W9-APX-898

An American National Standard

IEEE Recommended Practice for Electric Power Distribution for Industrial Plants

Published by
The Institute of Electrical and Electronics Engineers, Inc

Distributed in cooperation with
Wiley-Interscience, a division of John Wiley & Sons, Inc

ANSI/IEEE
Std 141-1986
(Revision of IEEE
Std 141-1976)

An American National Standard

IEEE Recommended Practice for Electric Power Distribution for Industrial Plants

Sponsor

Power Systems Engineering Committee
of the
IEEE Industry Applications Society

Approved December 12, 1985

IEEE Standards Board

Approved June 4, 1986

American National Standards Institute

First Printing
October 1986

ISBN 0-471-85687-8

Library of Congress Catalog Number 86-81947

The Institute of Electrical and Electronics Engineers, Inc
345 East 47th Street, New York, NY 10017, USA

October 10, 1986 *SH10710*

Foreword

(This Foreword is not a part of ANSI/IEEE Std 141-1986, IEEE Recommended Practice for Electric Power Distribution for Industrial Plants.)

This publication was prepared by the Systems Design Subcommittee of the IEEE Power Systems Engineering Committee, which is a technical committee of the IEEE Industry Applications Society. It has been approved by the IEEE Standards Board as an IEEE standards document to provide current information and recommended practices for the design, construction, operation, and maintenance of electric power systems in industrial plants.

It was forty-one years ago that *Electric Power Distribution for Industrial Plants* was first published by the AIEE. It was given the nickname the "Red Book" because of the color of its cover, and it became the first of the present IEEE Color Book series.

The second edition of the Red Book was produced in 1956, and identified as AIEE Std 952-1956. This was followed by the third edition in 1964; it was also identified as IEEE Std 141-1964. The fourth edition was produced in 1969; it was approved as a Recommended Practice of the Institute and identified as IEEE Std 141-1969. The fifth edition was produced in 1976 and identified as IEEE Std 141-1976. This, the sixth edition, is identified as ANSI/IEEE Std 141-1986. It was initiated in 1970 with the participation of more than fifty electrical engineers from industrial plants, consulting firms, and equipment manufacturers.

This IEEE Recommended Practice continues to serve as a companion publication to the following other Recommended Practices prepared by the IEEE Power Systems Engineering Committee:

Recommended Practice for Grounding of Industrial and Commercial Power Systems (IEEE Green Book), ANSI/IEEE Std 142-1982.

Recommended Practice for Electric Power Systems in Commercial Buildings (IEEE Gray Book), ANSI/IEEE Std 241-1983.

Recommended Practice for Industrial and Commercial Power System Analysis (IEEE Brown Book), ANSI/IEEE Std 399-1980.

Recommended Practice for Emergency and Standby Power for Industrial and Commercial Applications (IEEE Orange Book), ANSI/IEEE Std 446-1980.

Recommended Practice for the Design of Reliable Industrial and Commercial Power Systems (IEEE Gold Book), ANSI/IEEE Std 493-1980.

Recommended Practice for Electric Systems in Health Care Facilities (IEEE White Book), ANSI/IEEE Std 602-1986.

Recommended Practice for Energy Conservation and Cost-Effective Planning in Industrial Facilities (IEEE Bronze Book), ANSI/IEEE Std 739-1984.

Comments, corrections, and suggestions for the next revision of this standard are welcome and should be submitted to:

Secretary
IEEE Standards Board
345 East 47th Street
New York, NY 10017

The following persons were on the balloting committee that approved this document for submission to the IEEE Standards Board:

L. G. Ananian
D. S. Baker
A. J. Banes
G. M. Bauer
J. H. Beall
C. Becker
W. Bloomquist
J. M. Boyd
G. R. Bracey
T. F. Brandt
W. P. Burt
K. W. Carrick
K. Chen
L. S. Corey
J. M. Daly
F. A. Denbrock
L. C. Drapela
A. Freund
E. O. Gaylon
D. L. Goldberg
A. B. Gospel
L. E. Griffith
M. S. Griffith
H. L. Harkins
J. R. Harvey
C. R. Heising
E. Hesla
W. C. Huening
D. N. Hunder
L. Ilgen
R. W. Ingham
D. R. Kanitz
P. E. Kaup
A. M. Killin
J. T. Lambert
D. P. LaMorte
C. H. LaPlatney
S. A. Larson
R. H. Lee
J. R. Linders
R. E. Long
B. K. Mathur
F. Meyers
M. W. Migliaro
H. C. Miles
W. J. Moylan
J. F. Perkins
D. A. Pomering
M. D. Robinson
R. L. Simpson
D. S. Slaughter
E. R. Smith
R. L. Smith, Jr
W. L. Stebbins
G. A. Terry
C. P. Tsung
G. W. Walsh
J. Warren
S. J. Wells
J. I. Ykema
D. W. Zipse

When the IEEE Standards Board approved this standard on December 12, 1985, it had the following membership:

Electric Power Distribution for Industrial Plants

Sixth Edition

Working Group Members and Contributors

Lucas G. Ananian, *Working Group Chairman*

Chapter 1—Introduction: Donald T. Michael, *Chairman*; Daniel L. Goldberg, R. Gerald Irving

Chapter 2—System Planning: Donald W. Zipse, *Chairman*; Lucas G. Ananian, Graydon M. Bauer, Barry L. Christen, Arthur B. Gospel, Richard H. Kaufmann (deceased), Carl Noel

Chapter 3—Voltage Considerations: Donald T. Michael, *Chairman*; Walter C. Bloomquist, Arthur B. Gospel, Ralph H. Lee

Chapter 4—Surge Voltage Protection: David S. Baker, *Co-Chairman*; M. Shan Griffith, *Co-Chairman*; Richard L. Nailen, George W. Walsh, Glenn E. Word

Chapter 5—Application and Coordination of System Protective Devices: David S. Baker, *Chairman*; M. Shan Griffith, Richard H. Kaufmann (deceased), Clifton A. LaPlatney

Chapter 6—Fault Calculations: Walter C. Huening, *Chairman*; M. Shan Griffith, William R. Haack, Richard H. Kaufmann (deceased), Bal K. Mathur

Chapter 7—Grounding: Bal K. Mathur, *Chairman*; Barry L. Christen, K. V. Patel, Donald W. Zipse

Chapter 8—Power Factor and Related Considerations: Walter C. Bloomquist, *Chairman*; Lucas G. Ananian, Graydon M. Bauer, Kao Chen, M. Shan Griffith, Bal K. Mathur, William J. Moylan, Ray P. Stratford, George W. Walsh, Myron Zucker

Chapter 9—Power Switching, Transformation, and Motor-Control Apparatus: Harold C. Miles, *Chairman*; Lucas G. Ananian, J. Michael Boyd, William P. Burt, Kao Chen, Jerome Frank, Paul W. Gulick, Douglas R. Kanitz, Steven A. Larson, Joseph F. McCormick, Edward F. Troy, Jr

Chapter 10—Instruments and Meters: William J. Moylan, *Chairman*; Kao Chen, James R. Harvey

Chapter 11—Cable Systems: James M. Daly, *Chairman*; Alex Banes, Graydon M. Bauer, Raymond F. Bilinski, James R. Carey, David N. Hunder, Ralph H. Lee, Milton D. Robinson, Barry Schuster, Clarence P. Tsung, Dale E. Wagner

Chapter 12—Busways: Russell O. Ohlson, *Chairman*; Lucas G. Ananian, Robert W. Ingham, Frank C. Johnston, Robert H. Wood

Chapter 13—Electrical Energy Conservation: Carl E. Becker, *Chairman*; Charles S. Claar, Wayne L. Stebbins

Chapter 14—Cost Estimating of Industrial Power Systems: Erling C. Hesla, *Chairman*; Dan Agustin, Lucas G. Ananian, James M. Daly, Robert D. Giese, William L. Holveck, Douglas R. Kanitz, Harold C. Miles, Russell O. Ohlson

Contents

CHAPTER PAGE

CHAPTER PAGE

CHAPTER PAGE

CHAPTER PAGE

CHAPTER PAGE

CHAPTER PAGE

CHAPTER PAGE

FIGURES

FIGURES PAGE

FIGURES PAGE

FIGURES PAGE

FIGURES PAGE

FIGURES PAGE

FIGURES PAGE

TABLES

TABLES PAGE

TABLES PAGE

TABLES PAGE

1. Introduction

1.1 Institute of Electrical and Electronics Engineers (IEEE). ANSI/IEEE Std 141-1986, IEEE Recommended Practice for Electric Power Distribution for Industrial Plants, commonly called the IEEE Red Book, is published as a recommended practice by IEEE, a nonprofit, transnational professional society divided into 36 societies and councils covering the various sectors of electrical engineering. One of these is the Industry Applications Society, comprising 27 technical committees involved in the application of electric systems to industry. Of these, the Power Systems Engineering Committee is concerned with electric systems for industrial plants and is responsible for the development and maintenance of this recommended practice. Approval as an IEEE standard is conferred by the IEEE Standards Board.

1.2 IEEE Meetings and Publications. The IEEE and its constituent societies hold more than 100 technical conferences each year for the presentation of technical papers. Most of these are published in the *IEEE Transactions* or in conference records. Local sections of the IEEE and local chapters of the component societies hold meetings on various subjects of interest to members. Electrical engineers find these meetings and publications invaluable in keeping up-to-date on the latest technical developments in their particular areas of specialization. Many of these papers are referred to in this document, and copies may be obtained from the IEEE.[1]

1.3 Standards, Recommended Practices, and Guides. Electrical engineers use standards, recommended practices, and guides extensively in designing, installing, operating, and maintaining electric systems for industrial plants. Standards establish specific requirements such as definitions of electrical terms, methods of measurement and test procedures, and dimensions and ratings of equipment. Recommended practices suggest the best method of accomplishing an objective for

[1] All IEEE publications are available from Publication Sales, IEEE Service Center, 445 Hoes Lane, Piscataway, NJ 08854.

specified conditions. Guides specify the factors that must be considered in accomplishing a specific objective. All are grouped together as standards documents.

All standards carry the year of publication as a part of their title. If a copy of a standard referred to in this document is obtained, the year should always be checked. If it is older than five years, the standard may be obsolete; the publisher should be consulted to see if a more recent edition exists and, if so, a copy should be obtained. If the standard has been updated, the material referred to in the reference may have been revised and should be checked. If the date on a standard is followed by a later date preceded by the letter *R*, the standard has been reaffirmed as of the *R* date without change and the standard, therefore, is still in effect.

1.4 IEEE Standards Documents. The IEEE publishes several hundred standards documents covering various fields of electrical engineering. Appropriate IEEE standards are routinely submitted to the American National Standards Institute (ANSI) for consideration as American National Standards. Those that have been approved by ANSI carry the ANSI letters in their designation. A few that have also been approved by the Canadian Standards Association carry CSA letters. Basic standards of general interest include the following:

(1) ANSI/IEEE Std 91-1984, IEEE Standard Graphic Symbols for Logic Diagrams.
(2) ANSI/IEEE Std 100-1984, IEEE Standard Dictionary of Electrical and Electronics Terms.
(3) ANSI/IEEE Std 260-1978, IEEE Standard Letter Symbols for Units of Measurement.
(4) ANSI/IEEE Std 268-1982, American National Standard Metric Practice.
(5) ANSI/IEEE Std 280-1985, IEEE Standard Letter Symbols for Quantities Used in Electrical Science and Electrical Engineering.
(6) ANSI/IEEE Std 315-1975 (CSA Z99-1975), Graphic Symbols for Electrical and Electronics Diagrams.

The IEEE publishes several standards documents of special interest to electrical engineers involved with industrial plant electric systems, which are sponsored by the Power Systems Engineering Committee of the IEEE Industry Applications Society:

(1) ANSI/IEEE Std 142-1982, IEEE Recommended Practice for Grounding of Industrial and Commercial Power Systems (IEEE Green Book).
(2) ANSI/IEEE Std 241-1983, IEEE Recommended Practice for Electric Power Systems in Commercial Buildings (IEEE Gray Book).
(3) ANSI/IEEE Std 242-1986, IEEE Recommended Practice for Protection and Coordination of Industrial and Commercial Power Systems (IEEE Buff Book).
(4) ANSI/IEEE Std 446-1980, IEEE Recommended Practice for Emergency and Standby Power Systems for Industrial and Commercial Applications (IEEE Orange Book).
(5) ANSI/IEEE Std 399-1980, IEEE Recommended Practice for Power Systems Analysis (IEEE Brown Book).
(6) ANSI/IEEE Std 493-1980, IEEE Recommended Practice for the Design of Reliable Industrial and Commercial Power Systems (IEEE Gold Book).

(7) ANSI/IEEE Std 602-1986, IEEE Recommended Practice for Electric Systems in Health Care Facilities (IEEE White Book).
(8) ANSI/IEEE Std 739-1984, IEEE Recommended Practice for Energy Conservation and Cost-Effective Planning in Industrial Facilities (IEEE Bronze Book).

1.5 National Electrical Manufacturers Association (NEMA) Standards. NEMA[2] prepares standards that establish dimensions, ratings, and performance requirements for electric equipment for manufacturers. Their standards are widely used in the preparation of purchase specifications.

1.6 National Fire Protection Association (NFPA) Standards Documents. NFPA[3] publishes standards documents specifying requirements for fire protection and safety. Of interest to industrial plant electrical engineers are the following:
(1) ANSI/NFPA 70-1984, National Electrical Code (NEC).
(2) The NFPA Handbook of the National Electrical Code, sponsored by the NFPA and published by McGraw-Hill, contains the complete NEC text plus explanations edited to correspond with each edition of the NEC.
(3) ANSI/NFPA 70B-1983, Recommended Practice for Electrical Equipment Maintenance.

1.7 Underwriters Laboratories, Inc (UL) Standards. UL[4] prepares safety standards for electric equipment including appliances and tests equipment for compliance with these standards. Manufacturers who have their products approved by UL as meeting the standards are authorized to use the UL label on the equipment. UL periodically publishes lists of approved equipment.

1.8 American National Standards Institute (ANSI). ANSI[5] does not write standards. It promotes and coordinates the development of American National Standards and approves as American National Standards those documents that have been prepared in accordance with ANSI regulations.

Standards that have been approved by other organizations and then approved as American National Standards carry the identification numbers of both organizations and may be purchased from either. The sponsoring organization retains the responsibility for keeping the standard up-to-date. Standards carrying only an ANSI number were prepared by American National Standards committees organized and administered by other organizations in accordance with ANSI regulations. These ANSI committees are generally used to coordinate participation by a large number of organizations.

ANSI standards of interest to industrial plant electrical engineers include the following:
(1) ANSI Y1.1-1972, American National Standard Abbreviations for Use on Drawings and in Text.

---

[2] 2101 L Street, NW, Washington, DC 20037.
[3] Batterymarch Park, Quincy, MA 02269.
[4] 333 Pfingsten Road, Northbrook, IL 60020.
[5] 1430 Broadway, New York, NY 10018.

(2) ANSI Y32.9-1972, American National Standard Graphic Symbols for Electrical Wiring and Layout Diagrams Used in Architecture and Building Construction.

1.9 Occupational Safety and Health Administration (OSHA). Legislation by the US Federal Government has had the effect of giving standards, such as those of ANSI, the impact of law. The Occupational Safety and Health Act, administered by the US Department of Labor, permits federal enforcement of codes and standards. OSHA established *Design Safety Standards for Electrical Systems*, published in the Federal Register as 29CFR Part 1910, Subpart S. The regulations became effective April 16, 1981, and some articles and sections apply to all electrical installations and utilization equipment, and thus are retroactive.

1.10 Environmental Considerations. In all branches of engineering, an increasing emphasis is being placed on social, ecological, and environmental concerns. Today's engineer should consider air, water, noise, and all other items that have an environmental impact. The limited availability of energy sources and the steadily increasing cost of electric energy require a concern with energy conservation on the part of the engineer.

The electrical engineer may participate in studies such as total energy required compared to utility power available, electric heating versus fossil fuel, a comparison of boilers, purchased steam, and the heat pump. The use of steam turbines compared to absorption units and electric drives for air-conditioning may be evaluated. In these studies the effects of noise, vibration, exhaust gases, cooling methods, and energy requirements must be considered in relationship to the immediate and sometimes general environment.

1.11 Edison Electric Institute (EEI). The Edison Electric Institute,[6] the trade association of the privately owned electric utilities, publishes the following handbooks:
(1) *Electric Heating and Cooling Handbook*
(2) *A Planning Guide for Architects and Engineers*
(3) *Electric Space-Conditioning*
(4) *Industrial and Commercial Power Distribution*
(5) *Industrial and Commercial Lighting*
(6) *Underground Systems Reference Book*

1.12 Handbooks. The following handbooks have, over the years, established reputations in the electrical field. This list is not intended to be all-inclusive; other excellent references are available, but are not listed here because of space limitations.
(1) CROFT, T., CARR, C., and WATT, J. *American Electricians' Handbook* (McGraw-Hill, 1970). The practical aspects of equipment, construction, and installation are covered.

---

[6] 1111 19th Street, NW, Washington, DC 20036.

(2) *ASHRAE Handbook* (American Society of Heating, Refrigerating, and Air Conditioning Engineers[7] [ASHRAE]). This series of reference books in four volumes, which are periodically updated, details the electrical and mechanical aspects of space conditioning and refrigeration.
(3) *Electrical Transmission and Distribution Reference Book* (Westinghouse Electric Corporation,[8] 1964). The design and application of electric systems are outlined.
(4) *Electric Utility Engineering Reference Book, vol 3: Distribution Systems* (Westinghouse Electric Corporation, 1965). The application of high-voltage equipment and the design of high-voltage and network systems are covered in detail.
(5) McPARTLAND, J. and the editors of *Electrical Construction and Maintenance* magazine. *How to Design Electrical Systems* (McGraw-Hill, 1968); McPARTLAND, J. and NOVAK, W. *Electrical Design Details* (McGraw-Hill, 1960). These references offer the younger engineer an insight into the systems approach to electrical design.
(6) BEEMAN, D.L., editor. *Industrial Power Systems Handbook* (McGraw-Hill, 1955). A text on electrical design with emphasis on equipment, including that which is applicable to commercial buildings.
(7) *IES Lighting Handbook* (Illuminating Engineering Society[9] [IES]: Application Volume, 1981; Reference Volume, 1984). All aspects of lighting, including seeing, recommended lighting levels, lighting calculations, and design, are included in extensive detail in this comprehensive text.
(8) FINK, D.G. and CARROLL, J.M., editors. *Standard Handbook for Electrical Engineers* (McGraw-Hill, 1968). Virtually the entire field of electrical engineering is treated, including equipment and systems design.
(9) *Transformer Connections, Including Auto-transformer Connections*, Publication GET-2G (General Electric Company, 1960).
(10) *Underground Systems Reference Book* (Edison Electric Institute, 1957). The principles of underground construction and the detailed design of vault installations, cable systems, and related power systems are fully illustrated; cable splicing design parameters are thoroughly covered. Unfortunately, it has not been updated for some time.
(11) *Electrical Maintenance Hints* (Westinghouse Electric Corporation, 1975).
(12) SHAW, E.T. *Inspection and Test of Electrical Equipment* (Westinghouse Electric Corporation, 1967).

1.13 Periodicals. *Spectrum*, the basic monthly publication of the IEEE, covers all aspects of electrical and electronic engineering. It contains references to IEEE books and other publications, technical meetings and conferences, IEEE group, society, and committee activities, abstracts of papers and publications of the IEEE

---

[7] Publication Sales, 1791 Tullie Circle, NE, Atlanta, GA 30329.
[8] Printers' Division, Forbes Road, Trafford, PA 15085.
[9] 345 East 47 Street, New York, NY 10017.

and other organizations, and other material essential to the professional advancement of the electrical engineer.

Following are some other well-known periodicals:

(1) *Actual Specifying Engineer*, 205 East 42 Street, New York, NY 10017

(2) *Electrical Construction and Maintenance*, 1221 Avenue of the Americas, New York, NY 10020

(3) *Electrical Consultant*, 1760 Peachtree Road NW, Atlanta, GA 30309

(4) *Lighting Design and Application*, Illuminating Engineering Society, 345 East 47 Street, New York, NY 10017

(5) *Plant Engineering*, 1301 South Grove Avenue, Barrington, IL 60010

(6) *Power*, 1221 Avenue of the Americas, New York, NY 10020

(7) *Power Engineering*, 1301 South Grove Avenue, Barrington, IL 60010

1.14 Manufacturers' Data. The electrical industry, through its assocations and individual manufacturers of electrical equipment, issues many technical bulletins and data books. While some of this information is difficult for the individual to obtain, copies should be available to each major design unit. The advertising sections of electrical magazines contain excellent material, usually well-illustrated and presented in a clear and readable form, concerning the construction and application of equipment. Such literature may be promotional; it may present the advertiser's equipment or methods in a *best light* and should be carefully evaluated. Manufacturers' catalogs are a valuable source of equipment information. Some of the larger manufacturers' complete catalogs are very extensive, covering dozens of volumes; however, these companies may issue abbreviated or condensed catalogs that are adequate for most applications. Data sheets referring to specific items are almost always available from the sales offices. Some technical files may be kept on microfilm at larger design offices for use either by projection or by printing. Manufacturers' representatives, both sales and technical, can do much to provide complete information on a product.

2. System Planning

2.1 Introduction. The continuity of production in an industrial plant is only as reliable as its electric power distribution system. This chapter outlines the procedures for system planning and presents a guide to the use of the succeeding chapters.

No standard electric distribution system is adaptable to all industrial plants because two plants rarely have the same requirements. The specific requirements must be analyzed qualitatively for each plant and the system designed to meet its electrical requirements. Due consideration must be given to both the present and future operating and load conditions.

2.2 Basic Design Considerations. Any approach to the problem should include several basic considerations that will affect the overall design.

2.2.1 Safety. Safety of life and preservation of property are two of the most important factors in the design of the electric system. Safety to personnel involves no compromise; only the safest system can be considered. Following established codes in the selection of material and equipment is imperative.

2.2.2 Reliability. The continuity of service required is dependent on the type of manufacturing or process operation of the plant. Some plants can tolerate interruptions while others require the highest degree of service continuity. The system should be designed to isolate faults with a minimum disturbance to the system and should have features to give the maximum dependability consistent with the plant requirements and justifiable cost.

The majority of utilities today are supplying energy at 34.5, 69, and 138 kV. Some industrial plants accept supplies directly from utility distribution voltage substations at 4.16, 12.5, 13.8 kV, etc. In most instances, the utility substations also serve other customers so that there are usually several distribution lines connected to the same bus as the plant supply line(s).

For the most part, plant personnel feel secure with this type of supply, especially if the supply substation is nearby and if multiple supply lines are provided to meet *firm power* requirements. However, this tends to be a false sense of security when the facts are known regarding the reliability of distribution lines and the impact this has on plant operations. For example, when a fault occurs on the supply system near the plant there is an accompanying voltage dip on all of the plant primary distribution and utilization voltage buses. This lowered voltage persists while the utility relays operate and until the utility breaker trips, at which point the voltage will be reduced to zero on the faulted line and will be restored to near normal on the unfaulted portions of the system. Experience has shown that these short time voltage dips are severe enough and persist long enough to cause motor starter contactors to open automatically. When this occurs in a plant there will be parts spoilage, tool breakage, etc, all of which cause major disruptions in plant operations even though the supply may be lost only temporarily. Furthermore, new customers added to the supply substation could reduce the quality and reliability of the service.

Statistically, 138 kV lines may have outage rates of four or five outages per hundred miles per year. On the other hand, distribution lines in the 8-23 kV voltage range may have outage rates of 100 or more outages per hundred miles per year. Thus the probability of damaging voltage dips is at least 20 times as great on low voltage lines as on transmission voltage lines. This difference in probabilities is magnified even more when it is realized that the exposure to voltage dip incidents is not limited to only the plant supply line that may be extremely short. Instead, because distribution lines are not looped from substation to substation, the exposure is represented by the total miles of all lines connected to the utility substation.

2.2.3 System Reliability Analysis. One of the questions often raised during the design of the power distribution system is how to make a quantitative comparison of the failure rate and the forced downtime in hours per year for different circuit arrangements including radial, primary-selective, secondary-selective, simple spot network, and secondary-network circuits. This quantitative comparison could be used in tradeoff decisions involving the initial cost versus the failure rate and forced downtime per year. The estimated cost of power outages at the various distribution points could be considered in deciding which type circuit arrangement to use. The decisions could thus be based upon total owning cost over the useful life of the equipment rather than the first cost.

2.2.4 Reliability Data for Electrical Equipment. In order to calculate the failure rate and the forced downtime per year, it is necessary to have reliability data on the electric utility supply and each piece of electrical equipment used in the power distribution system. One of the best sources for this type of data are the extensive IEEE surveys on the reliability of electrical equipment in industrial plants and commercial buildings. See ANSI/IEEE Std 493-1980 [3].[10]

2.2.5 Reliability Analysis and Total Owning Cost. Statistical analysis methods involving probability of failure may be used to make calculations of the failure rate and the forced downtime for the power distribution system. The methods and formulas used in these calculations are given in ANSI/IEEE Std 493-1980 [3]. This includes the minimum revenue requirements method for calculating the total own-

---

[10] The numbers in brackets correspond to those of the references listed at the end of this chapter; when preceded by B, they correspond to the bibliography at the end of this chapter.

ing cost over the useful life of the equipment. Data on the cost of power outages are also given in ANSI/IEEE Std 493-1980 [3].

2.2.6 Simplicity of Operation. Simplicity of operation is very important in the safe and reliable operation and maintenance of the industrial power system. The operation should be as simple as possible to meet system requirements.

2.2.7 Voltage Regulation. Poor voltage regulation is detrimental to the life and operation of electrical equipment. Voltage at the utilization equipment must be maintained within equipment tolerance limits under all load conditions.

2.2.8 Maintenance. The distribution system should include preventive maintenance requirements in the design. Accessibility and availability for inspection and repair with safety are important considerations in selecting equipment. Space should be provided for inspection, adjustment, and repair in clean, well-lighted, and temperature-controlled areas.

2.2.9 Flexibility. Flexibility of the electric system means the adaptability to development and expansion as well as to changes to meet varied requirements during the life of the plant. Consideration of the plant voltages, equipment ratings, space for additional equipment, and capacity for increased load must be given serious study.

2.2.10 First Cost. While first costs are important, safety, reliability, voltage regulation, maintenance, and the potential for expansion should also be considered in selecting the best from alternate plans.

2.3 Planning Guide for Distribution Design. The following procedure will guide the engineer in the design of an electric distribution system for any industrial plant. The system designer should have or acquire knowledge of the plant's processes in order to select the proper system and its components.

2.3.1 Load Survey. Obtain a general layout, mark it with the major loads at various locations, and determine the approximate total plant load in kilowatts or kilovolt-amperes. Initially the amount of accurate load data may be limited. Some loads such as lighting and air conditioning may be estimated from generalized data. The majority of industrial plant loads are a function of the process equipment, and such information will have to be obtained from process and equipment designers. Since their design is often concurrent with power system design, initial information will be subject to change. It is therefore important that there be continuing coordination with the other design disciplines. For example, a change from electric powered to absorption refrigeration or a change from electrostatic to high-energy scrubber air-pollution control can change the power requirements for these devices by several orders of magnitude. The power system load estimates will require continual refinement until job completion.

2.3.2 Demand. The sum of the electrical ratings of each piece of equipment will give a total connected load. Because some equipment operates at less than full load and some intermittently, the resultant demand upon the power source is less than the connected load.

Standard definitions for these load combinations and their ratios have been devised.

demand. The electric load at the receiving terminals averaged over a specified interval of time.

NOTE: Demand is expressed in kilowatts, kilovolt-amperes, amperes, or other suitable units. The interval of time is generally 15 min, 30 min, or 1 h.

peak load. The maximum load consumed or produced by a unit or group of units in a stated period of time. It may be the maximum instantaneous load or the maximum average load over a designated period of time.

maximum demand. The greatest of all demands that have occurred during a specified period of time.

NOTE: For utility billing purposes the period of time is generally one month.

demand factor. The ratio of the maximum demand of a system to the total connected load of the system.

diversity factor. The ratio of the sum of the individual maximum demands of various subdivisions of the system to the maximum demand of the complete system.

load factor. The ratio of the average load over a designated period of time to the peak load occurring in that period.

coincident demand. Any demand that occurs simultaneously with any other demand, also the sum of any set of coincident demands.

Information on these factors for the various loads and groups of loads is useful in designing the system. For example, the sum of the connected loads on a feeder, multiplied by the demand factor of these loads, will give the maximum demand that the feeder must carry. The sum of the individual maximum demands on the circuits associated with a load center or panelboard, divided by the diversity factor of those circuits, will give the maximum demand at the load center and on the circuit supplying it. The sum of the individual maximum demands on the circuits from a transformer, divided by the diversity factor of those circuits, will give the maximum demand on the distribution transformer. The sum of the maximum demand on all distribution transformers, divided by the diversity factor of the transformer loads, will give the maximum demand on their primary feeder. By the use of the proper factors, as outlined, the maximum demands on the various parts of the system from the load circuits to the power source can be estimated.

2.3.3 Systems. Investigate the various types of distribution systems and select the system or systems best suited to the requirements of the plant.

A variety of basic circuit arrangements is available for industrial plant power distribution. Selection of the best system or combination of systems will depend upon the needs of the manufacturing process. In general, system costs increase with system reliability if component quality is equal. Maximum reliability per unit investment will be secured by using properly applied and well-designed components.

The first step is the analysis of the manufacturing process to determine its reliability need and potential losses in the event of power interruption. Some processes are little affected by interruption. Here a simple radial system may be satisfactory. Other processes may sustain long-term damage by even a brief interruption. Here a more complex system with an alternate power source for critical load may be justified.

A need for circuit redundancy may exist in continuous-process industries to allow equipment maintenance. Although the reliability of electric power distribution equipment is high, optimum reliability and safety of operation require routine maintenance. A system that cannot be maintained because of the needs of a continuous process is improperly designed.

Far more can be accomplished by the proper selection of the circuit arrangement than by economizing on equipment details. Cost reductions should never be made at the sacrifice of safety and performance by using inferior apparatus. Reductions should be obtained by using a less expensive distribution system with some sacrifice in reserve capacity and reliability.

2.3.3.1 Simple Radial System (Fig 1). Distribution is at the utilization voltage. A single primary service and distribution transformer supply all the feeders. There is no duplication of equipment. System investment is the lowest of all circuit arrangements.

Operation and expansion are simple. If quality components are used, reliability is high. Loss of a cable, primary supply, or transformer will cut off service. Equipment must be shut down to perform routine maintenance and servicing.

This system is satisfactory for small industrial installations where process allows

Fig 1
Simple Radial System

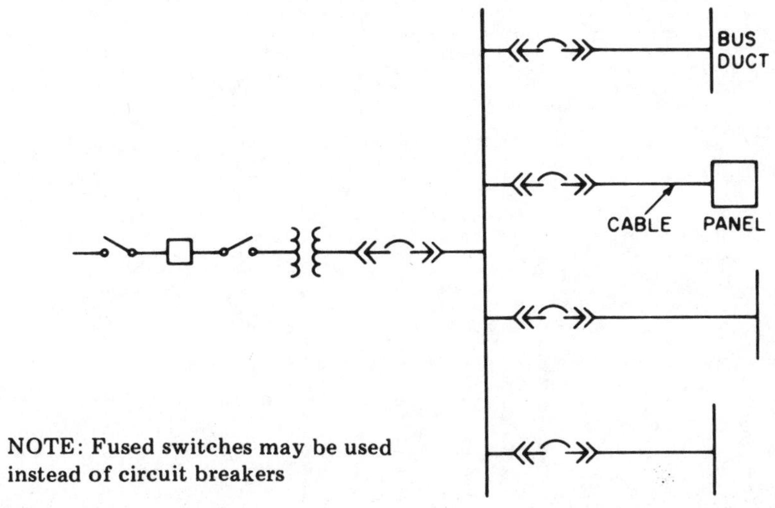

BUS DUCT

CABLE PANEL

NOTE: Fused switches may be used instead of circuit breakers

sufficient down time for adequate maintenance and the plant can be supplied by a single transformer.

2.3.3.2 Expanded Radial System (Fig 2). The advantages of the radial system may be applied to larger loads by using a radial primary distribution system to supply a number of unit substations located near the centers of load supplying the load through radial secondary systems.

The advantages and disadvantages are the same as those described for the simple radial system.

2.3.3.3 Primary Selective System (Fig 3). Protection against loss of a primary supply can be gained through use of a primary selective system. Each unit substation is connected to two separate primary feeders through switching equipment to provide a normal and an alternate source. Upon failure of the normal source, the distribution transformer is switched to the alternate source. Switching can be either manual or automatic, but there will be an interruption until load is transferred to the alternate source.

If the two sources can be paralleled during switching, some maintenance of primary cable and, in certain configurations, switching equipment may be performed with little or no interruption of service. Cost is somewhat higher than that for a radial system because of duplication of primary cable and switchgear.

2.3.3.4 Primary Loop System (Fig 4). This system offers the same advantages and disadvantages as the primary selective system. The failure of the normal source of a primary cable fault can be isolated and service restored by sectionalizing. However, finding a cable fault in the loop may be difficult. The system may be dangerous because the quickest way to find a fault is to sectionalize the loop and

Fig 2
Expanded Radial System

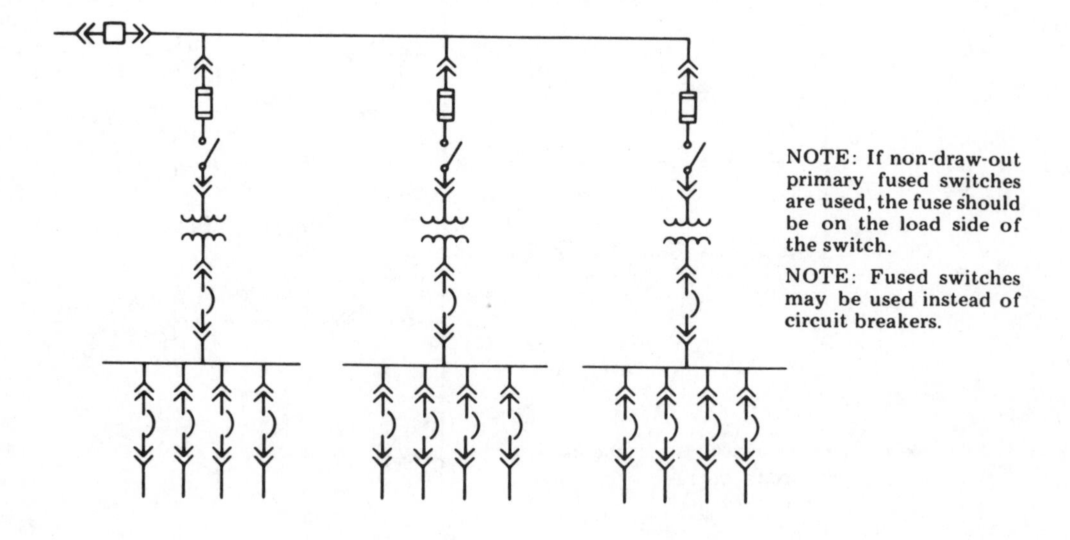

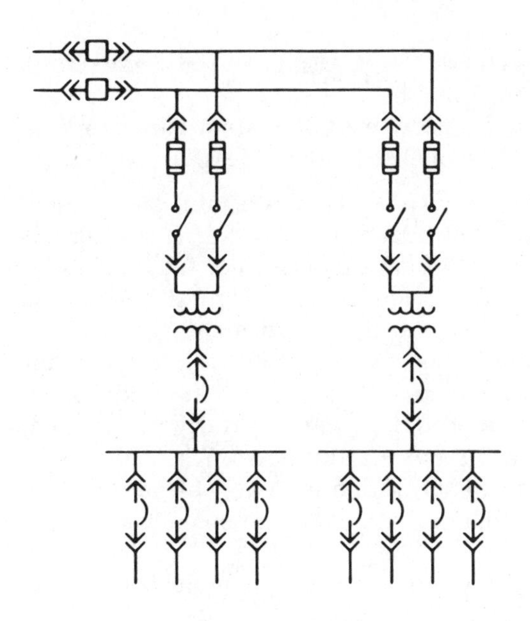

NOTE: If non-draw-out fused switches are used, the fuse should be on the load side of the switch.

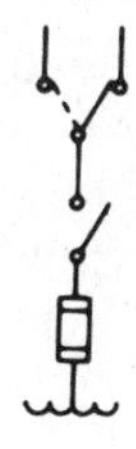

NOTE: An alternate arrangement uses a primary selector switch with a single fused interrupter switch (which may not have certified current-switching ability).

Fig 3
Primary Selective System

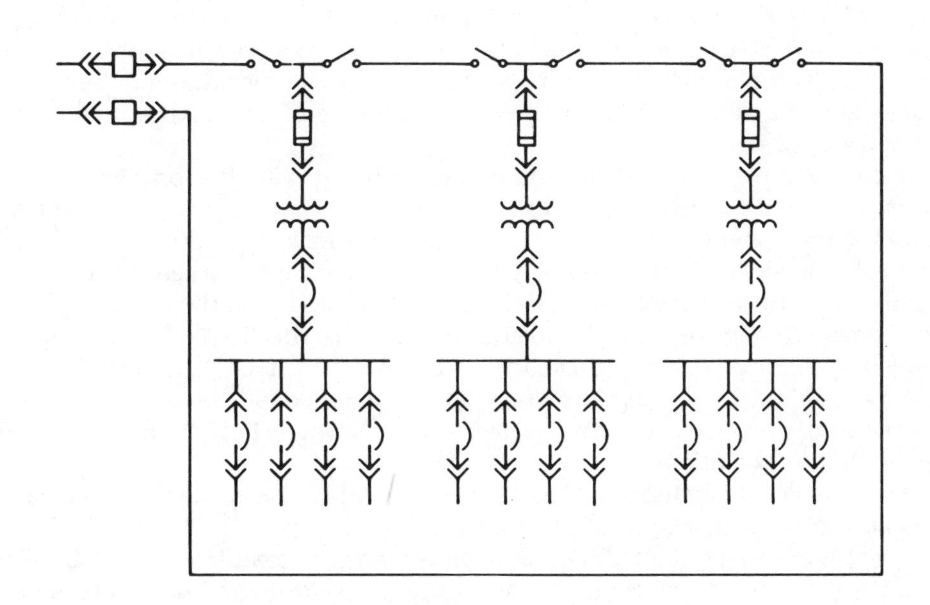

Fig 4
Primary Loop System

reclose. This may involve several reclosings on the fault. Also potentially dangerous is the fact that a section may be energized from both ends.

Cost may be somewhat less than for the primary selective system. However, it is doubtful that the savings are justified in view of the disadvantages.

2.3.3.5 Secondary Selective System [Fig 5(a)]. If pairs of unit substations are connected through a normally open secondary tie circuit breaker, the result is a secondary selective system. If the primary feeder or a transformer fails, the main secondary circuit breaker on the affected transformer is opened and the tie circuit breaker closed. Operation may be manual or automatic. Normally the stations operate as radial systems. Maintenance of primary feeders, transformer, and main secondary circuit breakers is possible with only momentary power interruption or no interruption if the stations may be operated in parallel during switching, although complete station maintenance will require a shutdown. With the loss of one primary circuit or transformer, the total substation load may be supplied by one transformer. To allow for this condition, one (or a combination) of the following should be considered.

(1) Oversizing both transformers so that one transformer can carry the total load

(2) Providing forced-air cooling to the transformer in service for the emergency period

(3) Shedding nonessential load for the emergency period

(4) Using the temporary overload capacity in the transformer and accepting the loss of transformer life

A distributed secondary selective system has pairs of unit substations in different locations connected by a tie cable and a normally open circuit breaker in each substation. The designer should balance the cost of the additional tie circuit breaker and the tie cable against the cost advantage of putting the unit stations nearer the load center.

The secondary selective system may be combined with the primary selective system to provide a high degree of reliability. This reliability is purchased with additional investment and addition of some operating complexity.

In Fig 5(a), while adhering to the *firm* capacity concept, it is seen that the total load allowable to the substation must be equal to or less than the capability of one transformer or one low side breaker, whichever is the least. By contrast, the arrangement of Fig 5(b), using a sparing transformer, can be developed for a firm capacity equal to $(n-1)$ transformers or low side breakers.

Operations, protection, etc, for configurations shown by Fig 5(a) and (b) are the same with two exceptions:

(1) Automatic transfer initiated by loss of voltage on a low side bus is not applicable in the sparing transformer scheme.

(2) Feeder breaker fault duties are almost always greater in the double-end scheme due to emergency condition when the tie is closed and one main is open.

2.3.3.6 Secondary Spot Network (Fig 6). In this system two or more distribution transformers are each supplied from a separate primary distribution feeder. The secondaries of the transformers are connected in parallel through a special

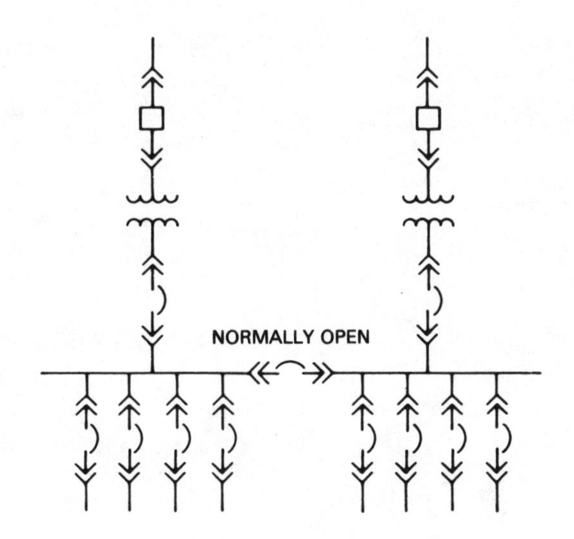

(a) Secondary Selective System

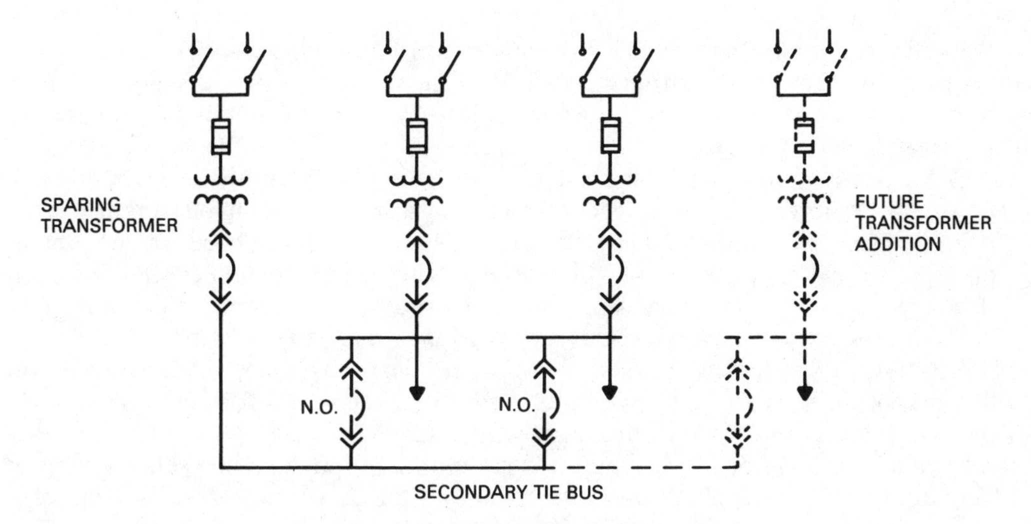

(b) Sparing Transformer Scheme

Fig 5
Typical Configurations Load Center Substations

type of circuit breaker, called a network protector, to a secondary bus. Radial secondary feeders are tapped from the secondary bus to supply utilization equipment.

If a primary feeder fails, or a fault occurs on a primary feeder or distribution transformer, the other transformers start to feed back through the network pro-

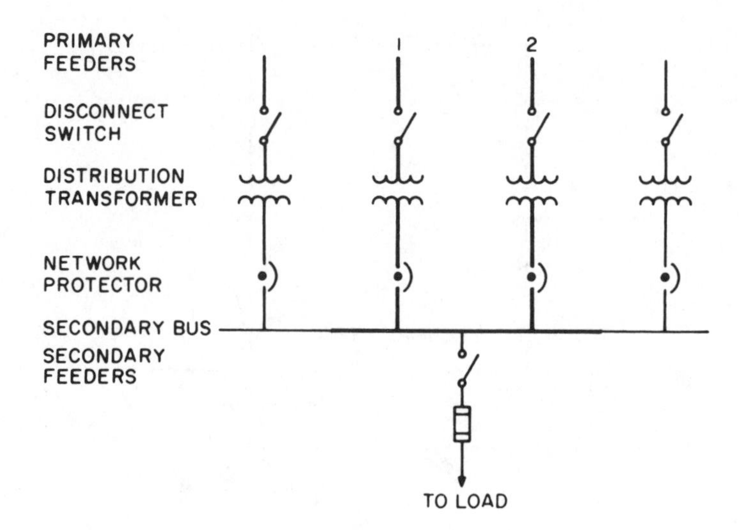

Fig 6
Secondary Spot Network

tector on the faulted circuit. This reverse power causes the network protector to open and disconnect the supply circuit from the secondary bus. The network protector operates so fast that there is a minimal exposure of secondary equipment to the associated voltage drop.

The secondary spot network is the most reliable power supply for large loads. A power interruption can only occur when there is a simultaneous failure of all primary feeders or when a fault occurs on the secondary bus. There are no momentary interruptions caused by the operation of the transfer switches that occur on primary selective, secondary selective, or loop systems. Voltage dips caused by faults on the system or large transient loads are materially reduced.

Networks are expensive because of the extra cost of the network protector and duplication of transformer capacity. In addition, each transformer connected in parallel increases the short-circuit-current capacity and may increase the duty ratings of the secondary equipment. It should be noted that this scheme is used only in low-voltage applications with a very high load density. Also, it requires a special bus construction to reduce the potential of arcing fault escalation. This scheme has not been applied to industrial power distribution.

2.3.3.7 Ring Bus (Fig 7). The ring bus offers the advantage of automatically isolating a fault and restoring service. Should a fault occur in Source 1, Device A and D would operate to isolate the fault while Source 2 would feed the loads. A fault anywhere in the ring results in two devices opening to isolate the fault.

Manual isolating switches are installed on each side of the automatic device. This allows maintenance to be performed safely and without interruption of service. This will also allow the system to be expanded without interruption.

2.3.4 Equipment Locations. The architect, in cooperation with process personnel, selects locations for distribution transformers and major utilization voltage

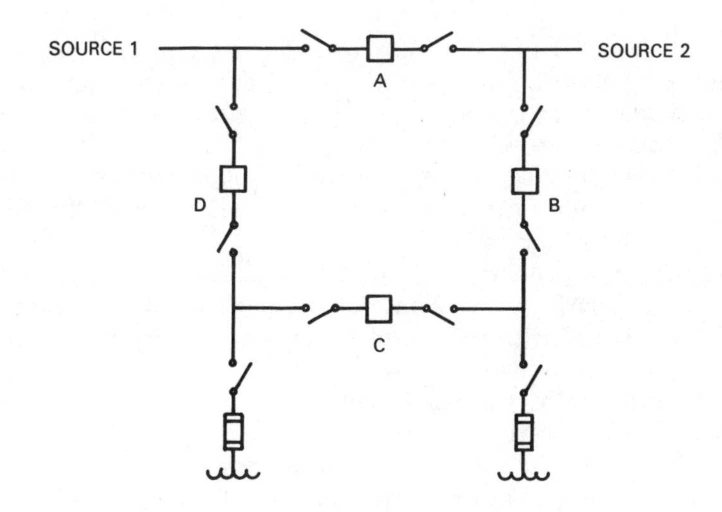

Fig 7
Ring Bus System

switching centers. In general, the closer the transformer to the load center of the area served, the lower the distribution system cost.

2.3.5 Voltage. Select the best voltages for the various system voltage levels. The most common utilization voltage in United States industrial facilities is 480 V. Other voltage levels depend upon motor rating, utility voltage available, total load served, potential expansion requirements, voltage regulation, and cost. Chapter 3 is a guide for correct voltage selection. The system should be capable of providing power to all equipment within published voltage limits under all normal operating conditions. Additional voltage considerations will include motor voltage drop during starting and restrictions placed on the user by the utility company to prevent disturbances to their system when starting large motors.

2.3.6 Utility Service. Whenever it is anticipated that a new service or a change in the existing utility supply is required, sufficient time should be allowed, prior to the scheduled in-service date, to permit all parties to act. While it is obvious to most that time must be allowed for the specification, procurement, manufacture, and installation of facilities, it is not always recognized that these efforts must frequently be preceded by negotiations and preparation and execution of a contract between the company and the utility.

It is important to note that the utility's response seemingly requires an inordinate amount of time. It should be recognized that utility company rules preclude any engineering, right-of-way acquisition, preparation of environmental impact statements, or filing of permit applications until there is a valid contractual arrangement with their customer. In some instances, an exchange of letters between the company and the utility can serve as a formal contract and the utility start date can be advanced.

Recent experience indicates that a utility requires anywhere from 18–30 months to place a new or expanded substation or transmission line in service, depending upon the complexity of the project. Moreover, this time interval begins only after the aforementioned contract has been executed. Therefore, it behooves the plant engineer to start developing plans early so details involving the utility will have been decided prior to the start of utility negotiations.

As soon as is practical in the design sequence, a conference should be arranged with the utility to determine the requirements for electric service. The utility should be furnished with the following data:

(1) Plan of the plant property showing buildings and other structures
(2) Plant load in kilowatts and maximum demand in kilovolt-amperes
(3) Preferred point of delivery of electric service
(4) Preferred service voltage
(5) Preferred utility supply arrangement
(6) Construction and startup schedule
(7) Unusual requirements such as high-speed reclosing
(8) Any unusually large motors on the system and starting duty
(9) Anticipated power factor
(10) Nature of connected load

The utility will be able to furnish the following data:

(1) Supply voltage or voltages available
(2) Anticipated voltage spread (maximum and minimum steady-state voltage magnitude)
(3) Point of delivery and line route
(4) Billing rate or rates available
(5) Options available on ownership of supply transformers
(6) Space requirements for the distribution substation if supplied by the utility
(7) Space required for the distribution transformer if plant is supplied at utilization voltage
(8) Maximum and minimum three-phase short-circuit duty at the point of delivery and system characteristics
(9) Requirements for metering
(10) Type of grounding on the supply system including maximum and minimum ground fault currents at point of delivery
(11) Requirements for coordination with the utility protection system
(12) Performance data on reliability of supply as necessary
(13) Back-up supply circuitry as necessary

Consideration should also be given to the need for temporary power during construction. It may be necessary to develop two sets of data, one for construction power and one for permanent power.

In the discussion with the utility company, one should explore the utility's future plans for expansion and the possible effects on plant loads. Consideration should be given to the utility's expected growth for the area.

2.3.7 Generation. Determine whether parallel, standby, or emergency generation will be required. If so, the following will have to be considered:

(1) Generator load in kilovolt-amperes
(2) Generator voltage

(3) Relaying and generator protection
(4) Metering
(5) Voltage regulation
(6) Synchronizing
(7) Grounding
(8) Cost
(9) Maintenance requirements
(10) Largest motor to be started

Consideration should be given to the load imposed on the generator when groups of motors are started in place of one large motor. These groups of motors may be arranged for staggered starting so that a smaller generator can be specified.

The complete design must be coordinated with the utility if parallel operation with the utility's system is anticipated. With today's electrical utility cost, it may be advantageous to utilize generator equipment to shave peak load.

2.3.8 One-Line Diagram. A complete one-line or single-line diagram, in conjunction with a physical plan of the installation, should present sufficient data to plan and evaluate the electric power system. Figures 66 and 99 in Chapters 5 and 6 represent one-line diagrams containing information required for system-protection design and fault-current analysis.

Symbols commonly used in one-line diagrams are defined in ANSI/IEEE Std 315-1975 [2].

The following items should be shown on any one-line diagram:

(1) Power sources including voltages and available short-circuit currents

(2) Size, type, ampacities, and number of all conductors in kilovolt-ampere rating

(3) Capacities, voltages, impedance, connections, and grounding methods of transformers

(4) Identification and quantity of protective devices (relays, fuses, and circuit breakers)

(5) Instrument transformer ratios

(6) Type and location of surge arresters and capacitors

(7) Identification of all loads

(8) Identification of any other distribution system equipment

(9) Type of relays and relay settings on utility supply lines for both phase and ground faults

The one-line diagram should show proposed future additions, and the effect of such additions should be a part of the original system planning.

The actual drawing should be kept as simple as possible. It is a schematic diagram and need not show geographical relationship. Duplication should be avoided.

2.3.9 Short-Circuit Analysis. Calculate short-circuit currents available at all system components. Chapter 6 provides a detailed guide to making these calculations.

2.3.10 Protection. Using the data presented in Chapter 5, design the required protective systems. System protection design must be an integral part of the total system design and not superimposed on a system after the fact.

Coordination of critical loads such as uninterruptible power supplies (UPS) with their fast-acting fuses to protect solid-state devices should not be overlooked.

2.3.11 Expansion. At every step, whether for an all-new plant or for an existing plant, consideration of the future is absolutely necessary. When this is done properly, subsequent increases in plant load can be accommodated by adding capacity to the initial system instead of necessitating a redesign of the whole primary system. In short, no plant primary system nor its supply should be designed in a manner that will make it difficult or impossible to expand its capacity.

More important than the timing of capital investment is the need to permanently allocate space for installing power supply and distribution apparatus that may be required for the ultimate plant development at the site. Thus, it is seen that an estimate of the maximum ultimate demand is necessary in order to establish the number and nature of future required facilities.

Experience has shown that when insufficient space is allocated for expanding either the high voltage outdoor substation or the indoor main distribution switchgear, there is a tendency to compromise both safety and convenience. The latter usually translates into a depreciation of reliability because space for equipment removal or maintenance, or both, is sacrificed. This being the case, it usually becomes necessary to remove additional equipment in order to repair or replace another piece of apparatus that has failed.

Figure 8 is presented to illustrate some of the considerations that are a necessary part of planning the power supply. Although the example suggests a totally new plant, similar forward planning is equally necessary at an existing facility *before* the demand exceeds the firm supply capacity.

Every plant either has or should have an up-to-date five- or six-year forecast of production or projects, or both, that may or may not actually be implemented. These, along with longer term projections, should be weighed in terms of their impact upon electric demand. In similar fashion the prospect of future expanded utilization of the site must be recognized in terms of electric demand. Thus, even if long-range projections are in error, it will be possible to develop an array of possible plans for the capacity to keep pace with the demands. In short, if the full range of future possibilities is explored generally both as to size and timing, long-range plans can be developed that at least offer a chance of coping successfully with demands that may actually occur.

Admittedly, forecasting is not an exact science and everyone realizes it is economic folly to construct or provide capacity that is never used. However, there are opportunities in planning and designing electrical systems for selecting apparatus and arranging these in schemes that minimize the probabilities of early retirements due to improper ratings and of reconstructing major portions of the system.

By using various combinations of systems, one can design a system to meet the load requirements. It may be necessary to design a system that can be expanded for future growth. As an example, one can start with a radial system supplied at 13.8 kV from the utility. See Fig 1. As the load at the site grows, one can convert to a double-ended substation such as is shown in Fig 5(a). At the time of a major expansion and when the load requires additional power, one can go to a ring bus.

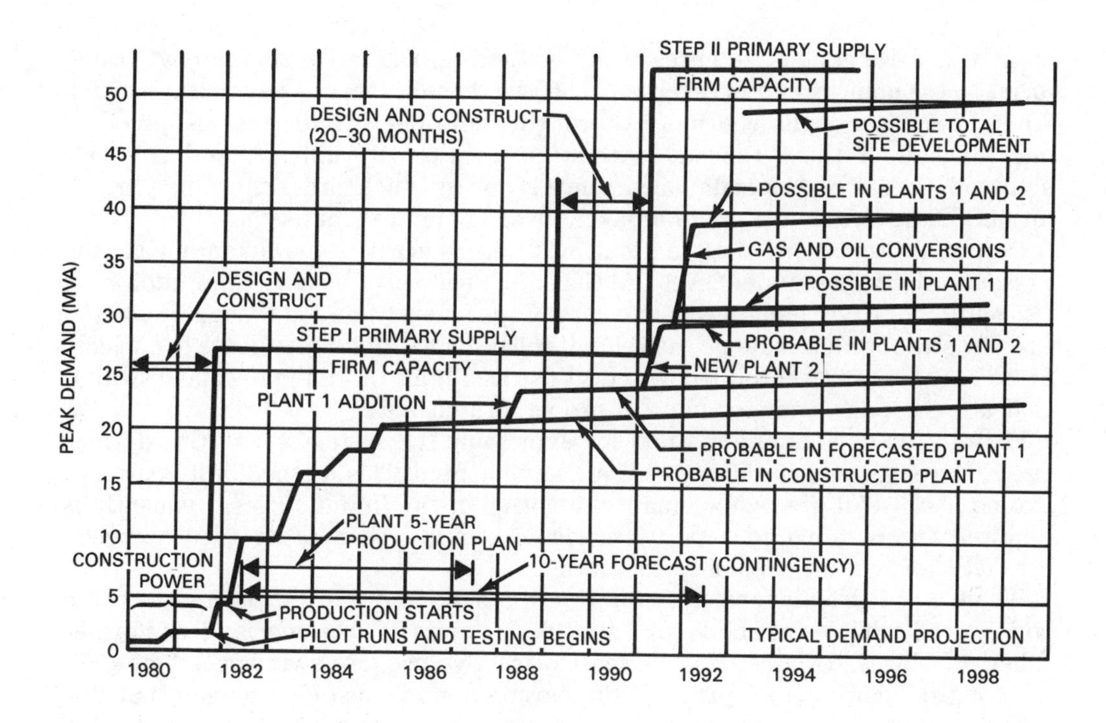

Fig 8
Power Supply Planning Considerations

See Fig 7. This ring bus through transformation can supply an intermediate distribution system at the original utility voltage of 13.8 kV. With proper planning and system selection, one can expand to meet any load requirement.

The following discussion is intended to stimulate serious thought about future expansion problems and methods at the time of the initial development. At the same time, it is hoped that engineering staffs will recognize that there are many ways to implement future expansion. It remains for the engineer to weigh the alternatives and to determine the most effective plan for the specific situation at hand.

Selecting the number of main distribution buses and the method of interconnecting them is not an arbitrary decision. Instead, many factors must be considered, such as the sizes of the immediate and ultimate plant loads, the primary distribution voltage, the availability of suitable supply lines, and other factors relating to the utility system.

Figure 9 is presented for the purpose of illustrating a number of main primary substation configurations that can be considered for use. The significance of Fig 9 is to demonstrate that there are many ways in which to develop the main primary distribution configuration while meeting specific requirements or constraints.

The two-bus arrangement (Fig 9), shown as the initial configuration, represents the lowest cost arrangement. It can also be seen that to maximize the *firm* capability of this arrangement requires that the thermal ratings of the supply (transformer), the main circuit breaker, and the switchgear main bus bar must be equal. While later discussions will more thoroughly consider specific ratings, precisely matching all of these thermal capacities is extremely difficult and is rarely achieved. Therefore, the usual occurrence is one in which either the supply transformer or the switchgear, main breaker, or bus bars are limiting.

Despite protestations that the load will never exceed some amount near the present or initially planned level and that there will *never* be a need for additional capacity, this is virtually never the case. Consequently, as Fig 9 suggests, initial sizing of switchgear should be such that it will be compatible with future expansion. *Future* in this case is intended to reflect the useful life of well-maintained switchgear at perhaps 40 years or more, if properly rated.

Figure 9(a) depicts one method for expanding the system where transformer size is increased and initial switchgear is augmented. This method, if it is used to extend the useful life of too-small switchgear in the initial stages, frequently is required to operate under fairly restrictive conditions in order to not exceed breaker short-circuit capability limitations.

Where a third transformer (supply) is possible and feasible, there are three widely used schemes as shown by Fig 9(b), (c), and (d). The sparing transformer scheme of Fig 9(d) is usually the least attractive because switchgear limits are unchanged. Figure 9(c) represents the most commonly installed scheme, but this scheme is nearly always capacity-limited by switchgear even when current-limiting reactors are installed in bus tie circuits. Even so, 9(c) is usually the preferred configuration because reactive losses are incurred only when the bus loading is unbalanced and during emergency periods. Figure 9(b) is more expensive than either 9(c) or 9(d) and nearly always its capacity is limited by the transformer rating, particularly when special duplex reactors are installed between transformers and main breakers to limit breaker fault current duty.

Although it does not affect the substation configuration, substation expansion is also possible by exchanging initially installed transformers for larger units. This is rarely done in instances where the plant owns the supply transformers. However, when transformers are supplied by the utility and when the initial transformers can be utilized by the utility, it is frequently more economical to exchange units than to add a third unit and all of the associated switching apparatus. Usually when a utility exchanges transformers, the customer receives credit for the retired equipment on the basis of *replacement cost new less depreciation*. Thus, for example, it may be financially attractive to begin a supply substation with two 15/20/25 MVA units and later exchange them for two 30/40/50 MVA units. However, in order to be technically feasible, the main plant primary switchgear must be capable of handling the higher load current as well as the higher available fault current.

When expanding an existing plant, determine if all the existing equipment is adequate for additional load and short-circuit requirements. Check such ratings as voltage, interrupting capacity, short-circuit withstand ratings, momentary ratings, switch close and latch requirements, and current-carrying capacity. Coordinate new and existing system protective devices. Carefully review the new design to ensure that maximum safety levels are maintained.

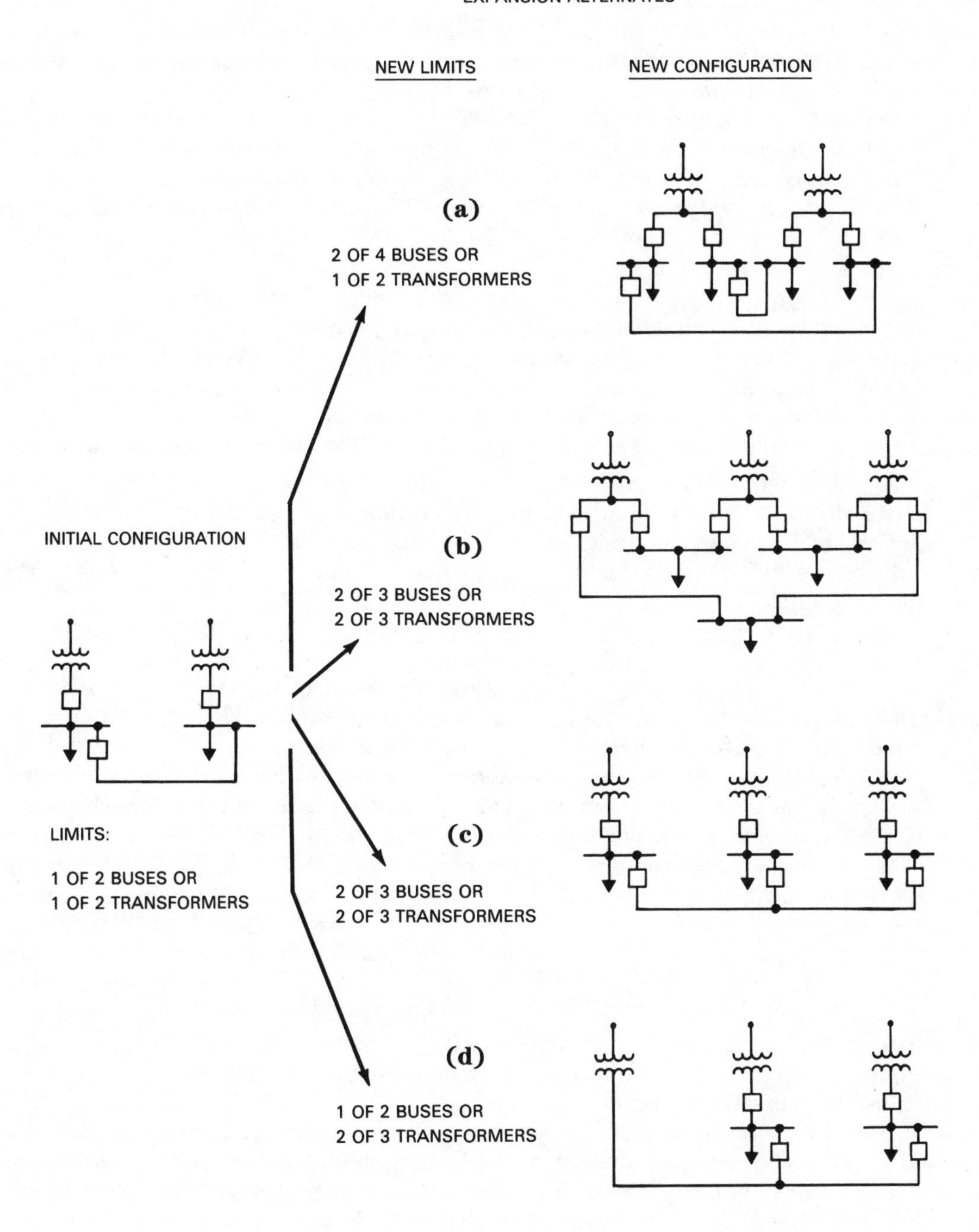

Fig 9
Typical Main Primary Distribution Arrangements

Determine the best method of connecting the new part of the power system to the existing system to minimize production loss and construction cost.

2.3.12 Other Requirements. Investigate unusual loads or conditions such as:

(1) Starting requirements of large motors

(2) Arc furnace operation

(3) Welder operation

(4) Loads that must be kept in operation under all conditions

(5) Sensitive loads such as computers and high-gain laboratory equipment that may be affected by voltage and frequency transients that do not affect other equipment

(6) Equipment producing high noise levels. (Federal law requires that personnel not be exposed continuously for an eight-hour period to noise pressure levels of more than 90 dBA. Higher levels are allowed up to 115 dBA with a reduction in exposure time.)

(7) The possibility of demand limiting to reduce power cost

(8) Coordination of the electric power system with other energy systems

2.3.13 Safety. Make sure that adequate safety features are incorporated into all parts of the system. Safety of life and preservation of property are two of the most important factors in the design of the electric system.

Listed below are limits that should be considered in order to provide safe working conditions for personnel:

(1) Interrupting devices must be able to function safely and properly under the most severe duty to which they may be exposed.

(2) Protection must be provided against accidental contact with energized conductors by enclosing the conductors, installing protective barriers, or installing the conductors at sufficient height to avoid accidental contact.

(3) Isolating switches must not be operated while they are carrying current, unless designed to do so. They should be equipped with interlocks or warning signs if load break or fault closing capability is not provided.

(4) In many instances it is desirable to isolate a power circuit breaker using disconnect switches. In such cases the circuit breaker must be opened first, then the disconnect switches. Interlocks to ensure this sequence are usually desirable.

(5) The system should be designed so that maintenance work on circuits and equipment can be accomplished with the particular circuits and equipment de-energized and grounded. System design should provide for locking out circuits or equipment for maintenance. A written procedure should be established to provide instructions on tagging or locking out circuits during maintenance and re-energizing after completion.

(6) Electric equipment rooms, especially those containing apparatus over 600 V such as transformers, motor controls, or motors, should be equipped and located to eliminate or minimize the need for access by nonelectrical maintenance or operating personnel. Convenient remotely located exits should be provided to allow quick exit during an emergency.

(7) Electric apparatus located outside special rooms should be provided with protection against mechanical damage. The area should be accessible to mainte-

nance and operating personnel for emergency operation of protective devices.

(8) Hazardous areas should be considered. A nonhazardous area may be set aside for electrical equipment; or it may be advantageous to locate explosion-proof equipment in the hazardous area. The advantages and disadvantages of not only initial cost but the maintenance cost and the ability to maintain the integrity of the equipment should be considered.

(9) Warning signs should be installed on electric equipment accessible to unqualified personnel, on fences surrounding electric equipment, on doors giving access to electrical rooms, and on conduits or cables above 600 V in areas that include other equipment or pipelines.

(10) An adequate grounding system must be installed.

(11) Emergency lights should be provided where necessary to protect personnel against sudden lighting failure. In areas with high populations, exit routes, process control locations, and electric switching centers are particularly important.

(12) Operating and maintenance personnel should be provided with complete operating and maintenance instructions including wiring diagrams, equipment ratings, and protective device settings. Spare fuses of the correct ratings should be stocked.

2.3.14 Communications. Any plan for the protection of a plant must include a reliable communication system. This can be accomplished either by a self-contained and self-maintained interplant system of telephones, alarms, etc, and may include modern radio and television equipment, or by a joint system tied into the existing communication services.

Fire and smoke alarm circuits, whether self-contained or connected to city alarm systems, should be installed in a manner that will be least affected by faults and changes in buildings or plant operations. Circuits should be arranged to provide easy means of testing and of isolating portions of the system without interference with the remainder of the system.

Watchman circuits, including television and radio equipment, are used in many plants for the purpose of providing a ready means for the individual watchman to report unusual circumstances to his supervisor without delay. Such systems are frequently combined with loudspeaker paging systems and other alarm methods.

Annunciator systems are available for alerting operations to abnormal situations in critical areas. The operator can dispatch others to investigate the malfunction or disorder or take corrective action with controls furnished to him.

2.3.15 Maintenance. Electric equipment must be selected and installed with attention to adequacy of performance, safety, and reliability. To preserve these features, a maintenance program must be established, tailored to the type of equipment and the details of the particular installation. Some items require daily attention, some weekly, and others can be tested or checked annually or less frequently.

Requirements of a maintenance program should be incorporated in the electrical design to provide working space, easy access for inspection, facilities for sampling and testing, and disconnecting means for protection of the workmen, lighting, and standby power. The maintenance program should have the following objectives.

2.3.15.1 Cleanliness. Dirt and dust accumulation affects the ventilation of equipment and causes excess heat, which reduces the life of the insulation. Dirt and dust also build up on the surfaces of insulators to form paths for leakage that may result in arcing faults. Insulated surfaces should be regularly cleaned to minimize these hazards.

2.3.15.2 Moisture Control. Moisture reduces the dielectric strength of many insulating materials. Unnecessary openings should be closed and necessary openings should be baffled or filtered to prevent the entrance of moisture, especially light snow. Also, even though equipment is adequately housed and indoors, condensation from weather changes should be minimized by supplying heat, usually electric, to the enclosure interiors. From 5–7.5 W/ft^2 of enclosure external surface is usually effective when placed at the bottom of each space affected. A small amount of ventilation (or breathing) outdoors is necessary even with heating to avoid condensation damage and insulation failure.

2.3.15.3 Adequate Ventilation. Much electric equipment is designed with paths for ventilating air to pass over insulating surfaces to dissipate heat. Filters must be changed, fans inspected, and equipment cleaned often enough to keep such ventilating systems operating properly.

2.3.15.4 Reduced Corrosion. Corrosion destroys the integrity of equipment and enclosures. As soon as evidence of corrosion is noted, action should be taken to clean the affected surfaces and inhibit future deterioration.

2.3.15.5 Maintenance of Conductors. Conducting surfaces reveal problems caused by overheating, wear, or misalignment of contact surfaces. These conditions should be corrected by tightening bolts, correcting excessive operations, aligning contacts, or whatever action is necessary.

2.3.15.6 Regular Inspections. Inspections should be scheduled on a regular basis depending on equipment needs and process requirements. External inspection can often be made and reveal significant information without process shutdown. However, a complete inspection will require a shutdown. Plans for repairs should be based on such inspections so that necessary manpower, tools, and replacement parts will be available as needed during the shutdown.

2.3.15.7 Regular Testing. Performance of protective devices depends on the accuracy of the sensing devices and the integrity of the control circuits. Periodic tests of such devices as well as of the dielectric strength of insulating systems and the color and acidity of the insulating oils, etc, will reveal deteriorating conditions that cannot be determined by visual inspection. Necessary adjustments or corrections can be made before failure occurs.

2.3.15.8 Adequate Records. An organized system of records of inspection, maintenance, tests, and repairs provides a basis for trouble-shooting, predicting equipment failures, and selecting future equipment.

2.3.16 Codes and Standards. Throughout the design the engineer must comply with all national and local laws, codes, and standards.

2.4 References. This standard shall be used in conjunction with the following publications:

[1] ANSI/IEEE Std 277-1983, IEEE Recommended Practice for Cement Plant Power Distribution.[11]

[2] ANSI/IEEE Std 315-1975 (CSA Z99-1975), Graphic Symbols for Electrical and Electronics Diagrams.

[3] ANSI/IEEE Std 493-1980, IEEE Recommended Practice for the Design of Reliable Industrial and Commercial Power Systems.

[4] ANSI/NFPA 70B-1983, Recommended Practice for Electrical Equipment Maintenance.[12]

2.5 Bibliography

[B1] BJORNSON, N.R. How Much Redundancy and What It Will Cost. *IEEE Transactions on Industry and General Applications*, vol IGA-6, May/June 1970, pp 192-195.

[B2] GANNON, P.E. Cost of Interruptions; Economic Evaluation of Reliability. IEEE Industrial and Commercial Power Systems Technical Conference, Los Angeles, May 10-13, 1976.

[B3] HEISING, C.R. Examples of Reliability and Availability Analysis of Common Low-Voltage Industrial Power Distribution Systems, IEEE Industrial and Commercial Power Systems Technical Conference, Los Angeles, May 10-13, 1976.

[B4] HEISING, C.R. Reliability and Availability Comparison of Common Low-Voltage Industrial Power Distribution Systems. *IEEE Transactions on Industry and General Applications*, vol IGA-6, Sept/Oct 1970, pp 416-424.

[B5] IEEE Committee Report. Report on Reliability Survey of Industrial Plants, Parts I-III. *IEEE Transactions on Industry Applications*, vol IA-10, Mar/Apr 1974, pp 213-252.

[B6] JOHNSON, G.T. and BRENIMAN, P.E. Electrical Distribution System for a Large Fertilizer Complex. *IEEE Transactions on Industry and General Applications*, vol IGA-5, Sept/Oct 1969, pp 566-577.

[B7] McFADDEN, R.H. Power-System Analysis: What It Can Do for Industrial Plants. *IEEE Transactions on Industry and General Applications*, vol IGA-7, Mar/Apr 1971, pp 181-188.

---

[11] ANSI/IEEE publications can be obtained from the Sales Department, American National Standards Institute, 1430 Broadway, New York, NY 10018, or from the Institute of Electrical and Electronics Engineers, Service Center, Piscataway, NJ 08854.

[12] ANSI/NFPA publications can be obtained from the Sales Department, American National Standards Institute, 1430 Broadway, New York, NY 10018, or from Publication Sales, National Fire Protection Association, Batterymarch Park, Quincy, MA 02269.

[B8] PATTON, A.D. Fundamentals of Power System Reliability Evaluation, IEEE Industrial and Commercial Power Systems Technical Conference, Los Angeles, May 10-13, 1976.

[B9] REGOTTI, A.R. and TRASKY, J.G. What to Look for in a Low-Voltage Unit Substation. *IEEE Transactions on Industry and General Applications*, vol IGA-5, Nov/Dec 1969, pp 710-719.

[B10] SHAW, E.T. *Inspection and Test of Electrical Equipment.* Pittsburgh, PA: Westinghouse Electric Corporation. Electric Service Division, Pub MB3051, 1967.

[B11] YUEN, M.H. and KNIGHT, R.L. On-Site Electrical Power Generation and Distribution for Large Oil and Gas Production Complex in Libya. *IEEE Transactions on Industry and General Applications*, vol IGA-7, Mar/Apr 1971, pp 273-289.

3. Voltage Considerations

3.1 General. An understanding of system voltage nomenclature and the preferred voltage ratings of distribution apparatus and utilization equipment is essential to ensure proper voltage identification throughout a power distribution system. The dynamic characteristics of the system need to be recognized and the proper principles of voltage control applied so that satisfactory voltages will be supplied to all utilization equipment under all normal conditions of operation.

3.1.1 Definitions. The following terms and definitions from the proposed revision of ANSI C84.1-1982 [2] [13] are used to identify voltages and voltage classes used in electric power distribution.

3.1.1.1 System Voltage Terms

system voltage. The root-mean-square phase-to-phase voltage on an ac electric system.

nominal system voltage. The root-mean-square phase-to-phase voltage by which the system is designated and to which certain operating characteristics of the system are related. (The nominal system voltage is near the voltage level at which the system normally operates. To allow for operating contingencies, systems generally operate at voltage levels about 5–10% below the maximum system voltage for which system components are designed.)

maximum system voltage. The highest root-mean-square phase-to-phase voltage that occurs on the system under normal operating conditions, and the highest root-mean-square phase-to-phase voltage for which equipment and other system components are designed for satisfactory continuous operation without derating of any kind. (When defining **maximum system voltage**, voltage transients and temporary overvoltages caused by abnormal system conditions, such as faults, load rejection, etc, are excluded. However, voltage transients and temporary overvoltages may affect equipment life and operating performance as well as conductor insulation and are considered in equipment application.)

[13] The numbers in brackets correspond to those of the references listed at the end of this chapter; when preceded by B, they correspond to the bibliography at the end of this chapter.

service voltage. The root-mean-square phase-to-phase or phase-to-neutral voltage at the point where the electric system of the supplier and the user are connected.

utilization voltage. The root-mean-square phase-to-phase or phase-to-neutral voltage at the line terminals of utilization equipment.

3.1.1.2 System Voltage Classes

low voltage. A class of nominal system voltages less than 1000 V.

medium voltage. A class of nominal system voltages equal to or greater than 1000 V and less than 100 000 V.

high voltage. A class of nominal system voltages equal to or greater than 100 000 V.

3.1.2 Standard Nominal System Voltages for the United States. These voltages and their associated tolerance limits are listed in ANSI C84.1-1982 [2] for voltages from 120–230 000 V and in ANSI C92.2-1981 [3] for voltages above 230 kV nominal. Table 1, reprinted from ANSI C84.1-1982 [2] and containing information from ANSI C92.9-1981 [3], provides all the standard nominal system voltages and their associated tolerance limits for the United States. Preferred nominal system voltages and voltage ranges are shown in boldface type while other systems in substantial use that are recognized as standard voltages are shown in regular type. Other voltages may be encountered on older systems but they are not recognized as standard voltages. The transformer connections from which these voltages are derived are shown in Fig 10.

Two sets of tolerance limits are defined: range A, which specifies the limits under most operating conditions, and range B, which allows minor excursions outside the range A limits.

3.1.3 Application of Voltage Classes

(1) Low-voltage class voltages are used to supply utilization equipment.

(2) Medium-voltage class voltages are used as primary distribution voltages to supply distribution transformers which step the medium voltage down to a low voltage to supply utilization equipment. Medium voltages of 13 800 V and below are also used to supply utilization equipment such as large motors (see 3.5.2, Table 8).

(3) High-voltage class voltages are used to transmit large amounts of electric power between transmission substations. Transmission substations located adjacent to generating stations step the generator voltage up to the transmission voltage for transmission-to-transmission substations in the load area, which transform the transmission voltage down to a primary distribution voltage in order to supply distribution transformers that step the primary distribution voltage down to a utilization voltage. Transmission lines also interconnect transmission substations to provide alternate paths for power transmission to improve the reliability of the transmission system.

3.1.4 Voltage Systems Outside of the United States. Voltage systems in other countries generally differ from those in the United States. For example, 416Y/240 V is widely used as a utilization voltage even for residential service. Also the frequency in many countries is 50 Hz instead of 60 Hz, which affects the operation of some

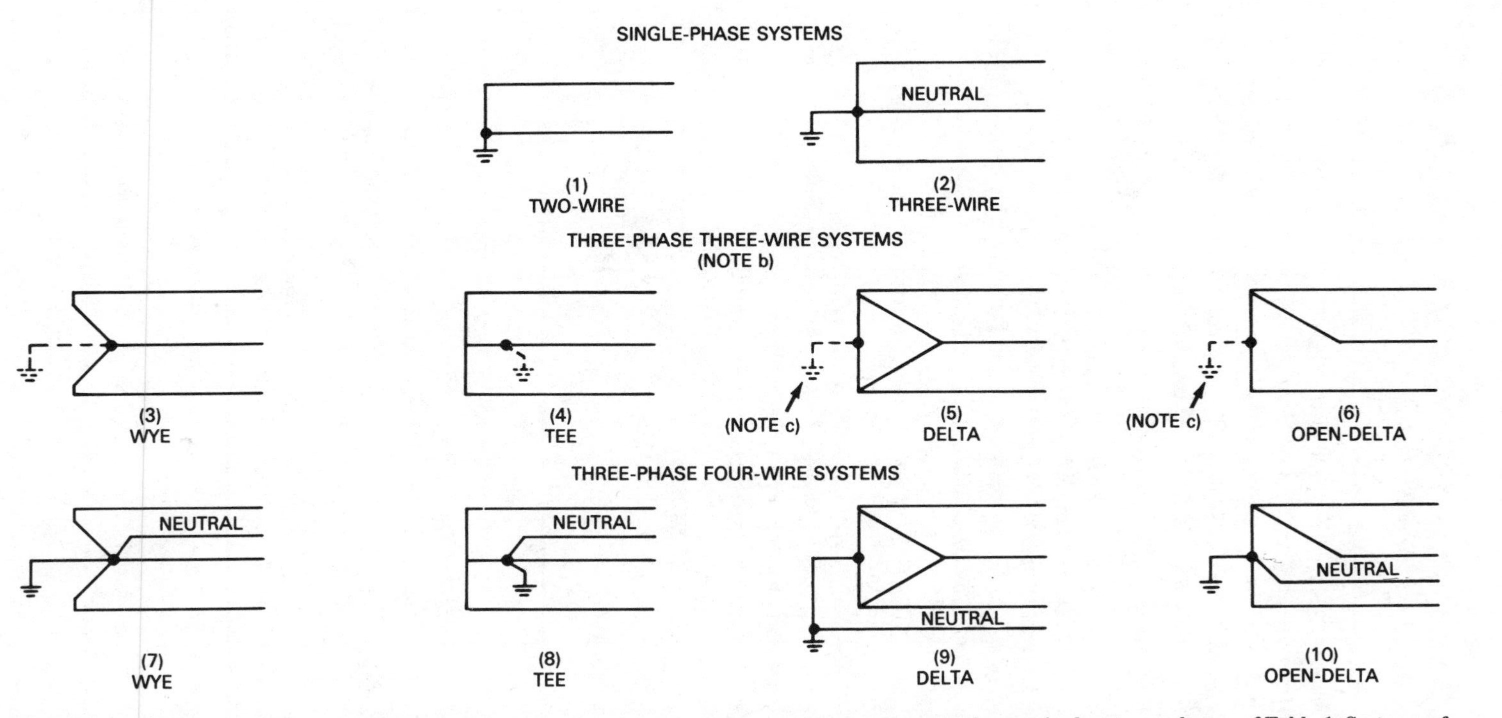

NOTES: (a) The above diagrams show connections of transformer secondary windings to supply the nominal system voltages of Table 1. Systems of more than 600 V are normally three-phase and supplied by connections (3), (5) ungrounded, or (7). Systems of 120–600 V may be either single-phase or three-phase and all of the connections shown are used to some extent for some systems in this voltage range.

(b) Three-phase, three-wire systems may be solidly grounded, impedance grounded, or ungrounded, but are not intended to supply loads connected phase-to-neutral (as the four-wire systems are).

(c) In connections (5) and (6), the ground may be connected to the midpoint of one winding as shown (if available), to one phase conductor (*corner* grounded), or omitted entirely (ungrounded).

(d) Single-phase services and single-phase loads may be supplied from single-phase systems or from three-phase systems. They are connected phase-to-phase when supplied from three-phase, three-wire systems and either phase-to-phase or phase-to-neutral from three-phase, four-wire systems.

Fig 10
Principal Transformer Connections to Supply the System Voltages of Table 1

Table 1
Standard Nominal System Voltages and Voltage Ranges

VOLTAGE CLASS	NOMINAL SYSTEM VOLTAGE (Note a)			VOLTAGE RANGE A (Note b)			VOLTAGE RANGE B (Note b)		
				Maximum	Minimum		Maximum	Minimum	
	Two-wire	Three-wire	Four-wire	Utilization and Service Voltage (Note c)	Service Voltage	Utilization Voltage	Utilization and Service Voltage	Service Voltage	Utilization Voltage
Low Voltage (Note 1)				**Single-Phase Systems**					
	120			126	114	110	127	110	106
		120/240		**126/252**	**114/228**	**110/220**	**127/254**	**110/220**	**106/212**
				Three-Phase Systems					
			208Y/120 (Note d)	**218Y/126**	**197Y/114**	**191Y/110**	**220Y/127**	**191Y/110** (Note 2)	**184Y/106** (Note 2)
			240/120	**252/126**	**228/114**	**220/110**	**254/127**	**220/110**	**212/106**
		240		252	228	220	254	220	212
			480Y/277	**504Y/291**	**456Y/263**	**440Y/254**	**508Y/293**	**440Y/254**	**424Y/245**
		480		**504**	**456**	**440**	**508**	**440**	**424**
		600 (Note e)		630 (Note e)	570	550	635 (Note e)	550	530
Medium Voltage		2 400		2 520	2 340	2 160	2 540	2 280	2 080
			4 160Y/2 400	4 370/2 520	4 050Y/2 340	3 740Y/2 160	4 400Y/2 540	3 950Y/2 280	3 600/2 080
		4 160		**4 370**	**4 050**	**3 740**	**4 400**	**3 950**	**3 600**
		4 800		5 040	4 680	4 320	5 080	4 560	4 160
		6 900		7 240	6 730	6 210	7 260	6 560	5 940
			8 320Y/4 800	8 730Y/5 040	8 110Y/4 680	(Note f)	8 800Y/5 080	7 900Y/4 560	(Note f)
			12 000Y/6 930	12 600Y/7 270	11 700Y/6 760		12 700Y/7 330	11 400Y/6 580	
			12 470Y/7 200	**13 090Y/7 560**	**12 160Y/7 020**		**13 200Y/7 620**	**11 850Y/6 840**	
			13 200Y/7 620	**13 860Y/8 000**	**12 870Y/7 430**		**13 970Y/8 070**	**12 504Y/7 240**	
			13 800Y/7 970	14 490Y/8 370	13 460Y/7 770		14 520Y/8 380	13 110Y/7 570	
		13 800		**14 490**	**13 460**	**12 420**	**14 520**	**13 110**	**11 880**
			20 780Y/12 000	21 820Y/12 600	20 260Y/11 700	(Note f)	22 000Y/12 700	19 740Y/11 400	(Note f)
			22 860Y/13 200	24 000Y/13 860	22 290Y/12 870		24 200Y/13 970	21 720Y/12 540	
		23 000		24 150	22 430		24 340	21 850	
			24 940Y/14 400	**26 190Y/15 120**	**24 320Y/14 040**		**26 400Y/15 240**	**23 690Y/13 680**	
			34 500Y/19 920	**36 230Y/20 920**	**33 640Y/19 420**		**36 510Y/21 080**	**32 780Y/18 930**	
		34 500		36 230	33 640		36 510	32 780	
				Maximum Voltage (Note g)					
		46 000		48 300					
		69 000		**72 500**					
High Voltage		**115 000**		**121 000**					
		138 000		**145 000**					
		161 000		169 000					
		230 000		**242 000**					
		(Note h)							
		345 000		**362 000**					
		500 000		**550 000**					
		765 000		**800 000**					
		1 100 000		**1 200 000**					

(Preferred system voltages in bold-face type)

NOTES: (1) Minimum utilization voltages for 120-600 volt circuits not supplying lighting loads are as follows:

	Nominal System Voltage	Range A	Range B
	120	108	104
	120/240	**108/216**	**104/208**
(Note 2)	**208Y/120**	**187Y/108**	**180Y/104**
	240/120	**216/108**	**208/104**
	240	216	208
	480Y/277	**432Y/249**	**416Y/240**
	480	**432**	**416**
	600	540	520

(2) Many 220 volt motors were applied on existing 208 volt systems on the assumption that the utilization voltage would not be less than 187 volts. Caution should be exercised in applying the Range B minimum voltages of Table 1 and Note (1) to existing 208 volt systems supplying such motors

Table 1
Standard Nominal System Voltages and Voltage Ranges *(continued)*

NOTES:

(a) Three-phase three-wire systems are systems in which only the three-phase conductors are carried out from the source for connection of loads. The source may be derived from any type of three-phase transformer connection, grounded or ungrounded. Three-phase four-wire systems are systems in which a grounded neutral conductor is also carried out from the source for connection of loads. Four-wire systems in Table 1 are designated by the phase-to-phase voltage, followed by the letter Y (except for the 240/120 volt delta system), a slant line, and the phase-to-neutral voltage. Single-phase services and loads may be supplied from either single-phase or three-phase systems. The principal transformer connections that are used to supply single-phase and three-phase systems are illustrated in Appendix A.

(b) The voltage ranges in this table are illustrated in Appendix B.

(c) For 120-600 volt nominal systems, voltages in this column are maximum service voltages. Maximum utilization voltages would not be expected to exceed 125 volts for the nominal system voltage of 120, nor appropriate multiples thereof for other nominal system voltages through 600 volts.

(d) A modification of this three-phase, four-wire system is available as a 120/208Y volt service for single-phase, three-wire, open-wye applications.

(e) Certain kinds of control and protective equipment presently available have a maximum voltage limit of 600 volts; the manufacturer or power supplier or both should be consulted to assure proper application.

(f) Utilization equipment does not generally operate directly at these voltages. For equipment supplied through transformers, refer to limits for nominal system voltage of transformer output.

(g) For these systems Range A and Range B limits are not shown because, where they are used as service voltages, the operating voltage level on the user's system is normally adjusted by means of voltage regulation to suit their requirements.

(h) Standard voltages reprinted from American National Standard C92.2-1981 for convenience only.

equipment such as motors; they run approximately 17% slower than in the United States. Plugs and receptacles are generally different, and this helps to prevent utilization equipment from the United States from being connected to the wrong voltage.

In general, equipment rated for use in the United States cannot be used outside of the United States, and equipment rated for use outside of the United States cannot be used in the United States. If electric equipment made for use in the United States must be used outside the United States, information on the voltage, frequency, and type of plug required should be obtained. If the difference is only in the voltage, transformers are generally available to convert the supply voltage to the equipment voltage.

3.1.5 Voltage Standard for Canada. The voltage standard for Canada is CAN3-C235-83 [10]. This standard differs from the United States standard in both the list of standard nominal voltages and the tolerance limits.

3.2 Voltage Control in Electric Power Systems

3.2.1 Principles of Power Transmission and Distribution in Utility Systems. To understand the principles of voltage control required to provide satisfactory voltage to utilization equipment, a general understanding of the principles of power transmission and distribution in utility systems is necessary since most industrial plants obtain most of their electric power from the local electric utility. Fig 11 shows a one-line diagram of a typical utility power generation, transmission, and distribution system.

Most utility generating stations are located near sources of water. Generated power, except for station requirements, is transformed in a transmission substation located at the generating station to a transmission voltage generally 69 000 V or higher for transmission to major load areas. Transmission lines are controlled only to keep the lines operating within normal voltage limits and to facilitate power flow. ANSI C84.1-1982 [2] specifies only the nominal and maximum values for systems over 34 500 V.

Transmission lines supply distribution substations equipped with transformers that step the transmission voltage down to a primary distribution voltage generally

Fig 11
Typical Utility Generation, Transmission, and Distribution System

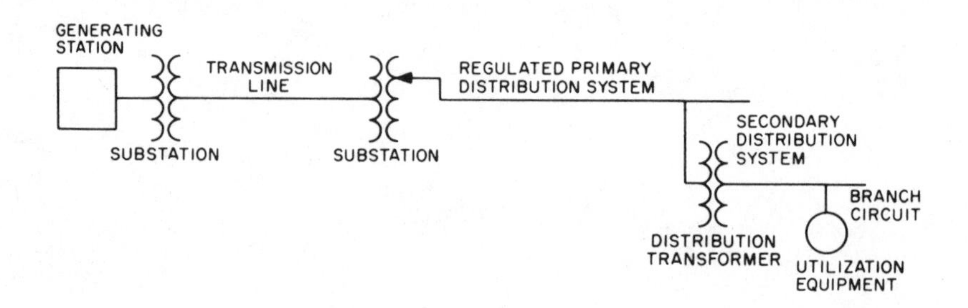

in the range from 4160–34 500 V with 12 470, 13 200, and 13 800 V in widest use.

There is a trend in the electric utility industry to use 23 kV and 34.5 kV subtransmission voltages for distribution. The majority of the utilities have 34.5 kV, a minority have 23 kV, and a few have both. If the supplying utility offers one of these voltages for primary distribution within a building, competent electricians experienced in making splices and terminations for 23 kV or 34.5 kV cable must be secured to obtain a good installation. Suitable liquid-cooled and dry-type distribution transformers, cable, and protective equipment are available for these voltages.

NOTE: As of this revision of this standard, the availability of US manufactured 35 kV indoor metal-clad switchgear is very limited.

Voltage control is applied when necessary for the purpose of supplying satisfactory voltage to the terminals of utilization equipment. The transformers used to step the transmission voltage down to the primary distribution voltage are generally equipped with tap-changing-under-load equipment, which changes the ratio of the transformer under load in order to maintain the primary distribution voltage at the substation at a specific value regardless of fluctuations in the transmission voltage or load. Separate step or induction regulators may also be used.

If the load is remote from the substation, the regulator controls are equipped with compensators that raise the voltage as the load increases and lower the voltage as the load decreases to compensate for the voltage drop in the primary distribution system that extends out radially from the substation to supply the surrounding area. Thus a fixed voltage is maintained at the feed point of the primary distribution system, and the voltage is prevented from rising to peak values during light-load conditions when the voltage drop along the primary distribution system is low. This is illustrated in Fig 12. Note that plants close to the substation will receive voltages which, on the average, will be higher than those received by

Fig 12
Effect of Regulator Compensation on Primary Distribution System Voltage

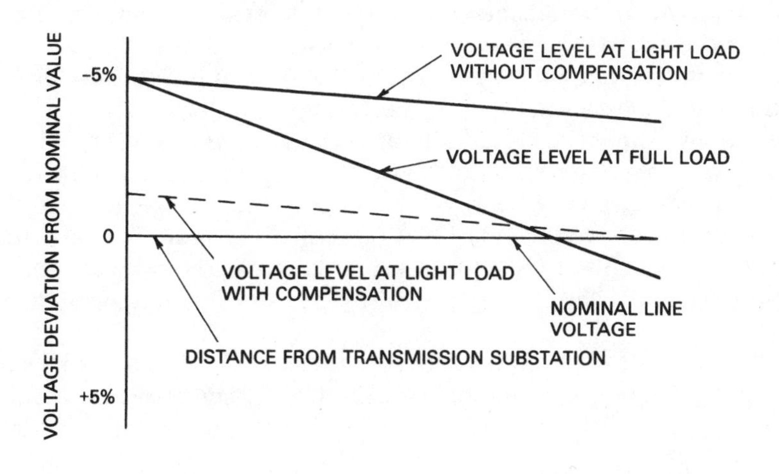

plants at a distance from the distribution substation. See 3.2.8 on the use of distribution transformer taps. Switched or fixed capacitors are also used to improve the voltage on primary feeders.

The primary distribution system supplies distribution transformers that step the primary distribution voltage down to utilization voltages generally in the range of 120-600 V to supply a secondary distribution system to which the utilization equipment is connected. Small transformers used to step a higher utilization voltage down to a lower utilization voltage such as 480 V to 208Y/120 V are considered part of the secondary distribution system.

The supply voltages available to an industrial plant depend upon whether the plant is connected to the distribution transformer, the primary distribution system, or the transmission system, which, in turn, depends on the size of the plant load.

Small plants with loads up to several hundred kilovolt-amperes and all plants supplied from low-voltage secondary networks are connected to the distribution transformer, and the secondary distribution system consists of the connections from the distribution transformer to the plant service and the plant wiring.

Medium-sized plants with loads of a few thousand kilovolt-amperes are connected to the primary distribution system, and the plant provides the portion of the primary distribution system within the plant, the distribution transformers, and the secondary distribution system.

Large plants with loads of more than a few thousand kilovolt-amperes are connected to the transmission system, and the plant provides the primary distribution system, the distribution transformers, the secondary distribution system, and it may provide the distribution substation.

Details of the connection between the utility system and the plant system will depend on the policy of the supplying utility. If power is supplied by in-plant generation, the generators will replace wholly or in part the utility supply system up to the distribution transformer if generation is over 600 V, and it will also replace the distribution transformer if generation is at 600 V or below.

3.2.2 Development of the Voltage Tolerance Limits for ANSI C84.1-1982 [2]. The voltage tolerance limits in ANSI C84.1-1982 [2] are based on ANSI/NEMA MG1-1978 [8], which established the voltage tolerance limits of the standard induction motor at ±10% of nameplate ratings of 230 V and 460 V. Since motors represent the major component of utilization equipment, they were given primary consideration in the establishment of the voltage standard.

The best way to show the voltages in an electric power distribution system is in terms of a 120 V base. This cancels the transformation ratios between systems so that the actual voltages vary solely on the basis of the voltage drops in the system. Any voltage may be converted to a 120 V base by dividing the actual voltage by the ratio of transformation to the 120 V base. For example, the ratio of transformation for the 480 V system is 480/120 or 4, so 460 V in a 480 V system would be 460/4 or 115 V.

The tolerance limits of the 460 V motor in terms of the 120 V base become 115 V plus 10%, or 126.5 V, and 115 V minus 10%, or 103.5 V. The problem is to decide how this tolerance range of 23 V should be divided between the primary distribution system, the distribution transformer, and the secondary distribution system, which make up the regulated distribution system. The solution adopted by ANSI Accredited Committee C84 is shown in Table 2.

The tolerance limits of the standard motor on the 120 V base of 126.5 V maximum and 103.5 V minimum were raised 0.5 V to 127 V maximum and 104 V minimum to eliminate the fractional volt. These values became the tolerance limits for range B in the standard. An allowance of 13 V was allotted for the voltage drop in the primary distribution system. Deducting this voltage drop from 127 V establishes a minimum of 114 V for utility services supplied from the primary distribution system. An allowance of 4 V was provided for the voltage drop in the distribution transformer and the connections to the plant wiring. Deducting this voltage drop from the minimum primary distribution voltage of 114 V provides a minimum of 110 V for utility secondary services from 120-600 V. An allowance of 6 V, or 5%, for the voltage drop in the plant wiring, as provided in ANSI/NFPA 70-1987 [9] (the National Electrical Code [NEC]), partials the distribution of the 23 V tolerance zone down to a minimum utilization voltage of 104 V.

The range A limits for the standard were established by reducing the maximum tolerance limits from 127 V to 126 V and increasing the minimum tolerance limits from 104 V to 108 V. The spread band of 18 V was then allotted as follows: 9 V for the voltage drop in the primary distribution system to provide a minimum primary service voltage of 117 V; 3 V for the voltage drop in the distribution transformer and secondary connections to provide a minimum utility secondary service voltage of 114 V; and 6 V for the voltage drop in the plant wiring to provide a minimum utilization voltage of 108 V.

Four additional modifications were made in this basic plan to establish ANSI C84.1-1982 [2]. The maximum utilization voltage in range A was reduced from 126 V to 125 V for low-voltage systems in the range from 120-480 V because there should be sufficient load on the distribution system to provide at least 1 V drop on the 120 V base under most operating conditions. This maximum voltage of 125 V is also a practical limit for lighting equipment because the life of the 120 V incandes-

Table 2
Standard Voltage Profile for
Low-Voltage Regulated Power Distribution System, 120 V Base

	Range A (V)	Range B (V)
Maximum allowable voltage	126 (125*)	127
Voltage drop allowance for primary distribution line	9	13
Minimum primary service voltage	117	114
Voltage drop allowance for distribution transformer	3	4
Minimum secondary service voltage	114	110
Voltage drop allowance for plant wiring	6 (4†)	6 (4†)
Minimum utilization voltage	108 (110†)	104 (106†)

* For utilization voltage of 120-600 V.
† For building wiring circuits supplying lighting equipment.

cent lamp is reduced by 42% when operated at 125 V (see 3.5.4, Table 9). The voltage drop allowance of 6 V on the 120 V base for the drop in the plant wiring was reduced to 4 V for circuits supplying lighting equipment. This raised the minimum voltage limit for utilization equipment to 106 V in range B and 110 V in range A because the minimum limits for motors of 104 V in range B and 108 V in range A were considered too low for satisfactory operation of lighting equipment. The utilization voltages for the 6900 V and 13 800 V systems in range B were adjusted to coincide with the tolerance limits of ±10% of the nameplate rating of the 6600 V and 13 200 V motors used on these respective systems.

The tolerance limits for the service voltage provide guidance to the supplying utility for the design and operation of its distribution system. The service voltage is the voltage at the point where the utility conductors connect to the user conductors, although in practice it is generally measured at the service switch for services of 600 V and below and at the billing meter potential transformers for services over 600 V. The tolerance limits for the voltage at the point of connection of utilization equipment provide guidance to the user for the design and operation of the user distribution system, and to utilization equipment manufacturers for the design of utilization equipment.

Electric supply systems are to be designed and operated so that most service voltages fall within the range A limits. User systems are to be designed and operated so that, when the service voltages are within range A, the utilization voltages are within range A. Utilization equipment is to be designed and rated to give fully satisfactory performance within the range A limits for utilization voltages.

Range B is provided to allow limited excursions of voltage outside the range A limits that necessarily result from practical design and operating conditions. The supplying utility is expected to take action within a reasonable time to restore service voltages to range A limits. The user is expected to take action within a reasonable time to restore utilization voltages to range A limits. Insofar as practicable, utilization equipment may be expected to give acceptable performance at voltages outside range A but within range B. When voltages occur outside the limits of range B, prompt corrective action should be taken.

To convert the 120 V base voltage to equivalent voltages in other systems, the voltage on the 120 V base is multiplied by the ratio of the transformer that would be used to connect the other system to a 120 V system. In general, oil-filled distribution transformers for systems below 15 000 V have nameplate ratings that are the same as the standard system nominal voltages; so the ratio of the standard nominal voltages may be used to make the conversion. However, for primary distribution voltages over 15 000 V, the primary nameplate rating of oil-filled distribution transformers is not the same as the standard system nominal voltages. Also most distribution transformers are equipped with taps that can be used to change the ratio of transformation. So if the primary distribution voltage is over 15 000 V, or taps have been used to change the transformer ratio, then the actual transformer ratio must be used to convert the base voltage to another system.

For example, the maximum tolerance limit of 127 V on the 120 V base for the service voltage in range B is equivalent, on the 4160 V system, to $4160/120 \cdot 127 = 4400$ V to the nearest 10 V. However, if the 4160–120 V transformer is set on the

+2½% tap, the voltage ratio would be 4160 + (4160 · 0.025) = 4160 + 104 = 4264 to 120. The voltage on the primary system equivalent to 127 V on the secondary system would be 4264/120 · 127 = 35.53 · 127 = 4510 V to the nearest 10 V. If the maximum primary distribution voltage of 4400 V is applied to the 4264–120 V transformer, the secondary voltage would be 4400/4260 · 120 =124 V. So the effect of using a +2½% tap is to lower the secondary voltage by 2½%.

3.2.3 System Voltage Tolerance Limits. The voltage range for all standard nominal system voltages in the utilization and distribution range of 120–34 500 V is specified in ANSI C84.1-1982 [2] for two critical points on the distribution system: the point of delivery by the supplying utility and the point of connection to utilization equipment. For transmission voltages over 34 500 V only the nominal and maximum voltage is specified, because the voltages are normally unregulated and only a maximum voltage is required to establish the design insulation level for the line and associated apparatus.

The actual voltage measured at any point on the system will vary depending on the location of the point of measurement and the system load at the time the measurement is made. Fixed voltage changes take place in transformers in accordance with the transformer ratio, while voltage variations occur from the operation of voltage control equipment and the changes in voltage drop between the supply source and the point of measurement as a result of changes in the current flowing in the circuit.

3.2.4 Voltage Profile Limits for a Regulated Distribution System. Figure 13 shows the voltage profile of a regulated power distribution system using the limits of range A in Table 1. Assuming a nominal primary distribution voltage of 13 800 V, range A in Table 1 shows that this voltage should be maintained by the supplying

Fig 13
Voltage Profile of Limits of Range A, ANSI C84.1-1982 [2]

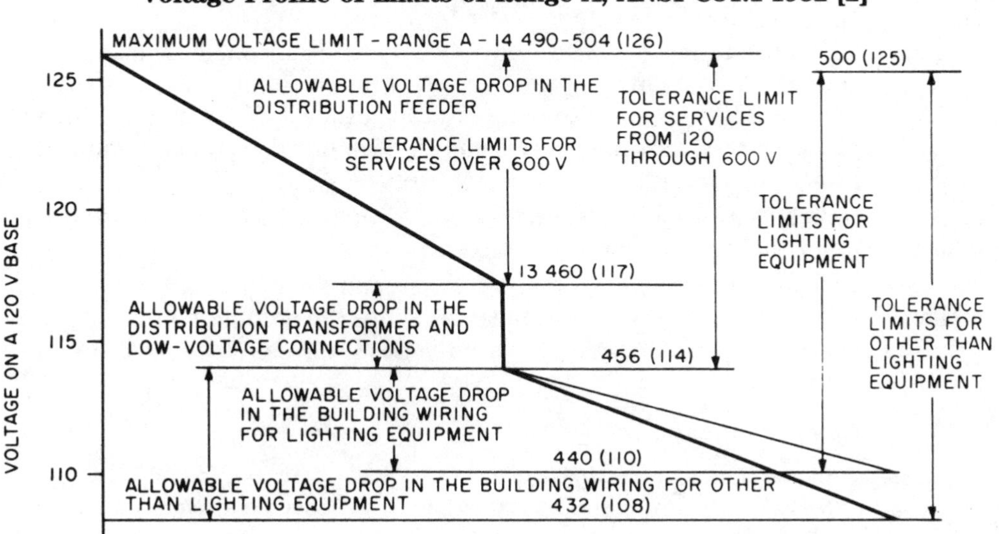

utility between a maximum of 126 V and a minimum of 117 V on a 120 V base. Since the base multiplier for converting from the 120 V system to the 13 800 V system is 13 800/120 or 115, the actual voltage limits for the 13 800 V system are 115 · 126 or 14 490 V maximum and 115 · 117 or 13 460 V minimum.

If a distribution transformer with a ratio of 13 800 to 480 V is connected to the
13 800/120 or 115, the actual voltage limits for the 13 800 V system are 115 · 126 or
secondary service be maintained by the supplying utility between a maximum of 126 V and a minimum of 114 V on the 120 V base. Since the base multiplier for the 480 V system is 480/120 or 4, the actual values are 4 · 126 or 504 V maximum and 4 · 114 or 456 V minimum.

Range A of Table 1 as modified for utilization equipment of 120–600 V provides for a maximum utilization voltage of 125 V and a minimum of 110 V for lighting equipment and 108 V for other than lighting equipment on the 120 V base. Using the base multiplier of 4 for the 480 V system, the maximum utilization voltage would be 4 · 125 V or 500 V and the minimum for other than lighting equipment would be 4 · 108 V or 432 V. For lighting equipment connected phase to neutral, the maximum voltage would be 500 V divided by the square root of 3 or 288 V and the minimum voltage would be 4 · 110 V or 440 V divided by the square root of 3 or 254 V.

3.2.5 System Voltage Nomenclature. The nominal system voltages in Table 1 are designated in the same way as on the nameplate of the transformer for the winding or windings supplying the system.

(1) *Single-Phase Systems*

120 — Indicates a single-phase, two-wire system in which the nominal voltage between the two wires is 120 V.

120/240 — Indicates a single-phase, three-wire system in which the nominal voltage between the two phase conductors is 240 V, and from each phase conductor to the neutral it is 120 V.

(2) *Three-Phase Systems*

240/120 — Indicates a three-phase, four-wire system supplied from a delta-connected transformer. The midtap of one winding is connected to a neutral. The three-phase conductors provide a nominal 240 V three-phase system, and the neutral and the two adjacent phase conductors provide a nominal 120/240 V single-phase system.

Single number — Indicates a three-phase, three-wire system in which the number designates the nominal voltage between phases.

Two numbers separated by a Y/ — Indicates a three-phase, four-wire system from a wye-connected transformer in which the first number indicates the nominal phase-to-phase voltage and the second number indicates the nominal phase-to-neutral voltage.

NOTES: (1) All single-phase systems and all three-phase, four-wire systems are suitable for the connection of phase-to-neutral load.

(2) See Chapter 7 for methods of system grounding.

(3) Figure 14 gives an overview of voltage relationships for 480 V three-phase systems and 120/240 V single- and three-phase systems.

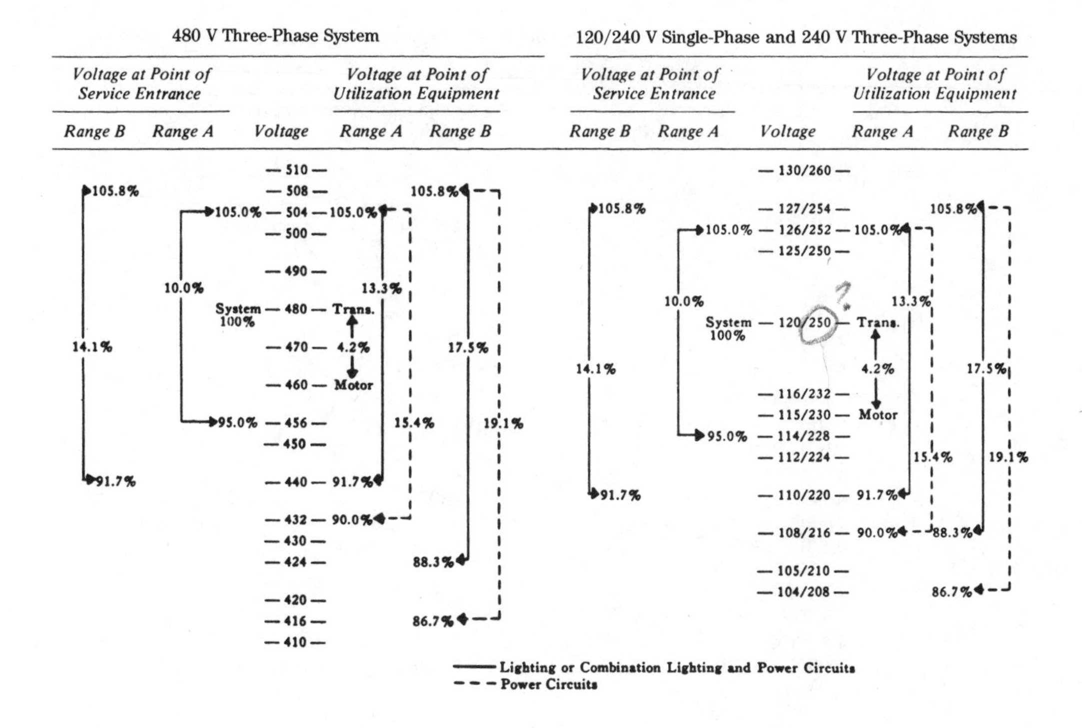

Fig 14
Voltage Relationships Based on Voltage Ranges in ANSI C84.1-1982 [2]

3.2.6 Nonstandard Nominal System Voltages. Since ANSI C84.1-1982 [2] lists only the standard nominal system voltages in common use in the United States, system voltages will frequently be encountered that differ from the standard list. A few of these may be so widely different as to constitute separate systems in too limited use to be considered standard. However, in most cases the nominal system voltages will differ by only a few percent as shown in Table 3. A closer examination of the table shows that these differences are due mainly to the fact that some voltages are multiples of 110 V, others are multiples of 115 V, some are multiples of 120 V, and a few are multiples of 125 V.

The reasons for these differences go back to the original development of electric power distribution systems. The first utilization voltage was 100 V. However, the supply voltage had to be raised to 110 V in order to compensate for the voltage drop in the distribution system. This led to overvoltage on equipment connected close to the supply, so the utilization equipment rating was also raised to 110 V. As generator sizes increased and distribution and transmission systems developed, an effort to keep transformer ratios in round numbers led to a series of utilization voltages of 110, 220, 440, and 550 V, a series of primary distribution voltages of 2200, 4400, 6600, and 13 200 V, and a series of transmission voltages of 22 000, 33 000, 44 000, 66 000, 110 000, 132 000, and 220 000 V.

Table 3
Nominal System Voltages

Standard Nominal System Voltages	Associated Nonstandard Nominal System Voltages
Low voltages	
120	110, 115, 125
120/240	110/220, 115/230, 125/250
208Y/120	216Y/125
240/120	
240	230, 250
480Y/277	460Y/265
480	440
600	550, 575
Medium Voltages	
2400	2200, 2300
4160Y/2400	
4160	4000
4800	4600
6900	6600, 7200
8320Y/4800	11 000, 11 500
12 000Y/6930	
12 470Y/7200	
13 200Y/7620	
13 200	
13 800Y/7970	14 400
13 800	
20 780Y/12 000	
22 860Y/13 200	
23 000	
24 940Y/14 400	
34 500Y/19 920	
34 500	33 000
46 000	44 000
69 000	66 000
High Voltages	
115 000	110 000, 120 000
138 000	132 000
161 000	154 000
230 000	220 000
Ultra-High Voltages	
345 000	
500 000	
765 000	
1 100 000	

As a result of the effort to maintain the supply voltage slightly above the utilization voltage, the supply voltages were raised again to multiples of 115 V, which resulted in a new series of utilization voltages of 115, 230, 460, and 575 V, a new series of primary distribution voltages of 2300, 4600, 6900, and 13 800 V, and a new series of transmission voltages of 23 000, 34 500, 46 000, 69 000, 115 000, 138 000, and 230 000 V.

As a result of continued problems with the operation of voltage-sensitive lighting equipment and voltage-insensitive motors on the same system, and the development of the 280Y/120 V network system, the supply voltages were raised again to multiples of 120 V. This resulted in a new series of utilization voltages of 120, 208Y/120, 240, 480, and 600 V, and a new series of primary distribution voltages of 2400, 4160Y/2400, 4800, 12 000, and 12 470Y/7200 V. However, most of the existing primary distribution voltages continued in use and no 120 V multiple voltages developed at the transmission level.

3.2.7 Standard Nominal System Voltages in the United States. The nominal system voltages listed in the left-hand column of Table 3 are designated as *standard nominal system voltages* in the United States by ANSI C84.1-1982 [2]. In addition, those shown in boldface type in Table 1 are designated as preferred standards to provide a long-range plan for reducing the multiplicity of voltages.

In the case of utilization voltages of 600 V and below, the associated nominal system voltages in the right-hand column are obsolete and should not be used. Where possible, manufacturers are encouraged to design utilization equipment to provide acceptable performance within the utilization voltage tolerance limits specified in the standard. Some numbers listed in the right-hand columnn are used in equipment ratings, but these should not be confused with the numbers designating the nominal system voltage on which the equipment is designed to operate.

In the case of primary distribution voltages, the numbers in the right-hand column may designate an older system in which the voltage tolerance limits are maintained at a different level than the standard nominal system voltage, and special consideration should be given to the distribution transformer ratios, taps, and tap settings.

3.2.8 Use of Distribution Transformer Taps to Shift Utilization Voltage Spread Band. Except for small distribution transformers in sizes of 50–100 kVA or less, power and distribution transformers are normally provided with four taps on the primary winding in 2½% steps, generally ± 2–2½% taps. These taps permit the transformer ratio to be changed to raise or lower the secondary voltage spread band to provide a closer fit to the tolerance limits of the utilization equipment. There are two situations requiring the use of taps:

(1) Where the primary voltage spread band is above or below the limits required to provide a satisfactory secondary voltage spread band. This occurs under two conditions:

(a) The primary voltage has a nominal value that is slightly different from the transformer primary nameplate rating. For example, if a 13 200–480 V transformer is connected to a nominal 13 800 V system, the nominal secondary voltage would be (13 800/13 200) · 480 = 502 V. However, if the 13 800 V system were connected to the +5% tap of the 13 200–480 V transformer at 13 860 V, the nominal secondary voltage would be (13 860/13 800) · 480 = 482 V, which is practically the same as would be obtained from a transformer having the proper ratio of 13 800–480 V.

(b) The primary voltage spread is in the upper or lower portion of the tolerance limits provided in ANSI C84.1-1982 [2]. For example, a 13 200–480 V transformer is connected to a 13 200 V primary distribution system close to the substation so that the primary voltage spread band falls in the upper half of the tolerance zone for range A in the standard, or 13 200–13 860 V. This would result in a nominal secondary voltage under no-load conditions of 480–504 V. By setting the transformer on the +2½% tap at 13 530 V, the secondary voltage would be lowered 2½% to a range of 468–491 V. This would materially reduce the overvoltage on utilization equipment.

(2) Adjusting the utilization voltage spread band to provide a closer fit to the tolerance limits of the utilization equipment. For example, Table 4 shows the shift in the utilization voltage spread band for the +2½% and +5% taps as compared to the utilization voltage tolerance limits for range A of ANSI C84.1-1982 [2] for the 480 V system. Table 5 shows the voltage tolerance limits of standard 460 V and 440 V three-phase induction motors. Table 6 shows the tolerance limits for standard 277 V and 265 V fluorescent lamp ballasts. A study of these three tables shows that a tap setting of *normal* will provide the best fit with the tolerance limits of the 460 V motor and the 277 V ballast, but a setting on the +5% tap will provide the best fit for the 440 V motor and the 265 V ballast. For buildings having appreciable numbers of

Table 4
Tolerance Limits for Lighting Circuits from Table 1, Range A, in Volts

Nominal System Voltage (volts)	Transformer Tap	Minimum Utilization Voltage (volts)	Maximum Utilization Voltage (volts)
480Y/277	Normal	440Y/254	500Y/288
468Y/270	Plus 2½%	429Y/248	488Y/281
456Y/263	Plus 5%	418Y/241	475Y/274

Table 5
Tolerance Limits for Low-Voltage Three-Phase Motors, in Volts

Motor Rating (volts)	− 10 Percent	+ 10 Percent
460	414	506
440	396	484

Table 6
Tolerance Limits for Low-Voltage Standard Fluorescent Lamp Ballasts, in Volts

Ballast Rating (volts)	− 10 Percent	+ 10 Percent
277	249	305
265	238	292

both ratings of motors and ballasts, a setting on the +2½% tap may provide the best compromise.

Note that these examples assume that the tolerance limits of the supply and utilization voltages are within the tolerance limits specified in ANSI C84.1-1982 [2]. This may not be true, so the actual voltages should be measured, preferably with a seven-day chart, to obtain readings during the night and over weekends when maximum voltages occur. These actual voltages can then be used to prepare a voltage profile similar to Fig 13 to check the proposed transformer ratios and tap settings.

Where a plant has not yet been built so that actual voltages cannot be measured, the supplying utility should be requested to provide the expected spread band for the supply voltage, preferably supported by a seven-day graphic chart from the nearest available location. If the plant furnishes the distribution transformers, recommendations should also be obtained from the supplying utility on the transformer ratios, taps, and tap settings. With this information a voltage profile can be prepared to check the expected voltage spread at the utilization equipment.

Where the supplying utility offers a voltage over 600 V that differs from the standard nominal voltages listed in ANSI C84.1-1982 [2], the supplying utility should be asked to furnish the expected tolerance limits of the supply voltage, preferably supported by a seven-day chart from a nearby location, and the recommended distribution transformer ratio and tap settings to be used to obtain a satisfactory utilization voltage range. With this information, a voltage profile for the supply voltage and utilization voltage limits can be constructed for comparison with the tolerance limits of utilization equipment. If the supply voltage offered by the utility is one of the associated nominal system voltages listed in Table 1, the taps on a standard distribution transformer will generally be sufficient to adjust the distribution transformer ratio to provide a satisfactory utilization voltage range.

Since the taps are on the primary side of the transformer, raising the tap setting by using the +2½% tap increases the transformer ratio by 2½% and lowers the secondary voltage spread band by 2½% minus the voltage drop in the transformer. Taps only serve to move the secondary voltage spread band up or down in the steps of the taps. They cannot correct for excessive spread in the supply voltage or excessive drop in the plant wiring system. If the voltage spread band at the utilization equipment falls outside the satisfactory operating range of the equipment, then action must be taken to improve voltage conditions by other means (see 3.7).

In general, transformers should be selected with the same primary nameplate voltage rating as the nominal voltage of the primary supply system, and the same secondary voltage rating as the nominal voltage of the secondary system. Taps should be provided at +2½% and +5% and at -2½% and -5% to allow for adjustment in either direction.

3.3 Voltage Selection

3.3.1 Selection of Low-Voltage Utilization Voltages. The preferred utilization voltage for industrial plants is 480Y/277 V. Three-phase power and other 480 V loads are connected directly to the system at 480 V, and gaseous discharge lighting is connected phase-to-neutral at 277 V. Small dry-type transformers rated

480-208Y/120 V are used to provide 120 V, single-phase, for convenience outlets, and 208 V, single-phase and three-phase, for small tools and other machinery. Where requirements are limited to 120 or 240 V, single-phase, 480-120/240 V single-phase transformers may be used. However, single-phase transformers should be connected in sequence to the individual phases in order to keep the load on each phase balanced (see 3.8).

For small industrial plants supplied at utilization voltage by a single distribution transformer, the choice of voltages is limited to those the utility will supply. However, most utilities will supply most of the standard nominal voltages listed in ANSI C84.1-1982 [2] with the exception of 600 V, although all voltages supplied may not be available at every location.

The built-up downtown areas of most large cities are supplied from secondary networks. Originally only 208Y/120 V was available, but most utilities now provide spot networks at 480Y/277 V for large installations.

3.3.2 Utility Service Supplied from a Medium-Voltage Primary Distribution Line. When an industrial plant becomes too large to be supplied at utilization voltage from a single distribution transformer normally furnished by the supplying utility and located outdoors, a tap from the primary distribution line is brought into the plant as a primary service to supply distribution transformers, which are generally dry type, epoxy-filled cast type, high fire point liquid, or nonflammable fluid-insulated types suitable for indoor installation. Generally these distribution transformers are combined with primary and secondary switching and protective equipment to become unit substations. They are designated as primary unit substations when the secondary voltage is over 1000 V and secondary unit substations when the secondary voltage is 1000 V and below. Primary distribution may also be used to supply large industrial plants or plants involving more than one building. In this case, the primary distribution line may be run overhead or underground and may supply distribution transformers located outside the building or unit substations inside the building.

Original primary distribution voltages were limited to the range from 2400-14 400 V, but the increase in load densities in recent years has forced utilities to limit expansion of primary distribution voltages below 15 000 V and to begin converting transmission voltages in the range from 15 000-50 000 V (sometimes called subtransmission voltages) to primary distribution. ANSI C84.1-1982 [2] provides tolerance limits for primary supply voltages up through 34 500 V. ANSI C57.12.20-1981 [1] lists overhead distribution transformers for primary voltages up through 69 000 V.

In case an industrial plant, supplied at utilization voltage from a single primary distribution transformer, contemplates an expansion that cannot be supplied from the existing transformer, a changeover to primary distribution will be required, unless a separate supply to the new addition is permitted by the local elecrical code enforcing authority and the higher cost resulting from separate bills from the utility is acceptable. In any case, the proposed expansion needs to be discussed with the supplying utility to determine whether the expansion can be supplied from the existing primary distribution system or whether the entire load can be transferred to another system. Any utility charges and the plant costs associated with the changes need to be clearly established.

In general, primary distribution voltages between 15 000 and 25 000 V can be brought into a plant and handled like the lower voltages. Primary distribution voltages from 25 000–35 000 V will require at least a preliminary economic study to determine whether they can be brought into the plant or transformed to a lower primary distribution voltage. Voltages above 35 000 V will require transformation to a lower voltage.

In most cases, plants with loads of less than 10 000 kVA will find that 4160 V is the most economical plant primary distribution voltage, and plants with loads over 20 000 kVA will find 13 800 V the most economical considering only the cost of the plant wiring and transformers. If the utility supplies a voltage in the range from 12 000–15 000 V, a transformation down to 4160 V at plant expense cannot normally be justified. For loads of 10 000–20 000 kVA, an economic study including consideration of the costs of future expansion needs to be made to determine the most economical primary distribution voltage.

Where overhead lines are permissible on plant property, an overhead primary distribution system may be built around the outside of the building or to separate buildings to supply utility-type outdoor equipment and transformers. This system is especially economical at voltages over 15 000 V. However, any plans for such an installation should be approved by the insurance company providing fire insurance because of rules governing the installation of oil-filled transformers close to buildings, and should comply with the local NEC [9] enforcing authority because some jurisdictions consider each transformer as a separate service.

Utility primary distribution systems are almost always solidly grounded wye systems, and this should be considered in the design of the plant primary distribution system.

3.3.3 Utility Service Supplied from Medium-Voltage or High-Voltage Transmission Lines. Voltages on transmission lines used to supply large industrial plants range from 23 000–230 000 V. There is an overlap with primary distribution system voltages in the range from 23 000–69 000 V, with voltages of 34 500 V and below tending to fall into the category of regulated primary distribution voltages and voltages above 34 500 V tending to fall into the category of unregulated transmission lines. The transmission voltage will be limited to those voltages the utility has available in the area. A substation is required to step the transmission voltage down to a primary distribution voltage to supply the distribution transformers in the plant.

3.3.3.1 Substation Is Supplied by the Industrial Plant. Most utilities have a low rate for service from unregulated transmission lines; this condition requires the plant to provide the substation. This permits the plant designer to select the primary distribution voltage but requires the plant personnel to assume the operation and maintenance of the substation. The substation designer should obtain from the supplying utility the voltage spread on the transmission line, and recommendations on the substation transformer ratio, tap provisions, and tap setting, and whether regulation should be provided.

With this information a voltage profile similar to Fig 13 is obtained using the actual values for the spread band of the transmission line and the estimated maximum values for the voltage drops in the substation transformer, primary distribution system, distribution transformers, and secondary distribution system to obtain the voltage spread at the utilization equipment. If this voltage spread is not within satisfactory limits, then regulations are required in the substation, preferably by equipping the substation transformer or transformers with tap changing under load.

For plants supplied at 13 800 V, the distribution transformers or secondary unit substations should have a ratio of 13 800-480Y/277 V with two ± 2½% taps. Where medium-sized motors in the 200 hp or larger range are used, a distribution transformer stepping down to 4160 V or 2400 V may be more economical than supplying these motors from the 480 V system.

For plants supplied at 4160 V, the distribution transformers or secondary unit substations should have a ratio of 4160-480Y/277 V with two ± 2½% taps. Medium-sized motors of a few hundred horsepower may economically be connected directly to the 4160 V system, preferably from a separate primary distribution circuit, to avoid the lighting dip that might occur when the large motors are started if distribution transformers supplying lighting equipment were connected to the same circuit.

3.3.3.2 Distribution Substation Is Supplied by the Utility. Most utilities have a rate for power purchased at the primary distribution voltage that is higher than the rate for service at transmission voltage because the utility provides the substation. The choice of the primary distribution voltage is limited to those supplied by the particular utility, but the utility will be responsible for keeping the limits specified for service voltages in ANSI C84.1-1982 [2]. The utility should be requested to provide recommendations for the ratio of the distribution transformers or secondary unit substations, provisions for taps, and the tap settings. With this information, a voltage profile similar to Fig 13 can be constructed using the estimated maximum values for the voltage drops in the primary distribution system, the transformers, and the secondary distribution system to make sure that the utilization voltages fall within satisfactory limits.

3.4 Voltage Ratings for Low-Voltage Utilization Equipment. Utilization equipment is defined as electric equipment that uses electric power by converting it into some other form of energy such as light, heat, or mechanical motion. Every item of utilization equipment is required to have, among other things, a nameplate listing the nominal supply voltage for which the equipment is designed. With one major exception, most utilization equipment carries a nameplate rating that is the same as the voltage system on which it is to be used; that is, equipment to be used on 120 V systems is rated 120 V (except for a few small appliances rated 117 or 118 V), for 208 V systems, 208 V, and so on. The major exception is motors and equipment containing motors. These are also about the only utilization equipment used on systems over 600 V. Single-phase motors for use on 120 V systems have been rated 115 V for many years. Single-phase motors for use on 208 V single-phase systems are rated 200 V, and for 240 V single-phase systems are rated 230 V.

Prior to the late 1960s, low-voltage three-phase motors were rated 220 V for use on both 208 and 240 V system, 440 V for use on 480 V systems, and 550 V for use on 600 V systems. The reason was that most three-phase motors were used in large industrial plants where relatively long circuits resulted in voltages considerably below nominal at the ends of the circuits. Also, utility supply systems had limited capacity and low voltages were common during heavy load periods. As a result, the average voltage applied to three-phase motors approximated the 220, 440, and 550 V nameplate ratings.

In recent years supplying electric utilities have made extensive changes to higher distribution voltages. Increased load density has resulted in shorter primary distribution systems. Distribution transformers have been moved inside buildings to be closer to the load. Lower impedance wiring systems have been used in the secondary distribution system. Capacitors have been used to improve power factors. All of these changes have contributed to reducing the voltage drop in the distribution system which raised the voltage applied to utilization equipment. By the mid-1960s surveys indicated that the average voltage supplied to 440 V motors on 480 V systems was 460 V, and there were increasing numbers of complaints of overvoltages as high as 500 V during light-load periods.

At about the same time the Motor and Generator Committee of the National Electrical Manufacturers Assocation (NEMA) decided that the improvements in motor design and insulation systems would allow a reduction of two frame sizes for standard induction motors rated 600 V and below. However, the motor voltage tolerance would be limited to ±10% of the nameplate rating. As a result, the nameplate voltage rating of the new motor designated as the T-frame motor was raised from the 220/440 V rating of the U-frame motor to 230/460 V. Subsequently, a motor rated 200 V for use on 208 V systems was added to the program. Table 7

Table 7
Nameplate Voltage Ratings of Standard Induction Motors

Nominal System Voltage	Nameplate Voltage
Single-phase motors	
120	115
240	230
Three-phase motors	
208	200
240	230
480	460
600	575
2400	2300
4160	4000
4800	4600
6900	6600
13 800	13 200

shows the nameplate voltage ratings of standard induction motors, as specified in ANSI/NEMA MG1-1978 [8].

The question has been raised why the confusion between equipment ratings and system nominal voltage cannot be eliminated by making the nameplate rating of utilization equipment the same as the nominal voltage of the system on which the equipment is to be used. However, manufacturers say that the performance guarantee for utilization equipment is based on the nameplate rating and not the system nominal voltage. For utilization equipment such as motors where the performance peaks in the middle of the tolerance range of the equipment, better performance can be obtained over the tolerance range specified in ANSI C84.1-1982 [2] by selecting a nameplate rating closer to the middle of this tolerance range.

3.5 Effect of Voltage Variations on Low-Voltage and Medium-Voltage Utilization Equipment

3.5.1 General Effects. When the voltage at the terminals of utilization equipment deviates from the value on the nameplate of the equipment, the performance and the operating life of the equipment are affected. The effect may be minor or serious depending on the characteristics of the equipment and the amount of the voltage deviation from the nameplate rating. Generally, performance conforms to the utilization voltage limits specified in ANSI C84.1-1982 [2] but it may vary for specific items of voltage-sensitive equipment. In addition, closer voltage control may be required for precise operations.

3.5.2 Induction Motors. The variation in characteristics as a function of the applied voltage is given in Table 8—(a) for the older U-frame motors and (b) for the current T-frame motors. Motor voltages below nameplate rating result in reduced starting torque and increased full-load temperature rise. Motor voltages above nameplate rating result in increased torque, increased starting current, and decreased power factor. The increased starting torque will increase the accelerating forces on couplings and driven equipment. Increased starting current causes greater voltage drop in the supply circuit and increases the voltage dip on lamps and other equipment. In general, voltages slightly above nameplate rating have less detrimental effect on motor performance than voltages slightly below nameplate rating.

3.5.3 Synchronous Motors. Synchronous motors are affected in the same manner as induction motors, except that the speed remains constant (unless the frequency changes) and the maximum or pull-out torque varies directly with the voltage if the field voltage remains constant, as in the case where the field is supplied by a generator on the same shaft with the motor. If the field voltage varies with the line voltage as in the case of a static rectifier source, then the maximum or pull-out torque varies as the square of the voltage.

3.5.4 Incandescent Lamps. The light output and life of incandescent filament lamps are critically affected by the impressed voltage. The variation of life and light output with voltage is given in Table 9. The figures for 125 V and 130 V lamps are also included because these ratings are useful in signs and other locations where long life is more important than light output.

Table 8
General Effect of Voltage Variations on Induction-Motor Characteristics

(a) U-Frame Motors

Characteristic	Function of Voltage	Voltage Variation: 90% Voltage	Voltage Variation: 110% Voltage
Starting and maximum running torque	(Voltage)^2	Decrease 19%	Increase 21%
Synchronous speed	Constant	No change	No change
Percent slip	1/(Voltage)^2	Increase 23%	Decrease 17%
Full-load speed	Synchronous speed slip	Decrease 1½%	Increase 1%
Efficiency			
Full load	—	Decrease 2%	Increase ½-1%
¾ load	—	Practically no change	Practically no change
½ load	—	Increase 1-2%	Decrease 1-2%
Power factor			
Full load	—	Increase 1%	Decrease 3%
¾ load	—	Increase 2-3%	Decrease 4%
½ load	—	Increase 4-5%	Decrease 5-6%
Full-load current	—	Increase 11%	Decrease 7%
Starting current	Voltage	Decrease 10-12%	Increase 10-12%
Temperature rise, full load	—	Increase 6-7 °C	Decrease 1-2 °C
Maximum overload capacity	(Voltage)^2	Decrease 19%	Increase 21%
Magnetic noise—no load in particular	—	Decrease slightly	Increase slightly

(b) T-Frame Motors

Characteristic	Function of Voltage	Voltage Variation: 90% Voltage	Voltage Variation: 110% Voltage*
Starting and maximum running torque	(Voltage)^2	Decrease 19%	Increase 21%
Percent slip	1/(Voltage)^2	Increase 20-30%	Decrease 15-20%
Full-load speed	Synchronous speed slip	Slight decrease	Slight increase
Efficiency			
Full load	—	Decrease 0-2%	Decrease 0-3%
¾ load	—	Practically no change	No change to slight decrease
½ load	—	Increase 0-1%	Decrease 0-5%
Power factor			
Full load	—	Increase 1-7%	Decrease 5-15%
¾ load	—	Increase 2-7%	Decrease 5-15%
½ load	—	Increase 3-10%	Decrease 10-20%
Full-load current	—	Increase 5-10%	Slight decrease to 5% increase
Starting current	Voltage	Decrease ≈ 10%	Increase ≈ 10%
Temperature rise, full load	—	Increase 10-15%	Increase 2-15%
Maximum overload capacity	(Voltage)^2	Decrease 19%	Increase 21%
Magnetic noise—no load in particular	—	Slight decrease	Slight increase

*There may be wide variations depending upon type of motor, such as dripproof (DP) or totally enclosed fan cooled (TEFC), and horsepower rating, with the smaller ratings showing the greater variations. Some data will vary according to manufacturers.

Table 9
Effect of Voltage Variations on Incandescent Lamps

Applied Voltage (volts)	Lamp Rating 120 V Percent Life	120 V Percent Light	125 V Percent Life	125 V Percent Light	130 V Percent Life	130 V Percent Light
105	575	64	880	55	—	—
110	310	74	525	65	880	57
115	175	87	295	76	500	66
120	100	100	170	88	280	76
125	58	118	100	100	165	88
130	34	132	59	113	100	100

3.5.5 Fluorescent Lamps. Fluorescent lamps, unlike incandescent lamps, operate satisfactorily over a range of ±10% of the ballast nameplate voltage rating. Light output varies approximately in direct proportion to the applied voltage. Thus a 1% increase in applied voltage will increase the light output by 1% and, conversely, a decrease of 1% in the applied voltage will reduce the light output by 1%. The life of fluorescent lamps is affected less by voltage variation than that of incandescent lamps.

The voltage-sensitive component of the fluorescent fixture is the ballast, a small reactor or transformer that supplies the starting and operating voltages to the lamp and limits the lamp current to design values. These ballasts may overheat when subjected to above normal voltage and operating temperature, and ballasts with integral thermal protection may be required. See NEC [9], Article 410.

3.5.6 High-Intensity Discharge (HID) Lamps (Mercury, Sodium, and Metal Halide). Mercury lamps using the conventional unregulated ballast will have a 30% decrease in light output for a 10% decrease in terminal voltage. If a constant wattage ballast is used, the decrease in light output for a 10% decrease in terminal voltage will be about 2%.

Mercury lamps require 4-8 min to suitably vaporize the mercury in the lamp and reach full brilliancy. At about 20% undervoltage, the mercury arc will be extinguished and the lamp cannot be restarted until the mercury condenses; this takes from 4-8 min unless the lamps have special cooling controls. The lamp life is related inversely to the number of starts so that, if low-voltage conditions require repeated starting, lamp life will be reduced. Excessively high voltage raises the arc temperature, which could damage the glass enclosure if the temperature approaches the glass softening point.

Sodium and metal halide lamps have similar characteristics to mercury lamps although the starting and operating voltages may be somewhat different so that the lamps and ballasts may not be interchangeable. See the manufacturers' catalogs for detailed information.

3.5.7 Infrared Heating Processes. Although the filaments in the lamps used in these installations are of the resistance type, the energy output does not vary with the square of the voltage because the resistance varies at the same time. The energy output does vary roughly as some power of the voltage slightly less than the square, however. Voltage variations can produce unwanted changes in the process heat available unless thermostatic control or other regulating means are used.

3.5.8 Resistance Heating Devices. The energy input and, therefore, the heat output of resistance heaters varies approximately as the square of the impressed voltage. Thus a 10% drop in voltage will cause a drop of approximately 19% in heat output. This, however, holds true only for an operating range over which the resistance remains essentially constant.

3.5.9 Electron Tubes. The current-carrying ability or emission of all electron tubes is affected seriously by voltage deviation from nameplate rating. The cathode life curve indicates that the life is reduced by half for each 5% increase in cathode voltage. This is due to the reduced life of the heater element and to the higher rate of evaporation of the active material from the surface of the cathode. It is extremely important that the cathode voltage be kept near rating on electron tubes for satisfactory service. In many cases this will necessitate a regulated power source. This may be located at or within the equipment, and often consists of a regulating transformer having constant output voltage or current.

3.5.10 Capacitors. The reactive power output of capacitors varies with the square of the impressed voltage. A drop of 10% in the supply voltage, therefore, reduces the reactive power output by 19%, and where the user has made a sizable investment in capacitors for power factor improvement, he loses the benefit of almost 20% of this investment.

3.5.11 Solenoid-Operated Devices. The pull of ac solenoids varies approximately as the square of the voltage. In general, solenoids are designed to operate satisfactorily on 10% overvoltage and 15% undervoltage.

3.5.12 Solid-State Equipment. Thyristors, transistors, and other solid-state devices have no thermionic heaters. Thus they are not nearly as sensitive to long-time voltage variations as the electron tube components they are largely replacing. Internal voltage regulators are frequently provided for sensitive equipment such that it is independent of supply system regulation. This equipment as well as power solid-state equipment is, however, generally limited regarding peak reverse voltage, since it can be adversely affected by abnormal voltages of even microsecond duration. An individual study of the maximum voltage of the equipment, including surge characteristics, is necessary to determine the effect of maximum system voltage or whether abnormally low voltage will result in malfunction.

3.6 Voltage Drop Considerations in Locating the Low-Voltage Secondary Distribution System Power Source. One of the major factors in the design of the secondary distribution system is the location of the power source as close as possible to the center of the load. This applies in every case, from a service drop from a distribution transformer on the street to a distribution transformer located outside the building or a secondary unit substation located inside the building. Frequently building esthetics or available space require the secondary distribution system power supply to be installed in a corner of a building without regard to what this adds to the cost of the building wiring to keep the voltage drop within satisfactory limits.

Figure 15 [14] shows that if a power supply is located in the center of a horizontal floor area at point 0, the area that can be supplied from circuits run radially from point 0 with specified circuit constants, and voltage drop would be the area enclosed by the circle of radius 0–X. However, conduit systems are run in rectangular coordinates so, with this restriction, the area that can be supplied is reduced to the square X–Y–X′–Y′ when the conduit system is run parallel to the axes X–X′and Y–Y′. But the limits of the square are not parallel to the conduit system. Thus to fit the conduit system into a square building with walls parallel to the conduit system, the area must be reduced to F–H–B–D.

If the supply point is moved to the center of one side of the building, which is a frequent situation when the transformer is placed outside the building, the area that can be served with the specified voltage drop and specified circuit constants is E–A–B–D. If the supply station is moved to a corner of the building—a frequent location for buildings supplied from the rear or from the street—the area is reduced to O–A–B–C.

Every effort should be made to place the secondary distribution system supply point as close as possible to the center of the load area. Note that this study is based on a horizontal wiring system and any vertical components must be deducted to establish the limits of the horizontal area which can be supplied.

Using an average value of 30 ft/V drop for a fully loaded conductor, which is a good average figure for the conductor sizes normally used for feeders, the distances in Fig 15 for 5% and 2½% voltage drops are shown in Table 10. For a distributed load, the distances will be approximately twice the values shown.

Fig 15
Effect of Secondary Distribution System Power Source Location on Area Which Can Be Supplied Under Specified Voltage Drop Limits

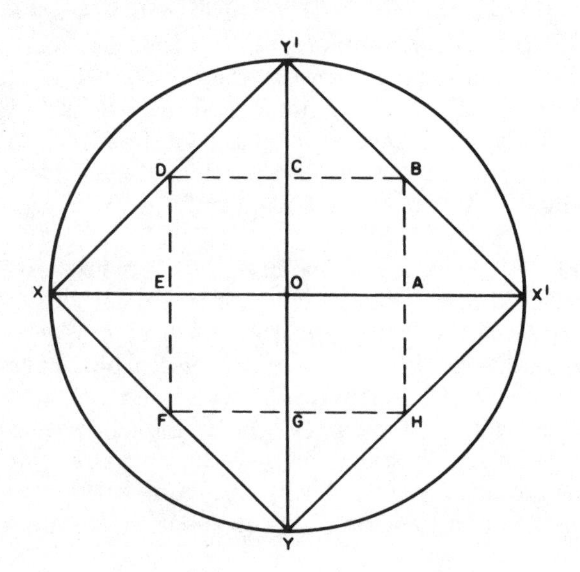

3.7 Improvement of Voltage Conditions. Poor equipment performance, overheating, nuisance tripping of overcurrent protective devices, and excessive burnouts are signs of unsatisfactory voltage. Abnormally low voltage occurs at the end of long circuits. Abnormally high voltage occurs at the beginning of circuits close to the source of supply, especially under lightly loaded conditions such as at night and over weekends.

In cases of abnormally low voltage, the first step is to make a load survey to measure the current taken by the affected equipment, the current in the circuit supplying the equipment, and the current being supplied by the supply source under peak-load conditions to make sure that the abnormally low voltage is not due to overloaded equipment. If the abnormally low voltage is due to overload, then corrective action is required to relieve the overloaded equipment.

If overload is ruled out or if the utilization voltage is excessively high, a voltage survey should be made, preferably by using graphic voltmeters, to determine the voltage spread at the utilization equipment under all load conditions and the voltage spread at the utility supply. This survey can be compared with ANSI C84.1-1982 [2] to determine if the unsatisfactory voltage is caused by the plant distribution system or the utility supply. If the utility supply exceeds the tolerance limits specified in ANSI C84.1-1982 [2], the utility should be notified. If the industrial plant is supplied at a transmission voltage and furnishes the distribution substation, the operation of the voltage regulators should be checked.

If excessively low voltage is caused by excessive voltage drop in the plant wiring (over 5%), then additional circuits are required in parallel with the affected circuits, or to supply portions of the affected equipment in order to reduce the voltage drop to suitable values. If the load power factor is low, capacitors may be installed to improve the power factor and reduce the voltage drop. Where the excessively low voltage affects a large area, the best solution may be to go to primary distribution if the building is supplied from a single distribution transformer, or to install an additional distribution transformer in the center of the affected area if the plant has primary distribution. Plants wired at 208Y/120 or 240 V may be changed over economically to 480Y/277 V if an appreciable portion of the wiring system is rated 600 V and motors are dual rated 220 : 440 V or 230 : 460 V.

Table 10
Areas That Can Be Supplied for Specific Voltage Drops and Voltages at Various Secondary Distribution System Power Source Locations

Nominal System Voltage (volts)	Distance (feet) 5% Voltage Drop		2½% Voltage Drop	
	0X	0A	0X	0A
120/240	360	180	180	90
208	312	156	156	78
240	360	180	180	90
480	720	360	360	180

3.8 Phase-Voltage Unbalance in Three-Phase Systems

3.8.1 Causes of Phase-Voltage Unbalance. Most utilities use four-wire grounded-wye primary distribution systems so that single-phase distribution transformers can be connected phase-to-neutral to supply single-phase load such as residences and street lights. Variations in single-phase loading cause the currents in the three-phase conductors to be different, producing different voltage drops and causing the phase voltages to become unbalanced. Normally the maximum phase-voltage unbalance will occur at the end of the primary distribution system, but the actual amount will depend on how well the single-phase loads are balanced between the phases on the system.

Perfect balance can never be maintained because the loads are continually changing, causing the phase-voltage unbalance to vary continually. Blown fuses on three-phase capacitor banks will also unbalance the load and cause phase-voltage unbalance.

Industrial plants make extensive use of 480Y/277 V utilization voltage to supply lighting loads connected phase-to-neutral. Proper balancing of single-phase loads among the three phases on both branch circuits and feeders is necessary to keep the load unbalance and the corresponding phase-voltage unbalance within reasonable limits.

3.8.2 Measurement of Phase-Voltage Unbalance. The simplest method of expressing the phase-voltage unbalance is to measure the voltages in each of the three phases:

$$\text{phase-voltage unbalance} = \frac{\text{maximum deviation from average phase voltage}}{\text{average phase voltage}}$$

The amount of voltage unbalance is better expressed in symmetrical components as the negative sequence component of the voltage:

$$\text{voltage-unbalance factor} = \frac{\text{negative-sequence voltage}}{\text{postive-sequence voltage}}$$

3.8.3 Effect of Phase-Voltage Unbalance. When unbalanced phase voltages are applied to three-phase motors, the phase-voltage unbalance causes additional negative-sequence currents to circulate in the motor, increasing the heat losses primarily in the rotor. The most severe condition occurs when one phase is opened and the motor runs on single-phase power. Table 11 [11] shows the increased temperature rise that occurs for specified values of phase-voltage unbalance for both U-frame and T-frame motors. Reference [13] provides a more comprehensive review of the problem.

Although there will generally be an increase in the motor load current when the phase voltages are unbalanced, the increase is insufficient to indicate the actual temperature rise that occurs because standard current-responsive thermal or magnetic overload devices only provide a trip characteristic that correlates with the motor thermal damage due to normal overload current (positive-sequence) and not negative-sequence current.

All motors are sensitive to phase-voltage unbalance, but sealed compressor motors used in air conditioners are most susceptible to this condition. These

Table 11
Effect of Phase-Voltage Unbalance on Motor Temperature Rise

Motor Type	Load	Percent Voltage Unbalance	Percent Added Heating	Insulation System Class	Temperature Rise (°C)
U-frame	Rated	0	0	A	60
	Rated	2	8	A	65
	Rated	3½	25	A	75
T-frame	Rated	0	0	B	80
	Rated	2	8	B	86.4
	Rated	3½	25	B	100

motors operate with higher current densities in the windings because of the added cooling effect of the refrigerant. Thus the same percent increase in the heat loss due to circulating currents, caused by phase-voltage unbalance, will have a greater effect on the sealed compressor motor than it will on a standard air-cooled motor.

Since the windings in sealed compressor motors are inaccessible, they are normally protected by thermally operated switches embedded in the windings, set to open and disconnect the motor when the winding temperature exceeds the set value. The motor cannot be restarted until the winding has cooled down to the point at which the thermal switch will reclose.

When a motor trips out, the first step in determining the cause is to check the running current after it has been restarted to make sure that the motor is not overloaded. The next step is to measure the three-phase voltages to determine the amount of phase-voltage unbalance. Table 11 indicates that where the phase-voltage unbalance exceeds 2%, the motor is likely to become overheated if it is operating close to full load.

Some electronic equipment such as computers may also be affected by phase-voltage unbalance of more than 2 or 2½%. The equipment manufacturer can supply the necessary information.

In general, single-phase loads should not be connected to three-phase circuits supplying equipment sensitive to phase-voltage unbalance. A separate circuit should be used to supply this equipment.

3.9 Voltage Dips and Flicker. The previous discussion has covered the relatively slow changes in voltage associated with steady-state voltage spreads and tolerance limits. However, certain types of utilization equipment such as motors have a high initial inrush current when turned on and impose a heavy load at a low power factor for a very short time. This sudden increase in the current flowing to the load causes a momentary increase in the voltage drop along the distribution system, and a corresponding reduction in the voltage at the utilization equipment. A voltage dip of ¼–½% will cause a noticeable reduction in the light output of an incandescent lamp and a less noticeable reduction in the light output of gaseous discharge lighting equipment.

In general, the starting current of a standard motor averages about 5 times the full-load running current. The approximate values for all ac motors over ½ hp are indicated by a code letter on the nameplate of the motor. The values indicated by these code letters are given in ANSI/NEMA MG1-1978 [8] and also in Article 430 of the NEC [9].

A motor requires about 1 kVA for each motor horsepower in normal operation, so the starting current of the average motor will be about 5 kVA for each motor horsepower. When the motor rating in horsepower approaches 5% of the secondary unit substation transformer capacity in kilovolt-amperes, the motor starting apparent power approaches 25% of the transformer capacity which, with a transformer impedance voltage of 6–7%, will result in a noticeable voltage dip on the order of 1%.

In addition, a similar voltage dip will occur in the wiring between the secondary unit substation and the motor when starting a motor with a full-load current which is on the order of 5% of the rated current of the circuit. This will result in a full-load voltage drop on the order of 4 or 5%. However, the voltage drop is distributed along the circuit so that maximum dip occurs only when the motor and the affected equipment are located at the far end of the circuit. As the motor is moved from the far end to the beginning of the circuit, the voltage drop in the circuit approaches zero. As the affected equipment is moved from the far end to the beginning of the circuit, the voltage dip remains constant up to the point of connection of the motor and then decreases to zero as the equipment connection approaches the beginning of the circuit.

The total voltage dip is the sum of the dip in the secondary unit substation transformer and the secondary circuit. In the case of very large motors of several hundred to a few thousand horsepower, the impedance of the supply system should be considered.

Where loads are turned on and off rapidly as in the case of resistance welders, or fluctuate rapidly as in the case of arc furnaces, the rapid fluctuations in the light output of incandescent lamps, and to a lesser extent, gaseous discharge lamps, is called flicker. When flicker continues over an appreciable period, voltage variations as low as ½% may be objectionable. If utilization equipment involving rapidly fluctuating loads is on the order of 10% of the capacity of the secondary unit substation transformer and the secondary circuit, accurate calculations should be made using the actual load currents and system impedances to determine the effect on lighting equipment.

Figure 16 [B4] may be used to determine whether voltage fluctuations will cause objectionable fluctuations in the light output of incandescent lamps. The borderline of irritation curve starts with a voltage change of 1% at a frequency of 7 fluctuations per second and increases to about 6% at 1 fluctuation per minute. The range between permissible flicker and objectionable flicker is due to the fact that some people are bothered more than others. Also, the effect of flicker depends upon lighting intensity and working conditions. Tests have indicated that flicker that is irritable to some people is hardly noticed by other people. Flicker is more of a problem with incandescent lighting than with fluorescent and high-intensity discharge types.

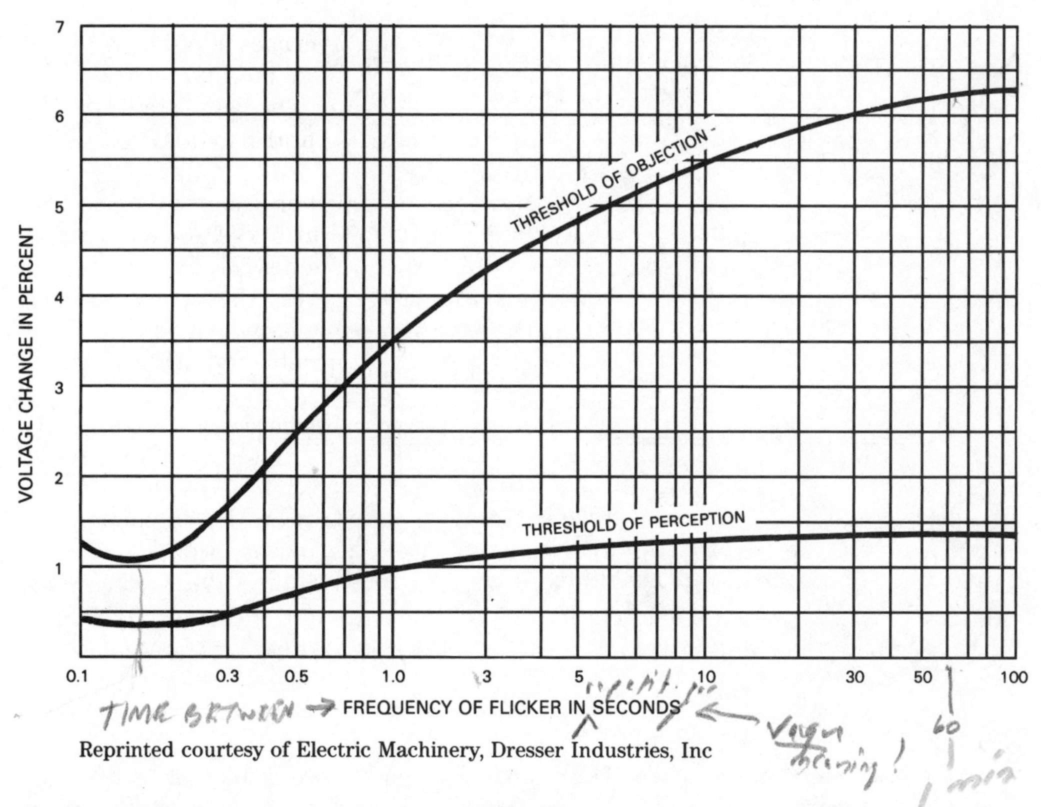

Reprinted courtesy of Electric Machinery, Dresser Industries, Inc

Fig 16
Range of Observable and Objectionable Voltage Flicker Versus Time

In using this curve, the purpose for which the lighting is provided needs to be considered. For example, lighting used for close work such as drafting requires flicker limits approaching the borderline of visibility curve. For general area lighting such as storage areas, the flicker limits may approach the borderline of the irritation curve. Note that the effect of voltage dips depends on the frequency of occurrence. An occasional dip, even though quite large, is rarely objectionable.

When objectionable flicker occurs, either the load causing the flicker should be reduced or eliminated, or the capacity of the supply system increased to reduce the voltage drop caused by the fluctuating load. In large plants, flicker-producing equipment should be segregated on separate transformers and feeders so as not to disturb flicker-sensitive equipment.

Special consideration should always be given when starting larger motors to minimize the voltage dip so as not to affect the operation of other utilization equipment on the system supplying the motor. Large motors (see Table 12) may be supplied at medium voltage such as 2400, 4160, 6900, or 13 200 V from a separate transformer to eliminate the voltage dip on the low-voltage system. However, consideration should be given to the fact that the maintenance electricians may not be

Table 12
Standard Voltages and Preferred Horsepower Limits for Polyphase Induction Motors

Motor Nameplate Voltage	Preferred Horsepower Limits
	Low-Voltage Motors
115	No minimum — 15 hp maximum
230	No minimum — 200 hp maximum
460 and 575	1 hp minimum — 1000 hp maximum
	Medium-Voltage Motors
2300	50 hp minimum — 6000 hp maximum
4000	100 hp minimum — 7500 hp maximum
4500	250 hp minimum — no maximum
6000	400 hp minimum — no maximum
13 200	1500 hp minimum — no maximum

qualified to maintain medium-voltage equipment. A contract with a qualified electrical firm may be required for maintenance. Standard voltages and preferred horsepower limits for polyphase induction motors are shown in Table 12 (based on [B5]).

Objectionable dips in the supply voltage from the utility should be reported to the utility for correction.

3.10 Harmonics

3.10.1 Nature of Harmonics. Harmonics are integral multiples of the fundamental frequency. For example, for 60 Hz power systems, the second harmonic would be 2 · 60 or 120 Hz and the third harmonic would be 3 · 60 or 180 Hz.

Harmonics are caused by devices that change the shape of the normal sine wave of voltage or current in synchronism with the 60 Hz supply. In general those include three-phase devices in which the three phase coils are not exactly symmetrical, and single- and three-phase loads in which the load impedance changes during the voltage wave produce a distorted current wave such as the magnetizing current in a coil with an iron core. It can be shown that a distorted wave can be made up of a fundamental and harmonics of various frequencies and magnitudes.

Inductive reactance varies directly as the frequency so that the current in an inductive circuit is reduced in proportion to the frequency for a given harmonic voltage. Conversely, capacitive reactance varies inversely as the frequency so that the current in a capacitive circuit is increased in proportion to the frequency for a given harmonic voltage. If the inductive reactance and the capacitive reactance in a series circuit are the same, they will cancel each other, and a given harmonic voltage will cause a large current to flow limited only by the resistance of the circuit. This condition is called resonance, and is more likely to occur at the higher harmonic frequencies.

3.10.2 Characteristics of Harmonics. The harmonic content and magnitude existing in any power system is largely unpredictable and effects will vary widely in different parts of the same system because of the different effects of different frequencies. Since the distorted wave is in the supply system, harmonic effects may occur at any point on the system where the distorted wave exists; this is not limited to the immediate vicinity of the harmonic-producing device. Where power is converted to direct current or some other frequency, harmonics will exist in any distorted alternating component of the converted power.

Harmonics may be transferred from one circuit or system to another by direct connection or by inductive or capacitive coupling. Since 60 Hz harmonics are in the low-frequency audio range, the transfer of these frequencies into communication, signaling, and control circuits employing frequencies in the same range may cause objectionable interference. In addition, harmonic currents circulating within a power circuit reduce the capacity of the current-carrying equipment and increase losses without providing any useful work.

3.10.3 Harmonic-Producing Equipment

(1) *Arc Equipment.* Arc furnaces and arc welders have a changing load characteristic during each half-cycle that demands harmonic currents from the supply system. Normally these do not cause very much trouble unless the supply conductors are in close proximity to communication and control circuits or there are large capacitor banks on the system.

(2) *Gaseous Discharge Lamps.* Fluorescent and mercury lamps produce small arcs and, in combination with the ballast, produce harmonics, particularly the third. Experience shows that the third-harmonic current may be as high as 30% of the fundamental in the phase conductors and up to 90% in the neutral where the third harmonics from each phase add directly, since they are displaced one third of a cycle. This is why the NEC [9] requires a full neutral for circuits supplying this type of load.

(3) *Rectifiers.* Half-wave rectifiers, which suppress alternate half-cycles of current, generate both even- and odd-numbered harmonics. Full-wave rectifiers tend to eliminate the even-numbered harmonics and usually diminish the magnitude of the odd-numbered harmonics.

The major producer of harmonics is the controlled rectifier, which chops the ac wave, particularly near the peak of the cycle. Since the wave shapes of both the input and the output depend upon the control setting to start rectification, both the shape of the input and output waves and, hence, the frequency and magnitude of the harmonics will vary with the setting of the control. Large rectifiers, which are frequently supplied from six- and twelve-phase transformer connections to produce smoother direct current, will produce different harmonics than those supplied from a three-phase system.

Phase-controlled rectifiers used to provide variable-speed drives for dc motors, or used as frequency changers to provide variable-speed drives for ac motors, are major sources of harmonics.

(4) *Rotating Machinery.* Normally the three phase coils of both motors and generators are sufficiently symmetrical that any harmonic voltages generated from lack of symmetry are too small to cause any interference. The nonlinear characteristics of the stator iron can produce appreciable harmonics, especially at high-flux densities.

(5) *Induction Heaters.* Induction heaters use 60 Hz or higher frequency power to induce circulating currents in metals to heat the metal. Harmonics are generated by the interaction of the magnetic fields caused by the current in the induction heating coil and the circulating currents in the metal being heated. Large induction heating furnaces may create objectionable harmonics.

(6) *Capacitors.* Capacitors do not generate harmonics. However, the reduced reactance of the capacitor to the higher frequencies magnifies the harmonic current in the circuit containing the capacitors. In cases of resonance, this magnification may be very large. High harmonic currents may overheat the capacitors. In addition, the high currents may induce interference with communication, signal, and control circuits.

Special capacitors prescribed by equipment manufacturers are required to perform satisfactorily under actual operating conditions. Thus the manufacturer must be furnished the harmonic voltage content of the power supply to determine the correct type to be used.

3.10.4 Reduction of Harmonic Effects. Where harmonic interference exists, the regular measures of increasing the separation between the power and communication conductors and the use of shielded communication conductors should be considered. Where capacitor banks magnify the harmonic current, the capacitors should be changed to suitable types or removed. Where resonant conditions exist, the capacitor bank should be changed in size to shift the resonant point to another frequency. Where harmonics pass from a power system to a communications, signal, or control circuit through a direct connection such as a power supply, filters may be required to suppress or short-circuit the harmonic frequencies.

During preliminary meetings with the supplying utility, the anticipated harmonic analysis of the power supply should be determined. This information, coupled with that provided by the manufacturer of any equipment to be installed that may generate a voltage distortion (along with appropriate safety factors), can be used to govern the specifications or the application of other equipment that may be exposed to the harmonic voltage condition.

3.11 Calculation of Voltage Drops. Building wiring designers must have a working knowledge of voltage drop calculations, not only to meet NEC requirements, but also to ensure that the voltage applied to utilization equipment is maintained within proper limits. Due to the phasor relationships between voltage and current and resistance and reactance, voltage drop calculations require a working knowledge of trigonometry, especially for making exact computations. Fortunately, most voltage drop calculations are based on assumed limiting conditions, and approximate formulas are adequate.

3.11.1 General Mathematical Formulas. The phasor relationships between the voltage at the beginning of a circuit, the voltage drop in the circuit, and the voltage at the end of the circuit are shown in Fig 17.

The approximate formulas for the voltage drop is

$$V = IR \cos \phi + IX \sin \phi$$

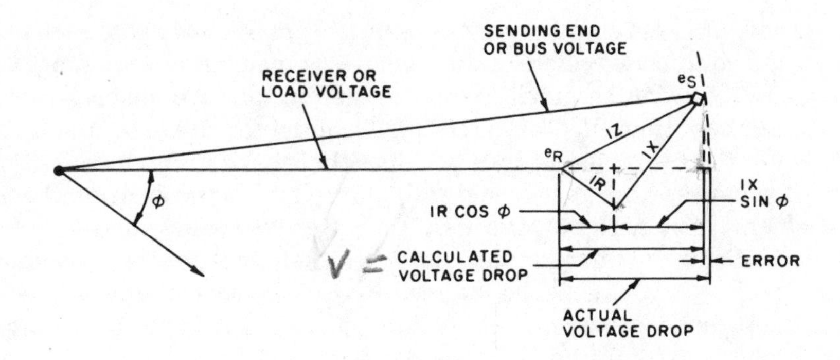

Fig 17
Phasor Diagram of Voltage Relations for Voltage-Drop Calculations

where

V	=	voltage drop in circuit, line to neutral
I	=	current flowing in conductor
R	=	line resistance for one conductor, in ohms
X	=	line reactance for one conductor, in ohms
ϕ	=	angle whose cosine is the load power factor
$\cos \phi$	=	load power factor, in decimals
$\sin \phi$	=	load reactive factor, in decimals

The voltage drop V obtained from this formula is the voltage drop in one conductor, one way, commonly called the line-to-neutral voltage drop. The reason for using the line-to-neutral voltage is to permit the line-to-line voltage to be computed by multiplying by the following constants:

Voltage system	Multiply by
Single-phase	2
Three-phase	1.732

In using this formula, the line current I is generally the maximum or assumed load current-carrying capacity of the conductor.

The resistance R is the ac resistance of the particular conductor used and of the particular type of raceway in which it is installed as obtained from the manufacturer. It depends on the size of the conductor measured in American Wire Gauge (AWG) for smaller conductors and in thousands of circular mils (kcmil) for larger conductors, the type of conductor (copper or aluminum), the temperature of the conductor (normally 75 °C for average loading and 90 °C for maximum loading), and whether the conductor is installed in magnetic (steel) or nonmagnetic (aluminum or nonmetallic) raceway. The resistance opposes the flow of current and causes the heating of the conductor.

The reactance X is obtained from the manufacturer. It depends on the size and material of the conductor, whether the raceway is magnetic or nonmagnetic, and on the spacing between the conductors of the circuit. The spacing is fixed for multiconductor cable but may vary with single-conductor cables so that an average value is required. Reactance occurs because the alternating current flowing in the conductor causes a magnetic field to build up and collapse around each conductor in synchronism with the alternating current. This magnetic field, as it builds up and falls radially, cuts across the conductor itself and the other conductors of the circuit, causing a voltage to be induced in each in the same way that current flowing in the primary of a transformer induces a voltage in the secondary of the transformer. Since the induced voltage is proportional to the rate of change of the magnetic field, which is maximum when the current passes through zero, the induced voltage will be a maximum when the current passes through zero, or, in vector terminology, lags the current wave by 90 degrees.

ϕ is the angle between the load voltage and the load current and is obtained by looking up the power factor expressed as a decimal (1 or less) in the cosine section of a trigonometric table or by using a scientific calculator.

Cos ϕ is the power factor of the load expressed in decimals and may be used directly in the computation of $IR \cos \phi$.

Sin ϕ is obtained by looking up the angle ϕ in a trigonometric table of sines, or by using a calculator. By convention, sin ϕ is positive for lagging power factor loads and negative for leading power factor loads.

$IR \cos \phi$ is the resistance component of the voltage drop and $IX \sin \phi$ is the reactive component of the voltage drop.

For exact calculations, the following formula may be used:

$$e_R = e_S + IR \cos \phi + IX \sin \phi - \sqrt{e_S^2 - (IX \cos \phi - IR \sin \phi)^2}$$

where the symbols correspond to those in Fig 17.

3.11.2 Cable Voltage Drop. Voltage drop tables and charts are sufficiently accurate to determine the approximate voltage drop for most problems. Table 13 contains four sections giving the three-phase line-to-line voltage drop for 10 000 circuit ampere-feet (A·ft) for copper and aluminum conductors in both magnetic and nonmagnetic conduit. The figures are for single-conductor cables operating at 60 °C. However, the figures are reasonably accurate up to a conductor temperature of 75 °C and for multiple-conductor cable. Although the length of cable runs over 600 V is generally too short to produce a significant voltage drop, Table 13 may be used to obtain approximate values. For borderline cases the exact values obtained from the manufacturer for the particular cable should be used. The resistance is the same for the same wire size, regardless of the voltage, but the thickness of the insulation is increased at the higher voltages, which increases the conductor spacing resulting in increased reactance causing increasing errors at the lower power factors. For the same reason, Table 13 cannot be used for open-wire or other installations such as trays where there is appreciable spacing between the individual phase conductors.

In using Table 13, the normal procedure is as follows: Look up the voltage drop for 10 000 A·ft and multiply this value by the ratio of the actual number of ampere-feet

Table 13
Three-Phase Line-to-Line Voltage Drop for 600 V Single-Conductor Cable per 10 000 A·ft
(60 °C Conductor Temperature, 60 Hz)

Load Power Factor Lagging	Wire Size (AWG or kcmil) 1000	900	800	750	700	600	500	400	350	300	250	4/0	3/0	2/0	1/0	1	2	4	6	8*	10*	12*	14*
Section 1: Copper Conductors in Magnetic Conduit																							
1.00	0.28	0.31	0.34	0.35	0.37	0.42	0.50	0.60	0.68	0.78	0.92	1.1	1.4	1.7	2.1	2.6	3.4	5.3	8.4	13	21	33	53
0.95	0.50	0.52	0.55	0.57	0.59	0.64	0.71	0.81	0.88	1.0	1.1	1.3	1.5	1.9	2.3	2.8	3.5	5.3	8.2	13	20	32	50
0.90	0.57	0.59	0.62	0.64	0.66	0.71	0.78	0.88	0.95	1.1	1.2	1.3	1.6	1.9	2.3	2.8	3.4	5.2	8.0	12	19	30	48
0.80	0.66	0.68	0.71	0.73	0.74	0.80	0.85	0.95	1.0	1.1	1.2	1.4	1.6	1.9	2.3	2.6	3.2	4.8	7.3	11	17	27	43
0.70	0.71	0.73	0.76	0.78	0.80	0.83	0.88	0.97	1.0	1.1	1.2	1.3	1.5	1.8	2.1	2.5	3.0	4.4	6.6	9.9	15	24	38
Section 2: Copper Conductors in Nonmagnetic Conduit																							
1.00	0.23	0.26	0.28	0.29	0.33	0.38	0.45	0.55	0.62	0.73	0.88	1.0	1.3	1.6	2.1	2.6	3.3	5.3	8.4	13	21	33	53
0.95	0.40	0.43	0.45	0.47	0.50	0.54	0.62	0.71	0.80	0.92	1.0	1.1	1.5	1.8	2.2	2.7	3.4	5.3	8.2	13	20	32	50
0.90	0.47	0.48	0.52	0.54	0.55	0.59	0.68	0.76	0.85	0.95	1.1	1.1	1.5	1.8	2.2	2.7	3.3	5.1	7.9	12	19	30	48
0.80	0.54	0.55	0.57	0.59	0.62	0.66	0.73	0.81	0.88	0.97	1.1	1.1	1.4	1.7	2.1	2.5	3.1	4.7	7.2	11	17	27	43
0.70	0.57	0.59	0.62	0.64	0.66	0.69	0.74	0.83	0.88	0.97	1.1	1.1	1.4	1.6	2.0	2.4	2.8	4.3	6.4	9.7	15	24	38
Section 3: Aluminum Conductors in Magnetic Conduit																							
1.00	0.42	0.45	0.49	0.52	0.55	0.63	0.74	0.91	1.0	1.2	1.4	1.7	2.1	2.6	3.3	4.2	5.2	8.4	13	21	33	52	—
0.95	0.62	0.65	0.70	0.73	0.76	0.83	0.94	1.1	1.2	1.4	1.6	1.8	2.3	2.7	3.4	4.2	5.3	8.2	13	20	32	50	—
0.90	0.69	0.72	0.76	0.79	0.82	0.88	0.99	1.2	1.3	1.4	1.6	1.9	2.3	2.7	3.4	4.1	5.1	7.9	12	19	30	48	—
0.80	0.76	0.80	0.83	0.85	0.88	0.95	1.0	1.2	1.3	1.4	1.6	1.8	2.2	2.6	3.2	3.9	4.7	7.3	11	17	27	43	—
0.70	0.80	0.83	0.87	0.89	0.92	0.98	1.1	1.2	1.3	1.4	1.6	1.7	2.1	2.4	2.9	3.6	4.3	6.5	10	15	24	37	—
Section 4: Aluminum Conductors in Nonmagnetic Conduit																							
1.00	0.36	0.39	0.44	0.47	0.51	0.59	0.70	0.88	1.0	1.2	1.4	1.7	2.1	2.6	3.3	4.2	5.2	8.4	13	21	33	52	—
0.95	0.52	0.56	0.60	0.63	0.67	0.74	0.85	1.0	1.1	1.3	1.5	1.8	2.2	2.7	3.4	4.2	5.2	8.2	13	20	32	50	—
0.90	0.57	0.61	0.65	0.68	0.71	0.79	0.89	1.1	1.2	1.3	1.5	1.8	2.2	2.6	3.3	4.1	5.0	7.9	12	19	30	48	—
0.80	0.63	0.66	0.71	0.73	0.76	0.83	0.92	1.1	1.2	1.3	1.5	1.7	2.1	2.5	3.1	3.8	4.6	7.2	11	17	27	42	—
0.70	0.66	0.69	0.73	0.75	0.78	0.83	0.92	1.1	1.1	1.3	1.4	1.6	1.7	2.3	2.8	3.4	4.2	6.4	9.9	15	24	37	—

*Solid Conductor. Other conductors are stranded.

To convert voltage drop to	Multiply by
Single-phase, three-wire, line-to-line	1.18
Single-phase, three-wire, line-to-neutral	0.577
Three-phase, line-to-neutral	0.577

to 10 000. Note that the distance in feet is the distance from the source to the load.

Example 1. 500 kcmil copper conductor in steel (magnetic) conduit; circuit length 200 ft; load 300 A at 80% power factor. What is the voltage drop?

Using Section 1 of Table 13, the intersection between 500 kcmil and 80% power factor gives a voltage drop of 0.85 V for 10 000 A·ft.

200 ft · 300 A = 60 000 circuit A·ft
(60 000/10 000) · 0.85 = 6 · 0.85 = 5.1 V drop
voltage drop, phase-to-neutral = 0.577 · 5.1
= 2.9 V

Example 2. AWG No 12 aluminum conductor in aluminum (nonmagnetic) conduit; circuit length 200 ft; load 10 A at 70% power factor. What is the voltage drop?

Using Section 4 of Table 13, the intersection between AWG No 12 aluminum conductor and 0.70 power factor is 37 V for 10 000 A·ft.

200 ft · 10 A = 200 circuit A·ft

voltage drop = (2000/10 000) · 37
= 7.4 V

Example 3. Determine the wire size in Example 2 to limit the voltage drop to 3 V. The voltage drop in 10 000 A·ft would be

(10 000/2000) · 3 = 15 V

Using Section 4 of Table 13, move along the 0.70 power factor line to find the voltage drop not greater than 15 V. AWG No 8 aluminum has a voltage drop of 15 V for 10 000 A·ft, so it is the smallest aluminum conductor in aluminum conduit that could be used to carry 10 A for 200 ft with a voltage drop of not more than 3 V, line-to-line.

3.11.3 Busway Voltage Drop. See Chapter 12 for busway voltage drop tables and related information.

3.11.4 Transformer Voltage Drop. Voltage-drop curves in Figs 18 and 19 may be used to determine the approximate voltage drop in single-phase and three-phase 60 Hz liquid-filled, self-cooled, and dry transformers. The voltage drop through a single-phase transformer is found by entering the chart at a kilovolt-ampere rating three times that of the single-phase transformer. Figure 18 covers transformers in the following ranges:

(1) *Single-Phase*
250–500 kVA, 8.6–15 kV insulation classes
833–1250 kVA, 5–25 kV insulation classes
(2) *Three-Phase*
225–750 kVA, 8.6–15 kV insulation classes
1000–10 000 kVA, 5–25 kV insulation classes

An example of the use of the chart is given in the following:

Example. Find the voltage drop in a 2000 kVA three-phase 60 Hz transformer rated 4160–480 V. The load is 1500 kVA at 0.85 power factor.

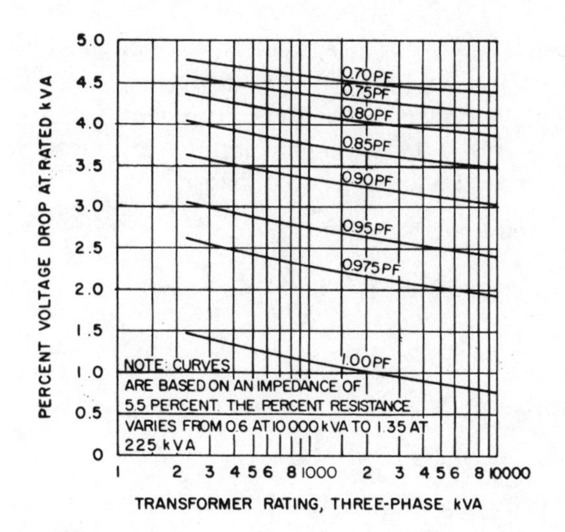

Fig 18
Approximate Voltage Drop Curves for Three-Phase Transformers, 225–10 000 kVA, 5–25 kV

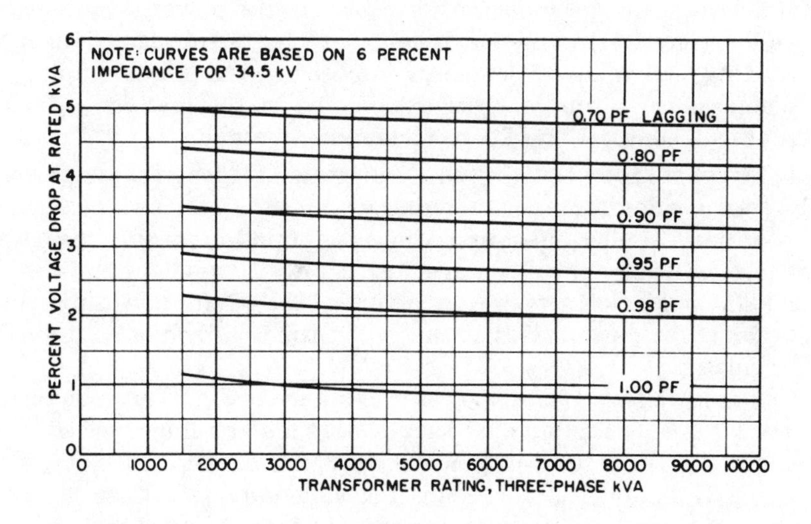

Fig 19
Approximate Voltage Drop Curves for Three-Phase Transformers, 1500–10 000 kVA, 34.5 kV

Solution. Enter the chart on the horizontal scale at 2000 kVA, extend a line vertically to its intersection with the 0.85 power factor curve. Extend a line horizontally from this point to the left to its intersection with the vertical scale. This point on the vertical scale gives the percent voltage drop for rated load. Multiply this value by the ratio of actual load to rated load:

$$\text{percent drop at rated load} = 3.67$$

$$\text{percent drop at 1500 kVA} = 3.67 \cdot \frac{1500}{2000}$$

$$= 2.75$$

$$\text{actual voltage drop} = 2.75\% \cdot 480$$

$$= 13.2 \text{ V}$$

Figure 19 applies to the 34.5 kV insulation class power transformer in ratings from 1500–10 000 kVA. These curves can be used to determine the voltage drop for transformers in the 46 and 69 kV insulation classes by using appropriate multipliers at all power factors except unity.

To correct for 46 kV, multiply the percent voltage drop obtained from the chart by 1.065, and for 69 kV, multiply by 1.15.

3.11.5 Motor-Starting Voltage Drop. It is characteristic of ac motors that the current that they draw on starting is much higher than their normal running current. Synchronous and squirrel-cage induction motors started on full voltage may draw a current as high as seven or eight times their full-load running current. This sudden increase in the current drawn from the power system may result in excessive drop in voltage unless it is considered in the design of the system. The motor-starting load in kilovolt-amperes, imposed on the power supply system, and the available motor torque are greatly affected by the method of starting used. Table 14 gives a comparison of several common methods.

3.11.6 Effect of Motor Starting on Generators. Figure 20 shows the behavior of the voltage of a generator when an induction motor is started. Starting a synchronous motor has a similar effect up to the time of pull-in torque. The case used for this illustration utilizes a full-voltage starting device, and the full-voltage motor starting load in kilovolt-amperes is about 100% of the generator rating. It is assumed for curves A and B that the generator is provided with an automatic voltage regulator.

The minimum voltage of the generator as shown in Fig 20 is an important quantity because it is a determining factor affecting undervoltage devices and contactors connected to the system and the stalling of motors running on the system. The curves of Fig 21 can be used for estimating the minimum voltage occurring at the terminals of a generator supplying power to a motor being started.

3.11.7 Effect of Motor Starting on Distribution System. Frequently in the case of purchased power, there are transformers and cables between the starting motor and the generator. Most of the drop in this case is within the distribution equipment. When all the voltage drop is in this equipment, the voltage falls immediately

Table 14
Comparison of Motor-Starting Methods

Type of Starter (Settings Given Are the More Common for Each Type)	Motor Terminal Voltage (Percent Line Voltage)	Starting Torque (Percent Full-Voltage Starting Torque)	Line-Current (Percent Full-Voltage Starting Current)
Full-voltage starter	100	100	100
Autotransformer			
80% tap	80	64	68
65% tap	65	42	46
50% tap	50	25	30
Resistor starter, single step (adjusted for motor voltage to be 80% of line voltage)	80	64	80
Reactor			
50% tap	50	25	50
45% tap	45	20	45
37.5% tap	37.5	14	37.5
Part-winding starter (low-speed motors only)			
75% winding	100	75	75
50% winding	100	50	50

NOTE: For a line voltage not equal to the motor-rated voltage, multiply all values in the first and last columns by the ratio (actual voltage)/(motor-rated voltage). Multiply all values in the second column by the ratio $[(\text{actual voltage})/(\text{motor-rated voltage})]^2$.

Fig 20
Typical Generator Voltage Behavior Due to Full-Voltage Starting of a Motor

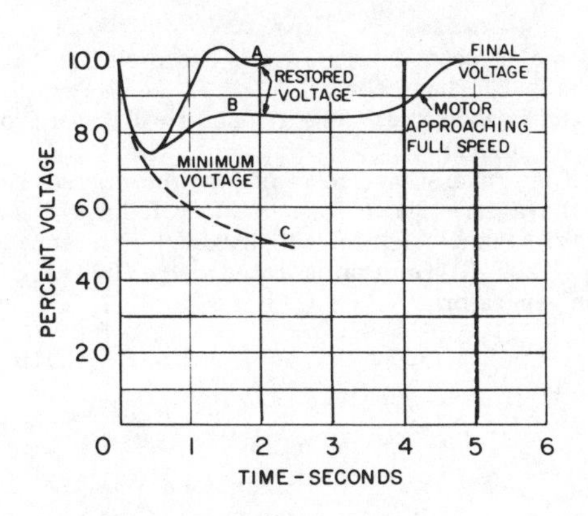

Motor-starting kVA = 100% of generator rating
A — No initial load on generator
B — 50% initial load on generator
C — No regulator

(because it is not influenced by a regulator as in the generator case) and does not recover until the motor approaches full speed. Since the transformer is usually the largest single impedance in the distribution system, it takes almost the total drop. Figure 22 has been plotted in terms of motor starting load in kilovolt-amperes that would be drawn if rated transformer secondary voltage were maintained.

Fig 21
Minimum Generator Voltage Due to Full-Voltage Starting of a Motor

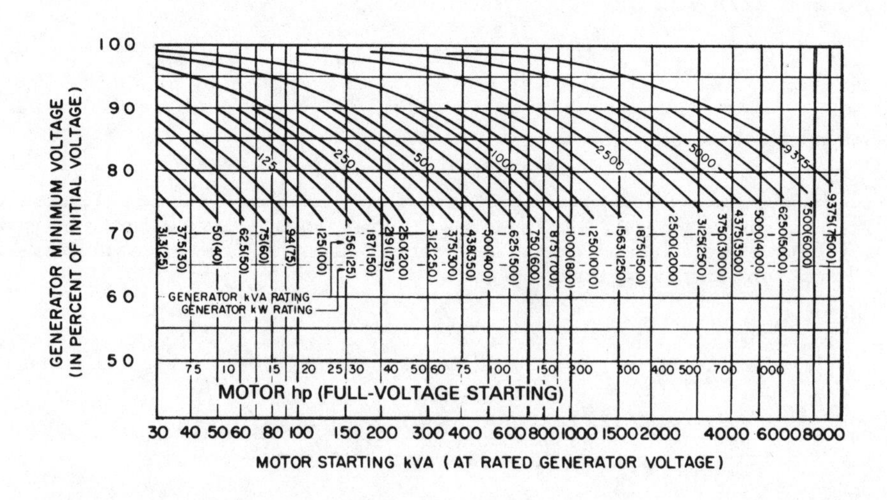

NOTES: (1) The scale of motor horsepower is based on the starting current being equal to approximately 5.5 times normal.

(2) If there is no initial load, the voltage regulator will restore voltage to 100% after dip to values given by curves.

(3) Initial load, if any, is assumed to be of constant-current type.

(4) Generator characteristics are assumed as follows: (a) Generators rated 1000 kVA or less: Performance factor k = 10; transient reactance X_d' = 25%; synchronous reactance X_d = 120%. (b) Generators rated above 1000 kVA: Characteristics for 3600 r/min turbine generators.

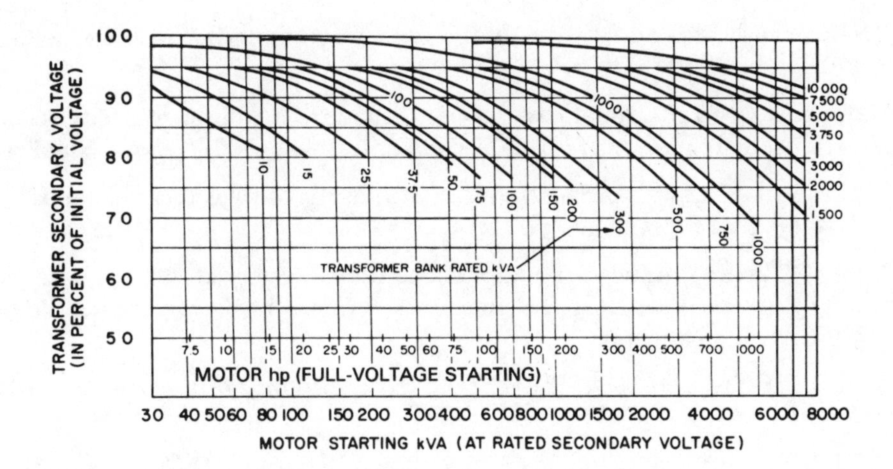

NOTES: (1) The scale of motor horsepower is based on the starting current being equal to approximately 5.5 times normal.

(2) Short-circuit capacity of primary supply is assumed to be as follows:

Transformer Bank Load (kVA)	Primary Short-Circuit Capacity (kVA)
0-300	25 000
500-1000	50 000
1500-3000	100 000
3760-10 000	250 000

(3) Transformer impedances are assumed to be as follows:

Transformer Bank Load (kVA)	Transformer Bank Impedance (percent)
10-50	3.0
75-150	4.0
200-500	5.0
750-2000	5.5
3000-10 000	6.0

(4) Representative values of primary system voltage drop, expressed as a fraction of total drop, for the assumed conditions, are as follows:

Transformer Bank Load (kVA)	System Drop / Total Drop
100	0.09
1000	0.25
10 000	0.44

Fig 22
Approximate Voltage Drop in a Transformer Due to Full-Voltage Starting of a Motor

3.12 References. This standard shall be used in conjunction with the following publications:

[1] ANSI C57.12.20-1981, American National Standard Requirements for Overhead-Type Distribution Transformers 67 000 Volts and Below, 500 kVA and Smaller.[14]

[2] ANSI C84.1-1982, American National Standard Voltage Ratings for Electric Power Systems and Equipment (60 Hz).

[3] ANSI C92.2-1981, American National Standard Preferred Voltage Ratings for Alternating-Current Electrical Systems and Equipment Operating at Voltages above 230 Kilovolts Nominal.

[4] ANSI/IEEE Std 100-1984, IEEE Standard Dictionary of Electrical and Electronics Terms.

[5] ANSI/IEEE Std 142-1982, IEEE Recommended Practice for Grounding of Industrial and Commercial Power Systems.

[6] ANSI/IEEE Std 242-1975, IEEE Recommended Practice for Protection and Coordination of Industrial and Commercial Power Systems.

[7] ANSI/IEEE Std 446-1980, IEEE Recommended Practice for Emergency and Standby Power Systems for Industrial and Commercial Applications.

[8] ANSI/NEMA MG1-1978, Motors and Generators.[15]

[9] ANSI/NFPA 70-1987, National Electric Code.

[10] CAN3-C235-83, Preferred Voltage Levels for AC Systems, 0 to 50 000 V (Canadian Standards Association).[16]

[11] ARNOLD, R.E. NEMA Suggested Standards for Future Design of AC Integral Horsepower Motors. *IEEE Transactions on Industry and General Applications*, vol IGA-6, Mar/Apr 1970, pp 110–114.

[12] *Electric Utility Engineering Reference Book*, vol 3: *Distribution Systems*. Trafford, PA: Westinghouse Electric Corporation, 1965.

[13] Linders, J.R., Effects of Power Supply Variations on AC Motor Characteristics. *Conference Record*, 6th Annual Meeting of the IEEE Industry and General Applications Group, 1971, IEEE 71 C1-IGA, pp 1055–1068.

---

[14] ANSI publications can be obtained from the Sales Department, American National Standards Institute, 1430 Broadway, New York, NY 10018.

[15] ANSI/NEMA publications can be obtained from the Sales Department, American National Standards Institute, 1430 Broadway, New York, NY 10018, or from the National Electrical Manufacturers Association, 2101 L Street, NW, Washington, DC 20037.

[16] In the US, CSA Standards are available from the Sales Department, American National Standards Institute, 1430 Broadway, New York, NY 10018. In Canada they are available at the Canadian Standards Association (Standards Sales), 178 Rexdale Blvd, Rexdale, Ontario, Canada M9W 1R3.

[14] MICHAEL, D.T. Proposed Design Standard for the Voltage Drop in Building Wiring for Low-Voltage Systems. *IEEE Transactions on Industry and General Applications*, vol IGA-4: Jan/Feb 1968, pp 30-32.

3.13 Bibliography

[B1] BRERETON, D.S. and MICHAEL, D.T. Developing a New Voltage Standard for Industrial and Commercial Power Systems. *Proceedings of the American Power Conference*, vol 30, 1968, pp 733-751.

[B2] BRERETON, D.S. and MICHAEL, D.T. Significance of Proposed Changes in AC System Voltage Nomenclature for Industrial and Commercial Power Systems: I—Low-Voltage Systems. *IEEE Transactions on Industry and General Applications*, vol IGA-3, Nov/Dec 1967, pp 504-513.

[B3] BRERETON, D.S. and MICHAEL, D.T. Significance of Proposed Changes in AC System Voltage Nomenclature for Industrial and Commercial Power Systems: II—Medium-Voltage Systems. *IEEE Transactions on Industry and General Applications*, vol IGA-3, Nov/Dec 1967, pp 514-520.

[B4] *Electrical Data Book*, Section 5, Fig 5-3, Electric Machinery Manufacturing Company, Minneapolis, Minnesota.

[B5] *Standard Handbook for Electrical Engineers*, 10 Ed, Table 18-5. New York: McGraw-Hill.

4. Surge Voltage Protection

4.1 Nature of the Problem. There are numerous mechanisms by which surge voltages can be generated on plant electrical distribution systems. Surges can originate on a system externally from lightning strokes on or near overhead power lines serving the plant, and internally from forced current zero switching, the blowing of current limiting fuses, restriking interruption of circuit switching devices, and a variety of other operating phenomena that result in sudden overvoltage conditions.

A lightning induced surge will have the form of a steep front wave that will travel away from the stricken point in both directions along the power system conductors (Fig 23). As the surge travels along the conductors, losses cause the magnitude of the voltage surge to constantly diminish. If the voltage magnitude is sufficient to produce corona, the decay of the voltage surge will be fairly rapid until below the corona starting voltage. Beyond this point the decay will be more deliberate. Properly rated arresters at the plant terminal of the incoming line will generally reduce the overvoltage to a level the terminal station apparatus can withstand.

In instances where the local industrial plant system is itself without lightning exposure except from the exposed high-voltage lines through stepdown transformers effectively protected with high-side surge arresters, any surges induced by

Fig 23
Two *Traveling* Bodies of Charge Result when Quantity of Charge is Deposited on Conducting Line by Lightning

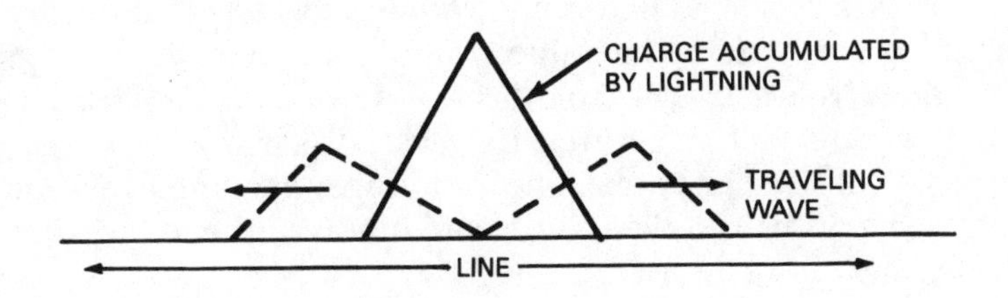

lightning are infrequent and quite moderate. Likewise, surges due to switching phenomena, although more common, are generally not as severe. Only occasionally would line-to-ground potentials on the local system reach arrester-protective levels. The multiplicity of radiating cable circuits with their array of connected apparatus acts to greatly curb the slope and magnitude of the voltage surge that reaches any particular item of connected apparatus. Transformers and other equipment items connected as a single load at the end of circuits are particularly vulnerable. Experience has indicated that certain types of apparatus are susceptible to voltage surges for almost any circuit connection arrangement, and it is advisable to fully investigate the subject of damaging voltage surges.

The appearance of abnormal applied voltage stresses, either transient, short-time, or sustained steady state, contributes to premature insulation failure. Electrical insulation deterioration to the point of failure is a result of an aggregate accumulation of insulation damage finally reaching the critical stage, where a conducting path is rapidly driven through the insulation sheath and failure (short circuit) takes place. Large amounts of current are driven through the faulted channel, liberating large amounts of heat. The excessive temperature rise rapidly expands the zone of insulation damage, and complete destruction can occur rather quickly unless the input of electric power supply is interrupted. Even small insulation punctures such as might be discovered after special nondestructive testing of apparatus require repair or rewinding.

An acceptable system of insulation protection will be influenced by a number of factors. Of prime importance is a knowledge of the insulation system withstand capability and endurance qualities. These properties are indicated by insulation-type designations and specified high-potential and surge-voltage test withstand capabilities. Another facet of the problem relates to the identification of likely sources of overvoltage exposure and the character, magnitude, duration, and repetition rate that are likely to be impressed on the apparatus and circuits. The appropriate application of surge protective devices will lessen the magnitude and duration, and is the most effective tool for achieving the desired insulation security. A working understanding of the behavior pattern of electric surge voltage propagation along electrical conductors is necessary to achieving the optimum solution.

The problem is complicated by the fact that insulation failure results not only because of excess-magnitude impressed overvoltages, but also because of the aggregate sum total duration of such overvoltages. No simple devices are available that can correctly integrate the cumulative effects of sequentially applied excessive overvoltage. The time factor must be estimated and then factored into the design and application of the protection system.

Lightning is a major source of transient overvoltages. Numerous industrial operations use open-wire overhead lines that are subject to direct exposure to lightning, allowing lightning surges to be introduced into the industrial distribution system.

Direct lightning strokes are rare to overhead outdoor plant wiring because of the shielding effect of adjacent structures. However, direct strokes to objects 25–50 ft away can induce substantial transient overvoltages into the overhead line so that surge arresters should be installed at the end of an overhead line that connects to the building wiring.

A sudden overvoltage can be imparted to system conductors when they come into contact with conductors from a higher voltage system.

ANSI C2-1987 [1][17] (The National Electrical Safety Code [NESC]), Rule 222C2, recommends that open conductors of different voltages installed on the same support must have the highest voltage on top and the lowest voltage below. When a higher voltage conductor breaks for any reason such as an automobile striking the supporting pole or a tree limb falling across the line, the higher voltage conductor may fall across the lower voltage conductor, impressing the higher voltage on the lower voltage circuit. Little can be done to protect against this condition other than to ensure that the protection equipment on the higher voltage line will disconnect the line as quickly as possible, generally as a result of a fault due to an insulation failure on the lower voltage line.

Steep wave-front transient overvoltages are also generated in plant wiring by switching actions that change the circuit operation from one steady-state condition to another. Switches that tend to chop the normal ac wave such as thyristors, vacuum switches, current-limiting fuses, and high-speed circuit breakers, force the current to zero, which accelerates the collapse of the magnetic field around the conductor, generating a transient overvoltage. The initial overvoltage *spike* resulting from the interrupting action of a current-limiting fuse is depicted in Fig 24.

Restriking current interruption of certain circuit configurations by the circuit switching device can also cause high-frequency transient overvoltages. The circuit arrangement and response for one such switching action, involving restriking interruption of capacitance current, is illustrated in Fig 25.

Prior to the initial interruption, E_{cap} remained solidly referenced to E_A. At a capacitor current zero, an initial interruption of current is assumed to occur, at which time E_{cap} continues at fixed potential while E_A proceeds to reverse according to normal system operation at fundmental frequency. During the first half-cycle E_A completely reverses its potential, which would cause twice the normal line-to-neutral crest voltage to appear across the open switching contact. Should the switch restrike, it suddenly changes from an insulator to a conductor. Since the capacitor voltage holds fast on voltage, the required transition snaps the "A" phase conductor to the capacitor voltage. This is a steep-front snap transition. The line and capacitor together begin a transition oscillation toward where line "A" will eventually be at *normal* potential. A corresponding transitory capacitor restrike current is involved in this process that crosses and recrosses zero. At one of these zero crossings, conditions may be such as to permit another interruption, perhaps with E_{cap} at a greater potential than at the first interruption as illustrated. This would increase the possible step-voltage transition at the next restrike.

In this manner there is opportunity to develop a steep-front step-voltage change of 6800 V on the first restrike in a 4160 V system with greater values possible on subsequent restrikes. Had the capacitor bank been ungrounded, as it usually would

[17] The numbers in brackets correspond to those of the references listed at the end of this chapter; when preceded by B, they correspond to the bibliography at the end of this chapter.

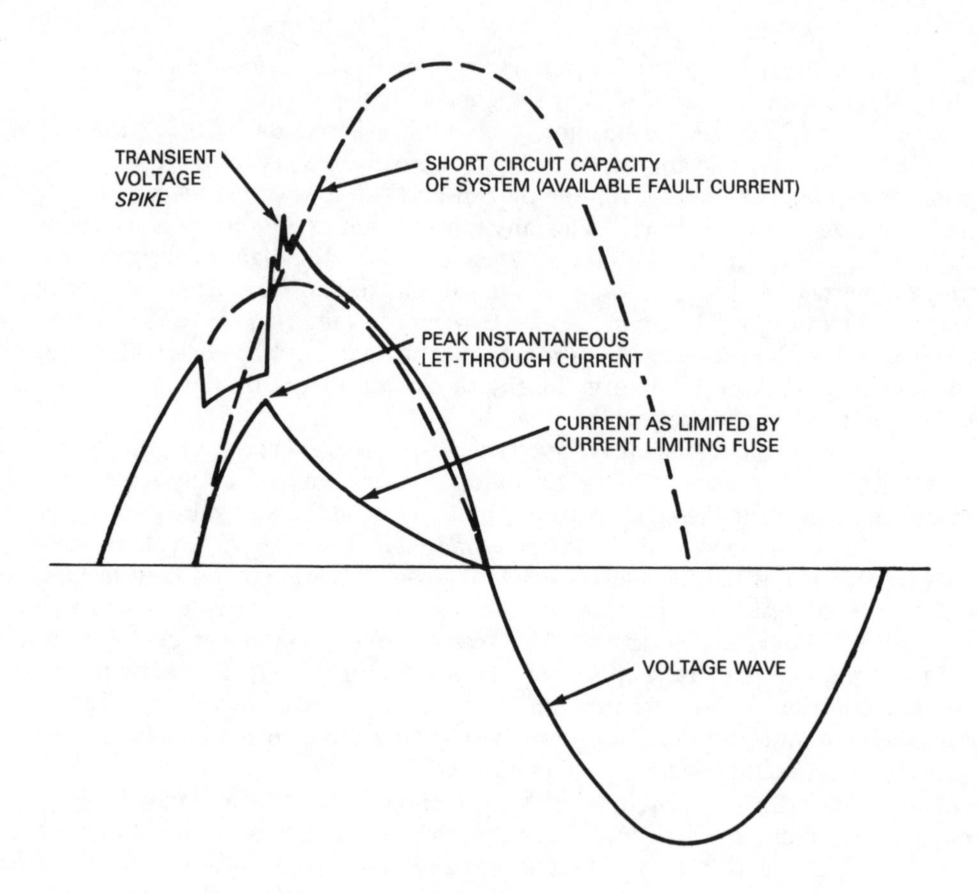

Fig 24
Oscillogram Showing Typical Short-Circuit Current-Limiting Action of Fuse to Produce Transient Overvoltage

be in industrial systems, there would be opportunity for more than twice the line-to-neutral crest voltage to appear across the first pole to clear.

In this example, only two restrikes occur. Additional restrikes would cause an even more dramatic overvoltage condition [37].

A short circuit (that is, insulation breakdown) is a switching action that creates a bypass around part of a circuit. The heat generated by the heavy flow of current across the short circuit may melt or even vaporize the conductor, creating a gap with an arc. Heated air rising from the arc creates a draft causing the arc resistance to fluctuate rapidly, which produces transient overvoltages. An insulation failure results in an arc through the failure path with similar results.

The nonlinear inductance of an iron-core transformer and a capacitor in the same circuit may go into oscillation and produce a condition of ferroresonance, which will generate overvoltages. Other sources of overvoltages are described in [36] and [37].

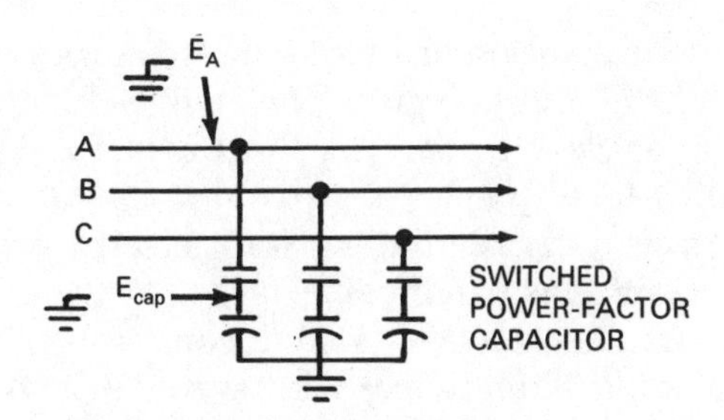

(a) Circuit Arrangement

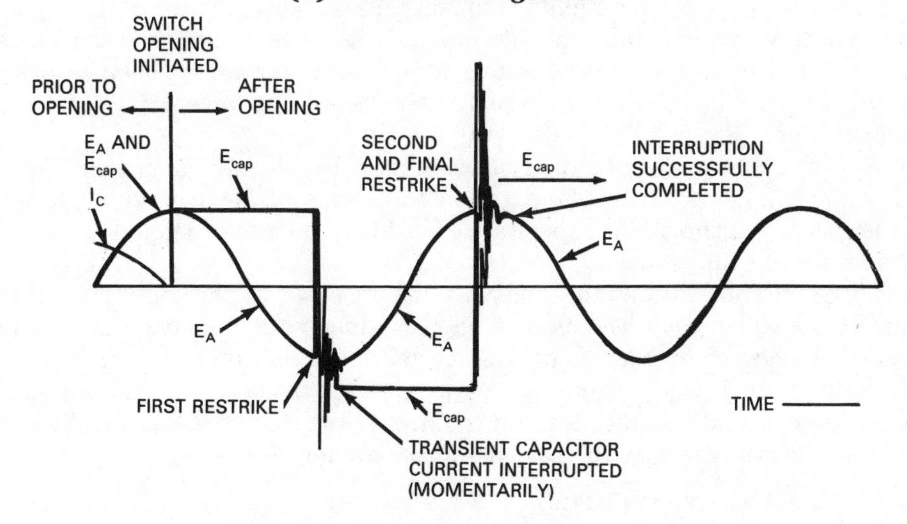

(b) Transient Overvoltage Character Due to Switch Restrike During Interruption

Fig 25
Equivalent Circuit and Transient Response for Capacitor Switching Restrike Phenomena

Transient overvoltages are propagated along the electric power conductors to create insulation distress far removed from the origin of the voltage surge. Furthermore, the voltage stress imposed on insulation far removed from the point of surge origin may exceed that appearing at the source point.

4.2 Traveling-Wave Behavior

4.2.1 Surge-Voltage Propagation. The electric power circuits transmit undesired surge voltages equally as well as normal-frequency voltages and can do so efficiently, even for frequencies into the megahertz range. When circuit geometry is short compared to wavelength, lumped circuit constants (L,R,C) often suffice for the particular analysis at hand. The usual concepts of line impedance, expressed as resistance and reactance in ohms, used in power-frequency computations do not, however, apply for the solution of short-time transitory voltages such as lightning-produced waves traveling on typical power lines, cables, and other apparatus. When wavelengths (or wave fronts) are short compared to the lengths of circuitry involved, then it may be necessary to use a distributed-constant representation.

Figure 26 illustrates a distributed-constant electric transmission line, either an open-wire line or a solid-insulated cable.

Such a transmission line can be viewed as consisting of a continued succession of small incremental series inductances with evenly distributed increments of shunt capacitance. When the switch SW is closed, the voltage E becomes connected to the line terminal. The first increment of capacitance is charged to a voltage E. Current begins to flow through the first increment of L to the next increment of shunt capacitance. The appearance of voltage along the line is always being impeded by the next incremental element of inductance. The voltage wave takes time to travel down the line.

The electrical behavior of a distributed-constant transmission line can, for practical surge-voltage problems, be expressed in terms of the series inductance per unit length L and the shunt capacitance per identical unit of length C. Consider L in henries and C in farads.

Each elementary inductance has a surge voltage impressed upon it by an assumed traveling wave. The associated electromagnetic energy ($\frac{1}{2}LI^2$) and electrostatic energy ($\frac{1}{2}CE^2$) are expressed in joules (wattseconds) when units are as defined in Fig 26. It is a profound property of traveling waves that the two forms of energy are of equal magnitudes and a surge voltage/current ratio will so result. Equating the two energy expressions and solving for E/I,

$$E/I = \sqrt{L/C} = \text{surge impedance} = Z_0$$

Fig 26
Distributed-Constant Transmission Circuit

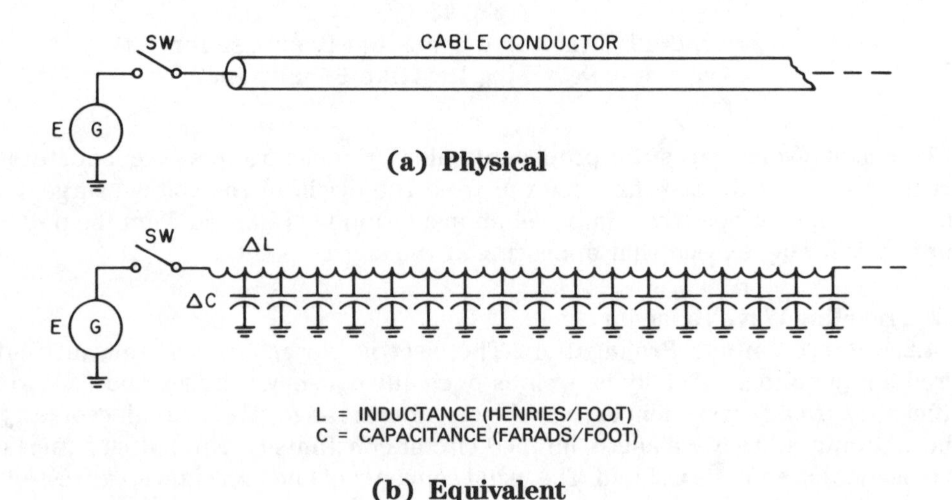

(a) Physical

(b) Equivalent

as depicted in Fig 27. The equivalent distributed-constant relationships exist in apparatus (transformers, rotating machines, etc) as well, but are somewhat more complex to analyze and visualize.

With the prescribed units, the quantity Z_0 has the dimensions of ohms and relates with E expressed in volts and I in amperes. The quantity $\sqrt{L/C}$ is called the surge impedance and assigned the reference symbol Z_0. This symbol has no relation to the zero-sequence impedance, which uses the same symbol.

Typical values of Z_0 for a single transmission line are (1) open-wire line, 400 Ω and (2) solid-dielectric metal-sheathed cable, 40 Ω. The applied voltage E and the surge current I are in phase. The current flow duplicates the wave shape of the impressed voltage and is in time phase with it. At this point it appears exactly as a resistor of ohmic value $\sqrt{L/C}$, but the behavior differs from that of a resistor. In a true resistor the I^2R line loss energy is converted to heat. In the distributed-constant line the electric energy is stored in the inductance and capacitance as the $LI^2/2$ and $CE^2/2$ of an electric surge existing on a finite length of the transmission line.

The transit of the surge along the line is propagated at a rate controlled by the quantity LC. The propagation velocity is expressed as $1/\sqrt{LC}$. An increased value of the LC product slows down the transit rate. With the units chosen, the propagation velocity will be in feet per second. Had L and C been expressed in henries/meter and farads/meter, respectively, the propagation velocity would have the units of meters per second.

Ignoring the influence of the ground circuit impedance and various second-order effects on an open-wire line with air dielectric, the propagation velocity is approximately that of the speed of light, 1000 ft/μs, or 1 ft/ns. A solid-insulation cable will display a propagation velocity about half that of the open-wire line, that is, 500 ft/μs or ½ ft/ns.

4.2.2 Surge-Voltage Reflection. Were the line of infinite length, the bundle of energy would continue to travel forever, never again to be observed at the point of origin. Since practical circuits have a finite length, problems develop as the surge reaches the end of the line. The equivalent circuit, shown in Fig 28, allows evaluation of the effect of continuing along a line of different surge impedance that terminates at an open circuit, at a short circuit, or with a network of lumped constants. An electric surge E traveling along a transmission line of surge impedance Z_0 toward a junction J can be replaced by the equivalent circuit shown, that

Fig 27
Surge-Voltage Wave in Transit Along a Line of Surge Impedance Z_0

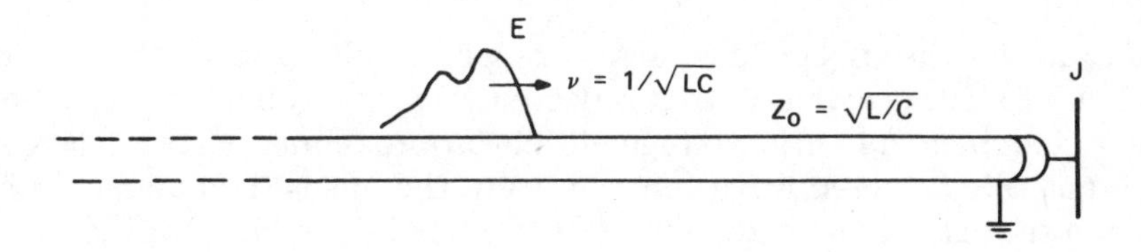

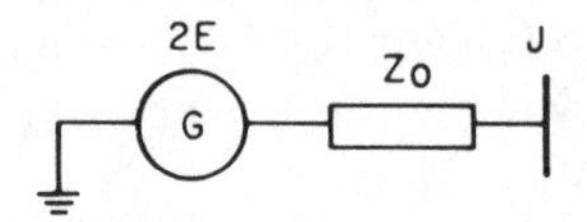

Fig 28
Equivalent Circuit Representing the Arrival of Surge at Junction Point J

is, a driving voltage of twice the actual traveling-wave surge voltage magnitude in series with a resistor of ohmic value Z_0.

If at the junction bus J every circuit that exists in the real system, whether a lumped impedance or distributed-constant line, is connected line-to-ground, the resulting network correctly satisfies the voltage-current relationships that will prevail at the junction bus J. Every distributed-constant line connected to bus J is represented by a line-to-ground-connected resistor of ohmic value Z_0 for each respective line.

An examination of some familiar line-termination cases will aid in developing a conviction that the equivalent circuit is indeed a valid one.

(1) An open-ended line at bus J [Fig 29(a)]. The equivalent circuit yields the following, which we know to be correct:

bus J voltage = $2E$
line terminal current = 0

(2) A short-circuited line at bus J [Fig 29(b)]. Therefore the equivalent circuit yields the following familiar relationships:

bus J voltage = 0
line terminal current = $2E/Z_0$

(3) A line joining another line of equal surge impedance at bus J [Fig 29(c)]. Again, the equivalent circuit correctly yields

bus J voltage = $2E(Z_0)/2Z_0$
line terminal current = $2E/2Z_0$

A verification of these solutions appears in Note 1.

The construction of an equivalent circuit for more complex combinations is accomplished using the techniques described. The simplest equivalent circuit used to accommodate distributed-constant lines is valid only until a returning reflected wave arrives at the junction under study. In many cases the entire critical voltage excursion at the bus under study will have passed before the first reflected wave returns.

To account for the effect of a returning reflected voltage wave, the correct equivalent circuit for surge arrival of that reflected wave at any bus must be created as if it were an independent surge voltage initially approaching the bus. The computed voltage that this reflected wave contributes to the bus is then added with proper polarity to that still being contributed at the bus by the initial surge. When more

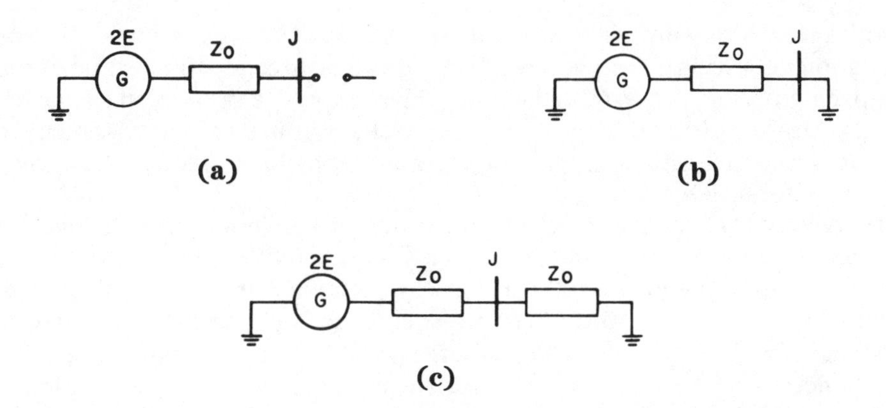

Fig 29
Equivalent Circuits Representing a Line Terminated in Different Ways
(a) Open Circuit at Junction Point J
(b) Grounded Circuit at Junction Point J
(c) Line Continuation of Equal Surge Impedance

than a few wave reflections must be accepted, *lattice diagram* techniques should be used to ensure correct results ([29], Chapter 9, p 215).

The energy of a traveling wave can be dissipated completely if the traveling wave is directed to a junction whose equivalent circuit displays a real resistance termination of ohmic value equal to the Z_0 value of the transmission line. The wave energy is disposed of as heat in the terminating resistor. Although such terminating devices are seldom applied, they are sometimes called transient snubbers.

4.2.3 Amplification Phenomena. A surge-voltage wave traveling along a distributed-constant line, upon encountering a junction terminated in a higher surge impedance Z_0, will increase in voltage to as much as double if the junction is terminated in an open circuit. A surge arrester installed a finite distance ahead of such a junction (Fig 30) can result in a voltage at the junction well above the voltage at the arrester. The junction voltage rise will depend on

(1) the steepness of the surge voltage wave
(2) the propagation velocity along the line
(3) the distance of the line extension ΔD
(4) the magnitude of surge impedance connected to the terminal junction

The rise in terminal voltage will be aggravated by

(1) a steeper front wave
(2) greater ΔD
(3) slower line propagation velocity
(4) greater magnitude of terminating surge impedance

The terminal voltage will not exceed twice the traveling-wave value with any possible value of parameters, and the mechanism associated with this amplification is discussed in Note 2.

Many application charts are available that display the maximum ΔD for specific application conditions. With voltage wave fronts no steeper than 0.5 μs, a separation spacing ΔD of 25 ft is quite generally allowable. The protection system design should locate the protective device as close to the terminals of the critical protected apparatus as is reasonable. Surge-voltage waves may have steeper fronts than the standard reference wave.

A traveling surge voltage, encountering in succession junctions with higher surge impedance, may have its voltage magnitude elevated to a value in excess of twice the magnitude of the initial voltage (Fig 31). Assume the surge impedance of line sections 1-4 to be 10, 20, 40, and 70 Ω, respectively. Next, assume each line section to be long enough to contain the complete wave front, distributed along its length. At the junction between sections 1 and 2, the refracted wave, which continues, will have a magnitude of $1.33E$. This wave, encountering the junction between sections 2 and 3, would create a refracted wave of $1.78E$. In like fashion, this voltage wave, in turn, encountering the junction of sections 3 and 4, would be increased to $2.27E$. This voltage wave, upon reaching the open-end terminal at section 4, would be doubled to $4.54E$.

Typically, the change in surge impedance might result from a different cable construction, which may be additionally modified by a different number of cables run in multiple. The example might well have been represented by four 500 kcmil conductors in parallel in section 1, two in parallel in section 2, one alone in section 3, and a section of bus duct in section 4. The presence of an open-wire line (400 Ω Z_0) extension from a cable feeder (40 Ω Z_0) could, at an open-end terminal, develop a voltage of 3.64 times the surge voltage traveling in the cable.

Fig 30
Transmission Line Extended Beyond a Surge-Voltage Arrester

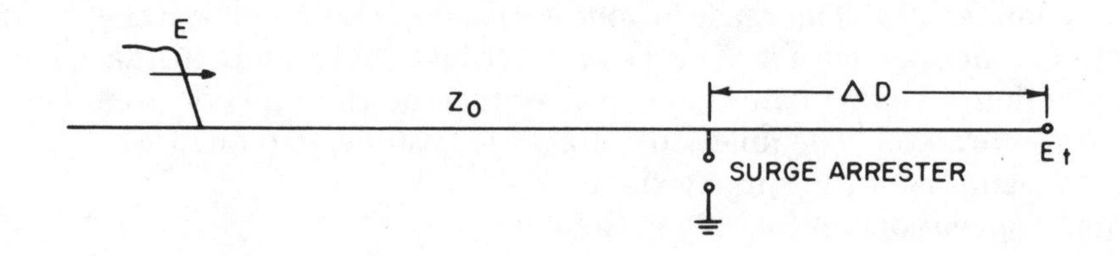

Fig 31
Voltage Amplification by a Series Chain of Line Sections of Progressively Higher Surge Impedance

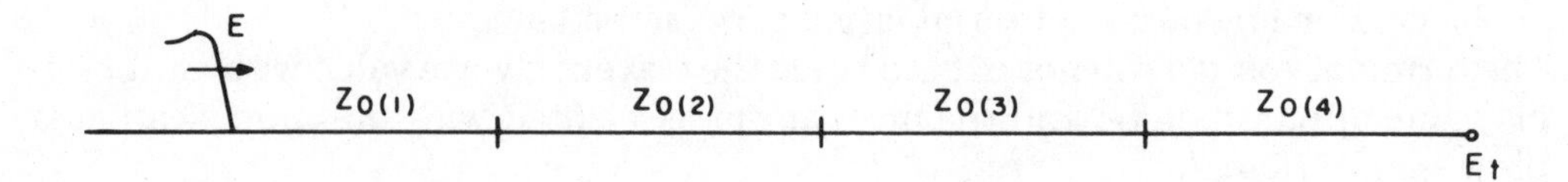

In most instances, a surge voltage approaching a junction bus will encounter a surge impedance of lower value, resulting in a stepdown rather than a stepup in voltage magnitude. Where stepup conditions exist, supplementary protective devices may be required.

The use of several insulation systems (of different character) operating in series may, under a transient surge condition, be broken down at far less voltage than what had been expected by a cascade chain type failure within an equipment insulation system, which actually amplifies the magnitude of the incident wave. Figure 32 illustrates the mechanics involved. Three insulation systems 1, 2, and 3 are in series, having individual breakdown strengths of 10, 15, and 25 kV. Dependence was placed mainly on insulation systems 2 and 3 to assure a withstand capability of 35 kV.

Suppose that a surge voltage E has charged capacitor C_1. The circuit geometry is such that when the voltage on C_1 reaches 12 kV, the voltage across gap 1 reaches 10 kV, the insulation fails, and a sparkover occurs. Conduction across gap 1 causes the 12 kV to be impressed across the L_2-C_2 circuit. The response will be oscillatory with a period of about 15 μs per cycle. The voltage across C_2 will crest in about 7 μs with perhaps 70% overshoot. The voltage across C_1 will drop about 5% during this excursion. The resulting crest voltage across C_2 will be 19.4 kV, more than enough to break down gap 2. Consider that the insulation system 2 fails when the voltage level reaches 17 kV on C_2. Conduction across gap 2 energizes the L_3-C_3 circuit, initiating an oscillatory response very similar to the L_2-C_2 circuit, except that the time period for a first half-cycle swing to crest voltage on capacitor C_3 will take only about 0.5 μs. A 70% overshoot would be reasonable with a 10% drop in voltage across capacitor C_2. The resulting first-crest voltage on capacitor C_3 would be expected to be 26 kV, sufficient to spark over gap 3. The series insulation system that was expected to withstand 35 kV failed with an applied voltage of 12 kV.

Fig 32
Vulnerability of a Chain of Insulation Systems in Series

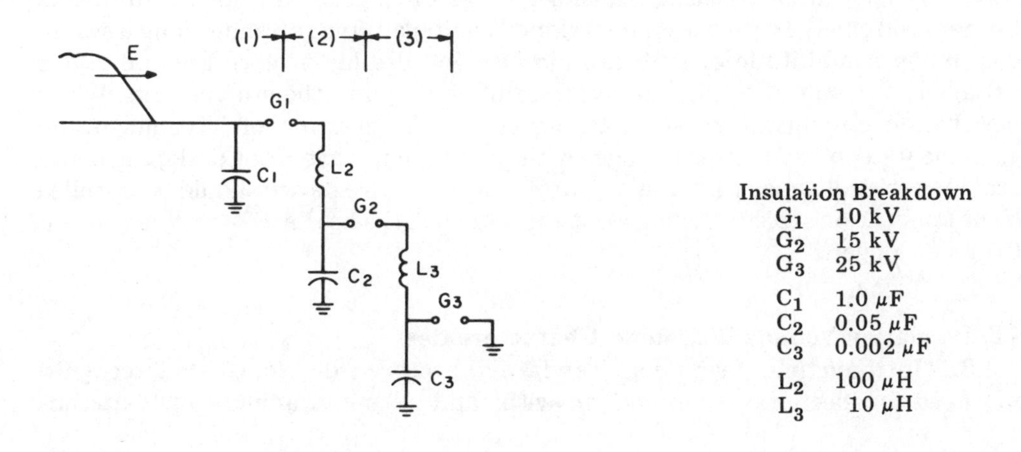

One solution is to introduce voltage-dividing impedances across the several individual insulation systems to force a proportional distribution of voltage across the individual systems under applied voltage conditions of direct-current, power-frequency, and high-frequency transient waves. Another solution would use an independent protective device across each insulation system. If the unbalanced-voltage distribution occurs only in the presence of a fast-front transient voltage wave, a wave-sloping capacitor network in the line connection of the same character used for turn-insulation protection of motor windings could be used (Note 3).

Surge voltages as they arrive at utilization apparatus usually display a substantial delay time in the buildup to crest value of perhaps 1-5 μs. Standard protection procedures recognize a wave front of 0.5-1.5 μs to crest. (In certain cases a chopped wave test is included that involves a fast-front voltage change of substantial magnitude.) Since standard test procedures are based on a 1.2 μs wave front, it follows that most of the available application data will be based on a surge having this characteristic.

There are practical circumstances that generate surge-voltage waves having extremely fast fronts. In the case of vehicular collisions with power poles that cause high- and low-voltage conductors to be brought together, the result can be a severe steep-front voltage surge to the low-voltage system. A lightning stroke to an overhead ground wire may create high voltages on the ground wire and tower structures that may flash back to a power conductor over the line insulator. The result is a step-like steep-front voltage transient on the power line conductor. The voltage rise on the ground mat of a high-capacity substation may be high enough to spark back to low-voltage conductors, resulting in a severe steep-front voltage transient on the low-voltage conductors. Switching interrupters, especially when switching capacitor banks, may make several unsuccessful attempts to interrupt after current zero. When a restrike occurs at the switching contacts, the voltage previously stored on the capacitor is abruptly impressed on the line conductors, creating a very steep-front surge voltage on the line. Multiple restrikes can cause the voltage stored on the capacitor to become elevated to a value much greater than the normal operating voltage ([25], Chapter 5, p 286).

Some conditions tend to reduce the slope of the wave front. A voltage surge, in traveling through an isolating transformer (as distinguished from an autotransformer) will emerge with a less steep slope. A voltage surge traveling along a system containing a multitude of stub tap circuits will display a more gradual rise in voltage as a result of the initial diversion of charge into the stub circuits. Where installation circumstances seem favorable for the creation of large-magnitude extremely steep-front surge voltages with no inherent wave-front desloping before reaching critical apparatus, a surge-capacitor protective device should be installed. Most application tables showing allowable lead lengths are based on wave fronts of 0.5 μs or longer.

4.3 Insulation Voltage Withstand Characteristics

4.3.1 Introduction. Insulation standards have been developed that recognize the need for electrical equipment to withstand a limited amount of temporary excess-voltage stress over and beyond the normal operating voltage. The ability of equipment insulation systems to survive these stresses (without unreasonable loss of life expectancy) is certified by overvoltage tests applied to electrical products at the completion of manufacture. A number of different tests have been developed and standardized for use in rating equipment. The physical structure of insulation systems determines the overvoltage withstand properties for electrical equipment. Some of the important physical considerations affecting the dielectric strength of insulation systems are given in 4.3.3.

4.3.2 Insulation Tests and Ratings. The most common standard factory tests are the 1 min, 60 Hz high-potential (*hi-pot*) test and the 1.2 × 50 μs full-wave impulse test. The 1.2 × 50 μs designation means that a voltage pulse increases from zero to peak value in 1.2 μs and declines to one-half peak value in 50 μs. The wave shape defined by this designation is indicated as the full-wave test in Fig 33. Electrical aparatus assigned a given insulation class should be capable of withstanding, without flashover or apparent damage, a 1.2 × 50 μs full-wave impulse test of specified crest kV. This specified crest voltage is the basic impulse insulation level (BIL) of the equipment. Typical values of test voltages in use today are shown in Tables 15-18.

Transformer insulation systems are generally required to be capable of withstanding other overvoltage tests besides the 60 Hz hi-pot and full-wave tests. Voltages for two of these tests, chopped-wave withstand and switching surge withstand, are indicated in Table 15. For the chopped-wave test, a 1.2 × 50 μs wave with a crest voltage 10% or 15% higher than the full-wave (BIL) test is chopped by a suitable gap after the specified minimum time to flashover (Table 15). The resulting wave shape, shown in Fig 33, has a steep negative gradient that establishes certain withstand capabilities such as associated with sparkover of gap-type arresters or bushing flashover. The chopped-wave test stresses the turn-to-turn insulation more than the line-to-ground insulation, which is checked primarily by the full-wave test. The switching surge level test certifies the capability of an insulation system to withstand the transient overvoltages produced by such conditions as arcing ground faults or the switching of capacitor banks, lines, or transformers.

Fig 33
Standard Impulse Test Waves

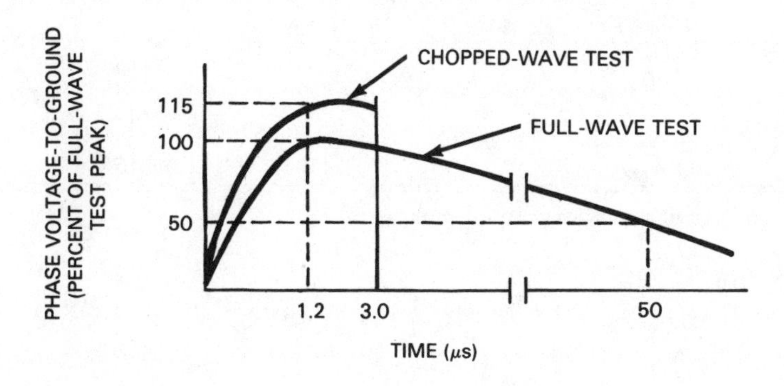

Table 15
Impulse Test Levels for Liquid-Filled Transformers

Insulation Class and Nominal Bushing Rating	Windings					Bushing Withstand Voltages		
	Hi-Pot Tests	Chopped Wave Minimum Time to Flashover		BIL Full Wave (1.2 × 50)	Switching Surge Level	60-cycle 1 min Dry	60-cycle 10 s Wet	BIL Impulse Full Wave (1.5 × 40)
kV (rms)	kV (rms)	kV (crest)	μs	kV (crest)	kV (crest)	kV (rms)	kV (rms)	kV (crest)
1.2	10	54 (36)	1.5 (1)	45 (30)	20	15 (10)	13 (6)	45 (30)
2.5	15	69 (54)	1.5 (1.25)	60 (45)	35	21 (15)	20 (13)	60 (45)
5.0	19	88 (69)	1.6 (1.5)	75 (60)	38	27 (21)	24 (20)	75 (60)
8.7	26	110 (88)	1.8 (1.6)	95 (75)	55	35 (27)	30 (24)	95 (75)
15.0	34	130 (110)	2.0 (1.8)	110 (95)	75	50 (35)	45 (30)	110 (95)
25.0	50	175	3.0	150	100	70	70 (60)	150
34.5	70	230	3.0	200	140	95	95	200
46.0	95	290	3.0	250	190	120	120	250
69.0	140	400	3.0	350	280	175	175	350
92.0	185	520	3.0	450	375	225	190	450
115.0	230	630	3.0	550	460	280	230	550
138.0	275	750	3.0	650	540	335	275	650
161.0	325	865	3.0	750	620	385	315	750

NOTE: Values in parentheses are for distribution transformers, instrument transformers, constant-current transformers, step- and induction-voltage regulators, and cable potheads for distribution cables. The switching surge levels shown are applicable only to power transformers (not distribution transformers). Test voltages defined in ANSI/IEEE C57.12.00-1980 [13].

Table 16
Basic Impulse Insulation Levels (BILs) of Power Circuit Breakers, Switchgear Assemblies, and Metal-Enclosed Buses

Voltage Rating (kV)	BIL (kV)	Voltage Rating (kV)	BIL (kV)	Voltage Rating (kV)	BIL (kV)
2.4	45	23	150	115	550
4.16	60	34.5	200	138	650
7.2	75*	46	250	161	750
13.8	95	69	350	230	900
14.4	110	92	450	345	1300

*95 for metal-clad switchgear with power circuit breakers.

Table 17
Impulse Test Levels for Dry-Type Transformers

Nominal Winding Voltage (Volts)		High-Potential Test	Standard BIL (1.2 × 50)
Delta or Ungrounded Wye	Grounded Wye	kV (rms)	kV (crest)
120-1200		4	10
	1200Y/693	4	10
2520		10	20
	4360Y/2520	10	20
4160-7200		12	30
	8720Y/5040	10	30
8320		19	45
12 000-13 800		31	60
	13 800Y/7970	10	60
18 000		34	95
	22 860Y/13 200	10	95
23 000		37	110
	24 940Y/14 400	10	110
27 600		40	125
	34 500Y/19 920	10	125
34 500		50	150

NOTE: Data from ANSI/IEEE C57.12.01-1979 [14]. Nominal voltages shown are exactly as tabulated in ANSI/IEEE C57.12.01-1979 [14] and are not, in all cases, in accordance with the classifications commonly encountered on industrial and commercial systems.

Table 18
Rotating Machine High-Potential Test and Winding Impulse Voltages, Phase-to-Ground

	Rated Motor Voltage (Volts)					
	460	2300	4000	4600	6600	13 200
60 Hz, 1 min high-potential test voltage						
Crest value (kilovolts)	2.71	7.92	12.73	14.43	20.1	38.8
Per unit of normal crest	7.21	4.22	3.9	3.84	3.73	3.6
Impulse Strength						
Crest value (kilovolts)	3.39	9.9	15.91	18.0	25.1	48.5
Per unit of normal crest	9.01	5.27	4.87	4.8	4.66	4.5

NOTE: From ANSI C50.10-1977 [3] and ANSI C50.13-1977 [4] for synchronous motors, ANSI/NEMA MG1-1978 [22] for induction motors, and [32].

The impulse voltage waves used in switching surge level tests are based on the characteristics of these voltage disturbances, which may be generally described as much *slower* than those caused by lightning. Since some switching phenomena may produce very *fast* front waves, to characterize all switching surges as *slow* can be misleading. ANSI/IEEE C62.1-1984 [19], 8.3.3, has adopted *slow-front* as being preferable to the designation *switching surge*. The test standards require that the switching impulse time to crest exceed 100 μs, and that the time to the first voltage zero on the tail of the wave exceed 1000 μs. The voltage of the wave must exceed 90% of the crest value for at least 200 μs. Typical switching surge test crest voltages, which are 83% of the BIL for transformers of 450 kV BIL and higher, are listed in Table 15. One other test which is sometimes specified as another check on the strength of turn-to-turn insulation is the front-of-wave test. The front-of-wave test is similar to the chopped-wave test except the voltage is higher, and the impulse is chopped on the rising front of the wave before the normal crest.

Tables 15-17 provide a general picture of the standardized impulse capabilities of transformers and switchgear. These voltage withstand characteristics are useful for coordinating the equipment capabilities with the protective characteristics of surge arresters, an analytical procedure known as insulation coordination. The subject of insulation coordination is discussed in 4.6.2. Comparison of Tables 15 and 17 shows that the BILs of dry-type transformers are relatively low. Also, the insulation strength of dry-type transformers does not increase appreciably as the duration of the applied impulse decreases. Open-wire lines vary somewhat in their impulse withstand capacity depending upon such factors as construction, maintenance, and weather, but are generally considered well above associated transformers in this respect. An open-wire 13.8 kV distribution circuit, for example, is typically considered to have a 400 kV BIL. While cables do not have assigned BILs, they too have impulse capability significantly higher than associated liquid-filled transformers.

Rotating machines, like dry-type transformers, have relatively low impulse strength and have no established, standardized BILs. Rotating machines do, however, have standard high-potential test voltage values (shown in Table 18), which have become important in the application of surge protection. A recent report [32] contains a proposed voltage-time boundary (Fig 45) where the maximum impulse voltage is 1.25 times the crest value of the standard high-potential test voltages. These values are also shown in Table 18.

4.3.3 Physical Properties Affecting Insulation Strength. For each item of electrical apparatus to be protected, the security of major insulation (line-to-ground) and, where applicable, turn insulation (turn-to-turn) should be considered separately.

Circumstances will exist in which one of these insulation systems may be overstressed, while the other one is not subjected to any abnormal stress at all. The security of each must be independently examined and protected as necessary.

One of the confusing aspects of an insulation system capability and its protection is the progressive accumulation of deterioration within the dielectric that results from the complete history of voltage stress exposure. An item of equipment subjected to a 60 Hz high-potential test may withstand the voltage application for 50 s and then break down. The device failed the test. It may have withstood the voltage application for the entire 60 s period and passed the test, but it might have failed had the test voltage been continued for another 10 cycles.

With certain types of electric apparatus having a combination liquid and solid insulation system, the cumulative stress failure mechanism only occurs within a narrow band of stress voltages just below the breakdown voltage. Exposure to a lesser overvoltage may initially cause an incomplete failure of the solid insulation, but the subsequent penetration by the liquid material will partially *repair* the deteriorating region.

In all cases a large fraction of the insulation system's capability to withstand applied voltage can be destroyed simply by the process of testing it. For this reason overtesting with dynamic ac voltage should be avoided. Direct-current testing is preferred.

The design of the electric system, including the use of surge suppression devices (to assure adequate insulation security), should correctly interpret the effect of the inverse relationship between imposed voltage magnitude and the allowable duration. A 30% increase in the applied ac voltage magnitude for most equipment will result in a tenfold reduction in insulation life. The high-magnitude surges require careful attention because of the very rapid loss of life. The system design engineer must, largely by judgment, set the margin of safety between the design controls of allowed overvoltage exposure and the certified withstand capability of the insulation system, based on his knowledge of the probable character and repetition rate of troublesome surge voltage transients.

The problems relating to the achievement of insulation security for turn-to-turn insulation in multiturn coils are many and complex. The normal 60 Hz voltage developed in a single turn will range from perhaps a small fraction of 1 V in a contactor magnet coil to 20 V in a medium-sized induction motor to several hundred volts in a large transformer. If it were necessary to only insulate for this normal operating voltage developed in a single turn, the problem would be simple. However, the voltage stress that appears across a single turn-to-turn insulation element when high rate-of-rise voltage surges occur may be much greater than the single-turn operating voltage.

This aggravated voltage stress is most pronounced at turn insulation adjacent to the coil terminals and is intensified by the increased shunt capacitance between winding sections and ground, such as exists inherently in motor windings as a result of each coil in the construction being surrounded by grounded stator core iron.

The controlling parameters (Fig 34) are the elemental values of coil series inductance ΔL, the elemental values of capacitance C_S shunting the above elemental coil segments, and the elemental capacitance to ground C_G.

Under normal 60 Hz excitation, the voltage distribution is controlled almost entirely by the series-connected chain of elemental coil inductances, creating equal division of the impressed voltage across all turns. A steep-front voltage applied to the line terminal creates an entirely different pattern of distributed voltage. Consider the incoming voltage wave to be a step voltage of infinite rate of rise. If only

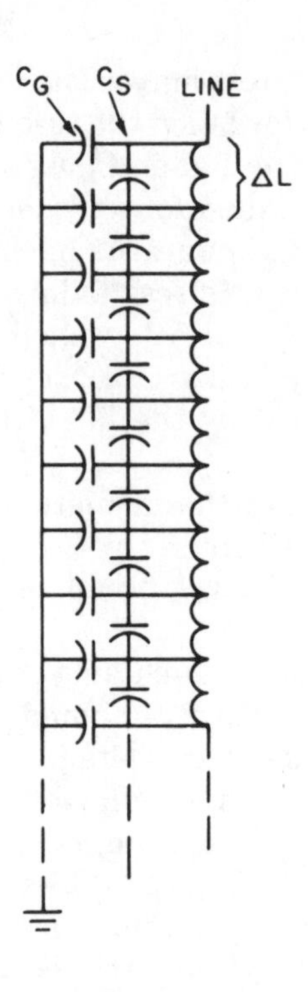

Fig 34
LC Network in a Multiturn Winding

the turn shunting capacitance C_S were present, the incoming transient overvoltage would be uniformly distributed. The distributed capacitive coupling to ground C_G is responsible for the nonuniform voltage distribution. Note that each elemental capacitance to ground C_G tends directly to hold the coil turns with which it is coupled at ground potential. If C_S were zero, a step voltage at the terminal would create full voltage at the terminal end of the first coil. The inner end of the first coil, and all coils deeper in the winding, would remain at ground potential as controlled by C_G. Only as current flow begins through the first coil inductance ΔL could any voltage appear across any C_G (except the one unit at the terminal). Thus the initial voltage distribution would display the full surge voltage across the terminal coil with zero voltage across all coils deeper in the winding.

Following the initial steep-front voltage application, current flow will build up in the ΔL, which will redistribute the voltage more uniformly. In the process, the internal L–C oscillations will create a voltage drop across some inner coil (or coils) substantially greater than the uniform-distribution value, but seldom as great as the initial voltage stress across the terminal coil.

In an actual motor winding, a very high percentage (upwards of 90%) of an applied surge voltage with a $^1\!/_{10}$ μs front can appear across the terminal coil.

Unfortunately, the internal electric network by which most apparatus can be represented is not commonly found in equipment specifications or industry standards. In the case of rotating machines, a guide to achieve turn-insulation security is presented in 4.6.3.6.

4.4 Arrester Characteristics and Ratings

4.4.1 Introduction. Historically, various types of surge arresters have been used for power system protection. The use of pellet and expulsion-type arresters diminished when the so-called valve-type arrester was introduced and refined. The basic elements of valve-type surge arresters are the gap unit (employed in some designs) and the so-called valve element. The gap unit is configured and augmented such as to achieve the most desirable balance between (low) sparkover and reseal capability consistent with cost. The valve element consists of a nonlinear resistance that exhibits a relatively high resistance at low voltages (and current) and a much lower resistance at high (surge) voltage and current. This nonlinear property greatly enhances overall arrester performance because it presents lowest resistance (low-surge impedance, in effect) at the important high-surge current condition while its high-resistance property at low surge and normal voltage levels assists the gap unit in resealing after-surge discharge to prevent continued flow of follow (power) current.

While the protective efficiency of today's conventional valve-type arrester permits most significant savings in the insulation levels of transmission and subtransmission equipment, it has not been possible until recently to achieve desired surge protective margins with these arresters for some very important industrial and commercial applications—specifically certain motors and certain dry-type transformers. This is due primarily to the arrester gap surge sparkover voltage level, which greatly exceeds the associated arrester discharge voltage characteristic in the vast majority of industrial and commercial exposures. Arrester gaps also introduce transients as they sparkover to engage the valve elements. Thus arrester gaps have been the principal impediment in securing a more ideal arrester function. However, the valve element material heretofore used, silicon carbide, does not possess sufficient nonlinearity to be self-protecting in the presence of continuous line-to-ground voltage, and series gaps are required to provide isolation.

A newly developed metal oxide valve-element material [42], specifically of zinc oxide base, is now available in an arrester design that has sufficient nonlinearity such that a series gap is not required. This metal oxide arrester can be designed to avoid the noted disadvantages of gaps, and the inherent, relatively constant voltage characteristic of its zinc oxide valve element initiates and completes its surge protective cycle in a more ideal manner as compared to valve-type arresters previously available. Certain commercially available metal oxide arresters actually employ the use of a modified gap design that retains the essential protective advantages of gapless construction. With the development of the gapless metal oxide arrester, such terms as *sparkover level*, *reseal voltage*, *power follow current*, which specifically relate to gap performance, are no longer pertinent and more general terms must be used to describe arrester prerformance. Instead of sparkover level, the terms *protective level*, *clamping level*, *clamping voltage*, and *discharge voltage* will henceforth be used to describe this important arrester characteristic. The terms *reseal voltage* and *power follow current* are not applicable to metal oxide arresters and apply in this text solely to gapped silicon carbide valve-type arresters. Traditionally, an arrester employing a gap in series with a valve element is referred to as a valve-type arrester or, referring to the material of the valve element, a silicon carbide arrester. The newer arrester may be generically referred to as a metal oxide, zinc oxide, or gapless arrester.

This section will present some comparisons of the zinc-oxide-based material with that of the traditional silicon carbide valve-element material used in modern valve-type arresters. Finally, some of the protective characteristics and ratings of both types of arresters will be presented.

4.4.2 Comparison of Metal Oxide and Silicon Carbide Volt-Ampere Characteristics. Both silicon carbide and zinc-oxide-based valve-element material is formulated by pressing their respective ingredients into discs and sintering at high temperature into a dense ceramic. The nonlinear volt-ampere characteristic of silicon carbide is primarily a temperature-activated phenomenon while that of zinc oxide is a voltage-gradient-related phenomenon, at least throughout much of its operating range. At higher currents, the nonlinearity of zinc-oxide-based valve-element material becomes moderated somewhat by its resistivity. Figure 35 illustrates the relative degrees of nonlinearity of the two materials by a normalized log-log plot of volts (per millimeter of disc thickness) versus amperes (per square centimeter of disc area). Note the following points in regard to the Figure 35 zinc oxide plot:

(1) A current increase of five orders of magnitude (from 10^{-1} to 10^{2} A/cm^2) has an associated voltage increase of only slightly greater than one-half. With such a high

Fig 35
Typical Volt-Ampere Characteristics of Zinc Oxide and Silicon Carbide Valve-Element Discs [42]

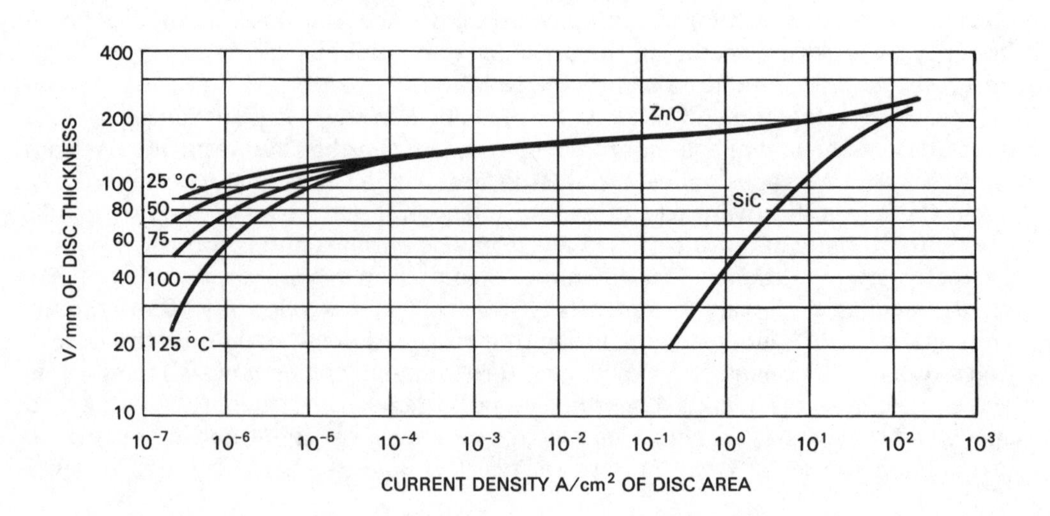

degree of nonlinearity it is feasible to use this material as a valve element without a series gap. By comparison, for the silicon carbide material an increase of only one order of magnitude in current is associated with a voltage increase by a factor ranging from approximately 1¾–3.

(2) The volt-ampere characteristics plotted over a wide temperature range blend together and are virtually coincident above approximately 0.0003 A/cm^2. This particular current density is well above leakage current density at normal operating voltage but well below surge current densities. Thus temperature does influence the thermal design of the arrester for normal voltage conditions but does not affect the surge protective characteristics of a given design. Presently used silicon carbide valve material exhibits a negative temperature coefficient of about 0.24%/°C.

(3) At moderately high current densities (greater than 10 A/cm^2) the volt-ampere characteristic develops a significant turn-up that is attributable to the resistivity of the zinc oxide grains. A similar turn-up occurs in silicon carbide valve elements that is somewhat greater in degree (than for zinc oxide valve elements) at the highest current densities encountered in arresters.

(4) Continuous application of ac voltage to zinc oxide valve elements will produce a very slow drift toward decreasing resistance at the low-current end of the volt-ampere characteristic. For example, the log-log plots of Fig 35 will gradually shift to the right (increasing current) at the low ends as voltage is applied for extended time. This low-current drift has associated increase in watts loss under normal operating conditions. This, coupled with the fact that zinc oxide also has a positive temperature coefficient of watts loss, makes it imperative that temperature limits and continuous voltage capabilities be assigned very carefully.

Figure 36 presents a comparison of the volt-ampere characteristics of zinc oxide and silicon carbide valve elements for particular designs wherein the 10 kA discharge voltage is essentially the same. Note the nearly constant-voltage characteristic of the zinc oxide valve element as compared to that for silicon carbide. The zinc oxide element draws very little current (exhibits very high resistance) until a voltage approaching the *protective level* is reached. Thereafter, only the current necessary to maintain this voltage is drawn, thus requiring the valve element to absorb a minimum of associated energy.

As a consequence of the differences in the volt-ampere characteristics of the two materials, metal oxide arresters will exhibit higher discharge voltages at low-discharge currents and lower discharge voltages at high-discharge currents as compared to silicon carbide arresters. The resistance of metal oxide valve elements is sufficiently high at very low currents to avoid the need for series gaps; although, as mentioned earlier, certain designs of metal oxide arresters do employ gaps.

Additional comments relating to comparative volt-ampere characteristics of metal oxide and silicon carbide valve elements are included in 4.5.1.

4.4.3 Basis of Arrester Rating. ANSI/IEEE C62.1-1984 [19], 2.46, defines *voltage rating* as "the designated maximum permissible operating voltage between its terminals at which an arrester is designed to perform its duty cycle. It is the voltage rating specified on the nameplate."

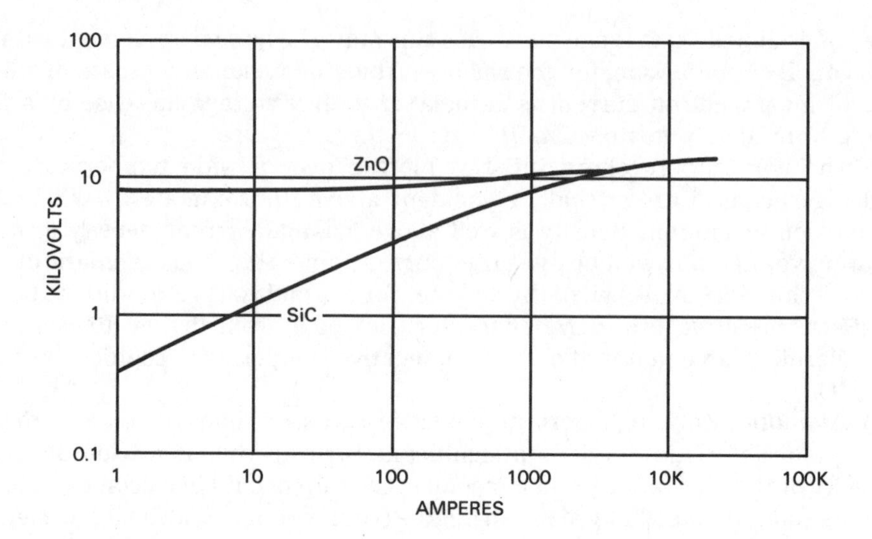

NOTE: These are particular designs that have essentially the same discharge voltage at 10 kA [42].

Fig 36
Typical Volt-Ampere Charcteristics of 6 kV Valve Elements

The duty cycle alluded to is the so-called *duty-cycle test* (defined in the same standard) which involves the application of 24 high-current, 8 × 20 μs impulses with a maximum interval of 1 min between impulses. For station-class arresters, the prescribed impulse current magnitude for the test is 10 000 A. Fundamental frequency voltage (60 Hz) at which the arrester is rated is applied continuously throughout the test series. This test verifies the ability of the arrester to accommodate the current impulses, the associated follow current, and the capability of the gap to reseal. Note that the arrester is rated on the basis of the associated applied 60 Hz voltage at which the arrester is capable of passing the duty-cycle test, and not on the basis of arrester protective characteristics. Actually, heretofore gap reseal for the gap-type arrester has been the most critical aspect of this test and, for this reason, it is often stated that arresters are rated on the basis of the maximum 60 Hz (rms) voltage at which the arrester is expected to reseal, that is to extinguish discharge current in the gap element after sparkover.

A gap-related basis of rating is not germane for a metal oxide arrester. The basic duty-cycle test noted above, however, is still retained as a basis of rating zinc oxide arresters as it represents a recognized and accepted criterion by which valve element adequacy may be demonstrated. The duty-cycle test does not establish all of the design criteria of the zinc oxide valve element. Protective characteristics and the continuous (line-to-ground) 60 Hz voltage impressed across the arrester establish important design parameters. As a result, the metal oxide arrester has a maximum circuit operating voltage (MCOV) in addition to its duty-cycle voltage rating, as shown in Table 19. Silicon carbide valve elements are designed to withstand rated

Table 19
Typical Metal-Oxide Arrester Ratings, Extended Time Capabilities, and Maximum Discharge Voltages at Indicated Impulse Currents

Arrester Rating[1] kV rms		Maximum Equivalent Front-of-Wave[2] Protective Level kV Crest			Maximum Discharge Voltage (kV Crest) at Indicated Impulse Current for an 8 × 20 μs Current Wave: 1.5 kA			5 kA			10 kA			20 kA			40 kA	Maximum Switching Surge Protection Level kV Crest[3]	
	MCOV[4]	Sta	Int	Dist	Sta	Int	Dist	Sta	Int	Dist	Sta	Int	Dist	Sta	Int	Dist	Sta	Sta	Int
2.7	2.2	8.1	—	—	6.1	—	—	6.7	—	—	7.3	—	—	8.5	—	—	10.3	5.6	—
3.0	2.54	9.3	9.9	—	6.9	7.1	—	7.7	8.0	—	8.3	8.8	—	9.7	10.3	—	11.7	6.4	—
4.5	3.70	13.5	14.3	—	10.0	10.3	—	11.0	11.5	—	11.9	12.6	—	13.8	14.7	—	16.4	9.2	—
5.1	4.2	15.2	—	—	11.3	—	—	12.5	—	—	13.4	—	—	15.5	—	—	18.4	10.4	—
6.0	5.08	18.4	19.6	—	13.6	14.1	—	15.0	15.7	—	16.1	17.2	—	18.6	20.0	—	22.0	12.6	—
7.5	6.10	22.1	23.5	—	16.4	16.9	—	18.0	18.8	—	19.3	20.6	—	22.4	24.0	—	26.5	15.1	—
8.5	6.9	25.0	—	—	18.5	—	—	20.4	—	—	21.8	—	—	25.2	—	—	29.7	17.0	—
9.0	7.62	27.6	29.4	27.6	20.4	21.1	20.4	22.5	23.4	22.4	24.0	25.7	24.0	27.7	29.9	27.7	32.7	18.8	—
10	8.47	30.6	32.6	30.6	22.7	23.4	22.7	24.9	26.0	24.9	26.7	28.5	26.7	30.7	33.1	—	36.1	20.9	—
12	10.16	36.7	39.1	—	27.2	28.1	—	29.9	31.1	—	32.0	34.1	—	36.9	39.5	—	43.4	25.0	—
15	12.7	45.9	48.9	—	34.0	35.1	—	37.3	38.9	—	39.9	42.6	—	45.9	49.4	—	53.9	31.3	—
18	15.24	55.1	58.7	55.1	40.7	42.1	40.7	44.7	46.6	44.7	47.8	51.0	47.8	54.9	59.0	54.9	64.4	37.5	—
21	17.1	61.8	65.8	—	46.7	47.2	—	50.2	58.3	—	53.6	57.2	—	61.6	66.1	—	72.1	42.1	—
24	19.5	70.5	75.1	—	52.1	53.9	—	57.2	59.7	—	61.2	65.3	—	70.3	75.6	—	82.4	48.0	—
27	21.4	79.1	84.3	83.5	58.5	60.5	61.4	64.3	67.0	67.4	68.6	73.3	72.0	78.9	84.8	82.7	92.3	53.9	—
30	24.4	88.2	93.9	—	65.2	67.4	—	71.5	74.6	—	76.4	81.5	—	87.7	94.1	—	102.5	60.1	—
36	29.3	106	113	—	78.2	80.9	—	85.9	89.6	—	91.7	97.9	—	105	113	—	123	72.1	—
39	31.7	115	122	—	84.6	87.6	—	92.9	96.9	—	99.2	106	—	114	123	—	133	78.0	—
45	36.5	132	141	—	97.4	101	—	107	112	—	114	122	—	131	141	—	153	89.8	—
48	36.9	141	150	—	108.8	107	—	114	119	—	122	130	—	139	150	—	163	95.7	—
54	43.7	142	—	—	105	—	—	115	—	—	122	—	—	135	—	—	157	106	—
60	48.6	158	188	—	117	134	—	128	149	—	136	162	—	150	188	—	174	118	122
72	58.3	189	226	—	140	161	—	153	178	—	163	195	—	180	225	—	209	142	147
90	72.9	237	281	—	175	201	—	192	223	—	204	244	—	225	281	—	261	177	183
96	77.8	253	300	—	187	214	—	204	237	—	218	259	—	241	300	—	279	189	195
108	87.5	284	338	—	210	242	—	230	268	—	245	293	—	271	338	—	313	213	220
120	97.2	315	376	—	233	269	—	255	298	—	272	325	—	300	376	—	348	236	245
132	105	347	—	—	256	—	—	281	—	—	299	—	—	331	—	—	383	260	—
144	110	380	—	—	280	—	—	307	—	—	327	—	—	362	—	—	419	284	—
168	119	441	—	—	326	—	—	357	—	—	380	—	—	420	—	—	487	331	—
180	146	474	—	—	350	—	—	383	—	—	408	—	—	451	—	—	523	355	—
192	152	506	—	—	374	—	—	410	—	—	436	—	—	482	—	—	558	379	—
228	164	600	—	—	443	—	—	486	—	—	517	—	—	527	—	—	662	449	—

[1] Ratings listed in this column are from ANSI/IEEE C62.1-1984 [19]. For ratings 48 kV and below the 2000 h capability is (.9) (arrester rating) and is suggested for use where ground-faults are not relayed off immediately (for example, high-resistance grounded and some ungrounded systems).

[2] Based on a 10 kA impulse that results in a discharge voltage with a 0.5 μs rise time.

[3] Standards have not established the basis of switching surge tests for arresters rated below 60 kV. For ratings 2.7 through 48 kV, the listed values are maximum 500 A discharge voltages.

[4] Maximum continuous operating voltage L–N kV rms.

arrester voltage for only very short periods of time, being nearly isolated from fundamental frequency voltage by the gap unit, except during sparkover.

4.4.4 Protective Characteristics. The surge protective characteristics of valve-type arresters are associated with the two basic units comprising the arrester:

(1) sparkover voltage characteristics relating to gap unit performance

(2) discharge voltage characteristics relating to valve unit performance

The sparkover characteristic, that is, the surge voltage at which the gap unit conducts to engage the valve unit, varies with the waveshape of the surge. A perfect sphere gap will spark over at markedly higher values of voltage on steeper front waves and at lower values of voltage on *slower* waves, that is, waves that take a longer time to reach a given crest value. Modern valve arrester gap units, however, are designed such as to produce a more uniform sparkover characteristic as a function of wavefront steepness.

Arrester standard ANSI/IEEE C62.1-1984 [19] specifies various waveshape tests whereby sparkover performance of gapped arresters is evaluated. Arrester manufacturers list performance characteristics based on these waveshapes. The two tests most frequently used for such listing are the front-of-wave and the 1.2×50 μs wave test. The front-of-wave sparkover is the voltage at which arrester-gap sparkover occurs on the front of a wave rising at the rate of 100 kV/μs for each 12 kV of arrester rating for arresters rated 3–144 kV, and 1200 kV/μs for arresters rated above 144 kV. For arresters rated less than 3 kV, the test wave rate-of-rise is 10 kV/μs. Also, all rotating-machine arresters have a specified front-of-wave test rate-of-rise of 10 ± 3 kV/μs to gap sparkover. The 1.2×50 μs sparkover is the crest value of a 1.2×50 μs wave that causes arrester sparkover at the time crest value is reached.

So-called *slow-front* or *switching-surge* sparkover values are also sometimes listed. These are similar in nature to the 1.2×50 μs test wave except a series of waveshapes are used with fronts varying from 30–200 μs.

Table 20 lists typical sparkover characteristics based on the foregoing standard test waves. Also note that the table includes so-called *discharge voltage values*. These are values of voltage that appear across the arrester when the indicated magnitude (crest) of a standardized 8×20 μs current wave as defined in ANSI/IEEE C62.1-1984 [19] is conducted through the arrester. This voltage is also sometimes referred to as the "IR" voltage. Note that very large increases in discharge current result in relatively small increases in discharge voltage. This exhibits the nonlinear nature of the valve unit.

Metal oxide arrester protective characteristics relate primarily to their valve element properties. Traditionally, valve element protective characteristics have been provided in terms of the maximum voltage associated with discharging a specified magnitude of surge current through them. Due to the nonlinear volt-ampere characteristics of valve elements, a given discharge current rises to its crest more slowly and decays from crest more rapidly than the associated impressed surge voltage. The standardized current waveshape that has been adopted, although not strictly accurate for metal oxide, approximates that resulting from the application of the standard full-wave impulse test voltage wave (1.2×50 μsec) to the valve elements of silicon carbide gap-type arresters.

Table 20
Surge Protective Characteristics of Valve-Type Silicon Carbide Surge Arresters[1]

Arrester Rating[3] kV rms	Maximum Impulse Sparkover Crest kV					Maximum Switching Surge—Crest kV	Maximum Discharge Crest kV at Indicated Impulse Current, 8 × 20 μs Wave													Arrester Rating kV rms
	Front-of-Wave			1.2 × 50 μs			1.5 kA			5 kA			10 kA			20 kA			40 kA	
	Sta	Int	Dist[2]	Sta	Int	Station	Sta	Int	Dist[2]	Sta	Int	Dist[2]	Sta	Int	Dist[2]	Sta	Int	Dist[2]	Sta	
3	12	11	14.5/11	12	11	12	5.0	6.5	9.5/9	6.4	7.4	11/11	7.3	8.3	12/13	8.3	9.5	13.5/15	10.2	3
4.5	16	16	—/17	15	15	15	7.4	9.5	—/13	9.5	10.8	—/17	10.8	12.0	—/19	12.3	14.0	—/23	15.1	4.5
6	20	21	28/21	18	19	18	9.8	12.5	19/17	12.6	14.5	22/22	14.3	16.0	24/26	16.3	18.5	27/30	19.9	6
7.5	25	26	—/26.5	22	23.5	23	12.2	15.8	—/21	15.7	17.8	—/28	17.7	20.0	—/32	0.3	22.8	—/38	24.8	7.5
9	30	31	39/32	25	27.5	27	14.6	19.0	28/26	18.8	21.0	33/32	21.2	23.5	36/39	24.3	27.0	40/45	29.6	9
10	—	35	43/32	—	31	—	—	21.8	28/29	—	24.5	33/38	—	27.2	36/45	—	31.5	40/53	—	10
12	39	40	61/39.5	32	35.5	35	19.4	24.7	34/34	24.9	28.0	43/43	28.1	31.0	51/51	32.1	36.0	60/60	39.2	12
15	48	50	76/46	39	43.5	43	24.2	31.0	42/42	31.0	35.0	53/53	35.0	39.0	63/63	40.0	45.0	74/74	48.4	15
18	57	—	91/54.5	47	—	51	28.9	—	50/50	37.6	—	63/63	41.8	—	75/75	47.8	—	90/90	58.5	18
21	66	68	106/—	54	59	59	33.7	43.0	—/58	43.2	49.0	—/73	48.7	55.0	—/87	55.5	63.0	—/104	68.0	21
24	76	78		61	67	67	38.4	49.0		49.2	56.0		55.5	62.0		63.5	72.0		77.5	24
30	95	97		75	81	84	47.8	61.5		61.5	70.0		69.5	78.0		79.0	90.0		96.6	30
36	113	116		90	95	100	57.5	74.0		73.5	83.0		83.0	94.0		94.5	108		115.0	36
39	123	126		97	102	108	62.5	80.0		79.5	90.0		89.5	102		102.0	117		125.0	39
48	151	154		118	123	132	76.0	98.0		97.5	111		110	125		125.0	145		153	48
60	180	190		136	153	142	95.0	123		122.0	139		137	156		156	180		190	60
72	213	228		166	180	170	114	148		146	166		164	187		187	216		227	72
78	231	245		183	195	184	123	160		158	180		178	203		202	234		246	78
84	247	262		198	209	198	133	172		170	194		191	218		217	252		265	84
90	267	282		214	223	213	142	184		182	208		204	233		232	270		283	90
96	280	300		231	236	227	151	197		194	222		218	249		248	288		302	96
108	315	335		262	263	253	170	221		218	249		245	281		278	324		333	108
120	347	370		294	290	284	188	246		241	277		272	311		309	360		376	120
132	380			320		312	207			262			294			333			402	132
144	413			350		304	226			287			321			363			439	144
168	495			396		397	263			334			374			422			510	168
180	530			400		400	281			358			400			452			550	180
192	560			427		426	300			382			427			482			585	192
228	640			510		506	355			452			510			575			695	228

[1] This table is based upon the offering of one manufacturer and may vary somewhat from other listings.
[2] Standard and low-sparkover models data listed—standard/low-sparkover.
[3] Arresters for rotating machine protection are available for ratings listed 3–24 kV, inclusive; also arrester rated 650 V available for low-voltage apparatus protection.
[4] Defined in ANSI C62.2-1981 [5].

Table 19 shows the maximum discharge voltages associated with various 8×20 μs discharge currents for arresters rated 2.7-192 kV. Virtually all discharge currents in effectively shielded industrial installations will be less than 10 kA, the vast majority being only a small fraction of this magnitude.

Table 19 also includes the maximum switching surge protective levels of the metal oxide arrester. Switching surge (sometimes called slow-front) impulse characteristics are derived on the basis of applying a prescribed array of test waves having front times ranging from 30-2000 μs. Switching surge tests have not as yet been established for arresters rated below 60 kV. For arrester ratings 2.7-60 kV in Table 19, the listed values of maximum switching-surge voltage is that associated with discharge of a 500 A (8×20 μs) impulse current. A discharge current of 500 A will encompass virtually all switching-produced surge currents at these voltage levels.

Figure 37 provides an insight to the very short-time protective characteristics of metal oxide valve elements as compared to silicon carbide valve elements. Fast-front impulse currents of 10 kA magnitude cresting in 0.12 μs, 1.3 μs, and 10 μs were used in developing these characteristics. The silicon carbide valve element voltage is shown to increase by 8-10% each time the time-to-crest is halved for the same current magnitude. The corresponding voltage increase for metal oxide valve elements is slightly less than one-half as much—approximately 4½%. For further

Fig 37
Typical Discharge Voltage Versus Current Wave Crest Time (μs) for 10 kA Discharge

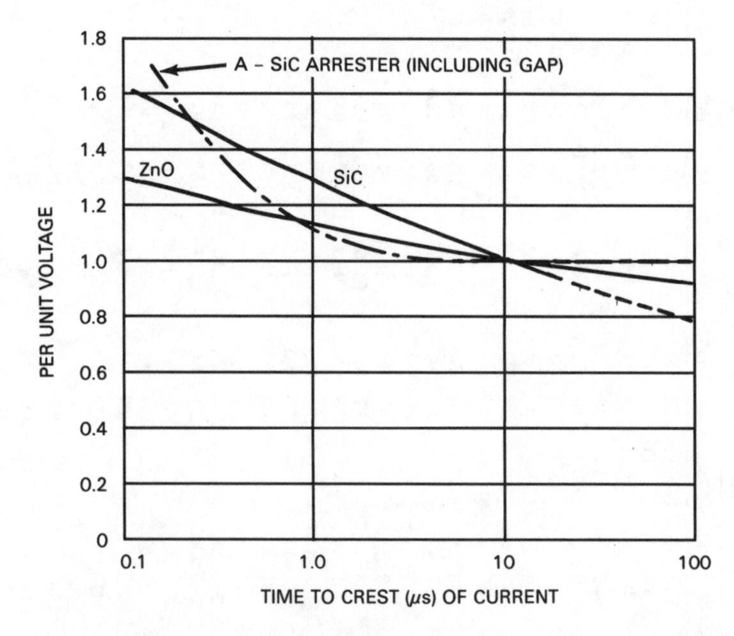

NOTE: The per unit voltage is based on discharge voltage associated with 10 μs to crest on current wave.

comparison, the volt-time sparkover characteristic of a silicon carbide high-voltage arrester is included as curve A. It is seen that the metal oxide arrester provides a better protective characteristic in relation to very fast-front surges, which include the so-called front-of-wave test voltage sometimes used to demonstrate fast-front surge insulation capability.

4.4.5 Arrester Classes. Three classes of valve-type arresters are recognized by industry standards (ANSI/IEEE C62.1-1984 [19]). In order of decreasing cost and overall protective quality and durability, they are

(1) station class
(2) intermediate class
(3) distribution class

Note from Table 20, which covers gapped silicon carbide arresters, that the station class, intermediate class, and special distribution class arresters all have comparable sparkover levels. The *standard* distribution class arrester of this design has a sparkover level on front-of-wave of approximately one-third to two-thirds greater. For a given discharge voltage, the silicon carbide station class arrester can discharge approximately twice as much impulse current (8×20 μs current wave) as the intermediate class arrester, which, in turn, can discharge one-half more impulse current than the distribution class.

Table 19 lists protective (discharge voltage) characteristics of the metal oxide arresters. It should be noted that the values listed in Tables 19 and 20 are representative of several manufacturers and may vary somewhat from actual listings. The nature of the zinc-oxide-based material used in the valve elements of this design is such that the protective characteristics among the three classes are relatively uniform. There are, however, distinct differences in design features, size, etc, among the three classes that enhance particularly the repetitive duty-cycle capability of the station class relative to the intermediate class and the intermediate class relative to the distribution class.

4.4.6 Arrester Discharge-Current Capability. To ensure that arresters have an acceptable capability to discharge lightning currents and line and cable capacitance, an array of discharge-current withstand and duty-cycle tests are specified by standards. Two of the tests relate to high-current, short-duration and to low-current, long-duration duties. The high-current, short-duration test consists of two discharges of a surge current (of 65 kA crest for intermediate and distribution class arresters, 100kA crest for station class) having a $(4-8) \times (10-20)$ μs waveshape. The low-current, long-duration tests require station and intermediate class arresters to display capability to discharge capacitance equivalent to specified transmission line lengths (150–200 mi for station class, depending upon arrester rating, and 100 mi for intermediate class). The procedure for determining cable length equivalent is included in ANSI/IEEE C62.1-1984 [19]. Distribution arresters must exhibit (in a specified series of discharges) the capability of withstanding an approximate rectangular waveshape of 75 A minimum surge current of a minimum time duration of 1000 μs.

Some arresters have discharge capabilities well in excess of these indicated minimums. Where high discharge currents are of concern, consult arrester manufacturer data to determine adequacy of arrester discharge capability.

The energy absorption capability of the metal oxide arrester upon discharge is limited, for a single event, by the thermal shock the valve element discs can sustain without cracking. In general, the metal oxide arrester *sudden* absorption capability is one to two orders of magnitude greater than the stored energy in the line used to perform the standard transmission line discharge test at these voltages (3–230 kV). After an interval of approximately 1 min to permit equalization of temperature throughout the discs, an additional approximately equal amount of energy absorption is permissible up to the transient thermal stability limit. This total (thermal stability) limit of energy absorption capability is approximately three times the energy absorbed in the standard (24 operation) duty-cycle test and is well above the capability of conventional arresters, an important consideration for severe-duty applications such as for installation near a large, switched capacitor bank.

4.5 Arrester Selection. The application of surge arresters depends upon three basic considerations:

(1) Selection of arrester rating
(2) Selection of arrester location
(3) Selection of arrester class

Whereas there is relatively little latitude in the selection of arrester ratings in most applications, there is somewhat more room for the exercise of judgment factors in the selection of arrester location and arrester class. These latter two basic considerations do provide for different degrees of risk that may be assumed by the owners and operators of a system.

4.5.1 Arrester Rating. As established in the foregoing discussion on silicon carbide surge arresters, they are rated on the basis of the maximum power system frequency rms voltage at which they may be expected to reseal against after having sparked over. The lower the arrester rating, the better the protection afforded circuit and apparatus insulation — also, the lower the cost. The basic philosophy of application of the metal oxide arrester is the same as for the silicon carbide arrester. Their respective application *rules* differ in only very limited areas. In any case, the lowest rated arrester capable of withstanding the maximum sustained fundamental frequency voltage that can occur across the arrester for the expected time of exposure is selected. The circumstance of a line-to-ground fault on the system, its duration, and the associated line-to-ground voltage on the unfaulted phases is the prime criterion for selection of arrester rating.

Line-to-ground faults tend to shift the system fundamental frequency phasor pattern from its normal position of symmetry with respect to ground. In the case of ungrounded systems, this shift is virtually complete, that is, the unfaulted (sound) phase arrester(s) will be subjected to 100% of the line-to-line operating voltage, resulting in the so-called *100% arrester requirement*. However, system *solid* grounding (depending upon degree) provides considerable restraint in voltage pattern shift and usually permits a considerable reduction in arrester rating requirement.

ANSI/IEEE C62.1-1984 [19] defines *coefficient of grounding* as "The ratio E_{LG}/E_{LL}, expressed as a percentage, of the highest root-mean-square line-to-ground power-frequency voltage E_{LG} on a sound phase, at a selected location, during a fault to ground affecting one or more phases to the line-to-line power-frequency voltage E_{LL} which would be obtained, at the selected location, with the fault removed." ANSI C62.2-1981 [5] provides a guide to facilitate the calculation and determination of coefficients of grounding. Such aids are often presented in terms of symmetrical component parameters, and surge arrester rating practices have evolved to a certain extent around symmetrical component terminology. Further, systems have been categorized as follows to aid in arrester rating selection:

(1) effectively grounded—coefficient of grounding not exceeding 80% (X_0/X_1 is positive and less than three, and R_0/X_1 is positive and less than one)

(2) noneffectively grounded or ungrounded when coefficient of grounding exceeds 80%

The vast majority of medium-voltage (2.4–13.8 kV) industrial power systems employ some form of resistance system grounding. For arrester application purposes, these are noneffectively grounded systems having coefficients of grounding of 100%. The same is true for the very infrequently used ungrounded systems. This simply means that such systems require arrester ratings of at least 100% of the maximum operating voltage of the system.

Some industrial complexes are served by medium-voltage systems that utilize *solid* system grounding only at the point of energy supply to the system. These systems exhibit a wide range of coefficents of grounding (about 75%), depending upon the system or location in the system. Therefore, these systems require individual study to ensure the most economical, secure, arrester rating selection.

Many high-voltage transmission systems may exhibit coefficients of grounding as low as 70%, and certain multigrounded four-wire distribution systems may be even slightly less. An IEEE working group [33] proposed open-wire four-wire multigrounded distribution systems using arresters with minimum rating of 1.25 times line-to-neutral (nominal) voltage, and similarly a 1.50 factor for same type systems utilizing spacer-cable construction. Based upon a 1.05 operating voltage regulation factor, this corresponds to coefficients of grounding of 69 and 84%, respectively.

The coefficient of grounding as a measure of the system grounding effectiveness is very important when applying arresters. On effectively grounded systems, for example, an arrester may be momentarily subjected to a voltage of up to 120–140% of normal line-to-ground voltage, during a ground fault involving another system phase. The most likely situation where this might occur would be for a lightning-induced overvoltage that caused a flashover (line-to-ground fault) on one phase of the system and a voltage surge of sufficient magnitude to simultaneously cause arrester protective action on the unfaulted phases. If these arresters were gap-type, they would then be called upon to reseal, that is, cease conduction, during the interval prior to fault interruption with the elevated voltage on the unfaulted phases and across the the arrester terminals. On noneffectively grounded systems, involving ungrounded and (high or low) resistance grounded systems, the abnormal steady-state voltage applied to the arrester is not only greater in magnitude, approaching line-to-line voltage as a limit, but is sometimes applied for substantially longer periods of time.

Neither the silicon carbide arrester nor the metal oxide arrester is capable of withstanding its rated voltage continuously. The silicon carbide arrester is capable of withstanding rated voltage for, say, several days and still be capable of passing the duty-cycle test (ANSI/IEEE C62.1-1984 [19]). Even in nonrelayed circumstances such as high-resistance grounded systems and some ungrounded systems, the silicon carbide arrester has had no difficulty in accommodating very extended periods of full line-to-line system voltage. The principal reason for this is simply that the arrester is not actually exposed to its rated voltage in the vast majority of such cases.

The metal oxide arrester cannot, however, be expected to have a similar extended-time capability at rated voltage. Its thermal capability (as noted in the foregoing) is well in excess of that required for immediate relayed systems but cannot be expected to sustain voltages at its rating for the periods of time associated with nonrelayed ground faults that may range up to many days.

Selection of appropriate arrester voltage ratings for these applications is discussed below. The minimum required voltage rating of a gap-type arrester is the maximum line-to-line operating voltage times the coefficient of grounding. Depending on voltage rating, there are three *classes* of arresters available with associated discharge characteristics as shown in Tables 19 and 20. For zinc oxide arresters, as the sustained voltage is reduced, the time capability of the arrester is increased and becomes unlimited at approximately 81% of arrester rating. This maximum circuit operating voltage is primarily established by arrester steady-state loss stability considerations. The arrester, therefore, possesses varying degrees of extended-time voltage capability between rated voltage and the MCOV. As an example, Table 21 identifies a 2000 h capability at 90% of arrester rated voltage for arresters of one particular design rated 48 kV and below. This 2000 h capability is provided to facilitate surge protection of the nonrelayed ground fault applications. Such zinc metal oxide arrester applications have been, therefore, classified as *110% applications*.

A final note on arrester rating selection, which is determined by maximum operating voltage and by coefficient of grounding, is as follows: Maximum operating voltage may be escalated by such factors as generator overspeeds, load rejections, and resonant possibilities. Also, emergency operating modes may significantly alter system grounding such as to increase coefficient of grounding. Recognition of these factors will enhance their proper evaluation in a particular installation. Table 21 shows the usually selected ratings of both gap-type and metal oxide arresters. The generally lower voltage ratings for the metal oxide arresters result in correspondingly lower clamping voltage levels (that is, improved protection) and reflect the absence of any reseal limitations in their gapless operating nature.

4.5.2 Arrester Class. The vagaries of lightning with its wide range of surge duties that may be imposed at random system locations introduces such a degree of probability or uncertainty that considerable latitude is found in actual lightning protective practices. Understandably then, selection of arrester class is no exception since there are wide differences in some of the protective characteristics and in the durability characteristics among the three classes.

Table 21
Voltage Ratings of Arresters Usually Selected for Three-Phase Systems in kV

Nominal System Voltage		Silicon Carbide Arrester		Metal Oxide Arrester			
		Ungrounded or Impedance Grounded[1]	Grounded Neutral Circuits[2]	Ungrounded or Impedance Grounded[3]		Grounded Neutral Circuits[4]	
E_{L-L}	E_{L-N}	Arrester Rating	Arrester Rating	Rating	MCOV[5]	Rating	MCOV
2.4	1.38	3.0	3	2.7	2.2	2.7	2.2
4.16	2.4	4.5	3	4.5	3.7	3.0	2.54
4.8	2.77	6.0	4.5	5.1,6.0	4.2,5.08	4.5	3.7
6.9	4.0	7.5	6	7.5	6.1	5.1,6.0	4.2,5.0
7.2	4.16	9	6	7.5	6.1	5.1,6.0	4.2,5.0
11.5	6.6	12	9	12	10.16	8.5,9.0	6.9,7.6
12.5	7.2	15	9,10	12,15	10.16,12.7	9.0	7.62
13.8	7.9	15	12	15	12.7	10	8.47
18	10.4	21	15	18	15.24	15	12.7
23	13.2	24	21	24	19.5	18	15.24
27	16.0	30	24	30	24.4	21	17.1
34.5	19.9	36	30	36	29.3	27	21.9

1 Arrester is rated for 100% of system E_{L-L}.
2 Arrester is selected for effectively grounded system where rating $\geq 0.80 \cdot$ system E_{L-L}.
3 Arrester is selected such that MCOV $\geq 0.90 \cdot$ system E_{L-L}. This corresponds to the 2000 h arrester capability.
4 Arrester is selected such that MCOV $\geq E_{L-L}/\sqrt{3}$.
5 MCOV — maximum continuous operating voltage.

As a general guide to arrester class usage versus equipment size, the following appears to prevail as typical practice:

station class	component protection of 7.5 MVA and above and large or essential rotating machines
intermediate class	component protection of 1–20 MVA substations and rotating machines
distribution class	distribution class apparatus, small rotating machines, and dry-type transformers

Considerable overlap of these categories prevails, tending toward the use of higher class arresters at higher voltages.

The effectiveness of the installation shielding can affect the selection of arrester class. Since excessive discharge currents are a prime cause of arrester failure, a knowledge of the discharge-current duties likely to prevail at a given location is necessary to a secure arrester application. Fig N1.3(a) and the associated equations in Note 1 provide a ready approach to the approximation of maximum discharge current for effectively shielded locations. The magnitude of surge voltage being delivered to an arrester location via a line surge impedance cannot exceed the actual impulse insulation strength of the line. Effective shielding of a line assures that this surge propagation mechanism is the process by which the line and connected apparatus are exposed to surges. Assuming the magnitude of such a surge voltage will not be more than 1.2 times the line BIL (1.2×50 μs wave, critical flashover), then the resulting maximum surge current through the arrester is

$$i_{\text{discharge}} = \frac{(2.4)\ (\text{line BIL})}{Z_L + Z_A}$$

where in Fig N1.3(a) E is the line BIL (1.2), $Z_{0(1)}$ is the line surge impedance Z_L, and $Z_{0(2)}$ is the arrester surge impedance Z_A.

At the higher discharge currents of concern, Z_A is generally small compared to Z_1 and is often neglected. For example, referring to Table 20, the 15 kV rated intermediate class arrester listed displays a maximum of 2.25 Ω at 20 kA discharge, similarly 18 Ω for the 120 kV rated arrester. This is very small compared to typical line surge impedance of 400 Ω. Since line BIL increases with increasing operating voltage but surge impedance does not (in fact, decreases modestly), it is to be expected that discharge currents will be greater at higher operating voltage levels. A 13.8 kV 400 kV BIL 400 Ω surge impedance line results in approximately 2400 A discharge current as derived by the foregoing expression. Similarily, a 138 kV line conservatively will result in 5000 A maximum discharge current since its BIL will be at least twice as great.

It is unlikely that the discharge voltage of silicon carbide arresters will exceed their sparkover in such installations. Accordingly, the sparkover levels of silicon carbide arresters may be used as a basis of their application in effectively shielded installations. For metal oxide arresters, coordination on the basis of a 2500–5000 A maximum discharge current range (as suggested above) appears reasonable in effectively shielded installations. Thus, in these installations that encompass the majority of in-plant systems, discharge current magnitude is relatively low and has little bearing on selection of arrester class.

It is in the noneffectively shielded installations where the capricious nature of lightning truly affects protective considerations and approaches. Obviously, nearby direct strokes to lines and apparatus subject them to extremely high surge currents and voltages. Voltage gradients in the order of 10^{12} V/s may be involved. Severe gradients also result from flashover. Extensive field measurements accumulated over many years have established a *firm* probability of surge contents in various areas of system exposure. In badly exposed rural areas (greater than 40 thunderstorm days per year), distribution class arrester discharge currents in excess of 65 kA may be found in, say, 0.1% of arrester locations. However, in urban areas where a greater density of arrester population exists along with more inherent shielding of surroundings, this figure drops to about 20 kA.

Similarly, arrester discharge currents associated with noneffectively shielded installations up to 138 kV are greatly reduced as compared to the surge currents for lightning at such stations. Statistics and experience are such that *standard* recommendations are to use 20 kA arrester discharge current for conservative lightning protection coordination for these installations. Less conservative practice may be acceptable in some cases (say coordination with arrester discharge current as low as 10 kA). However, the conservative approach would appear prudent for power systems serving sensitive high-investment industrial installations.

There has been a preference on the part of many to keep the discharge voltage of arresters below the sparkover levels of silicon carbide arresters. This has provided a more uniform margin of protection throughout the duration of a surge event. This, along with the added uncertainty entailed with nonshielded installations, has caused practice to evolve toward the use of higher class arresters. This would be particularly true in the higher isokeraunic levels in which many installations reside. The result has been to use station class and intermediate class arresters as a general practice in nonshielded installations where otherwise intermediate class and distribution class arresters, respectively, would be used. This same practice has carried over to the application of metal oxide arresters. Note that while the discharge voltage at 20 kA of both station class and intermediate class silicon carbide arresters may exceed their sparkover (1.2×50 μs), the station class may do so only slightly while the intermediate class may do so signficantly.

Switching surges, being of relatively long duration, are of particular concern in systems experiencing frequent switching. Particularly when capacitors (power factor) are connected, switching transients may impose a more severe time-current duty on arresters. Such applications should utilize station class arresters. Electric furnaces serving certain segments of industry involve very frequent switching and are classic applications of this type.

The fact that the metal-oxide arresters can provide superior protective levels with far less energy absorption than conventional arresters makes them ideal for frequently switched applications such as switched capacitor banks and arc furnace transformers. Capacitor switching, as noted, generally presents the most severe duty as related to a particular switching event. Arc furnaces tend to have a higher rate of switching events than switched capacitors, but the energy involvement with each event is likely to be somewhat less—probably less than 0.1 kW s/kV arrester rating per furnace no-load switching.

4.5.3 Arrester Location. The ideal location for surge arresters, from the standpoint of protection, is directly at the terminals of equipment to be protected. However, as noted previously, practical system circumstances and sound economics often dictate that arresters be mounted remotely from equipment to be protected. Sometimes, for example, one set of arresters is applied necessarily to protect more than one piece of apparatus. Even so, it will be found that the low BIL apparatus (certain dry-type transformers and rotating machines) will often require surge protective devices in direct shunt with associated insulation. It is necessary, therefore, to estimate the depreciation in protection occasioned by separation distances between arresters and protected equipment.

The traveling wave mechanics developed in Note 1 establish that the total (refracted) surge voltage (E_j) appearing at a junction of surge impedances is a function of the incident wave voltage magnitude (E) and the surge impedances

$$E_j = E\left(\frac{2Z_{0(2)}}{Z_{0(1)} + Z_{0(2)}}\right) = E\left(\frac{2}{1 + Z_{0(1)}/Z_{0(2)}}\right)$$

Therefore, for a given connected pair of surge impedances, the total junction surge voltage is dependent directly on the magnitude of E, instant by instant, as it (E) impinges on the junction. An arrester, in order to influence (reduce) the surge voltage at the junction, therefore must have a protective characteristic and location such as to reduce the incident wave voltage E.

Referring to Fig 38, the arrester is optimally located at the junction to hold its surge voltage to a minimum (of arrester clamping level). At *every* instant of time at this location the arrester experiences both the incident wave and reflected wave components. Necessarily then, E must be held to less than arrester clamping level to accommodate this circumstance. Equating the preceding expression to arrester clamping level, E is seen to be limited by arrester clamping to

$$E = \left(\frac{1 + Z_{0(1)}/Z_{0(2)}}{2} \right) \text{(arrester clamping)}$$

$Z_{0(2)}$ (representing protected equipment surge impedance) is usually much greater than $Z_{0(1)}$ (representing line surge impedance), and therefore the incident wave is usually limited to approximately slightly more than one-half the arrester clamping in this arrester location.

In order to hold the voltage of the wave incident to the protected junction to below arrester clamping voltage, the arrester must be located sufficiently close to ensure that the reflected wave from the junction participates in the clamping process. This will be found to be a distance ranging from zero up to the length commensurate with one half the clamping voltage on the front of the incident wave [see Fig 38(b)]. This distance in turn depends upon the rate of rise of incident wave voltage and travel velocity. Fig 39 illustrates the effect of separation distance between surge arrester and protected equipment for specific sets of surge impedances and wave travel velocities associated with line and cable circuits, and may be used for estimating purposes.

Fig 38(c) illustrates the case of an arrester remotely located from the protected junction. It is seen that the reflected wave is, in this case, delayed by travel time in getting to the arrester and does not contribute to arrester clamping until the incident wave has reflected fully at the protected junction. The incident wave arriving at the junction cannot be reduced below the clamping voltage at the arrester remote location. As a limit, the junction voltage will approach twice the arrester clamping voltage since the arrester at its location will limit the incident wave voltage to its clamping level.

More specific comments on recommended practices relating to arrester location are provided in 4.6.3 and in ANSI C62.2-1981 [5].

4.6 Application Concepts

4.6.1 General Considerations. Lightning is considered to be the most severe source of surge voltages and, for that reason, lightning protection is the main subject of the following discussion. It is to be understood, however, that many of the protection principles involved, particularly with regard to wave magnitude and wave shape control, apply to situations involving surges of nonlightning origin as well.

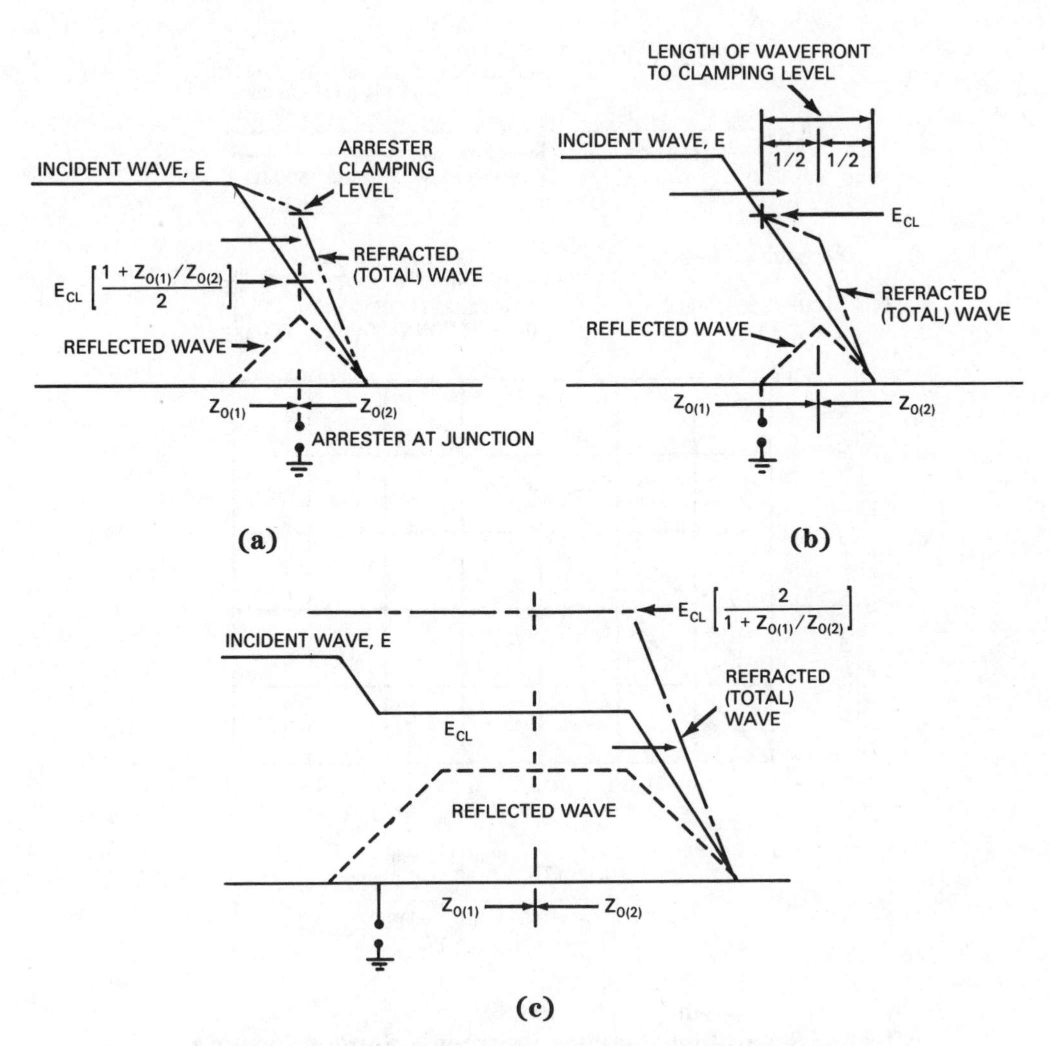

(a) Junction arrester limits incident wave voltage at junction to less than arrester clamping voltage (E_{CL}) depending upon ratio of surge impedances. This arrester location provides best protection for junction. (b) Arrester located remotely by a distance equal to length of front-of-incident wave at one-half clamping. This threshold location — up to this separation distance reflected wave returns in time to arrester to assist clamping and thus reduce incident wave at junction to less than arrester clamping. (c) Arrester is so remotely located form junction that reflected wave does not assist arrester clamping. Wave incident to junction is permitted to reach arrester clamping resulting in greater than arrester clamping voltage at junction. In limiting case where $Z_{0(1)}/Z_{0(2)}$ is very small, then junction voltage will approach twice the arrester clamping voltage.

Fig 38
Resulting Wave Phenomena for Various Arrester Locations

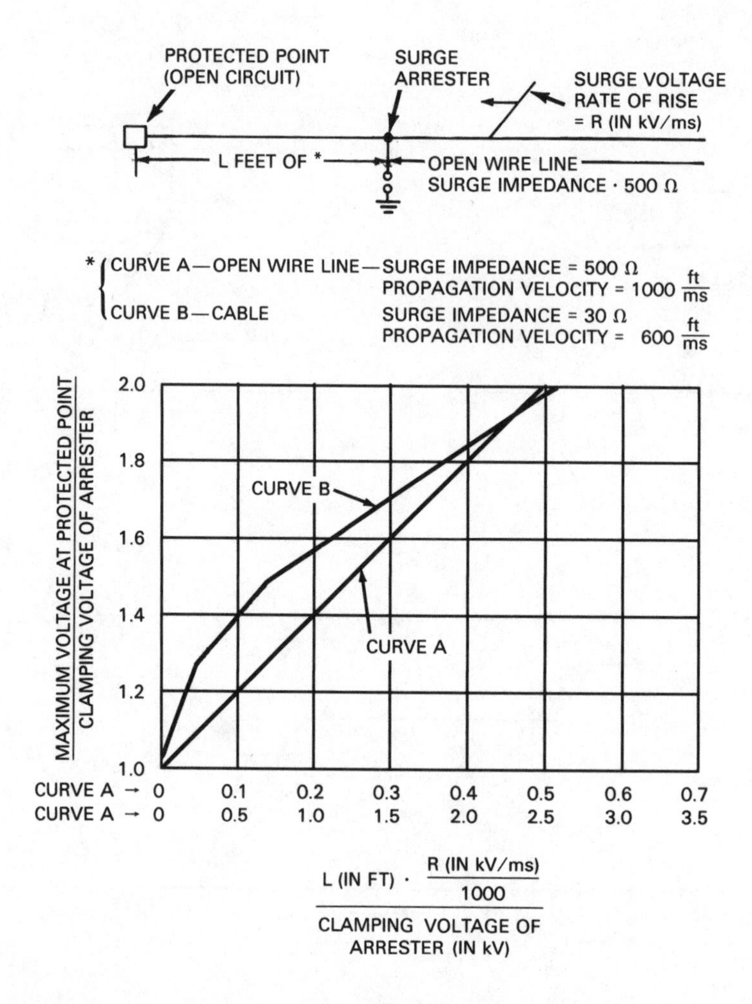

Fig 39
Effect of Separation Distance Between a Surge Arrester and the Protected Equipment on the Ratio of the Maximum Voltage at the Equipment to the Arrester A Voltage

In actual practice, lightning protection is achieved by the processes of *interception* of lightning-produced surges, *diverting* them to ground, and by *altering* their associated waveshapes [43]. Interception relates primarily to the prevention of direct strokes to lines and apparatus by shielding, which also functions as an energy diversion path to ground. However, an extremely low percentage of strokes may penetrate overhead line static wire shielding in addition to induced surges that will occur on the line in presence of lightning in the area. Also, of course, some lines are not shielded. Therefore, lightning-produced surges do become impressed on power system components due to imperfect or nonexistent shielding. Strategically located arresters are applied to divert most of this surge energy around sensitive apparatus insulation and thus afford the necessary protection. In the most ideal and simplest applications, the lightning arresters are connected in the closest practical shunt relationships with the insulation of the apparatus to be protected. Surge capacitors are applied to alter the shape of a steep incoming wavefront. The rate of rise of the surge voltage at the capacitor terminals is limited by the charging rate of the capacitor.

While most apparatus in the large majority of applications will tolerate the surge duties permitted by good shielding and proper arrester application, the associated gradients in particular may be damaging to rotating machines of multiturn coil construction. This includes virtually all motors and also generators up to approximately 35-40 MW at 13.8 kV. A discussion of motor protection in 4.6.3.6 will cover this aspect of surge protection and associated application of surge capacitors.

Internally generated surge sources should be recognized as well as lightning surges in order that the system is properly protected against all sources of hazardous overvoltage. These transient overvoltages can be produced by current-limiting fuse operation, vacuum and high-speed circuit breaker operations, thyristor switching, and ferroresonance. A more detailed description of these phenomena is given in 4.1.

4.6.2 Insulation Coordination. Insulation coordination is defined in ANSI C92.1-1982 [6] as "the process of correlating the insulation strength of electrical equipment with expected overvoltages and with the characteristics of surge protective devices." Fundamentally, insulation coordination involves checking to determine that an adequate margin of protection exists between the insulation withstand characteristic of the electrical apparatus and the protective characteristic of the applied surge arrester for any voltage impulse likely to be encountered. This is often demonstrated graphically, as shown in Fig 40, where the transformer insulation test-implied withstand curve is an attempt to simulate the actual withstand curve. The actual withstand curve is based on the insulation failure curve shown in Fig 41. This insulation failure curve is the envelope of the failure voltage crest at each time to failure for a series of impulse applications having a given wave shape (for example, a $1.2 \times 50\ \mu s$ wave) [28]. The actual withstand curve is then constructed at some safe margin beneath the insulation failure curve. It should be recognized that the technique of simulating the actual withstand curve from the results of the four (at most) generally available insulation tests discussed in 4.3.2 and graphically shown in Fig 40 is at best an approximation since different waveshapes are employed. Using such a curve on the same graph with a surge arrester protective characteristic curve that is based on tests performed using still other waveshapes is, therefore, not a truly accurate representation. For this reason, the calculation of three standardized protective margins (as described in the next paragraph) that may be compared to recommended minimums is preferred over graphical techniques for insulation coordination.

The degree of insulation coordination is determined by the magnitude of three protective margins. The fundamental definition of protective margin (PM) is

$$PM = \left[\frac{\text{insulation withstand level}}{\text{voltage at protected equipment}} - 1 \right] 100\%$$

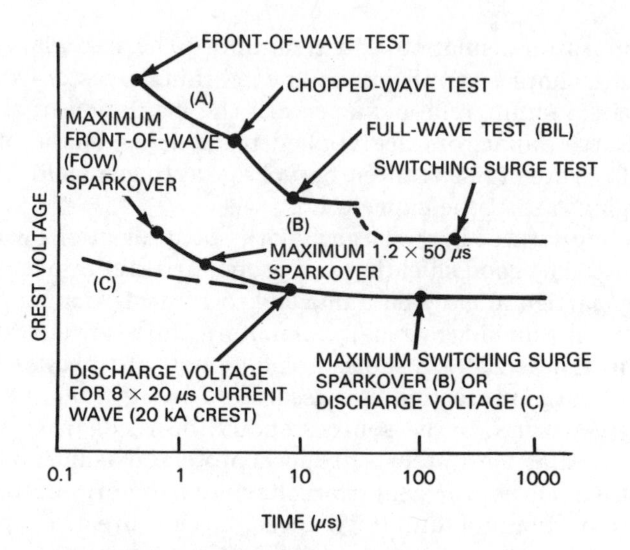

Fig 40
Insulation Coordination Based on Test-Implied Transformer Withstand Curve

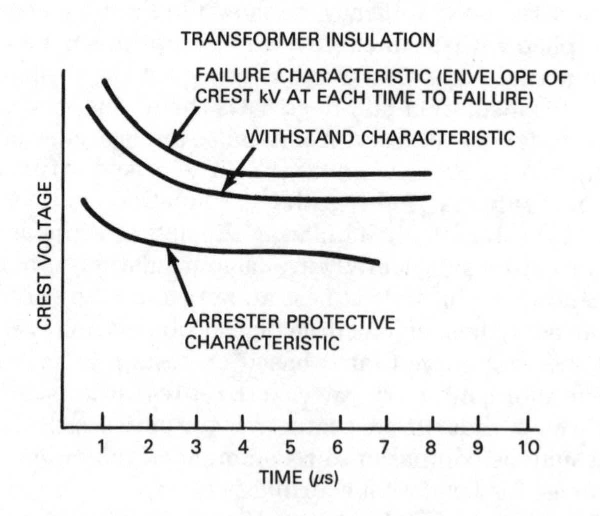

Fig 41
Coordination of Actual Transformer Insulation Withstand with Characteristic of a Surge Arrester (1.2 × 50 μs Wave)

For the basic case in which arresters are applied directly at the equipment being protected, the *voltage at protected equipment* is simply the appropriate arrester protective level. If the sum of the arrester lead length and the line length from arrester lead junction to the protected equipment is more than 10 ft (3 m), the *voltage at protected equipment* may be signficantly greater because of the effect of wave reflection as discussed in 4.5.3 and Note 1. ANSI C62.2-1981 [5] offers specific procedures for evaluating this so-called *separation effect*. The discussion that follows assumes that separation effect is negligible.

The following *insulation withstand levels* used in calculating the protective margins are discussed in 4.3.2:

(1) Chopped-wave withstand (CWW)
(2) Basic impulse insulation level (BIL)
(3) Switching surge level (SSL)

The following corresponding arrester protective levels to be used for *voltage at protected equipment* in calculating the protective margins are discussed in section 4.4.4:

(1) Front-of-wave sparkover (FOW), or equivalent for gapless metal oxide arresters.

(2) Lightning protective level (LPL), defined as the greater of the 1.2×50 μs voltage wave sparkover (if applicable) or the discharge voltage for a selected crest level 8×20 μs current impulse. The crest level used, generally 10 kA or 20 kA for exposed installations, is discussed in 4.5.2.

(3) Switching surge protective level (SPL), defined as the greater of the switching surge sparkover (if applicable) or switching surge discharge voltage.

Manufacturers' literature is always the best source for arrester protective characteristics, but Tables 19 and 20 may be used for approximate values.

The insulation withstand levels and arrester protective levels listed above define the three protective margins. These are listed along with the limits considered satisfactory for coordination:

$$PM\ (1) = [CWW/FOW - 1]\ 100\% \geq 20\%$$
$$PM\ (2) = [BIL/LPL - 1]\ 100\% \geq 20\%$$
$$PM\ (3) = [SSL/SPL - 1]\ 100\% \geq 15\%$$

Provided these margins are satisfied, it is usually assumed that the insulation will be protected over the full range of lightning and switching impulses. Distribution system insulation coordination is based only on PM (1) and PM (2), since switching surges imposed on distribution system apparatus are generally secondary in severity to the duty imposed by lightning surges. Also, standards do not generally provide for switching surge tests of distribution equipment.

4.6.3 Component Protection

4.6.3.1 Outdoor Substations. While actual lightning protective practices may necessarily vary from one type of installation to the next, the most basic categorical division relates to whether the installation is effectively shielded or noneffectively shielded. It is common practice to provide a safety factor of protective margin between established impulse capability of apparatus insulation and the protective level provided by arresters. The protective margin, defined in 4.6.2, that is generally recommended is 20% for impulse coordination (front-of-wave, full wave) and 15% for switching surge coordination.

4.6.3.1.1 Effectively Shielded Substations. The principle of shielding can be applied effectively to reduce the magnitude of lightning-derived surge voltages [26], [27], [31], [38], [40], [41]. The design pattern involves the installation of strategically located overhead grounded conductors as a sort of umbrella over the power conductors to be benefited. The presence of the grounded overhead conductors lessens the space voltage gradients beneath them which, in turn, lessens the induced voltage magnitude imparted to the power conductors by a nearby lighting stroke. The overhead shield wires will intercept a direct stroke ([38], Chapter 10). If the maximum discharge current is not too high, the stroke current will be diverted to ground without a flashover to the power conductors. At some level of discharge current the voltage gradients will become high enough to flash over to the power conductor. Even then the voltage on the power conductor will be no greater than if the initial stroke had contacted the power conductor. Since apparatus insulation destruction is the result of the aggregate accumulation of imposed transient overvoltage stress, any reduction in magnitude of the impressed surges is of direct benefit.

A single mast installed above the bare substation conductors is usually considered to provide a protective cone having its apex at the top of the mast and whose sides make an angle with the vertical of 30–45 degrees. With two or more masts the protective zone of each is increased somewhat in the area between them. This may be considered as an increase in the angle (made with the vertical) of the side of each protective cone that lies between two masts. With the usual spacings between masts, this angle may increase to 60 degrees. It is recommended that all overhead lines entering the substation be protected by a grounded shield conductor(s) for a distance of at least one-half mile from the substation. These shielding conductors should be grounded at each pole through as low a ground resistance as it is practicable to obtain, and they should be connected to the ground bus at the substation. Low ground resistance is particularly important for the ground connection at the first few poles adjacent to the substation.

A set of arresters is required on each overhead line as it enters the station to provide protection to disconnecting switches, buses, etc. Whether or not these arresters will also protect the transformer depends upon the system voltage, method of grounding, and circuit distance between the arresters and the transformer. It is often necessary to install an additional set of arresters at the transformer. Assessment of such need may be made with the assistance of aids such as Fig 39. It will be found that usually rather significant separation distances can be tolerated (say 75–200 ft—sometimes more) for station equipment 23 kV and above with full BIL insulation. For equipment in the 15 kV class and below, actual practice usually has been to avoid any appreciable separation distance. Low BIL dry-type transformers and rotating machines require special attention, even in shielded environments.

Finally, a set of arresters adequately rated for the service is recommended for installation at the remote end of the shielded section of the overhead line conductors. These arresters will intercept the severe surges and dissipate a large portion of their transient energy to ground. Only the attenuated voltage surge (perhaps one half or less of the original value) continues along the shielded one-half mile line section to the station. The lessened duty on the station surge protective devices results in a corresponding reduction in the surge-voltage magnitude arriving at the terminals of vulnerable apparatus.

4.6.3.1.2 Noneffectively Shielded Substations. These may be defined as those substations that do not have the overhead shielding described in the previous paragraphs. Such substations are likely to be small, low-voltage (up to and including 34.5 kV primary) installations entailing relatively simple circuit arrangements — often only one incoming exposed line or one secondary exposed circuit, or both. In such cases, incoming line arresters may suffice to protect the transformer if a minimum of circuit length, as defined in Fig 39, is devoted to associated overcurrent protection and switching equipment (breaker or fused switch, for example). Otherwise, arresters should be applied at the terminals of all transformers.

When a number of circuits are involved, the lightning-produced surge duties are divided among them in inverse proportion to their surge impedances and, in general, the hazard is reduced. Therefore, protective coordination should be established on the basis of the minimum number of circuits in service. Also, it is important to ensure that sensitive apparatus is not left isolated (from its surge protection) as a result of sectionalizing to accommodate an unusual operating condition.

Since noneffectively shielded applications entail a much higher surge exposure, it is most important that the lowest practical ground resistance be obtained and that circuitous connections between arrester ground and terminal and protected equipment conducting frame be avoided. Short ground interconnections between these two points are often employed to place arrester in closest practical shunt with insulation to be protected.

4.6.3.2 Metal-Clad Switchgear. Metal-clad switchgear installed in substations should be protected in similar fashion to that implied in the foregoing for transformers as BIL's of the two types of equipment are somewhat comparable. Often metal-clad switchgear has a limited exposure, that of a length of cable intervening between the metal-clad and exposed line. Where the cable is of continuous metallic sheath, Fig 42 illustrates this case and provides a guide as to the possible need for an arrester at the metal-clad switchgear. Note that an arrester is required at the line cable junction in any case to protect the cable. With arrester characteristics comparable to those listed in Tables 19 and 20, it will be found that arresters are not required at the switchgear for any length of continuous metallic sheath cable except in certain cases when standard distribution class arresters are used at the line cable junction. In these relatively few cases, a distribution class arrester at the switchgear will suffice.

Nonmetallic-sheathed cables have higher surge impedances than metallic-sheathed cables and their use necessitates the use of arresters at the switchgear (distribution class will suffice). However, the installation of a neutral or ground wire in the duct with each three-phase nonmetallic-sheathed cable provides very nearly the same surge impedance as continuous-metallic-sheathed cable and may be so considered for surge protective purposes.

In many industrial installations the only exposure of the metal-clad switchgear to lightning may be through a power transformer. When the power transformer has

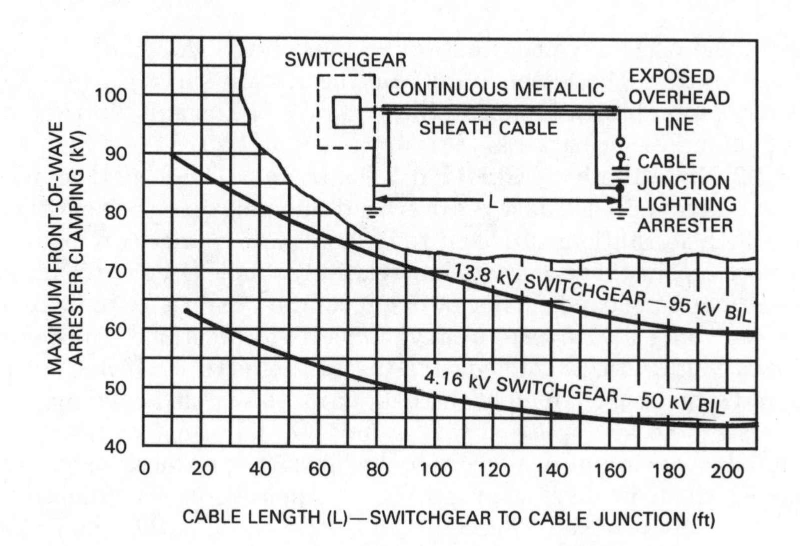

Fig 42
Curves Showing Maximum Permissible Length of Cable for Which Arresters Are Not Required in Metal-Clad Switchgear Versus Line-Cable Junction Arrester Clamping Voltage

adequate lightning protection on the exposed side opposite the switchgear, there is generally no necessity to provide arresters on the sheltered side of the transformer connected to the switchgear. Experience has shown that for the transformer sizes normally encountered in unit substations there is usually not enough surge transfer through the transformer to be harmful to the metal-clad switchgear.

4.6.3.3 Dry-Type Transformers. Dry-type transformers present relatively difficult lightning protective problems due to their usual low BIL's compared to liquid-filled transformers. When surge exposure is by direct-connected overhead lines, arresters are required in direct shunt with the dry-type transformer.

Regarding applications of surge exposure through cable, Fig 43 applies for dry-type transformers in the identical fashion that Fig 42 applies for metal-clad switchgear. With arresters comparable to those listed in Tables 19 and 20, it will be found that in many practical applications, even in this relatively shielded environment, the line-cable junction arrester will not protect dry-type transformers against lightning-produced traveling waves. Where an arrester is required at the transformer, a special low-sparkover distribution class valve-type arrester will suffice.

A somewhat less severe, though typical, surge exposure for dry-type transformers is through another (supply) transformer (see Fig 44). Any surges impinging on the primary side of the supply transformer will be mollified somewhat as they are transferred through the transformer to appear on its (the supply transformer's) secondary. For the most used wye-delta and delta-wye connected supply transformers, arresters are generally not required at the dry-type transformer.

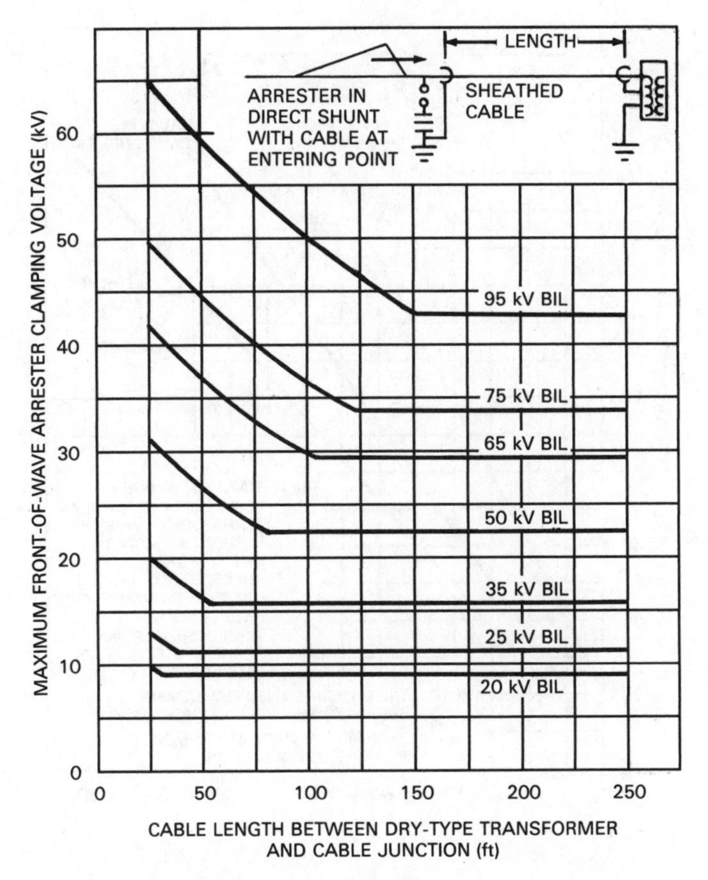

Note: Interpolate or extrapolate for BIL's not shown.

5333.F43

Fig 43
Curves for Determining Maximum Permissible Length of Cable for Which Arresters Are Not Required at Dry-Type Transformer Versus Line-Cable Junction Arrester Clamping Voltage

However, for those dry-type transformer applications where the transformer can be subjected to internally generated surges due to current-limiting fuse operation or chopping or switching effects of circuit breakers, surge protection should consist of arresters and possibly surge capacitors that reduce the effective load surge impedance and transient voltage rate-of-rise in those applications where an inadequate length of cable is present to achieve this.

4.6.3.4 Overhead Line Protection (4 kV–69 kV). Historically, relatively little consideration has been given to the protection of open-wire overhead line insulation. This often results in line insulator flashover which must be cleared by a protective device, which in turn results in a momentary or extended circuit inter-

-4.6.3.4

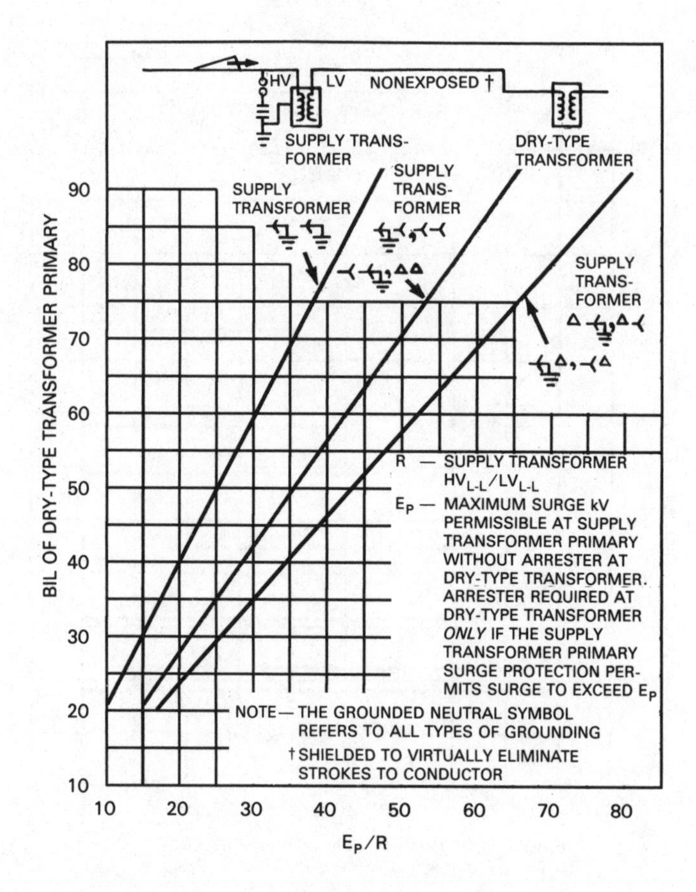

Fig 44
Curves Showing Maximum Surge Permissible at Supply Transformer Without Requiring Arrester at Dry-Type Transformer

ruption. Both the high gradient associated with insulator flashover and service interruption are associated disadvantages of considerable significance to sensitive petroleum and chemical plant equipment and loads.

Analytical and test-model studies relating to overhead transmission and distribution circuits [35] have disclosed a so-called predischarge current effect in association with strokes to lines which in effect tends to suppress surge overvoltages at midspan and concentrate them at grounded poles. At grounded poles there is opportunity to install surge arresters on all phases so that voltage stresses are relieved by the arrester protective characteristics and thus prevent flashover of line insulators. A comprehensive coverage of the study techniques and results are embodied in the preceding referenced task force reports. It will suffice to state here that study results and actual utility company experience show that arresters protecting each phase, at economically spaced intervals along the line, will often give improved protection and reliability of service over that of the overhead-wire-shield method. This new approach to line protection is also much less sensitive to footing resistance.

There is a growing practice among electric utility companies to use the foregoing approach in protecting overhead circuits in the range of 4-69 kV. This should certainly be a consideration toward improving protection on overhead circuits that serve sensitive industrial complexes.

4.6.3.5 Aerial Cable. Aerial cable is almost universally protected against direct lightning strokes by grounding the messenger and sheath at every pole through a low value of ground resistance. This is to allow a lightning stroke to the messenger to drain off by current flow to earth without causing the voltage of the messenger and sheath to rise excessively above the voltage of the cable conductors. If an aerial cable joins an open-wire line, lightning arresters should be installed at the junction to protect the cable insulation against lightning surges that arrive over the open line. The ground terminals of these arresters should be connected directly to the cable messenger and sheath as well as to ground.

Since the voltage and current surges produced in the messenger of aerial cable by lightning stroke to the messenger result in voltage and current surges in the cable conductors, it is generally recommended that aerial cable be considered the same as open-wire lines as far as the protection of terminal equipment is concerned.

4.6.3.6 Rotating Machine Protection. The basic winding design patterns of motors and generators involve rather large capacitance coupling between the conductor of the winding of each coil and the grounded core iron that surrounds it. A fast rising surge voltage at the motor terminal lifts the potential of the terminal turn, but the turns deeper in the winding are constrained (by this relatively large capacitance from coil to ground) and delayed in their response to the arriving voltage wave. The result is a greatly accentuated voltage gradient across the end-turns of the terminal coil that appears as severe voltage stress on the turn-to-turn insulation of the terminal coil. Although the major or *ground wall* insulation between conductors and ground is fairly thick, the turn insulation within the coils is thin. Economical design dictates a thickness of no more than 0.005-0.040 in, depending upon machine voltage rating. It is the protection of the turn insulation that becomes critical in avoiding failure in multiturn stator windings of ac motors and generators.

4.6.3.6.1 Machine Winding Impulse Strength. It has already been observed that there are no established impulse standards on the insulation structures for ac rotating machines. However, in the motor area there is evidence of unofficial endorsement of the acceptability of a voltage wave at the motor terminals that rises gradually to a maximum level at a rate not exceeding 1.25 times the crest value of the one minute acceptance (hi-pot) test voltage in 5 μs, Table 18.

A re-examination of the impulse voltage capability [32] has led to the recommendation that form wound coils of multiturn construction have the capability to withstand impulse voltages whose fronts and amplitude lie below the envelope of Fig 45. If a machine is subjected to impulse voltages of greater magnitude, it should be protected by arresters especially designed for rotating machines (and other sensitive equipment) in order to limit the magnitude of the impulse reaching the

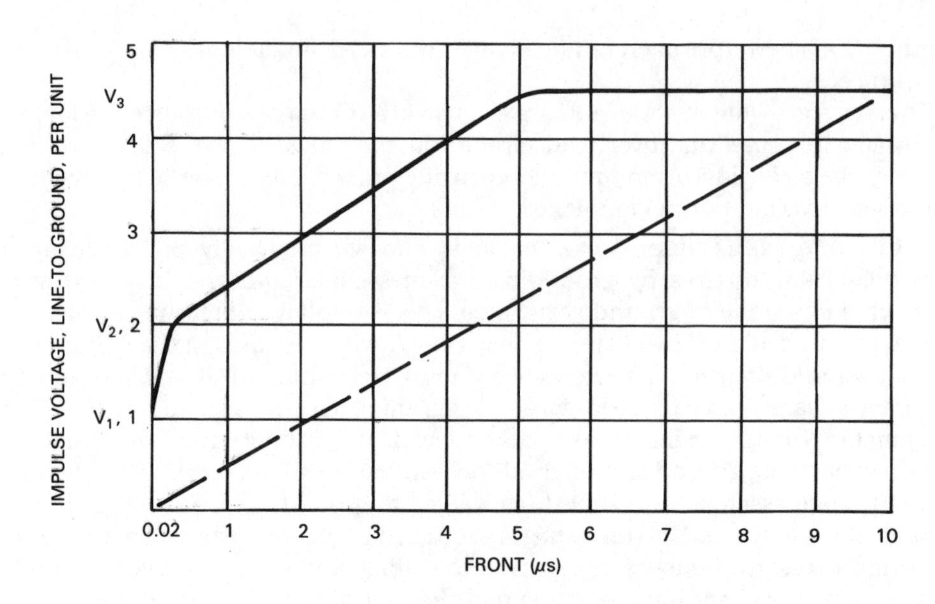

FOR V_L = MACHINE VOLTAGE, LINE-TO-LINE rms kV:
$V_1 = \sqrt{2/3}\ V_L$ = 1 PER UNIT CREST LINE-TO-GROUND
$V_2 = 2\ V_1$
$V_3 = 1.25\ \sqrt{2}\ (2\ V_L + 1)$, kV CREST VALUE

Fig 45
Machine Impulse Voltage Withstand Envelope

winding. The ratings of Table 21 are typical for these special arresters. One arrester should be connected between each machine line terminal and ground.

Impulse waves of lower magnitude, but having rise time less than 5 μs, will primarily endanger the turn insulation because of the nonuniform voltage distribution. Not only does 70–100% of the impulse magnitude appear across the first coil connected to the incoming line, but within that coil itself the voltage distribution is nonlinear, resulting in as much as half the total coil impulse voltage appearing between the first two adjacent turns [30], [39], [44]. By connecting a wave-sloping capacitor between each line terminal and ground affords protection against this condition [34]. By increasing the wave-front rise time to 10 μs or more, the capacitors safeguard the turn insulation as illustrated by the dotted line in Fig 45 [31].

The conventional method for wave-front rise protection of motors is shown in Fig 46. A capacitor is installed line-to-ground on each motor phase conductor. The rate of rise of the surge voltage across the motor winding terminals is limited by the charging rate of the capacitor. Special protective capacitor units are designed for this purpose with low internal inductance (Table 22) that control the rate of rise of incident overvoltages to protect the turn-to-turn insulation. Surge arresters of special design then complete the rotating machine insulation protection to ground by limiting the magnitude of the incident voltage wave.

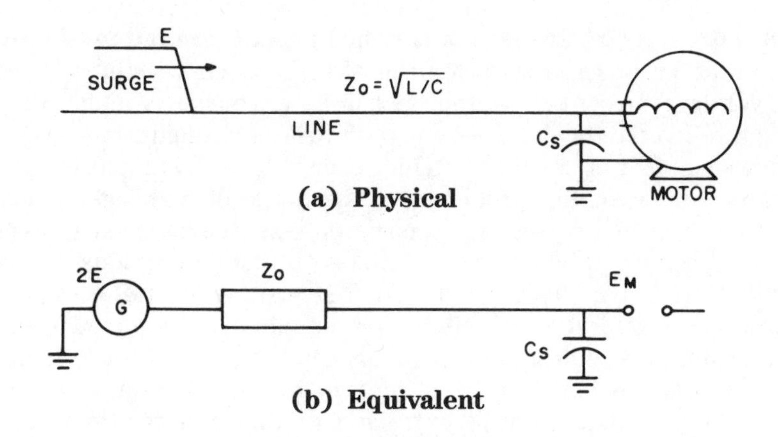

Fig 46
Application of Shunt-Connected Surge-Protective Capacitors for Wavefront Control

Table 22
Capacitance of Surge-Protective Capacitors per Line Terminal Connected Line to Ground

	Rated Motor Voltage		
	650 V and Less	2400-6900 V	11 500 V and Higher
Capacitance, μF	1.0	0.5	0.25

4.6.3.6.2 Rotating Machine Surge Protection Practice. Much documentation exists relating to the surge protection of rotating machines. An ideally protected installation requires

(1) a strictly effectively shielded environment
(2) arresters at terminals of machine
(3) surge capacitors at terminals of machine
(4) strict adherence to good grounding practices

Effective shielding requirements for stations have been defined previously. In case of rotating machines having overhead line exposure, either direct or through intervening equipment (such as reactors, transformers, or cables), arresters are also applied out on the exposed lines a distance of at least 1000–2000 ft to further reduce surge magnitude duties on the more immediate surge protection equipment (see Fig 39).

Actual practice indicates a very high percentage of motors above 4000 V are provided with arresters and surge capacitors. Similarly, at least half of the 4000 V motor installations are so equipped, while 2300 V motor installations are so equipped in only a minority of applications.

4.6.3.6.3 Special Care Required for Proper Installation of Surge Capacitors. Exploratory observations confirm the presence within *shielded* environments of voltage transients that approach arrester sparkover magnitudes and have exceedingly steep fronts (0.1 μs front-time). Although lightning does not usually entail such steep fronts, certain switching events do; for example, insulation breakdown, capacitor switching problems, or discharge of high lightning current-to-ground. As established previously, separation distance between protective equipment and apparatus to be protected invokes (sometimes serious) depreciation of protection. This is particularly true when steep wavefronts are involved. Surge capacitors, and preferably arresters also, should be connected directly to the machine terminals so that added inductance of the power cable circuit and of the surge capacitor lead will not interfere with their action. This limits the lead lengths to one or two feet, thus requiring extreme care in the motor terminal box equipment arrangement.

Each application should be reviewed on its own. If several machines are fed from a common bus, for example, it may be sufficient to connect arresters on the line side of the feeder circuit breaker, placing only the capacitors at the machine terminals. Such practice generally requires certainty that more than one feeder will be closed at the same time along with a careful analysis of the arrester protective level and the capacitor waveshaping action as a function of the feeder length involved.

A direct, low impedance path between machine winding and surge-protective devices must exist on both line and ground sides of the circuit. A good ground connection to the machine frame is essential. Published standards do not prescribe the size of such a connection, but ANSI/NFPA 70-1987 [23] (National Electrical Code [NEC]), Article 250-94, specifying the size of the grounding electrode conductor, could be used as a guide. But normally the size of the protective device ground terminal serves as a guide; for capacitors, this typically permits a ground wire up to AWG No 2.

When capacitor and arrester cases are solidly bolted to the conducting structure of a machine terminal box or frame, such a ground wire may seem superfluous. However, it should not be omitted. It should lead directly to a solid *frame ground* with the least possible number of bolted joints intervening between the protective device and the machine stator. Such joints risk having high resistance because of corrosion, bolt loosening, or paint. Furthermore, in some machines this wire may be the only ground provided, the reason being that users now often specify that the surge protective devices be mounted on ungrounded, insulated bases. This stems from the NEMA requirement for disconnection of surge protective devices from machine leads when winding insulation is tested (ANSI/NEMA MG1-1978 [22], Section 3.01H). There are at least three reasons for this recommendation:

(1) Over-potential winding tests may damage capacitors

(2) Such a test may falsely indicate bad insulation because the overvoltage is discharged to ground by an arrester

(3) Insulation resistance measurement by megohmmeter yields erroneous results because of the leakage current bypassed to ground through the discharge resistor built into every surge capacitor.

When a thorough preventive maintenance program includes such insulation tests once or twice a year, the necessity of disconnecting the surge protection devices from the line leads becomes an expensive part of the program. Since it is desirable to maintain the permanent line connections, isolation of the protective equipment can best be achieved by disconnecting the equipment ground conductor. A readily removable conductor or link is usually provided for ease in testing. After the test, this connection should be restored and tightened securely; otherwise protection may be lost.

4.7 References. This standard shall be used in conjunction with the following publications:

[1] ANSI C2-1987, American National Standard National Electrical Safety Code.

[2] ANSI C37.06-1979, American National Standard Preferred Ratings and Related Required Capabilities for AC High-Voltage Circuit Breakers Rated on a Symmetrical Current Basis.

[3] ANSI C50.10-1977, American National Standard General Requirements for Synchronous Machines.

[4] ANSI C50.13-1977, American National Standard Requirements for Cylindrical Rotor Synchronous Generators.

[5] ANSI C62.2-1981, American National Standard Guide for Application of Valve-Type Surge Arresters for Alternating-Current Systems.

[6] ANSI C92.1-1982, American National Standard on Insulation Coordination.

[7] ANSI/IEEE C37.04-1979, IEEE Standard Rating Structure for AC High-Voltage Circuit Breakers Rated on a Symmetrical Current Basis.

[8] ANSI/IEEE C37.13-1981, IEEE Standard for Low-Voltage AC Power Circuit Breakers Used in Enclosures.

[9] ANSI/IEEE C37.20-1969, IEEE Standard for Switchgear Assemblies Including Metal-Enclosed Bus.

[10] ANSI/IEEE C37.41-1981, IEEE Standard Design Tests for High-Voltage Fuses, Distribution Enclosed Single-Pole Air Switches, Fuse Disconnecting Switches, and Accessories.

[11] ANSI/IEEE C37.91-1985, IEEE Guide for Protective Relay Applications to Power Transformers.

[12] ANSI/IEEE C37.96-1976, IEEE Guide for AC Motor Protection.

[13] ANSI/IEEE C57.12.00-1980, IEEE Standard General Requirements for Liquid-Immersed Distribution, Power, and Regulating Transformers.

[14] ANSI/IEEE C57.12.01-1979, IEEE Standard General Requirements for Dry-Type Distribution and Power Transformers.

[15] ANSI/IEEE C57.12.90-1980 IEEE Standard Test Code for Liquid-Immersed Distribution, Power, and Regulating Transformers.

[16] ANSI/IEEE C57.12.91-1979, IEEE Standard Test Code for Dry-Type Distribution and Power Transformers.

[17] ANSI/IEEE C57.13-1978, IEEE Standard Requirements for Instrument Transformers.

[18] ANSI/IEEE C57.21-1981, IEEE Standard Requirements, Terminology, and Test Code for Shunt Reactors Over 500 kVA.

[19] ANSI/IEEE C62.1-1984, IEEE Standard for Surge Arresters for AC Power Circuits.

[20] ANSI/IEEE Std 100-1984, IEEE Standard Dictionary of Electrical and Electronics Terms.

[21] ANSI/IEEE Std 142-1982, IEEE Recommended Practice for Grounding of Industrial and Commercial Power Systems.

[22] ANSI/NEMA MG1-1978, Motors and Generators.

[23] ANSI/NFPA 70-1987, National Electrical Code.

[24] AIEE Committee Report. Impulse Testing of Rotating AC Machines. *AIEE Transactions (Power Apparatus and Systems)*, pt III, vol 79, 1960, pp 182-187.

[25] BEEMAN, D. L., Ed. *Industrial Power Systems Handbook*. New York: McGraw-Hill, 1955.

[26] BEWLEY, L. V. *Traveling Waves on Transmission Systems*, 2nd ed. New York: Wiley, 1951.

[27] DRAKE, C. W., Jr. Lightning Protection for Cement Plants, Part I—Surge Voltages on the Power System. *IEEE Transactions on Industry and General Applications*, vol IGA-4, Jan/Feb 1968, pp 57-61.

[28] *Electrical Transmission and Distribution Reference Book*, Westinghouse Electric Corporation, East Pittsburgh, PA, 1964.

[29] GREENWOOD, A. N. *Electrical Transients in Power Systems*. New York: Wiley, 1971.

[30] HUNTER, E. M. Transient Voltages in Rotating Machines. AIEE Transactions, vol 54, 1935, pp 599-603.

[31] IEEE Committee Report. Coordination of Lightning Arresters and Current-Limiting Fuses. *IEEE Transactions on Power Apparatus and Systems*, vol PAS-91, May/Jun 1972, pp 1075-1078.

[32] IEEE Working Group Progress Report. Impulse Voltage Strength of AC Rotating Machines. *IEEE Transactions on Power Apparatus and Systems*, vol PAS-100, No 8, Aug 1981, pp 4041-4053.

[33] IEEE Working Group Report. Voltage Rating Investigation for Application of Lightning Arresters on Distribution Systems. *IEEE Transactions on Power Apparatus and Systems*, vol PAS-91, May/Jun 1972, pp 1067-1074.

[34] JACKSON, D. W. Surge Protection of Rotating Machines. *IEEE Tutorial Course on Surge Protection in Power Systems*, Chapter 8, Pub 79EH0144-6 PWR, 1979.

[35] KAUFMAN, R. H. Nature and Causes of Overvoltages in Industrial Power Systems. *Iron and Steel Engineer*, Feb 1952.

[36] KAUFMANN, R. H. Overvoltages in Industrial Systems—How to Reduce Them by Neutral Grounding. *Industrial Engineering News*, May/Jun 1951.

[37] KAUFMANN, R. H. *Surge-Voltage Protection of Motors as Applied in Industrial Power Systems*. General Electric Company, Publication GET-3019, Dec 1971.

[38] LEWIS, W. W. *The Protection of Transmission Systems Against Lightning*. New York: Wiley, 1950.

[39] PETROV, G. N. and ABRAMOV, A. I. Overvoltage Stresses in the Turn Insulation of Electrical Machinery Windings During Electromagnetic Transients. *Electrichestvo*, NC 7, Mar 3, 1954, pp 24-31.

[40] SARRIS, A. E. Lightning Protection for Cement Plants, Part II—Surge Voltages in the Cement Plant. *IEEE Transactions on Industry and General Applications*, vol IGA-4, Jan/Feb 1968, pp 62-67.

[41] SHANKLE, D. F., EDWARDS, R. F., and MOSES, G. L. Surge Protection for Pipeline Motors. *IEEE Transactions on Industry and General Applications*, vol IGA-4, Mar/Apr 1968, pp 171-176.

[42] WALSH, G. W. A New-Technology Station Class Arrester for Industrial and Commercial Power Systems. *IEEE Industrial & Commercial Power Systems Technical Conference*, 1977 Conference Record 77CH1198-11A, pp 30-35.

[43] WALSH, G. W. A Review of Lightning Protection and Grounding Practices. *IEEE Transactions on Industry Applications*, vol IA-9, No 2, Mar/Apr 1973.

[44] WRIGHT, M. T., and McLEAY, K. Interturn Stator Voltage Distribution Due to Fast Transient Switching of Induction Motors. *IEEE Petroleum and Chemical Industry Conference*, Conference Paper PCI-81-14, pp 145-150.

4.8 Bibliography

[B1] ABETTI, P.A. Bibliography on the Surge Performance of Transformers and Rotating Machines. *AIEE Transactions (Power Apparatus and Systems)*, pt III, vol 77, 1958, pp 1150-1164 (first supplement, vol 81, 1962, pp 213-219; second supplement, vol 83, 1964, p 855-847).

[B2] ABETTI, P. A. Survey and Classification of Published Data on the Surge Performance of Transformers and Rotating Machines. *AIEE Transactions (Power Apparatus and Systems)*, pt III, vol 77, 1958, pp 1403-1414.

[B3] AIEE Committee Report. Power System Overvoltages Produced by Faults and Switching Operations. *AIEE Transactions*, vol 67, 1948, pp 912-922.

[B4] AIEE Committee Report. Switching Surges Due to Deenergization of Capacitative Circuits. *AIEE Transactions (Power Apparatus and Systems)*, pt III, vol 76, Aug 1957, p 562-564.

[B5] *Electric Utility Engineering Reference Book, vol 3: Distribution Systems*. Trafford, PA: Westinghouse Electric Corporation, 1965.

[B6] FINK, D. G. and CARROLL, J. M. *Standard Handbook for Electrical Engineers*. New York: McGraw-Hill, 1968.

[B7] HENDRICKSON, P. E., JOHNSON, L. B. and SCHULTZ, N. R. Abnormal Voltage Conditions Produced by Open Conductors on Three-Phase Circuits Using Shunt Capacitors. *AIEE Transactions (Power Apparatus and Systems)*, pt III, vol 72, Dec 1953, pp 1183-1193.

[B8] *Industrial Power Systems Data Book*. Schenectady, NY: General Electric Company, 1961.

[B9] JOHNSON, I. B., SCHULTZ, A. J., SCHULTZ, N. R., and SHORES, R. B. Some Fundamentals on Capacitance Switching. *AIEE Transactions (Power Apparatus and Systems)*, pt III, vol 74, Aug 1955, pp 727-736.

[B10] LIAO, T. W. and LEE, T. H. Surge Suppressors for the Protection of Solid-State Devices. *IEEE Transactions on Industry and General Applications*, vol IGA-2, Jan/Feb 1966, pp 44-52.

[B11] Lightning Protection of Metal-Clad Switchgear Connected to Overhead Lines. *General Electric Review*, Mar 1949.

[B12] MONTSINGER, V. M. Breakdown Curve for Solid Insulation. *Electrical Engineering*, vol 54, Dec 1935, pp 1300-1301.

[B13] NAILEN, R. L. Transient Surges and Motor Protection. *IEEE Transactions on Industry Applications*, vol IA-15, No 6, Nov/Dec 1979, pp 606-610.

[B14] NIEBUHR, W. D. Protection of Underground Systems Using Metal-Oxide Surge Arresters. *IEEE Rural Electric Power Conference*, 1981 Conference Record 81CH1654-3, pp 9-13.

[B15] RUDENBERG, R. *Electrical Shock Waves in Power Systems*. Cambridge, MA: Harvard University Press, 1968.

[B16] SAKSHAUG, E. C., KRESGE, J. S., and MISKE, S. A., Jr. A New Concept in Station Arrester Design, *IEEE Transactions on Power Apparatus and Systems*, Paper F76-393-9, Mar/Apr 1977, pp 647-656.

[B17] SCHULTZ, A. J., VAN WORMER, F. C., and LEE, A. R. Surge Performance of Aerial Cable, Part I—Surge Testing of the Aerial Cable and Analysis of the Test Oscillograms. *AIEE Transactions (Power Apparatus and Systems)*, pt III, vol 76, Dec 1957, pp 923-930.

[B18] SKEATES, W. F., TITUS, C. H., and WILSON, W. R. Severe Rates of Rise of Recovery Voltage Associated with Transmission Line Short Circuits. *AIEE Transactions (Power Apparatus and Systems)*, pt III, vol 76, Feb 1958, pp 1256-1266.

[B19] STACEY, E. M. and SELCHAU-HANSEN, P. V. SCR Drives—AC Line Disturbance, Isolation and Short-Circuit Protection. *IEEE Transactions on Industry Applications*, vol IA-10, Jan/Feb 1974, pp 88-105.

[B20] VAN WORMER, F. C. , SCHULTZ, A. J., and LEE, A. R. Surge Performance of Aerial Cable, Part II—Mathematical Analysis of the Cable Circuits and Synthesis of the Test Oscillograms. *AIEE Transactions (Power Apparatus and Systems)*, pt III, vol 76, Dec 1957, pp 930-942.

NOTE 1
Verifying the Validity of the Simplified Equivalent Circuits in Dealing with the Three Stated Line Terminations ([B15], Chapter 6)

A surge voltage E is considered to be traveling to the right along the circuit having a surge impedance of $Z_{0(1)}$ (Fig N1.1). Associated with this voltage wave is a current wave of magnitude $E/Z_{0(1)}$ traveling in the same direction. When this surge encounters a change in surge impedance at junction J, the impedance mismatch will cause changes in the resulting traveling surges of voltage and current. The effect of the impedance mismatch is resolved in terms of a forward-going refracted surge voltage E_F that will be associated with a forward-going surge current $E_F/Z_{0(2)}$, and a reflected surge voltage E_R associated with a current surge of $E_R/Z_{0(1)}$. The current created by a positive backward-going surge voltage will be a negative surge current in the forward reference direction: $-E_R/Z_{0(1)}$.

The relationships between these components as the surge passes junction J are (see Fig N1.2)

$$E_F = \left(\frac{2Z_{0(2)}}{Z_{0(1)} + Z_{0(2)}} \right)(E)$$

and associated current $I_F = E_F/Z_{0(2)}$, and

$$E_R = \left(\frac{Z_{0(2)} - Z_{0(1)}}{Z_{0(1)} + Z_{0(2)}} \right)(E)$$

and associated current $I_R = E_R/Z_{0(1)}$.

Fig N1.1
Surge Voltage Traveling on a Line of Surge Impedance $Z_{0(1)}$ Approaching a Junction where the Surge Impedance Changes to $Z_{0(2)}$

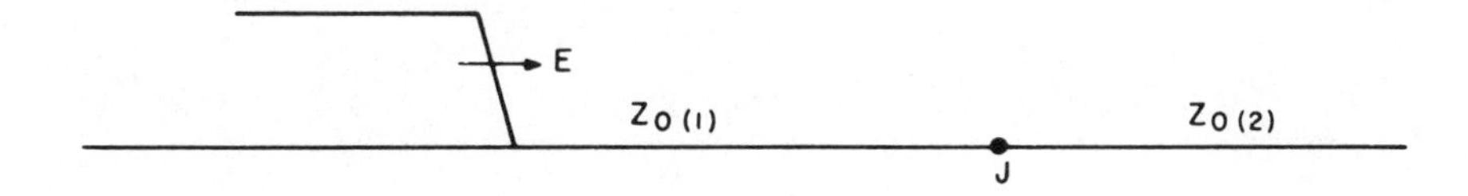

Note that when the reference direction of I_R is considered to be in the outgoing direction, the current expression becomes

$$I_R = -E_R / Z_{0(1)}$$

On the line section ahead of junction J where both the outgoing initial surge and the reflected returning surge exist simultaneously, the total value of surge voltage and surge current will be the sum of the two components.

For three specific cases A-C the surge impedance of the upstream line section remains at a fixed value $Z_{0(1)}$. The surge impedance of the downstream line section $Z_{0(2)}$ for an open-ended line at junction J (case A) will be $Z_{0(2)} = \infty$, for a short-circuited line at junction J (case B) it will be $Z_{0(2)} = 0$, and for a continuing line of the same surge impedance (case C) it will be $Z_{0(2)} = Z_{0(1)}$.

The component surge magnitudes for each of the cases are given in Table N1.1.

A first attempt at inserting $Z_{0(2)} = \infty$ leads to an indeterminate. This is avoided if the expressions for E_F and I_F are slightly revised by dividing both their numerators and denominators by $Z_{0(2)}$:

$$E_F = \left(\frac{2Z_{0(2)}}{Z_{0(1)} + Z_{0(2)}} \right)(E)$$

$$= \left(\frac{2}{Z_{0(1)}/Z_{0(2)} + 1} \right)(E)$$

$$E_R = \left(\frac{Z_{0(2)} - Z_{0(1)}}{Z_{0(1)} + Z_{0(2)}} \right)(E)$$

$$= \left(\frac{1 - Z_{0(1)}/Z_{0(2)}}{Z_{0(1)}/Z_{0(2)} + 1} \right)(E)$$

Note that the sign of the reflected surge current has been adjusted to a reference direction when *outwards* is positive.

The voltage of the refracted wave at the junction is the sum of the voltages of the incident and reflected waves, that is, it equals $(E)\ (2Z_{0(2)})/(Z_{0(2)} + Z_{0(1)})$.

Fig N1.2
Junction Mismatch Accounted for by a Forward-Going Refracted Surge Voltage E_F and a Backward-Going Reflected Surge Voltage E_R

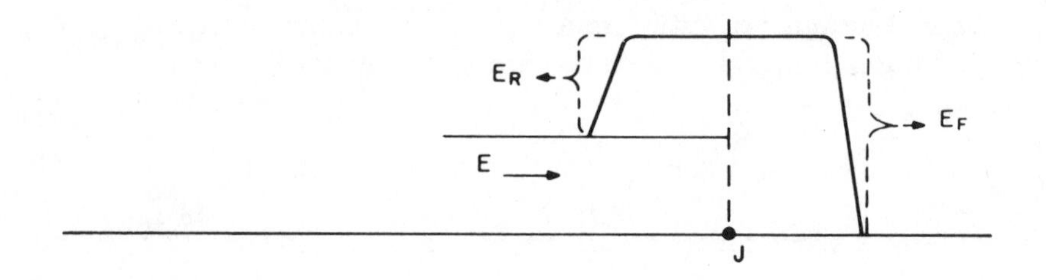

Combining the component values of voltage and current listed in Table N1.1 yields the values of bus J voltage and the line currents upstream and downstream of the junction bus shown in Table N1.2.

The simplified procedures presented in the text yield these results in a more direct manner with a better understanding of the behavior pattern involved.

Fig N1.3 is a frequently used simple dc equivalent circuit that may be derived from the foregoing to determine surge current and voltage at the junction of surge

Fig N1.3
Equivalent Circuits for Determining Junction Voltage (Maximum) for Different Terminations of Surge Impedance, $Z_{0(1)}$

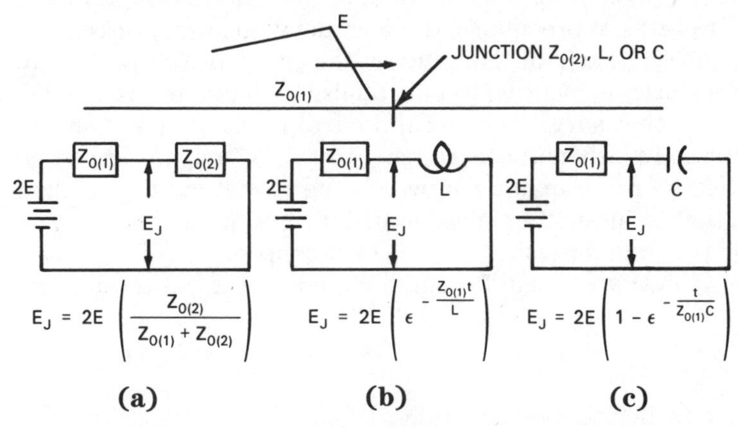

(a) (b) (c)

Table N1.1
Reflected and Refracted Surge Voltages and Currents for Different Line Termination Cases

Case	$Z_{0(2)}$	Initial Surge Voltage	Initial Surge Current	Refracted Surge Voltage	Refracted Surge Current	Reflected Surge Voltage	Reflected Surge Current
A	∞	E	$E/Z_{0(1)}$	$2E$	0	E	$-E/Z_{0(1)}$
B	0	E	$E/Z_{0(1)}$	0	0	$-E$	$E/Z_{0(1)}$
C	$Z_{0(1)}$	E	$E/Z_{0(1)}$	E	$E/Z_{0(1)}$	0	0

Table N1.2
Resulting Junction Voltage and Line Currents for Table N1.1 Line Termination Cases

Case	Bus Voltage	Line Current Upstream of J	Line Current Downstream from J
A	$2E$	0	0
B	0	$2E/Z_{0(1)}$	0
C	E	$E/Z_{0(1)}$	$E/Z_{0(1)}$

impedances $Z_{0(1)}$ and $Z_{0(2)}$. In this dc equivalent, E is a dc voltage equal to surge voltage magnitude, and the surge impedances display the characteristics of resistance of the same ohmic magnitude. This latter property greatly simplifies surge analysis. Further generalization is valid to the extent $Z_{0(2)}$ may also be a pure lumped inductance, capacitance, or combinations of these lumped circuit elements, providing the associated overall time constant is short compared to twice the surge travel time of the line, and E is a step function. See Fig N1.3 (b) and (c).

The foregoing has dealt with the simple case of a single incident wave as it impinges on a junction of two surge impedances. Obviously a practical system will experience on occasion a multitude of wave components traveling through its various surge impedances, producing a multitude of reflections and refractions at various junctions of surge impedances. Suffice it to state here, there are established methods and computer programs to evaluate these more complex cases if desired. However, as a practical procedure, the adequacy of surge protective measures is often judged on the basis of doubling at the junction (or protected apparatus terminal) of a single incident wave of established maximum probable value.

Thus it is seen that surge voltage (and current) at the junction may be determined by classical mathematical treatment of L, R, and C lumped constant elements depending on the manner in which the circuit surge impedance, $Z_{0(1)}$, is terminated (that is, in another surge impedance, R, L, or C).

Solving for the junction voltages (E_J) of arrangements of Fig N1.3 for various terminations of $Z_{0(1)}$ we obtain for surge impedance, $Z_{0(2)}$, termination

$$E_J = 2E \left(\frac{Z_{0(2)}}{Z_{0(1)} + Z_{0(2)}} \right)$$

for inductance (L henries) termination

$$E_J = 2E \left(\epsilon^{-\frac{Z_{0(1)}t}{L}} \right)$$

and for capacitance (C farads) termination

$$E_J = 2E \left(1 - \epsilon^{-\frac{t}{Z_{0(1)}C}} \right)$$

where E is the zero front-time step function.

The inductance termination responds initially like an open-circuited line but later behaves like a short-circuited line. This is important to the understanding of the detrimental effect of long interconnections to surge protective equipment. On the other hand the capacitance termination responds initially like a short-circuited line, but later behaves like an open-circuited line.

NOTE 2
The Degree of Voltage Amplification Produced by a Stub Line Extension Continuing Beyond the Arrester ([38], Chapter 10)

The situation being discussed is portrayed in Fig N2.1. A transient voltage surge is traveling along a transmission channel toward a utilization terminal. A surge-voltage arrester is installed on the transmission channel at a distance *d* ahead of the terminal junction. As the wave front passes the point of arrester installation, no action by the arrester occurs until the voltage at the arrester reaches the arrester sparkover value. Fig N2.2 illustrates the voltage distribution pattern that will exist along the line if *d* extends to the right of the wave tip when the arrester sparks over. (To simplify the analysis of the action under study, consider that the stub line section is terminated in an open circuit.) It is quite evident that the entire wave beyond the arrester will be doubled at the open-end terminal.

If the stub end is somewhat shorter, as pictured in Fig N2.3, the front tip of the wave will have encountered the line end and been doubled before the arrester acts.

Fig N2.1
30 kV Voltage Surge Traveling Toward a Line Termination Containing an Arrester Installation

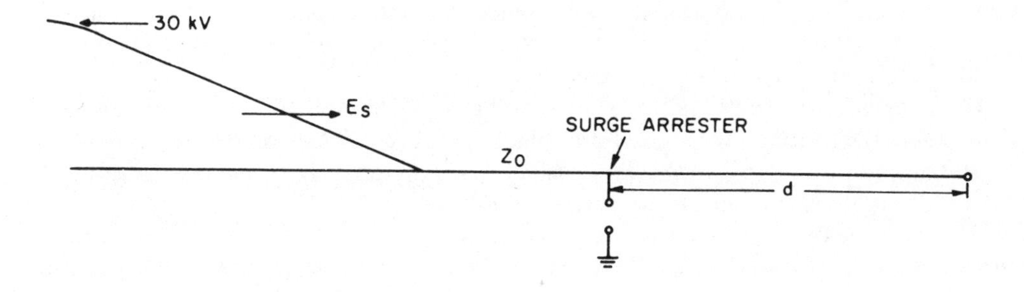

Fig N2.2
Entire Front Allowed by the Arrester Contained on a Stub Section ahead of the Termination

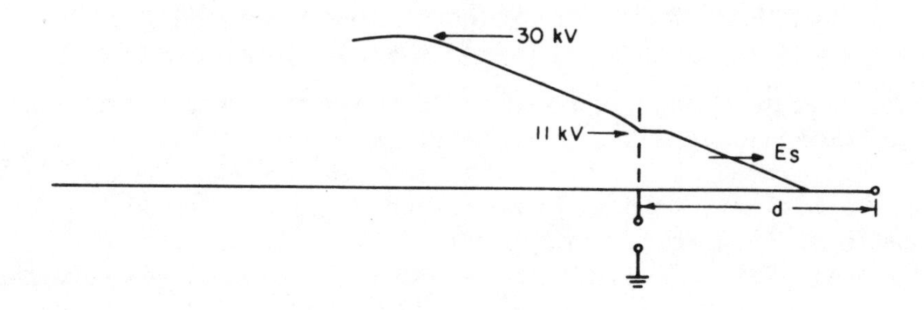

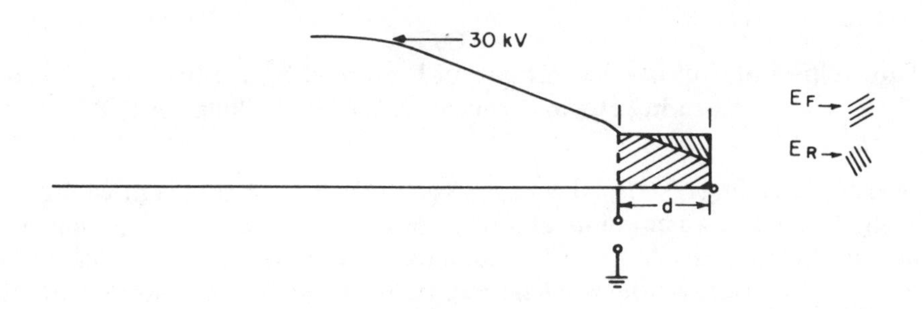

Fig N2.3
Forward Section of the Traveling Surge Already Reflected back from the Open End Before Arrester Conduction

That portion of the incident wave front that resides on the line between the arrester and the end terminal will also be doubled, unavoidably, upon reaching the open end. Hence, the top terminal voltage will be twice the voltage level that has already passed the arrester location.

If the stub line section is still shorter, such that the tip of the backward reflected voltage wave has just arrived at the arrester location when the arrester sparks over, the voltage at the open terminal will still go to twice the arrester sparkover voltage value. The first half of the wave front has already been doubled as it reaches the open end. The second half of the rising front is a forward-traveling wave already on the line that cannot be modified by any subsequent arrester action. Hence, it too will be doubled as it arrives at the open line terminal.

Thus, in general, the portion of the traveling wave front that can be contained on a line section of $2d$ length will be subject to full reflection amplification at the terminal. Line stubs of shorter lengths will proportionately reduce the amount of overshoot in voltage magnitude at the terminal.

With this basic understanding one can interpret the performance factors for other rates of rise of voltage wave fronts and for other values of propagation velocity. The adjustments needed to accommodate line terminals other than an *open circuit* are self-evident.

NOTE 3
How the Shunt-Connected Surge Capacitor Lessens the Slope of the Voltage Surge Front and Limits the Crest Voltage Magnitude

Fig N3.1 illustrates a voltage surge traveling along a branch cable circuit $Z_0 = 50\ \Omega$ to a 4160 V motor. At the line terminal, which is connected to the motor terminals, a set of surge protective capacitors (0.5 μF per phase) are installed. By the use of surge arresters, the voltage crest has already been clipped to 16 kV.

The electrical equivalent circuit applicable to the connection arrangement of Fig N3.1 is shown in Fig N3.2. The capacitor is charged by a 32 kV surge voltage through

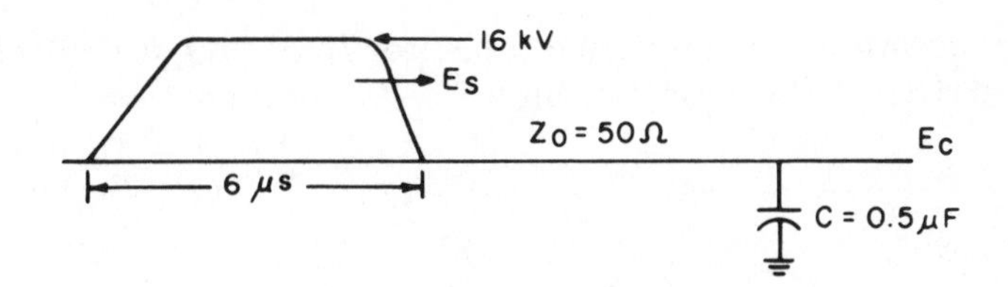

Fig N3.1
Surge Voltage Wave Traveling Toward a Motor Terminal on a 50 Ω Surge-Impedance Line

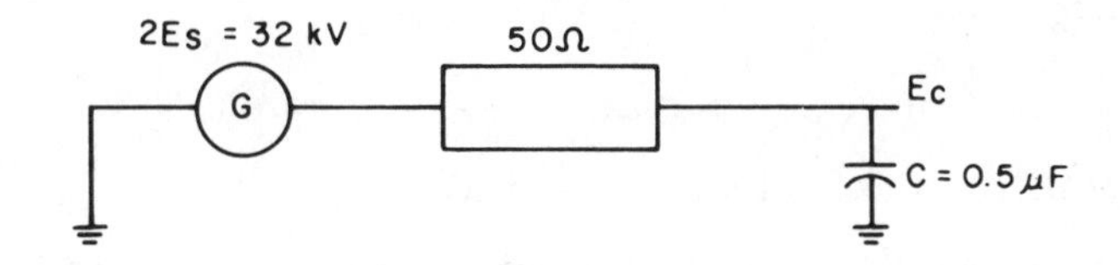

Fig N3.2
Accurate Lumped-Constant Equivalent Circuit for Analysis

a resistance of 50 Ω. The driving voltage is considered as a rectangular wave of 32 kV acting for a duration of 5 μs.

The motor major insulation security will be concerned primarily with the magnitude of the terminal voltage E_C, while the turn insulation security will be concerned primarily with the rate of rise of that voltage, dE_C/dt.

The fundamental current-voltage relationships associated with a capacitor, starting from a de-energized condition, are

$$E_C = \frac{Q}{C} = \frac{\int i\,dt}{C}$$

$$\frac{dE_C}{dt} = \frac{dQ}{dt} = \frac{I}{C}$$

When the capacitor is charged from a step-voltage source through a series resistor, as in Fig N3.2, the capacitor voltage builds up in accordance with

$$E_C = 2E_S\,(1 - e^{-t/t'})$$

where $t' = RC$, the circuit time constant, and

$$\frac{dE_C}{dt} \text{ (maximum)} = \frac{I}{C} \text{ (maximum)}$$

In this specific problem, $2E_S = 32$ kV, $Z_0 = 50$ Ω, and $C = 0.5\,(10)^{-6}$. The RC product is $50(0.5)(10)^{-6} = 25(10)^{-6}$s.

The maximum input current to the capacitor I_C occurs at $t = 0$ when the surge voltage first arrives at the capacitor and $E_C = 0$. We then have

$$I_C = \frac{2E_S}{Z_0} = \frac{32\,000}{50} = 640 \text{ A}$$

$$\frac{dE_C}{dt} \text{ (maximum)} = \frac{I_C}{C} = \frac{640}{0.5(10)^{-6}}$$

$$= 1280(10)^6 \text{ V/s} = 1280 \text{ V}/\mu\text{s}$$

For a surge-voltage duration of 5 μs, the quantity $\epsilon^{-t/t'}$ is equal to

$$\epsilon^{-0.2} = 0.8187$$

$$(1 - \epsilon^{-0.2}) = 0.1813$$

Thus at the end of the 5 μs interval, the capacitor voltage is

$$E_C = 2E_S(0.1813) = 32(0.1813) = 5.81 \text{ kV}$$

In comparison with the 13 kV crest value of the motor high-potential test, the voltage level developed across the capacitor (5.81 kV) is well below that level, and also well below the sparkover value of the special 4.16 kV surge arrester. The rate of rise of voltage at the motor terminal (maximum value) meets the criterion of at least 10 μs to reach the crest level of the nameplate voltage.

In conclusion the following should be noted:

(1) Had the circuit construction involved spaced conductors or open-wire lines, the surge impedance would have been substantially greater, making the surge current values lower, which in turn would account for lower values of capacitor voltage and lower values of the rate of rise of the capacitor voltage.

(2) Had the surge voltage been alternating, each subsequent half-cycle of surge current flow would create cancellation effects in the capacitor voltage created by the previous half-cycle.

(3) A greater duration of unidirectional voltage surge would account for a greater voltage across the capacitor, limited by the level $2E_S$ or the arrester sparkover voltage, whichever is lower.

(4) The presence of series inductance in the capacitor circuit acts to deteriorate the wave-sloping action of the surge capacitor, and even inductances of as little as a few microhenries can greatly impair performance ([31], Chapter 2).

5. Application and Coordination of System Protective Devices

5.1 Introduction

5.1.1 Purpose. The system and equipment protective devices guard the power system from the ever present threat of damage caused by overcurrents and transient overvoltages that can result in equipment loss and system failure. This chapter presents the principles of adequate system and equipment protection by introducing the many protective devices and their applications, special problems, system conditions associated with transient overvoltages, sound engineering techniques of protective device application and coordination, and suggested maintenance and testing procedures for circuit interrupting and protective devices.

5.1.2 Considering Plant Operation [31][18]. Industrial plants vary greatly in the complexity of their electric distribution systems. A small plant may have a simple radial design with low-voltage fuse protection only, whereas a large plant complex may incorporate an intricate network of medium- and low-voltage distribution substations, uninterruptible power sources, and in-plant generation required to operate in parallel with or isolated from local utility networks. At an early design stage, the plant engineering representatives should meet with the local power company to review and resolve the requirements of both the plant and the utility.

The need for higher production from industrial plants has created demands for greater industrial power system reliability. Trends to network systems and parallel operation with utilities have produced sources that have extremely high overcurrents during fault conditions and the development of new equipment standards.

The high costs of power distribution equipment and the time required to repair or replace damaged equipment, such as transformers, cable, high-voltage circuit breakers, etc, make it imperative that serious consideration be given to system protection design.

The losses associated with an electrical service interruption due to equipment or system failures vary widely with different types of industries. For example, a service

[18] The numbers in brackets correspond to those of the references listed at the end of this chapter; when preceded by B, they correspond to the bibliography at the end of this chapter.

interruption in a machining operation means loss of production, loss of tooling, and loss from damaged products. Likewise, a service interruption in a chemical plant can cause loss of product and create major clean-up and restart problems. To avoid a disorderly shutdown, which can be both hazardous and costly, it may be necessary to tolerate a short-time overload condition and the associated reduction in life expectancy of the affected electric apparatus. Other industries such as refineries, paper mills, automotive plants, textile mills, steel mills, and food processing plants are similarly affected, and losses can represent a substantial expense.

For some types of loads involving continuous processes and complex automation, even a momentary dip in voltage is as serious as a complete service interruption, whereas other types of loads can tolerate an interruption. Thus the nature of the industrial operation is a major consideration in determining the degree of protection that can be justified.

5.1.3 Equipment Capabilities. In addition to meeting the demands of overall system performance as dictated by the nature of the load, the protective devices must operate in conjunction with the associated circuit interrupters so as to afford protection to other power system equipment components. Transformers, cable, busway, circuit breakers, and otner switching apparatus all have short-circuit withstand limits as established by the National Electrical Manufacturers Association and the American National Standards Institute.

When a fault condition occurs, these devices, although perfectly intact, may be connected in series in the same circuit and subjected to severe thermal and magnetic stresses accompanying the passage of high-magnitude short-circuit current through their conducting parts. An important function of the system protective devices is to initiate operation of the circuit interrupter responsible for isolating the fault so that the other equipment connected in the same circuit is not stressed beyond safe limits. Otherwise the initial fault condition can affect far more than the specific circuit to be isolated, and a widespread outage can result.

The design engineer should examine the performance capability of all the individual system equipment components and not just the process sensitivity to local outages when justifying the protection to be applied. System and equipment protection is one of the most important items in the process of system planning, and enough time should be allowed in the early stages of a system design to properly investigate the selection and application of the protective devices.

5.1.4 Importance of Responsible Planning. Some industrial plants, because of their size or the nature of their operations, are able to maintain electrical engineering staffs capable of the design, installation, and maintenance of an efficient protective system; other plants probably find it more economical to engage competent engineering advice and services from consultants. This work is specialized and often very complex, and it is neither safe nor fair to the operating engineer to expect him to do it as a sideline. Modern computerized methods of calculating fault currents on complex systems are available from consulting firms and manufacturers. These provide accurate information essential for making decisions relative to the protection design in a short period of time ([57], Chapter 6).

Protection in an electric system is a form of insurance. It pays nothing as long as there is no fault or other emergency, but when a fault occurs it can be credited with reducing the extent and duration of the interruption, the hazards of property damage, and personnel injury. Economically, the premium paid for this insurance should be balanced against the cost of repairs and lost production. Protection, well integrated with the class of service desired, may reduce capital investment by eliminating the need for equipment reserves in the industrial plant or utility supply system.

While the protective devices are the watchmen of the power system, the industrial electrical engineer must be the custodian of the protection system. Adequate and regular maintenance and testing should be done, as well as reanalysis of the protection scheme when major system changes occur. Integrity of the system protection and, thereby, the system performance requires a continuing effort if it is to be preserved.

5.2 Analysis of System Behavior and Protection Needs

5.2.1 Nature of the Problem. It would be neither practical nor economical to build a fault-proof power system. Consequently, modern systems are designed to provide reasonable insulation, clearances, etc, but a certain number of faults must be tolerated during the life of the system. Even with the best design possible, materials deteriorate and the likelihood of faults increases with age. Every system is subject to short circuits and ground faults that should be removed quickly, and a knowledge of the effect of faults on system voltages and currents is necessary to design suitable protection, since these quantities are used to develop a protective plan. A reliable protective system is not only properly designed and maintained, but it should not be unnecessarily complicated or it can cause more problems than it solves in the form of undesired or incorrect operations.

Operating records show that the majority of electric circuit faults originate as a phase-to-ground failure. Protective devices should detect three-phase, phase-to-phase, double-phase-to-ground, as well as single-phase-to-ground short circuits. There are two general classifications of three-phase systems: (1) ungrounded systems, and (2) grounded systems where one conductor, generally the neutral, is grounded either solidly or through an impedance. Both classifications of systems are subject to all the aforementioned types of faults, but the severity of those faults involving ground depends to a large extent on the method of system grounding and the magnitude of the grounding impedance [42].

5.2.2 Grounded and Ungrounded Systems. The general subject of system grounding (see ANSI/IEEE Std 142-1982 [19]) is treated from the viewpoint of system design in Chapter 7, and it is only necessary to observe here the effect on basic relaying methods of the choice between a grounded and an ungrounded system.

In grounded systems, phase-to-ground faults produce currents of sufficient magnitude to operate ground-fault-responsive overcurrent relays, which automatically detect the fault, determine which feeder has failed, and initiate the tripping of the correct circuit breakers to de-energize the faulted portion of the system without interrupting service to healthy circuits. If the system neutral is grounded through a properly chosen impedance, the value of the ground-fault current can be restricted to a level that will avoid extensive damage at the point of the fault, yet be adequate for ground-fault relaying. In addition, the voltage dip caused by the flow of ground-fault current will be materially reduced.

In ungrounded systems, as shown in Fig 47(a), phase-to-ground faults produce relatively insignificant values of fault current. In a small, isolated-neutral industrial installation, the ground-fault current for a single line-to-ground fault may be well under 1 A, while the largest plant, containing miles of cable to provide electrostatic capacitance to ground, may produce not more than 20 A of ground-fault current. These currents usually are not of sufficient magnitude for the operation of overcurrent relaying to locate and remove such faults, not only because of the extreme sensitivity of the relays that would be required, but also because of the complexity

Fig 47
Analysis of Steady-State Conditions on an Ungrounded System Before and After the Occurrence of a Ground Fault

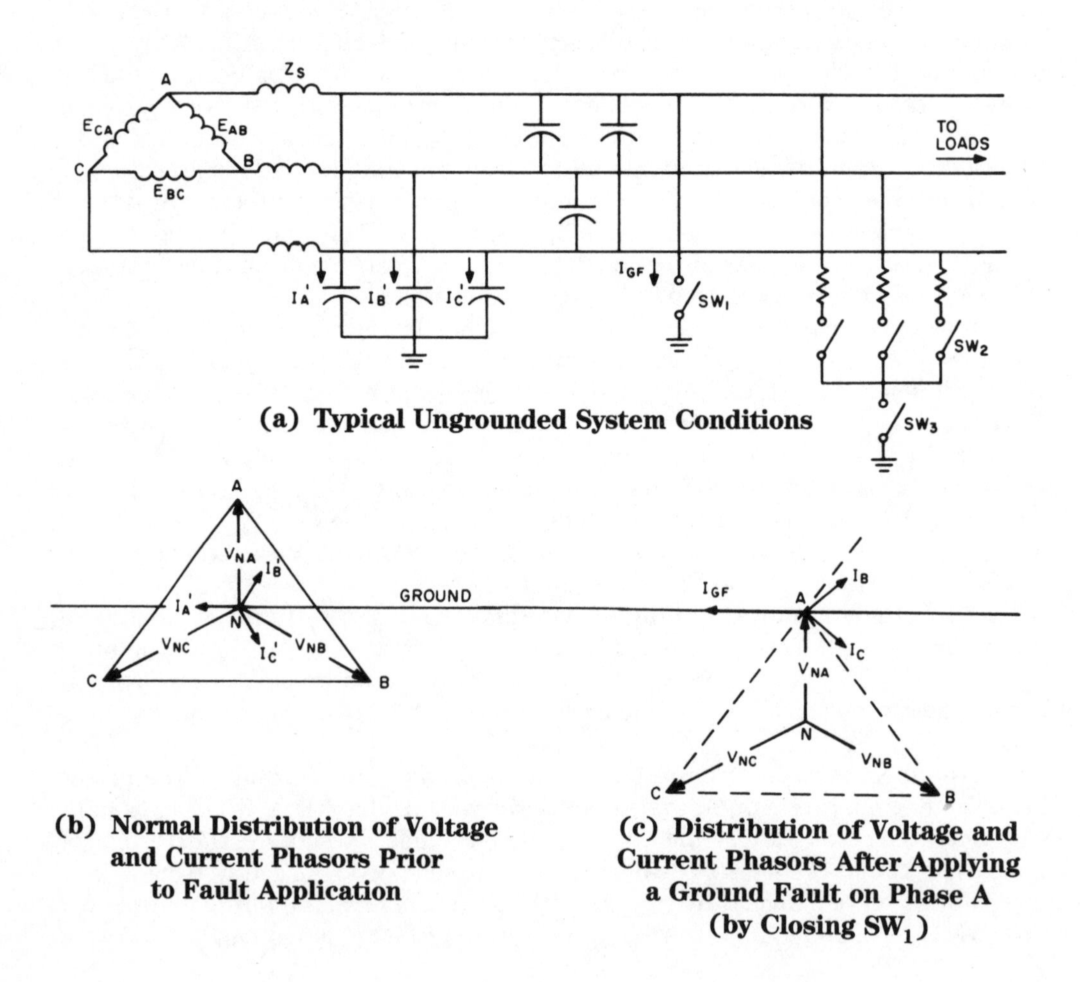

(a) Typical Ungrounded System Conditions

(b) Normal Distribution of Voltage and Current Phasors Prior to Fault Application

(c) Distribution of Voltage and Current Phasors After Applying a Ground Fault on Phase A (by Closing SW_1)

of the flow pattern resulting from the fact that the *source* of the ground current is the distributed capacitance to ground of the unfaulted conductors. It is possible, however, to provide phase-to-ground voltage relays that will operate an alarm on the occurrence of a ground fault, but that cannot provide any indication of its exact location. The voltage and current distribution for normal operation and for a single-phase-to-ground fault (phase A) condition for an ungrounded system are shown in Fig 47(b) and (c), respectively.

The one advantage of an ungrounded system lies in the possibility of maintaining service on the entire system, including the faulted section, until the fault can be located and the equipment shut down for repair. Against this advantage should be balanced such disadvantages as the impossibility of relaying the fault automatically, the difficulty of locating the fault, the continuation of burning and the escalation of damage at the point of the fault, the long-continued overstressing of the insulation of the unfaulted phases (1.73 times operating voltage in the case of solid ground faults and perhaps much more in the case of intermittent ground faults), and the hazard of multiple ground faults and transient overvoltages. High-resistance grounding should be considered as a preferred alternate.

5.2.3 Distortion of Phase Voltages and Currents During Faults. Balanced three-phase faults do not cause voltage distortion or current unbalance. Figure 48 shows the balanced conditions, both before and after a fault is applied on a system having an X/R ratio of approximately 1.7, which corresponds to an angle of 60 degrees between phase voltage and current or a 50% fault-circuit power factor. This

Fig 48
Voltage and Current Phasor Relationships for a Balanced Fault Condition

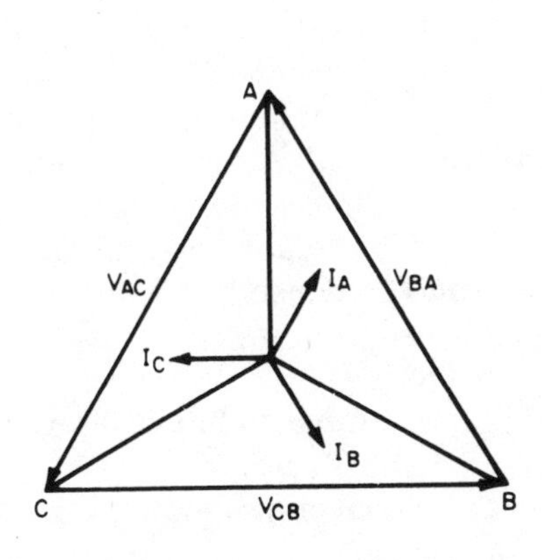

(a) Normal Conditions with Load Applied

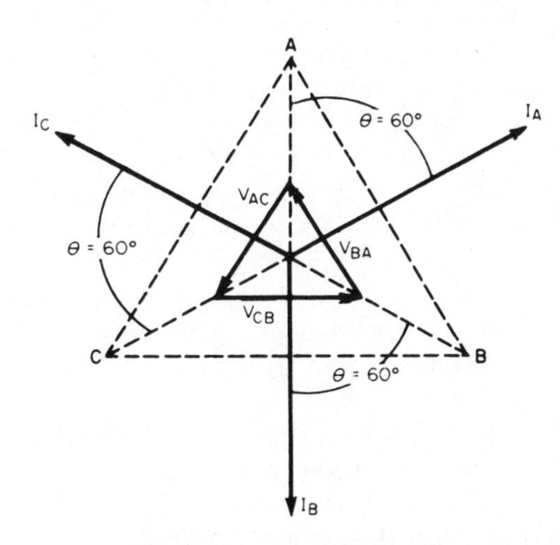

(b) Three-Phase Fault

condition would be realized by closing all three poles of switch SW_2 in Fig 47(a). Other types of faults, such as phase-to-phase, single-phase-to-ground, and two-phase-to-ground, cause distorted voltages and unbalanced currents. The voltage distortion is greatest at the fault and minimum at the generator or source.

Currents and voltages that exist during a fault vary widely for different systems, depending on type and location of the fault and the impedance of the system grounding connection. The vector diagrams of Fig 49 show voltage and current relations that exist for different types of faults on a solidly grounded system in which the currents lag the voltages by 60 degrees. Load currents are not included.

These diagrams are typical of the fault conditions that cause protective devices to operate. The characteristics of the voltage distortion that accompanies a fault are used to enable special types of relays to discriminate between different types of faults having otherwise similar current conditions. Some of these special devices will be discussed in further detail later in this chapter. The distortion can be greater or less than that shown, depending on the impedance of the fault and its distance from the relay. The small voltage drop shown between the faulted phases represents the fault impedance (arc) voltage drop plus the voltage drop in the system conductors due to the flow of fault current between the relay and the fault point.

5.2.4 Analytical Restraints. The one-line diagram commonly used to represent three-phase systems is a very useful analytical tool when its limitations are properly observed. Its validity is limited to symmetrical three-phase loading of symmetrical systems.

One-line diagrams, for example, provide no means for properly representing the effect of single-phase loads on the operation of the system or the protective devices. Likewise, the influence of surge protection equipment acting independently in any of the three phases cannot be correctly evaluated.

In the case of a line-to-ground fault on a three-phase system, the opening of that phase protector alone alters the system symmetry. Examination of a more precise three-phase diagram reveals that it is still possible for current to flow through the remaining phases via the line-to-line connected paths at the load apparatus, and then to ground at the fault point.

Determining how much current continues to flow to the ground fault after opening of one phase protector is a complex problem due to the alteration in system symmetry and introduction of additional variable impedances. However, it is generally a much lower value than the initial line-to-ground fault current and, hence, it can take the remaining protectors some time to sense and clear the circuit. Substantial heat with associated damage can be generated at the fault point before complete isolation of the circuit is accomplished.

By simultaneously opening all phases, the three-phase symmetry is not altered and the condition described above need not be given consideration.

5.2.5 Practical Limits of Protection. When the industrial power system is in normal operation, all parts should have some form of automatic protection. However, some fault possibilities may be legitimately considered too improbable to justify the cost of specific protection. Before accepting a risk on this basis alone, the magnitude of the probable damage should also be seriously considered. Too much protection might be provided for failures that occur frequently but cause only

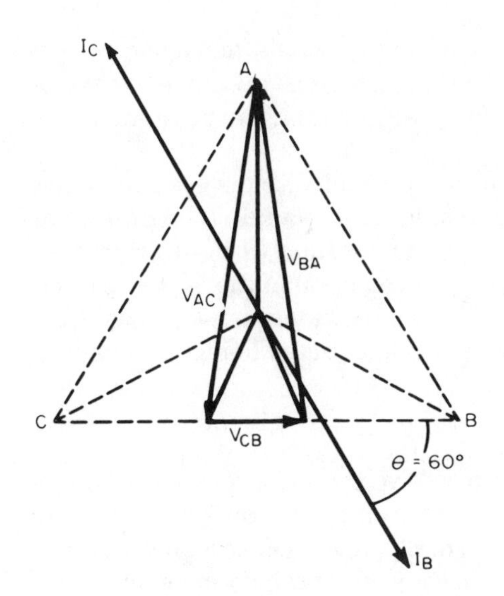

(a) Phase-to-Phase Fault Between Phases B and C on Ungrounded System [Close Poles B and C of SW_2 in Fig 47(a)]

(b) Two-Phase-to-Ground Fault Between Ground and Grounded System [Close SW_3 and Poles B and C of SW_2 in Fig 47(a)]

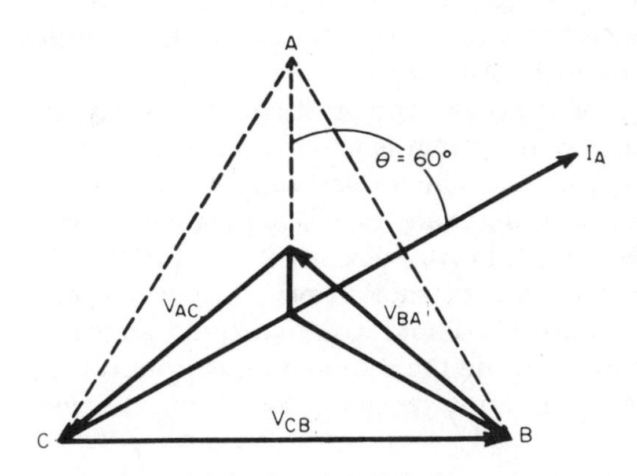

(c) Phase-to-Ground Fault Between Phase A and Ground on Grounded System

Fig 49
Voltage and Current Phasor Relationships for Various Unbalanced Fault Conditions (System X/R = 1.7)

minor difficulties, while rare but serious causes of trouble might be neglected. For example, internal transformer failures rarely occur, but the consequences may be very serious since such faults can cause fires and endanger personnel and equipment.

Most systems have some flexibility in the manner in which circuits are connected. The various possible arrangements should be considered in planning the protection system so as not to leave some emergency operating condition without protection. Some types of systems have so many possible operating combinations that protection cannot be applied to operate properly for all conditions. In such cases the operating connections for which the protection is inadequate should be avoided.

5.3 Protective Devices and Their Applications

5.3.1 General Discussion. Power system protective devices provide the intelligence and initiate the action that enables circuit switching equipment to respond to abnormal or dangerous system conditions. Normally, relays control power circuit breakers rated above 600 V and current-responsive self-contained elements operate multipole low-voltage circuit breakers to isolate circuits experiencing overcurrents on any leg. Similarly, fuses and single-pole interrupters function either alone or in combination with other suitable means to properly provide isolation of faulted or overloaded circuits. In other cases, special types of relays that respond to abnormal electric system conditions may cause circuit breakers or other switching devices to disconnect defective equipment from the remainder of the system.

The following is a brief description of the types and characteristics of relays and other protective devices most commonly used in industrial plant power systems along with some brief application considerations. A table of their relay device numbers referenced with respective device functions as given in ANSI/IEEE C37.2-1979 [8] appears in the Appendix of this book.

5.3.2 Overcurrent Relays. The most common relay for short-circuit protection of the industrial power system is the overcurrent relay, as shown in Fig 50. The overcurrent relays used in the industry are mostly of the electromagnetic attraction, induction, and solid-state types. Relays with bimetallic elements used for thermal overload protection are discussed in 5.3.16. The simplest overcurrent relay using the electromagnetic attraction principle is the solenoid type. The basic elements of this relay are a solenoid wound around an iron core and steel plunger or armature that moves inside the solenoid and supports the moving contacts. Other electromagnetic-attraction-type relays have hinged armatures or clappers of different shapes.

The construction of the induction disk-type overcurrent relay is similar to a watthour meter since it consists of an electromagnet and a movable armature, which is usually a metal disk on a vertical shaft restrained by a coiled spring. The relay contacts are operated by the movable armature (Fig 51).

The pickup or operating current for all overcurrent relays is adjustable. When the current through the relay coil exceeds a given setting, the relay contacts close and initiate the circuit breaker tripping operation. The relay usually operates on current from the secondary of a current transformer.

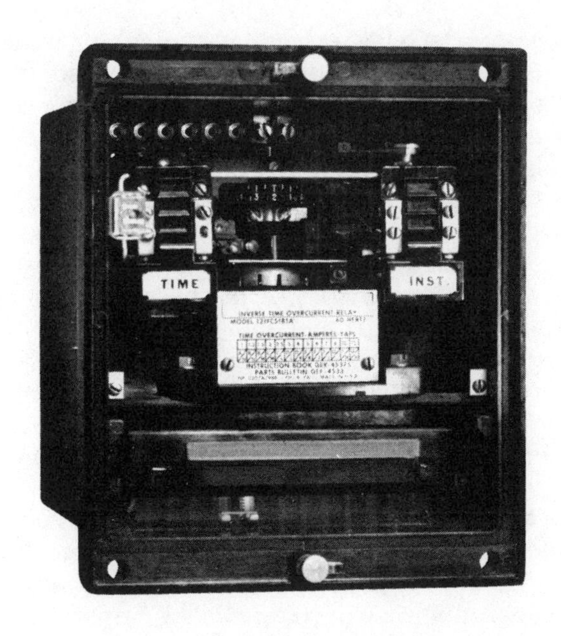

Fig 50
Typical Electromagnetic Overcurrent Relay

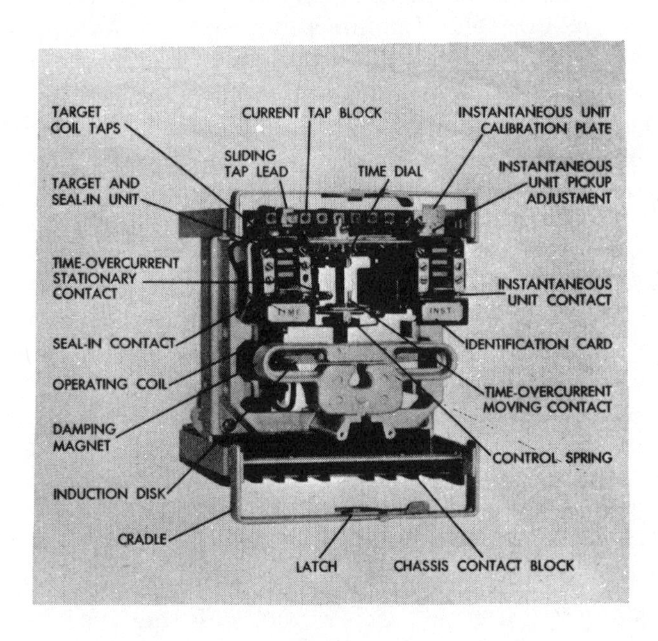

Fig 51
Induction-Disk Overcurrent Relay with Instantaneous Attachment
(Relay Removed from Drawout Case)

If the current operates the relay without intentional time delay, the protection is called instantaneous overcurrent protection. When the overcurrent is of a transient nature such as that caused by the starting of a motor or some sudden overload of brief duration, the circuit breaker should not open. For this reason most overcurrent relays are equipped with a time delay that permits a current several times in excess of the relay setting to persist for a limited period of time without closing the contacts. If a relay operates faster as current increases, it is said to have an inverse-time characteristic. Overcurrent relays are available with inverse, very inverse, and extremely inverse time characteristics to fit the requirements of the particular application. There are also definite minimum-time overcurrent relays that have an operating time that is practically independent of the magnitude of current after a certain current value is reached. Induction disk overcurrent relays have a provision for variation of the time adjustment and permit change of operating time for a given current. This adjustment is called the time lever or time dial setting of the relay.

Figure 52 shows the family of time-current operating curves available with a typical inverse-time overcurrent relay. Similar curves are published for other overcurrent relays having different time-delay characteristics. As is apparent, it is possible to adjust the operating time of relays. This is important since they are normally used to selectively trip circuit breakers that operate in series on the same system circuit. With increasing current values, the relay operating time will decrease in an inverse manner down to a certain minimum value. Figure 53 shows the characteristic curves of inverse (A), very inverse (B), and extremely inverse (C) time relays when set on their minimum and maximum time dial positions. It also shows the characteristics of the instantaneous element (D) that is usually supplied in these relays [39].

5.3.3 Overcurrent Relays with Voltage Restraint or Voltage Control [34]. A short circuit on an electric system is always accompanied by a corresponding voltage dip, whereas an overload will cause only a moderate voltage drop. Therefore a voltage-restrained or voltage-controlled overcurrent relay is able to distinguish between overload and fault conditions. A voltage-restrained overcurrent relay is subject to two opposing torques, an operating torque due to current and a restraining torque due to voltage. As such, the overcurrent required to operate the relay is higher at normal voltage than it is at reduced voltage. A voltage-controlled overcurrent relay operates by virtue of current torque only, the application of which is controlled by another relay element set to operate at some predetermined value of voltage. Such relay characteristics are useful where it is necessary to set the relay close to or below load current, while retaining certainty that it will not operate improperly on normal load current.

5.3.4 Directional Relays

5.3.4.1 Directional Overcurrent Relay. Directional overcurrent relays consist of a typical overcurrent unit and a directional unit that are combined to operate jointly for a predetermined phase-angle and magnitude of current. In the directional unit the current in one coil is compared in phase-angle position with a voltage or current in another coil of that unit. The reference current or voltage is called the polarization. Such a relay operates only for current flow to a fault in one direction and will be insensitive to current flow in the opposite direction. The overcurrent unit of the directional overcurrent relay is practically the same as for the usual overcurrent relay and has similar definite-minimum-time, inverse, and very inverse time-current characteristics. The directional overcurrent relays can be supplied with voltage restraint on the overcurrent element.

Fig 52
Time-Current Characteristics of a Typical Inverse-Time Overcurrent Relay

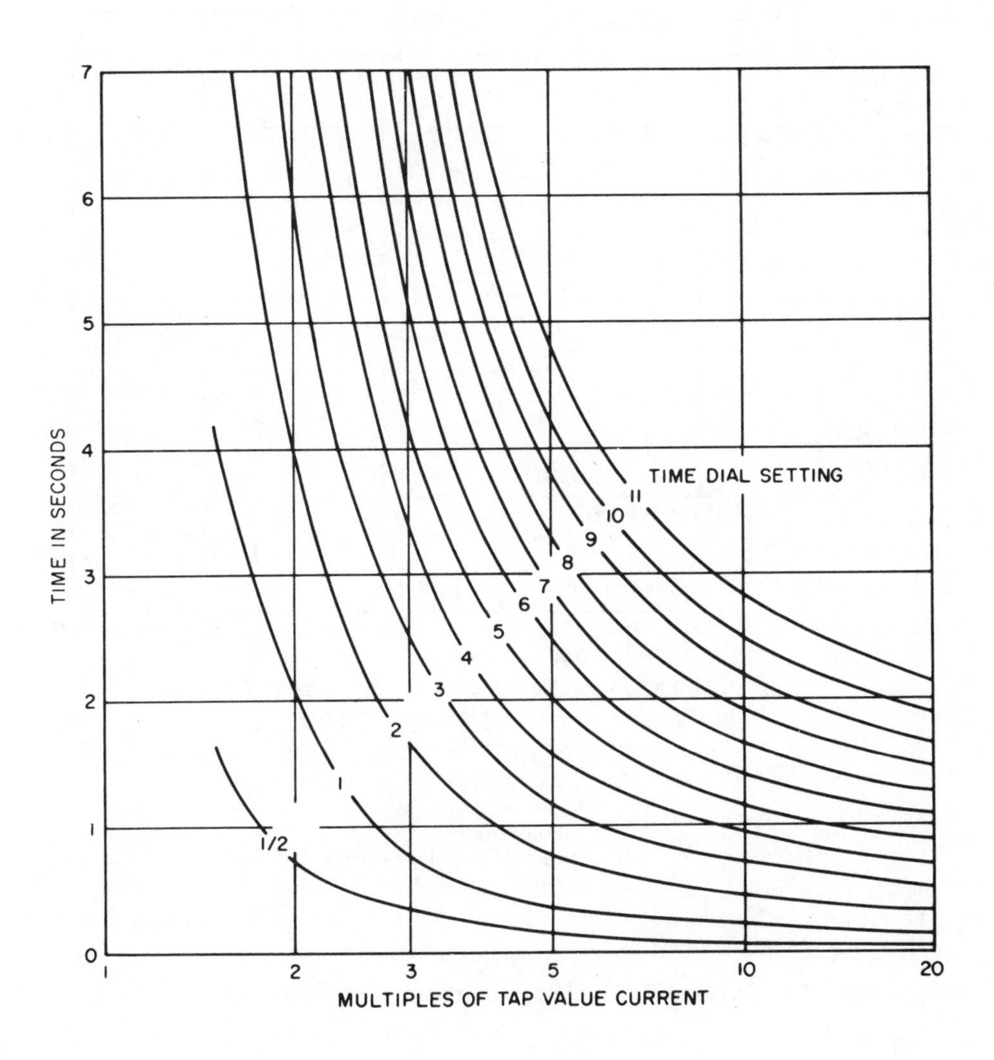

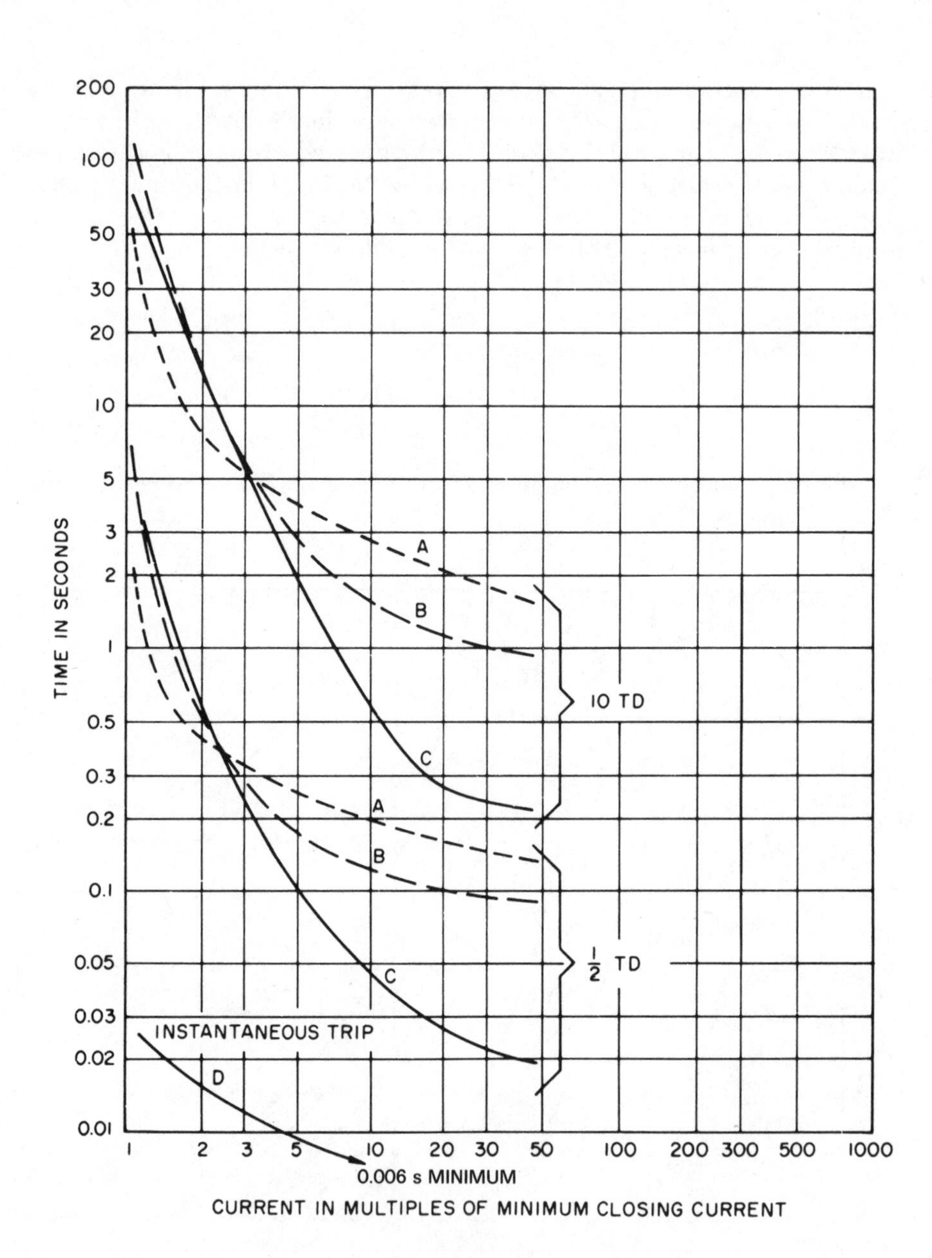

A Inverse
B Very Inverse
C Extremely Inverse
D Instantaneous
TD Relay Time Dial Setting

Fig 53
Typical Relay Time-Current Characteristics

The most commonly used directional relays are usually directionally controlled, that is, the overcurrent unit is inert until the directional unit detects the current in the tripping direction and releases or activates the overcurrent unit. Many directional relays are equipped with instantaneous elements, which in some cases operate nondirectionally, and unless it is possible to determine the direction of the fault by magnitude alone, the nondirectional instantaneous tripping feature should not be used.

5.3.4.2 Directional Ground Relay. The grounded-neutral industrial power system consisting of parallel circuits or loops may use directional ground relays, which are generally constructed in the same manner as the directional overcurrent relays used in the phase leads. In order to properly sense the direction of fault current flow, they require a polarizing source that may be either potential or current as the situation requires. Obtaining a suitable polarizing source requires special consideration of the system conditions during faults involving ground and a unique application of auxiliary devices.

5.3.4.3 Directional Power Relay. The directional power relay is in principle a single-phase or three-phase contact-making wattmeter and operates at a predetermined value of power. It is often used as a directional overpower relay set to operate if excess energy flows out of an industrial plant power system into the utility power system. Under certain conditions it may also be useful as an underpower relay to separate the two systems if the power flow drops below a predetermined value. Care should be used in the application of single-phase watt relays because at certain power factors they may cause a false trip operation.

5.3.5 Differential Relays. All the previously described relays have the common characteristic of adjustable settings to operate at a given value of some electrical quantity, such as current, voltage, frequency, power, or a combination of current and voltage or current and phase angle. There are other fault-protection relays that function by virtue of continually comparing two or more currents [Fig 54(a)]. Fault conditions will cause a change of these compared values with reference to each other and the resulting *differential* current can be used to operate the relay. However, current transformers have a small error in ratio and phase angle between the primary and secondary currents, depending upon variations in manufacture, the magnitude of current, and the connected secondary burden. These errors will cause a differential current to flow even when the primary currents are balanced. The error currents may become proportionately larger during fault conditions, especially when there is a dc component present in the fault current. The differential relays, of course, must not operate for the maximum error current which can flow for a fault condition external to the protected zone. To provide this feature, the percentage-type relay (Device 87[19]) illustrated at right of Fig 54(a) has been developed; it has special restraint windings to prevent improper operation due to the error currents on heavy *through* fault conditions while providing very sensitive detection of low-magnitude faults inside the differentially protected zone [53].

[19] ANSI device function numbers are defined in ANSI/IEEE C37.2-1979 [8].

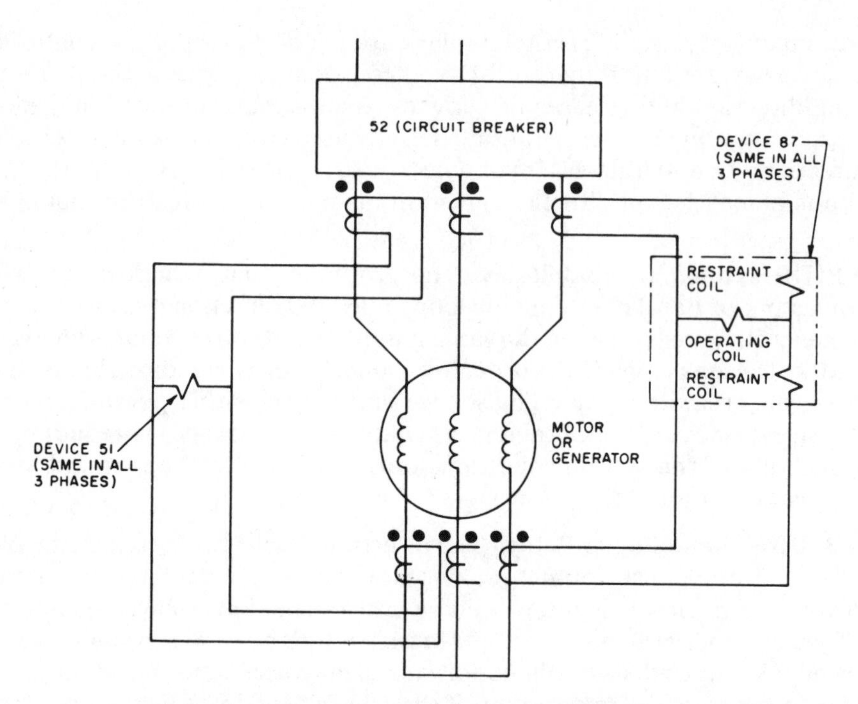

(a) Using Percentage-Type Differential Relays (Device 87) or Time-Delay Overcurrent Relays (Device 51)

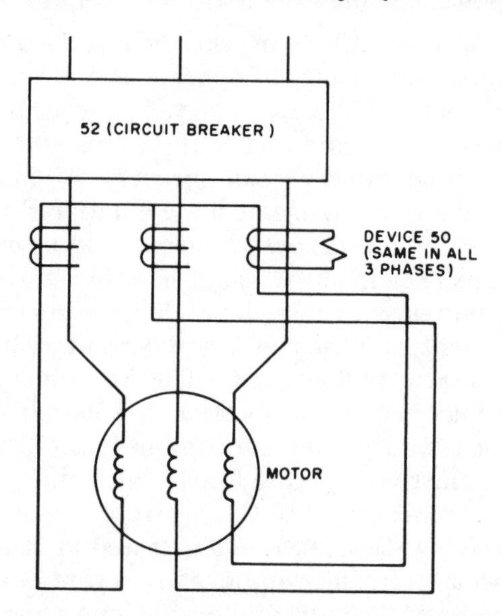

(b) Using Instantaneous Relays (Device 50), for Motors Only

Fig 54
Arrangements for Motor and Generator Differential Protection

5.3.5.1 Differential Protection of Motors and Generators. Using the connection arrangement illustrated at the left of Fig 54(a), overcurrent relays can be used for the differential protection of a motor or generator. As long as the current flowing into each winding of the motor is equal to the current flowing out of the same winding, no net current will flow through the relay operating coil (ignoring current transformer error currents). Any leakage of fault current to other phases or to ground will upset this balance and send differential current through the operating winding. When this current is greater than the pickup of the relay, its contacts close to activate tripping of the circuit breaker and disconnect the faulted apparatus.

Since no means is provided with this scheme to prevent false operation on current transformer error current, the overcurrent relays have to be set so that they will not operate on the maximum error current that can flow in the relay during an external fault. This results in a substantial sacrifice of sensitivity for low-magnitude internal faults. The percentage differential relay shown at right in Fig 54(a) overcomes this disadvantage. The restraint coils can be selected to provide a restraining torque of 10-25% of the through current on external faults, but produce zero restraint on internal faults.

Another form of motor differential protection involves a special routing of the machine phase and neutral leads of each leg through a common *window type* current transformer as shown in Fig 54(b). Under normal conditions the magnetizing flux produced by the phase and neutral currents adds to zero and no output is produced for the instantaneous relay. A fault in any winding will result in the fault current bypassing the current transformer, causing a differential current (and flux) that will in turn produce an output signal. The single relay and current transformer employed per phase in this scheme is less expensive, although additional machine terminal box space for neutral conductor cabling is required.

5.3.5.2 Differential Protection of a Two-Winding Transformer Bank. When differential relays are used for transformer protection, the inherent characteristics of power transformers introduce a number of problems that do not exist in the generators and motors. If the current transformer secondary currents on the two sides of the transformer differ in magnitude by more than the range provided by the relay taps, the relay currents can be altered by means of auxiliary current transformers or current-balancing autotransformers. If the high-voltage and low-voltage line currents are not in phase due to a delta-wye connection in the transformer, the secondary currents can be brought into phase by connecting the current transformers in delta on the wye side, and in wye on the delta side. The differential output signals of the current transformers are subject to the same errors as discussed above for generators. In addition, a significant trip current signal can be observed at the relay input due to primary magnetizing inrush current, which occurs upon transformer energization. This is why ordinary overcurrent relays cannot be given sensitive settings, and induction-type percentage differential relays are sometimes used instead. The best protection can be had by using the harmonic restraint-type relay. This relay typically has a filter to the operating coil that blocks the harmonic currents, and a filter to the restraint coil that passes only harmonic currents. Using this technique, undesired operation on magnetizing inrush currents is prevented while retaining good sensitivity for fault conditions.

5.3.5.3 Differential Protection of Buses. Large industrial power system buses often have sectionalizing circuit breakers so that a fault in one of the bus sections can be isolated without involving the remaining sections. Each of the bus sections or in some cases the whole bus (if not sectionalized) can be provided with differential relay protection, which in case of an internal fault isolates the bus section involved.

Differential bus protection distinguishes between internal and external fault by comparing the magnitudes of the currents flowing in and out of the protected bus. The major differences between bus protection and generator or transformer protection are in the number of circuits in the protected zone and in the magnitude of currents involved in the various circuits. Figure 55 shows phase and ground differential protection of an eight-circuit bus using overcurrent relays. This method, of course, is subject to the same disadvantages discussed in the preceding paragraphs. Several more acceptable types of bus protective relays are used, including the percentage differential relay, the linear coupler, and the differential voltage relay.

(1) *Percentage Differential Relay.* Where the number of circuits connected to the bus is relatively small, relays using the percentage differential principle similar to the transformer differential relay may be used. The problem of application of percentage differential relays for bus protection, however, increases with the number of circuits connected to the bus. All current transformers supplying the relays must have identical ratios and characteristics. Variations in the characteristics of the current transformers, particularly the saturation phenomena under short-circuit conditions, present the greatest problem to this type of protection and often limit it to applications where only a limited number of feeders are present.

Various relay-desensitizing schemes are used to avoid operation from the inrush of magnetizing current when switching transformers. The harmonic restraint type has features that distinguish between magnetizing inrush current and internal fault current. Another type utilizes external timing relays and shunting resistors during the switching interval.

(2) *Linear Coupler* [49]. The linear-coupler bus-protection scheme eliminates the difficulty due to differences in the characteristics of iron-core current transformers by using air-core mutual inductances. Since it does not contain any iron in its magnetic circuit, the linear coupler is free of any dc or ac saturation. The linear couplers of the different circuits are connected in series and produce voltages that are directly proportional to the currents in the circuits. For normal conditions, or for external faults, the sum of the voltages produced by linear couplers is zero. During internal (bus) faults, however, this voltage is no longer zero and operates a sensitive relay to trip all circuit breakers to clear the bus fault.

(3) *Differential Voltage Relay.* Another method of bus protection is the use of differential voltage relays. This scheme uses through-type iron-core current transformers. The problem of current-transformer saturation is overcome by using a voltage-responsive (high-impedance) operating coil in the relay.

Bus protection using linear couplers or differential voltage relays is not limited as to number of source and load feeders, and in general is faster in operation than

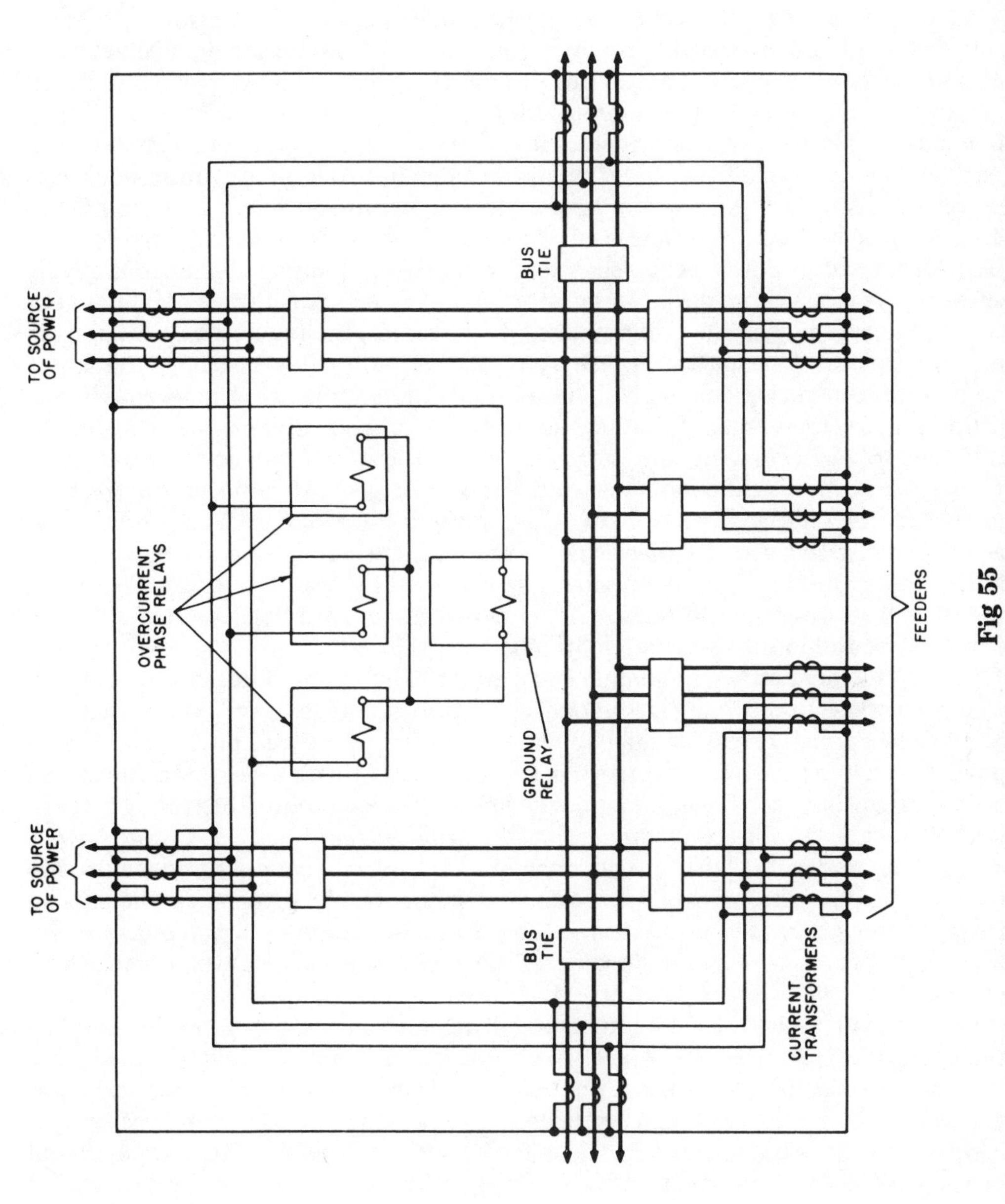

Fig 55
Phase and Ground Protection of an Eight-Circuit Bus Using Standard Induction-Disk Overcurrent Relays

protection using the percentage differential principle. It should be noted that linear couplers or current transformers used for differential voltage relays cannot be used for other purposes. Separate current transformers are required for line relaying and metering.

5.3.6 Current Balance Relay. The principle of a differential relay as applied to rotating machinery protection requires that the current transformers be available at both ends of the phase windings to permit the comparison between the current magnitudes at these ends. In some instances, particularly in smaller units, it may not be possible to justify the cost of installing these current transformers or of bringing out the extra winding terminals to make installation of differential relays possible. In such cases phase-balance current-comparison relays can provide an acceptable substitute for differential protection. A negative-sequence current relay is a more sensitive device that also detects unbalanced phase currents. In applying these relays it is assumed that under normal conditions the phase currents in the three-phase supply to the equipment and the corresponding output signals from each phase current transformer are balanced. Should the fault occur in the motor or generator involving one or two phases, or should an open circuit develop in any of the phases, the currents will become unbalanced and the relay will operate. In addition to protecting against winding faults, the phase-balance current relay affords protection against damage to the motor or generator due to single-phase operation. This type of protection is not provided for by the usual differential relays. Another current-balance type of differential protection for motors, which is both simple and relatively inexpensive, is provided by the use of a single current transformer zero-sequence relay scheme and is discussed further in 5.3.7.2.

5.3.7 Ground-Fault Relaying [43], [58]

5.3.7.1 Residually Connected Protection. Where the industrial power system neutral is intentionally grounded and ground-fault current can flow in the conductors, ground relaying may be used to provide improved protection. This is often an overcurrent relay connected in the common lead of the wye-connected secondaries of three line-current transformers. Figure 56 shows the typical current transformer and relay connections for this application. When used on four-wire systems, an additional current transformer in the neutral conductor is required to balance the residual signal of the normal line-to-neutral load currents. The ground relay can be set to pick up at a much lower current value than the phase relays because there is no current flowing in the residual circuit due to normal load current.

Overcurrent relays used for ground-fault protection are generally the same as those used for phase-fault protection, except that a more sensitive range of minimum operating current values is possible since they see only fault currents. Relays with inverse, very inverse, and extremely inverse time characteristics, as well as instantaneous relays, are all applicable for ground-fault relays. Precaution should be used, however, in applying this type of residually connected ground relay, since it is subject to nuisance operation due to error currents arising from current-transformer saturation and unmatched characteristics in the manner described for differential relays. Often the optimum speed and sensitivity of a residual ground relay must be compromised because of this.

5.3.7.2. Zero-Sequence Relay. An improved type of ground-fault protection can be obtained by a zero-sequence relay scheme in which a single window-type current transformer is mounted so as to encircle all three-phase conductors as illustrated in Fig 57. On four-wire systems with possible unbalanced line-to-neutral loads, the neutral conductor must also pass through the current transformer window. Only circuit faults involving ground will produce a current in the current-transformer secondary to operate the relay. Since only one current transformer is employed in this method of sensing ground faults, the relaying is not subject to current-transformer errors due to ratio mismatch or dc saturation effects; however, ac saturation can cause appreciable error. Therefore each relay-current-

Fig 56
Standard Arrangement for Residually Connected Ground Relay

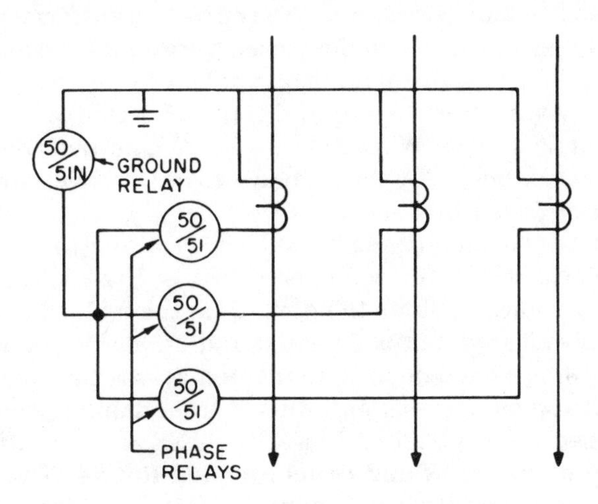

Fig 57
Relay and Current-Transformer Connection for Zero-Sequence Ground Relay

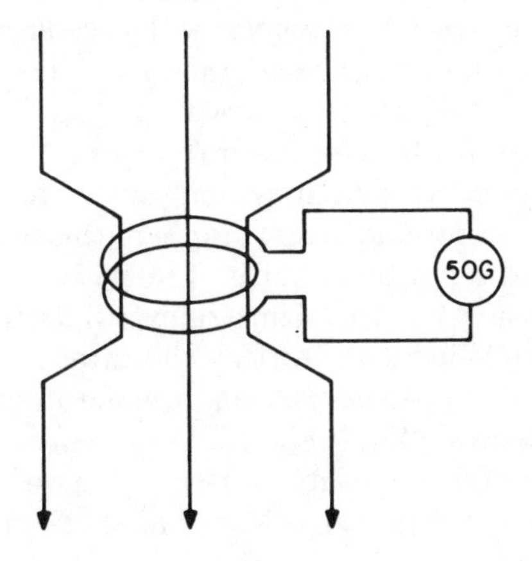

transformer combination should be tested before being applied in order to be assured of predictable performance. This scheme is widely applied on 5 kV and 15 kV systems and is also used on large low-voltage systems for improved protection. It is also often used as an economical alternative to differential protection for large motors on grounded systems.

5.3.7.3 Neutral Relaying. A time overcurrent relay, Device 51G, connected to a current transformer located in the grounded neutral of a transformer or generator, provides a convenient, low-cost method of detecting ground faults. Since only ground-fault currents will flow in this relay, it can be set to operate on very low values of current. This scheme is widely applied on 5 kV and 15 kV systems where low-resistance grounding is frequently used, and the fault current may be as low as 200 A. The relay can be set to minimum values of current pickup and time delay, which will be selective with the feeder ground-fault relays.

This scheme is also used on solidly grounded 480 V three-phase three-wire and four-wire systems. In four-wire systems the current transformer must be connected in the main bonding jumper located between the ground bus and the neutral bus to ensure that unbalanced load currents do not flow in the relay. The relay can be set to operate on low values of current and time delay that will be selective with the feeder ground-fault devices. When there is no ground-fault device on the feeders, the relay pickup and time delay should be set to be selective with the trip characteristic of the largest feeder breaker.

Another form of neutral relaying is used when the neutral resistor is sized to limit the ground-fault current to a few amperes, that is, 1-10 A. This method, known as high-resistance grounding, limits the damage at the fault site such that the fault is not automatically cleared, but is detected and an alarm initiated. An overvoltage relay, Device 59G, is used, which is connected across the resistor and senses the voltage that will appear across it only during ground faults. Additional information on grounding method can be found in ANSI/IEEE Std 142-1982 [19].

5.3.8 Synchronism-Check and Synchronizing Relays. The synchronism-check relay is used to verify when two ac circuits are within the desired limits of frequency and voltage phase angle to permit them to operate in parallel. These relays should be employed in switching applications on systems known to be normally paralleled at some other location so that they are only checking that the two sources have not become electrically separated or displaced by an unacceptable phase angle. The synchronizing relay, on the other hand, monitors two separate systems that are to be paralleled, automatically initiating switching as a function of the phase-angle displacement, frequency difference (beat frequency), and voltage deviation, as well as the operating time of the switching equipment, to accomplish interconnection when conditions are acceptable. An example of this application is a plant generating its own power with a parallel-operated tie with a utility system. The utility end of the tie line must have synchronizing relays that will check conditions on both systems prior to paralleling and initiate the interconnection so as to avoid any possibility of tying with the industrial plant generators out of phase.

5.3.9 Pilot-Wire Relays [33], [48], [56]. The relaying of tie lines, either between the industrial system and the utility system or between major load centers within the industrial system, often presents a special problem. Such lines should be capable of carrying maximum emergency load currents for any length of time, and they should be removable from service quickly should a fault occur. A type of differential relaying called pilot-wire relaying responds very quickly to faults in the protected line. Faults are promptly cleared, which minimizes line damage and disturbance to the system; yet the relay is normally unresponsive to load currents and to currents flowing to faults in other lines and equipment. The various types of pilot-wire relaying schemes all operate on the principle of comparing the conditions at the terminals of the protected line, the relays being connected to operate if the comparison indicates a fault in the line. The information necessary for this comparison is transmitted between terminals over a pilot-wire circuit, hence the designation of this type of relaying. Because, like all differential schemes, it is completely and inherently balanced within itself and completely selective, the pilot-wire relay scheme does not provide protection for faults of the adjacent station bus or beyond it. The new static wire pilot differential relay [48] operates on the circulating current principle. It offers an ideal form of cable differential protection for industrial plants.

5.3.10 Voltage Relays. Voltage relays function at predetermined values of voltage, which may be overvoltage, undervoltage, a combination of both, voltage unbalance (comparing two sources of voltage), reverse phase voltage, and excess negative-sequence voltage (that is, single phasing of a three-phase system). Plunger-type, induction-type, or solid-state-type relays are available. Adjustments of pickup or dropout voltage and operation timing are usually provided in these relays. Plunger-type relays are usually instantaneous in operation, although bellows, dash pots, or other delay means can be provided. The time-delay feature is often required in order that transient voltage disturbances will not cause nuisance relay operation. Some typical voltage relay applications are as follows:

(1) *Over- or Undervoltage Relays*

(a) capacitor switching control

(b) ac and dc overvoltage protection for generators

(c) automatic transfer of power supplies

(d) load shedding on undervoltage

(e) undervoltage protection for motors

(2) *Voltage Balance Relays*. Blocking the operation of a voltage-controlled current relay when a potential transformer fuse blows

(3) *Reverse-Phase Voltage Relays*

(a) detection of reverse phase connections of interconnecting circuits, transformers, motors, or generators

(b) prevention of any attempt to start a motor with one phase of the system open

(4) *Negative-Sequence Voltage Relays*. Detection of single phasing, damaging phase voltage unbalance, and reversal of phase rotation of supply for protection of rotating equipment.

5.3.11 Distance Relays [36], [56]. Distance relays comprise a family of relays that measure voltage and current, and the ratio is expressed in terms of impedance. Typically, this impedance is an electrical measure of the distance along a transmission line from the relay location to a fault. The impedance can also represent the equivalent impedance of a generator or large synchronous motor when a distance relay is used for loss-of-field protection.

The measuring element is usually instantaneous in action, with time delay provided by a timer element so that the delay, after operation of a given measuring element, is constant. In a typical transmission-line application three measuring elements are provided. The first operates only for faults within the primary protection zone of the line and trips the circuit breaker without intentional time delay. The second element operates on faults not only in the primary protection zone, but also in one adjacent or backup protection zone and initiates tripping after a short time delay. The third element is set to include a still more remote zone and to trip after a longer time delay. These relays have their greatest usefulness in applications where selective *stepped* operation of circuit breakers in series is essential, where changes in operating conditions cause wide variations in magnitudes of fault current, and where load currents may be large enough, in comparison with fault currents, to make overcurrent relaying undesirable.

The three main types of distance relay and their usual applications are as follows:

(1) *Impedance-Type*. Phase-fault relaying for moderate-length lines.

(2) *Mho-Type*. Phase-fault relaying for long lines or where severe synchronizing power surges may occur. Generator or large synchronous motor loss-of-field relaying.

(3) *Reactance-Type*. Ground-fault relaying and phase-fault relaying on very short lines and lines of such physical design that high values of fault arc resistance are expected to occur and affect relay *reach*, and on systems where severe synchronizing power surges are not a factor.

5.3.12 Phase-Sequence or Reverse-Phase Relays. Reversal of the phase rotation of a motor may result in costly damage to machines, long shutdown, and lost production. Important motors are frequently equipped with phase-sequence or reverse-phase relay protection. If this relay is connected to a suitable potential source, it will close its contacts whenever the phase rotation is in the opposite direction. It also can be made sensitive to unbalanced voltage or undervoltage conditions (see 5.3.10).

5.3.13 Frequency Relays [43]. Frequency relays sense under- or overfrequency conditions during system disturbances. Most frequency relays have provision for adjustment of operating frequency and voltage. The speed of operation depends on the deviation of the actual frequency from the relay setting. Some frequency relays operate if the frequency deviates from the set value. Others are actuated by the rate at which the frequency is changing. The usual application of this type of relay is to selectively drop system load based on the frequency decrement in order to restore normal system stability.

5.3.14 Temperature-Sensitive Relays. Temperature-sensitive relays usually operate in conjunction with temperature-detecting devices, such as resistance temperature detectors or thermocouples located in the equipment to be protected, and are used for protection against overheating of large motors (above 1500 hp), generator stator windings, and large transformer windings.

For generators and large motors, several temperature detectors are usually embedded in the stator windings, and the hottest (by test) reading detector is connected into the temperature relay bridge circuit. The bridge circuit is balanced at this temperature, and an increase in winding temperature will increase the resistance of the detector, unbalance the bridge circuit, and cause relay operation. Transformer temperature relays operate in a similar manner from detecting devices set in winding *hot spot* areas. Some relays are provided with a 10 °C differential feature that will prevent re-energizing of the equipment until the winding temperature has dropped 10 °C.

5.3.15 Pressure-Sensitive Relays. Pressure-sensitive relays used in power systems respond either to the rate of rise of gas pressure (sudden pressure relay) or to a slow accumulation of gas (gas detector relay), or a combination of both. Such relays are valuable supplements to differential or other forms of relaying on power, regulating, and rectifier transformers.

A sudden rise in the gas pressure above the liquid-insulating medium in a liquid-filled transformer indicates that a major internal fault has occurred. The *sudden-pressure* relay will respond quickly to this condition and isolate the faulted transformer. Slow accumulation of gas (in conservator tank-type transformers) indicates the presence of a minor fault, such as loose contacts, grounded parts, short-circuited turns, leakage of air into the tank, etc. The gas-detector relay will respond to this condition and either sound an alarm or isolate the faulted transformer.

5.3.16 Replica-Type Temperature Relays. Thermally activated relays respond to heat generated by current flow in excess of a certain predetermined value. The input to the relay is normally the output of the current transformer whose ratio should be carefully selected to match the available relay ratings. Many varied types are available, the most common being the bimetal strip and the melting alloy types. The relay should be checked for variations in operating characteristics as a function of ambient temperature.

Since the operating characteristics of this thermal *replica*-type relay closely match general-purpose motor heating curves in the light and medium overload areas, they are used almost exclusively for overload protection of motors up to 1500 hp.

5.3.17 Auxiliary Relays. Auxiliary relays are used in protection schemes whenever a protective device cannot in itself provide all the functions necessary for satisfactory fault isolation. This type of relay is available with a wide range of coil ratings, contact arrangements, and tripping functions, each suited for a particular application. Some of the most common applications of auxiliary relays are circuit breaker lockout, circuit breaker latching, targeting, multiplication of contacts, timing, circuit supervision, and alarming.

5.3.18 Direct-Acting Trip Devices for Low-Voltage Power Circuit Breakers

5.3.18.1 Electromechanical Trip Devices. Low-voltage power circuit breakers were for many years equipped with electromechanical series trip devices as the basic form of protection. State-of-the-art technology using solid-state devices has replaced electromechanical trip devices on low-voltage breakers; however, these older devices may still be available on replacement breakers. The electromechanical series trip is of the moving armature type, using a heavy copper coil carrying the full load current to provide the magnetizing force. Overload protection is provided by a dashpot restraining the movement of the armature. Short circuit protection is provided when the magnetic force suddenly overcomes a separate restraint spring. A separate adjustable unit is required for each trip rating.

Long-time, short-time, and instantaneous overcurrent tripping are available on these trip devices in any combination of the three forms of protection, each type having adjustable characteristics. These units do have some inherent disadvantages. The trip point will vary depending on age and severity of duty and the units have a limited calibration range. Because the trip characteristic curve of electromechanical devices has a very inverse shape with a somewhat unpredictable broad operating band (Fig 58), coordination of tripping with other devices is difficult.

5.3.18.2 Solid-State Trip Devices. In contrast to electromechanical devices, solid-state trip devices operate from a low-current signal generated by current sensors or current transformers in each phase. Output from the sensors is fed into the solid-state trip unit, which evaluates the magnitude of the incoming signal with respect to its calibration setpoints and acts to trip the circuit breaker if preset values are exceeded. In addition to phase protection, the solid-state trip device is available with integral ground-fault trip protection.

Solid-state trip devices are more accessible on the circuit breaker than are electromechanical trip devices and are much easier to calibrate since low values of currents can be fed through the device to simulate the effect of an actual fault-current signal. Special care or provisions are sometimes necessary to guarantee predictable operation when applying solid-state trip devices to loads having other than the pure sinusoidal current wave shapes. Vibration, temperature, altitude, and duty cycle have virtually no effect on the calibration of solid-state trip devices. Excellent reliability, therefore, is generally possible. The most important advantage of solid-state trip devices is the shape of the trip characteristic curve, which is essentially a straight line throughout its working portion (Fig 59). These devices have a very narrow and predictable operating band, which enables several such devices to be selectively coordinated.

5.3.19 Fuses [41], [52], [54]. The term *fuse* is defined by ANSI/IEEE Std 100-1984 [19] as "an overcurrent protective device with a circuit-opening fusible part that is heated and severed by the passage of overcurrent through it." From this definition it can be seen that a fuse is intended to be responsive to current and provide protection against system overcurrent conditions.

5.3.19.1 Power Fuses (over 600 V). Power fuses rated over 600 V are either of the current-limiting type or the expulsion type. The current-limiting type are categorized as either general purpose fuses (E-rated or non-E-rated) or R-rated fuses. These differ in that the general purpose fuse is designed to operate over a wider range of overcurrent levels than is the R-rated fuse, which is only intended to interrupt high-magnitude fault currents. Most general purpose current-limiting fuses and expulsion-type fuses comply with E rating requirements, as defined in ANSI C37.46-1981 [2], which are as follows:

(1) A fuse must be capable of carrying its rated current continuously.

(2) A fuse rated 100 A or less must open within 300 s at an rms current within the range of 200–240% of its continuous rating.

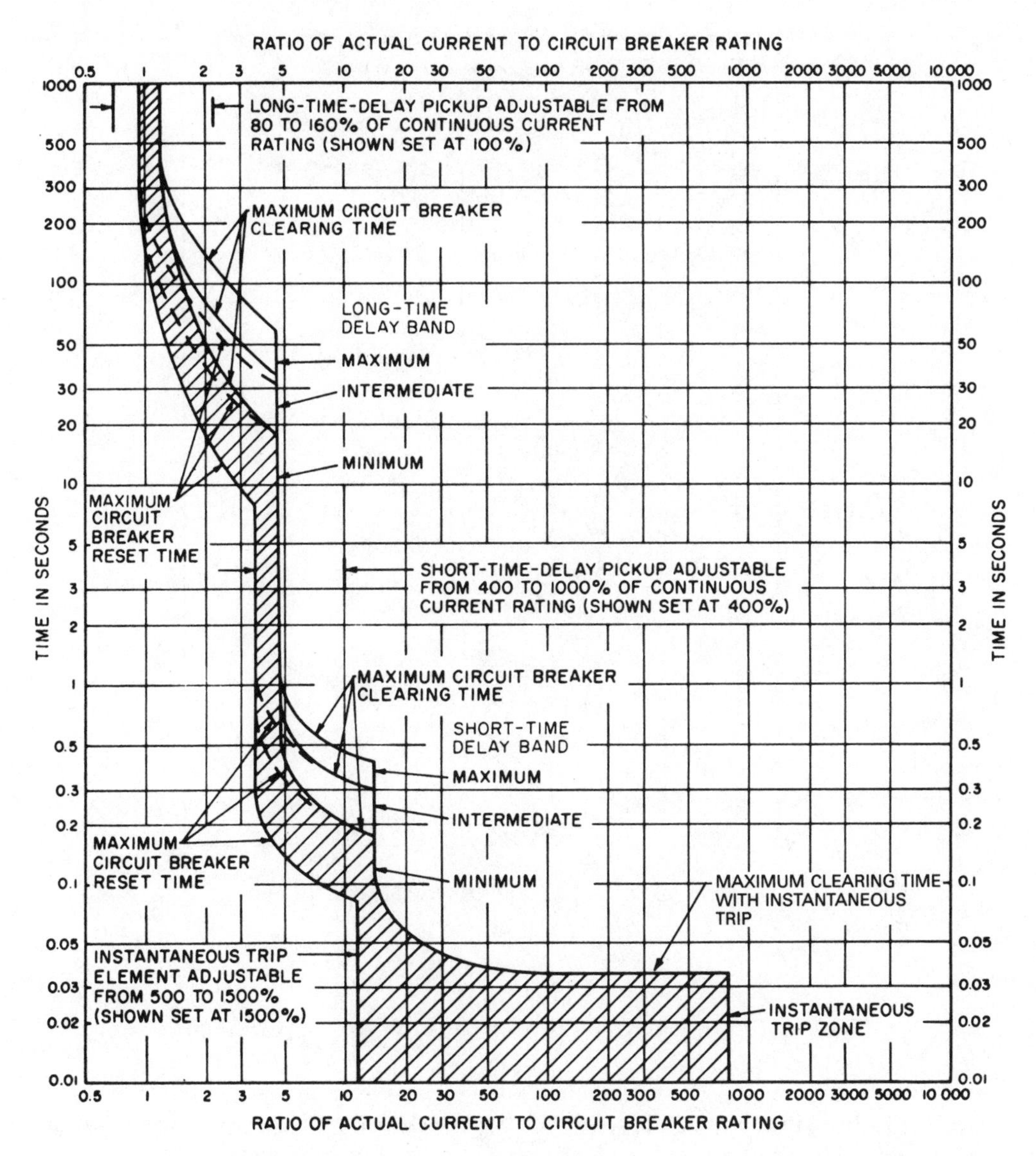

Fig 58
Typical Time-Current Plot for Electromechanical Trip Devices

(3) A fuse rated above 100 A must open within 600 s at an rms current within the range of 220–264% of its continuous rating.

5.3.19.1.1 Current-Limiting Power Fuses [41], [47], [52]. This type of fuse is designed so that the melting of the fuse element introduces a high arc resistance into the circuit in advance of the prospective peak current of the first half-cycle. If the fault current magnitude is sufficiently high, the arc voltage that rapidly escalates will forcibly limit the current to a peak value that is lower than the prospective

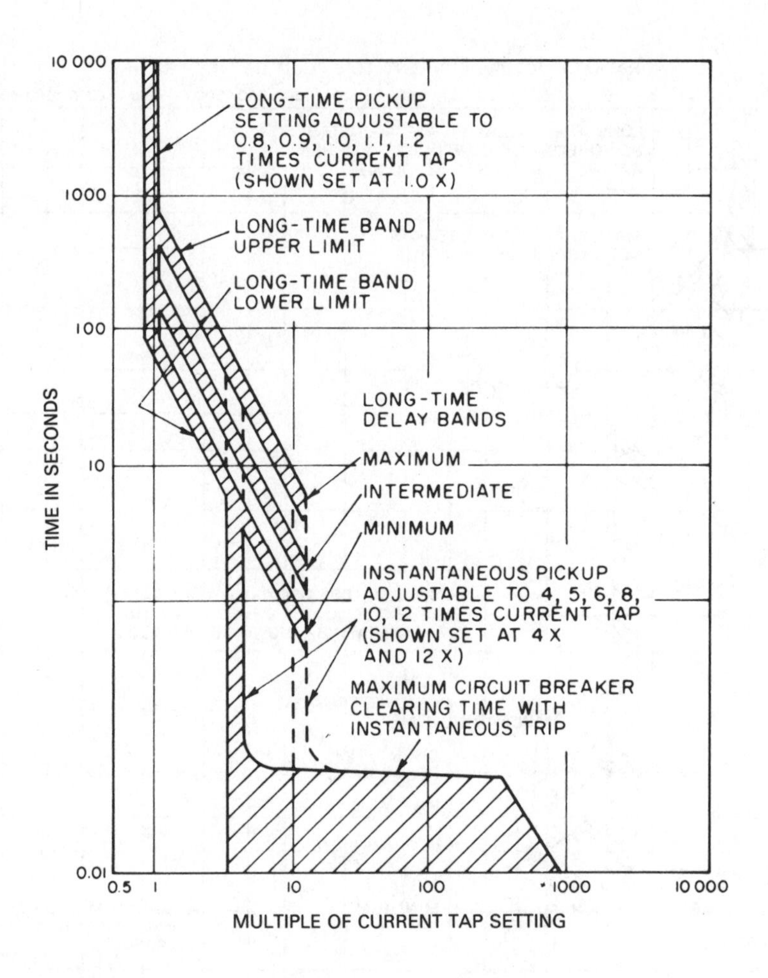

Fig 59
Typical Time-Current Plot for Solid-State Trip Devices

peak. This reduced peak value is referred to as the peak let-through current, which may be a small fraction of the peak current that would flow without the current-limiting action of the fuse.

A general purpose current-limiting fuse is defined as a fuse capable of interrupting all currents from the rated maximum interrupting current down to the current that causes melting of the fusible element in one hour. This type of fuse is not intended to provide protection against low-magnitude overload currents, since it can reliably interrupt only currents above approximately twice its continuous rating for E-rated fuses and usually above approximately three times its continuous rating for non-E-rated fuses. Typical applications are for the protection of power transformers, potential transformers, and feeder circuits. A typical time-current characteristic for this type of fuse is shown in Fig 60. Because of the fuse's nearly

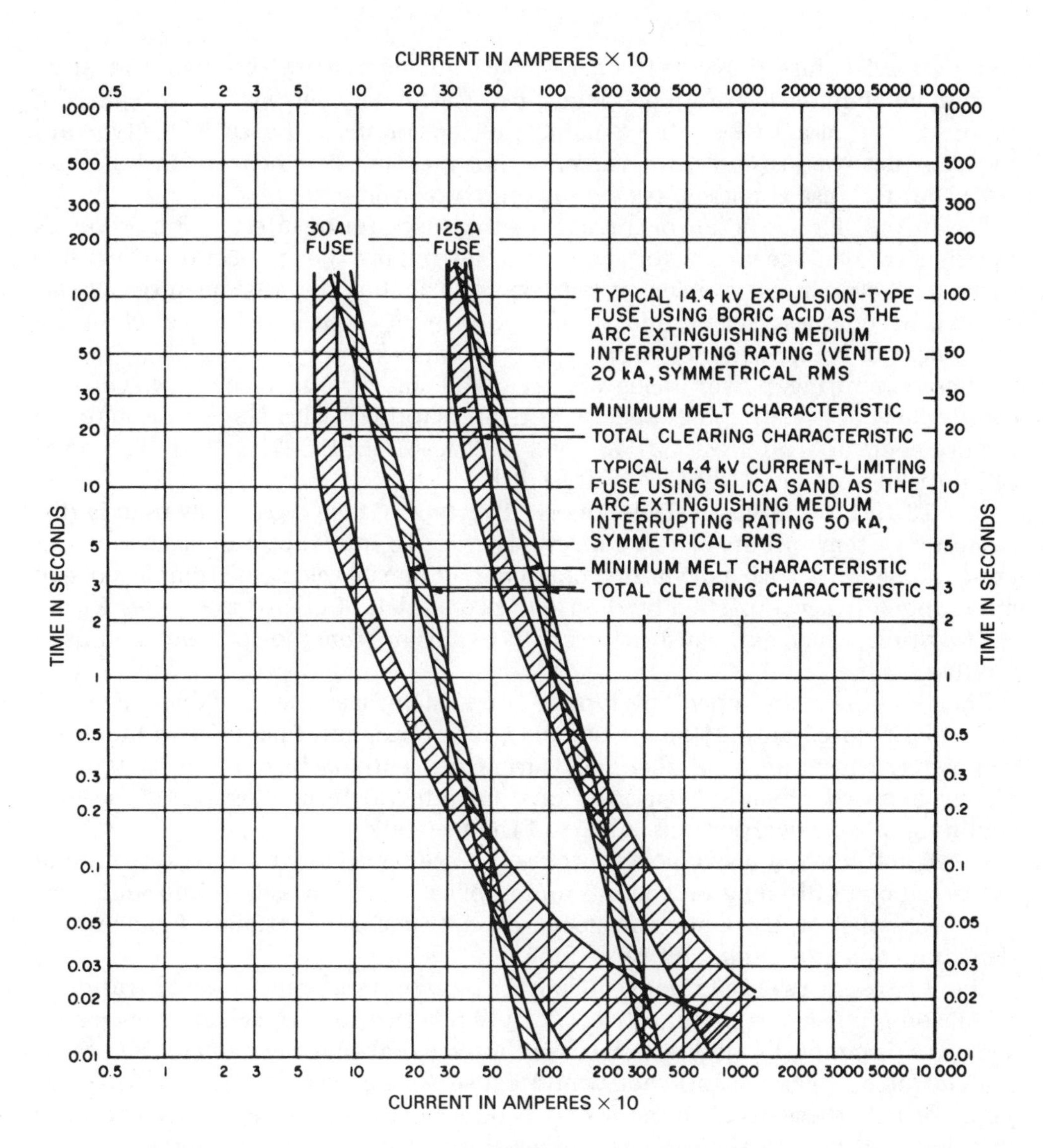

Fig 60
Time-Current Characteristic Curves Showing the Difference Between Boric-Acid Expulsion-Type and Current-Limiting Fuses

straight and vertical characteristic, it can be difficult to coordinate with overcurrent relays.

Current-limiting fuses of the R-rated type are most commonly applied in motor starters utilizing contactors that are not capable of interrupting high magnitudes of fault current. The "R" designation is not related to the continuous-current rating, although each fuse does have a permissible continuous current that is published by the manufacturer. The R number is $\frac{1}{100}$ of the amperes required to open the fuse in about 20 s. The fuse provides the necessary short-circuit protection, but must be used in combination with an overload protective device to sense lower values of overcurrent that are within the capability of the contactor. Fuses of this type are generally designed to interrupt currents that melt the fuse element in less than 100 s, but the fuse is not self-protecting on lower overcurrents.

The current-forcing action of current-limiting fuses during interruption produces transient overvoltage on the system, which may require the application of suitable surge-protective apparatus for proper control. The duty imposed on surge arresters can be relatively severe and should be carefully considered in selecting the equipment to be applied (see [51]).

Current-limiting power fuses are available in various frequency, voltage, continuous-current-carrying capacity, and interrupting ratings that conform to the requirements of ANSI/IEEE C37.40-1981 [10], ANSI/IEEE C37.41-1981 [11], ANSI C37.46-1981 [2], and ANSI C37.47-1981 [3].

5.3.19.1.2 Expulsion-Type Fuses. This type of fuse is generally used in distribution system cutouts or disconnect switches. To interrupt a fault current, an arc-confining tube with a deionizing fiber liner and fusible element is employed. Arc interruption is accomplished by the rapid production of pressurized gases within the fuse tube, which extinguishes the arc by expulsion from the open end or ends of the fuse.

Enclosed, open, and open-link types of expulsion fuses are available for use as cutouts. Enclosed cutouts have terminals, fuse clips, and fuse holders mounted completely within an insulating enclosure. Open cutouts have these parts completely exposed. Open-link cutouts have no integral fuseholders, and the arc-confining tube is incorporated as part of the fuse link.

Fused cutouts and disconnect switches are used outdoors for the protection of industrial plant distribution systems and applications such as line-fault and overload protection of distribution feeder circuits, transformer primary fault protection, and capacitor-bank fault protection.

Since gases are released rapidly during the interruption process, the operation of expulsion-type fuses is comparatively noisy. When they are applied in an enclosure such as a disconnect switch, special care should be taken to vent any ionized gases that might be released and that would cause a flashover between internal live parts. Despite these disadvantages, expulsion-type fuses are often applied because they have an inverse time-current characteristic that is more compatible with standard overcurrent relays (Fig 60).

5.3.19.2 Low-Voltage Fuses (600 V and Below). These fuses are covered by the following standards: ANSI C97.1-1972 [7] and by the ANSI/UL 198 series [24], [25], [26], [27], [28], [29].

Plug fuses are of three basic types, all rated 125 V or less to ground and up to 30 A maximum. Although they have no interrupting rating, they are subjected to one ac short-circuit test with an available current of 10 000 A. The three types are Edison base with no time delay in which all ratings are interchangeable; Edison base with a time delay and interchangeable ratings; and type S base available in three noninterchangeable current ranges: 0-15 A, 16-20 A, and 21-30 A. These last two types normally have a time-delay characteristic of at least 12 s at 200% of their rating, although time-delay plug fuses are no longer required by ANSI/NFPA 70-1987 [22] (National Electrical Code [NEC]).

Cartridge fuses may be either renewable or nonrenewable. Nonrenewable fuses are factory assembled and should be replaced after operating. Renewable fuses can be disassembled and the fusible element replaced. Renewal elements are usually designed to give a greater time delay than ordinary nonrenewable fuses, and in some designs the delay on moderate overcurrents is considerable. The renewable-type fuse is not available in the higher interrupting ratings.

(1) *Noncurrent-Limiting Fuses (Class H)*. These fuses interrupt overcurrents up to 10 000 A but do not limit the current that flows in the circuit to the same extent as do recognized current-limiting fuses. As a general rule, they should only be applied in circuits where the maximum available fault current is 10 000 A and the protected equipment is fully rated to withstand the peak available fault current associated with this fault duty, unless such fuses are specifically applied as part of an equipment combination that has been type-tested and designed for use at higher available fault current levels.

(2) *Current-Limiting Fuses*. Current-limiting fuses are intended for use in circuits where available short-circuit current is beyond the withstand capability of downstream equipment or the interrupting rating of ordinary fuses or standard circuit breakers. An alternating current-limiting fuse is a fuse which safely interrupts all available currents within its interrupting rating and, within its current-limiting range, limits the clearing time at rated voltage to an interval equal to or less than the first major or symmetrical current loop duration and limits peak let-through current to a value less than the peak current that would be possible with the fuse replaced with a solid conductor of the same impedance. A current-limiting fuse, therefore, places a definite ceiling on the peak let-through current and thermal energy, providing equipment protection against damage from excessive magnetic stresses and thermal energy.

These fuses are widely used in motor starters, fused circuit breakers, and fused switches of motors and feeder circuits for protection of busway and cable.

(3) *Let-Through Considerations*. Figure 61 illustrates typical current-limiting fuse operating characteristics during a high-fault-current interruption. In applications involving high available fault currents, the operating characteristics of the current-limiting fuse limits the actual current that is allowed to flow through the circuit to a level substantially less than the *prospective* maximum. The peak let-through current of a current-limiting fuse is the instantaneous peak value of current through the fuse during fuse opening. The let-through I^2t of a fuse is a measure of the thermal energy developed throughout the entire circuit during clearing of the fault. Both values are important in evaluating fuse performance and can be determined from peak let-through and I^2t curves supplied by the fuse manufacturer. A let-through current value considerably less than the available fault current will greatly reduce the magnetic stresses (which increase as the square of the current) and thus reduce fault damage in protected equipment. In some cases it becomes possible to use components (that is, motor starters, disconnect switches, circuit breakers, and bus duct) in the system that have fault capabilities much less than the maximum fault current available. The low peak let-through current and

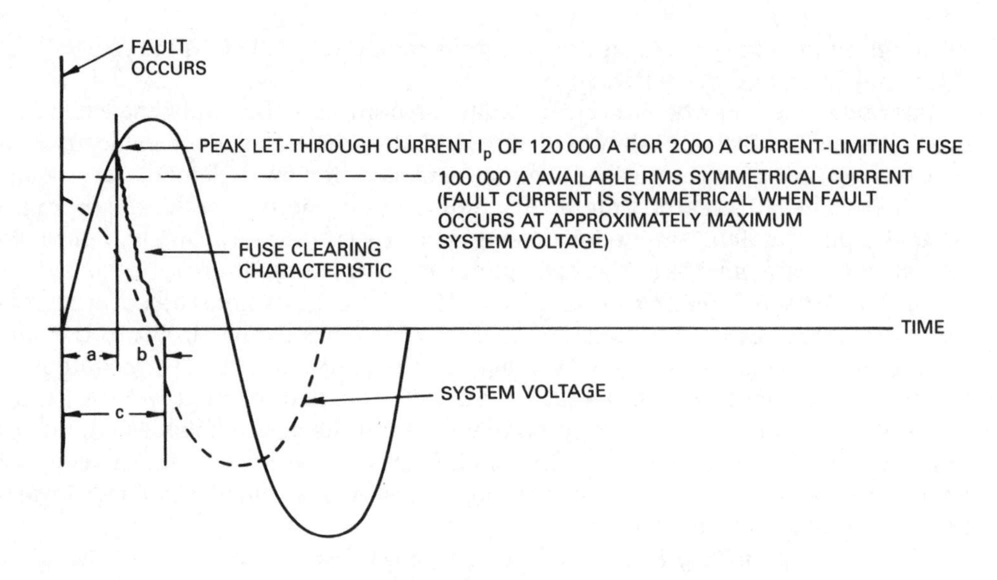

(a) Fault Occurring at Peak Voltage

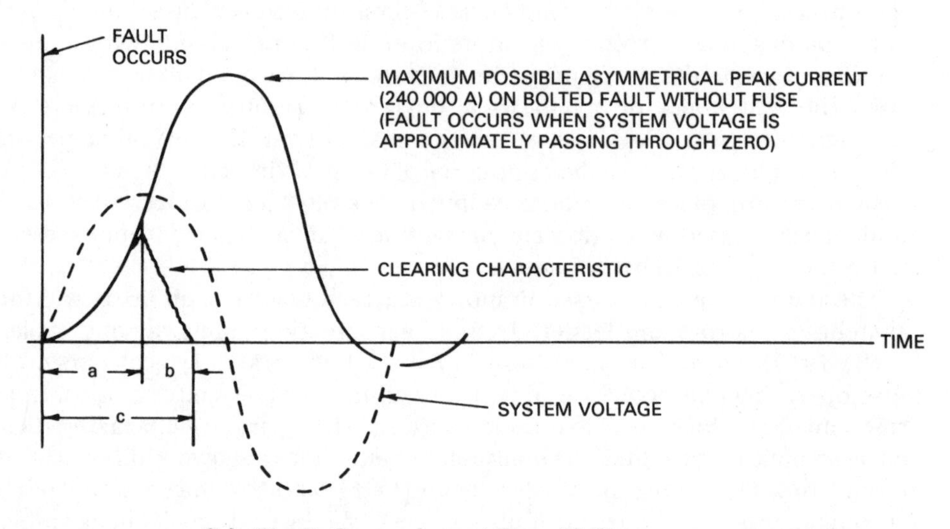

(b) Fault Occurring at Zero Voltage

a — Melting Time
b — Arcing Time
c — Total Clearing Time

Fig 61
Typical Current-Limitation Characteristics Showing Peak Let-Through and Maximum Prospective Fault Current As a Function of the Time of Fault Occurrence (100 kA Available rms Symmetrical Current)

I^2t levels can be achieved with current-limiting fuses because of their extremely fast (often less than one quarter-cycle) speed of response when subjected to high-fault current.

The speed of response is governed by fuse design. For highest speed, silver links with special configurations surrounded by quartz sand are used.

Peak let-through current values alone cannot determine the comparable effectiveness of current-limiting fuses. The product of the total clearing time and the effective value of the let-through current squared I^2t, or thermal energy, should be considered as well.

The melting I^2t of a fuse does not vary with voltage. However, arcing I^2t is voltage-dependent and the arcing I^2t at 480 V, for example, will not be as great as that at 600 V.

(4) *Dual-Element or Time-Delay Fuses*. A dual-element fuse has current-responsive elements of two different fusing characteristics in series in a single cartridge. The fuse is one-time in operation, and the fast-acting element responds to overcurrents that are in the short-circuit range. The time-delay element permits short-duration overloads, but melts if these overloads are sustained. The most important application for these fuses is motor and transformer protection. They do not open on motor starting or transformer magnetizing inrush currents, but protect the motor and branch circuits from damage by sustained overloads.

(5) *Fuse Standards (Cartridge)*. Cartridge fuses differ in dimensions according to voltage and current ratings. They have ferrule contacts in ratings of 60 A or less and knife-blade contacts in larger ratings. Cartridge fuses of varying types and characteristics have been classified by the ANSI/UL 198 series [24], [25], [26], [27], [28], [29] into the following classes.

(a) *Miscellaneous cartridge fuses*. These fuses are not intended for use in branch circuits but rather for use in control circuits, special electronic or automotive equipment, etc. To be UL-listed they must not be manufactured in the same dimensions as UL classes G, H, J, K, or L. They have ratings of 125, 250, 300, 500, and 600 V, and are applied in accordance with the NEC [22].

(b) *Class H fuses*. These fuses may be renewable or nonrenewable and are generally of zinc-link construction. They are rated and sized up to 600 A at either 250 V or less or 600 V or less. They may or may not be of dual-element construction, but if labeled as *time-delay* they must have a minimum delay of 10 s at 500% of rating. These fuses have no interrupting rating, but have been tested on an available alternating current of 10 000 A and are generally used where fault currents do not exceed this magnitude. The label shows neither the class letter nor any interrupting rating. Class H fuses are often referred to as NEC [22] fuses, and the nonrenewable class H fuse is sometimes referred to as a one-time fuse, although this may describe other fuse types as well. Renewable fuses are losing popularity because of their limited and uncertain interrupting ability and the dangers of *double-linking* or insecurely fastening renewal links.

(c) *Class K high-interrupting-capacity fuses*. These fuses are manufactured in identical physical sizes as class H fuses, with which they are interchangeable, and, therefore, cannot be labeled with the words *current-limiting*. However, they have been tested at various high available fault-current levels up to their maximum labeled ratings, which may be either 50 000, 100 000, or 200 000 A rms. Thus fuses in this class are referred to as high-interrupting-capacity fuses. Class K fuses must meet specified maximum values of instantaneous peak let-through currents and I^2t energy let-through maximum values for each physical case size. Each fuse label bears the ac interrupting rating, the class and subclass, and if the fuse meets the time-delay requirement of at least 10 s at 500% rating, it may be labeled as time-delay or with the letter D. According to ANSI/UL 198D-1982 [26], class K fuses are available in three distinct subclasses identified as class K1, class K5, and class K9. Class K1 uses have the lowest maximum values for peak let-through currents and I^2t; class K9 have the highest maximum values; and the class K5 values are between the K1 and K9 values. Most earlier class K9 types have been modified and now have class K5 characteristics.

(d) *Class R current-limiting fuses*. These fuses are a nonrenewable cartridge-type current-limiting fuse also manufactured to class H dimensional standards and have a 200 000 A rms symmetrical interrupting rating. The R designator signifies the fact that the fuse is built with a rejection feature that consists of grooves or notches provided in either the fuse ferrule or blade, depending on the size involved. Equipment rated and approved only for use with fuses having the current-limiting characteristics of class R fuses is then provided with fuse attachment hardware that will only permit the installation of the notched fuses. Since this rejection feature eliminates the possibility of interchanging a current-limiting fuse with a noncurrent-limiting fuse, the Class R fuses are labeled with the words *current-limiting*. These Class R fuses are available in either the single- or dual-element construction. The fuses are available in two subclasses identified as RK1 and RK5, which denote the fact that the fuses have let-through characteristics corresponding to class K1 and K5 fuses, respectively. Since equipment that is approved for class R service is always rated to withstand the higher let-through conditions of RK5 fuses, either the RK1 or the RK5 fuse can be safely applied.

(e) *Class J current-limiting fuses*. These fuses are manufactured in ratings up to 600 A and in specified dimensions per ANSI/UL 198C-1981 [25], which are noninterchangeable with class H and class K fuses. They are labeled as current-limiting. There is no 250 V or less rating; all are labeled 600 V or less and may be used only in fuseholders of suitable class J dimensions. Each case size has specified maximum values of peak let-through current values and maximum thermal (I^2t) values. Fuses having a time-delay characteristic are available in class J dimensional sizes. UL-listed class J fuses with time delay are available from several manufacturers. Class J fuses have a 200 000 A rms interrupting rating.

(f) *Class L current-limiting fuses*. These are the only UL-labeled fuses available with current ratings in excess of 600 A. Per ANSI/UL 198C-1981 [25], their ratings range from 601 to 6000 A rms, all at 600 V or less. There is no 250 V size. Class L fuses have an interrupting rating of 200 000 A rms and will safely interrupt any overcurrent up to this value. Like class J fuses, each case size or mounting dimension (mounting holes drilled in the blades) has a maximum allowable peak let-through current and I^2t value.

5.3.19.3 Fuse-Selection Considerations. For each fuse classification, the corresponding UL standards specify the following design features that are particularly important to fuse application: current rating, voltage rating, frequency rating, interrupting rating, maximum peak let-through current, and maximum clearing thermal energy, I^2t. Standards also specify maximum opening times at certain overload values such as 135 and 200% of rating and for time-delay qualification, minimum opening time at a specific overload percentage. Within these parameters and from various other overcurrent test data, manufacturers construct time-current curves. Normally such curves are based on available rms currents 0.01 s and above, and on either the average melting, minimum melting, or total clearing time. Caution should be exercised in the use of such curves to be certain that equivalent characteristics are being compared.

A fuse should be selected for voltage, current-carrying capacity, and interrupting rating. When fuses must be coordinated with other fuses or circuit breakers, the time-current characteristic curves, peak let-through curves, and I^2t curves may be useful. The load characteristics will dictate the time-delay performance required of the fuse. If fuses are applied in series in a circuit, it is essential for short-circuit coordination to verify that the clearing I^2t of the downstream fuse during a fault will be less than the melting I^2t of the upstream fuse. Fuse manufacturers publish fuse-ratio tables that provide listings of fuses that are known to operate selectively. Use of these tables permits coordination without the need for detailed analysis, provided the fuses being applied are all of the same manufacturer.

When coordinating an upstream circuit breaker with a downstream fuse, the let-through energy of the fuse (clearing I^2t) must be less than the required amount to release the circuit breaker trip latch mechanism. This is not easily accomplished with many types of circuit breakers. Critical operation occurs in the region for currents greater than the circuit breaker instantaneous trip device pickup for periods of time less than 0.01 s, even though a normal time-current plot would suggest that selective performance exists. Similar problems exist when attempting to coordinate a downstream circuit breaker with an upstream fuse. The clearing time of the circuit breaker can often exceed the minimum melting time of the fuse. Overload coordination for low-magnitude or moderate faults can be established with standard time-current curve overlays (for details see ANSI/IEEE Std 242-1986 [20].

For protection of a downstream circuit breaker with an upstream fuse during high-fault currents, the peak let-through current of the fuse must be compatible with the momentary withstand rating of the circuit breaker. Manufacturers' tables for the selection of fuses to protect circuit breakers are an easy solution provided that such tables are based upon current styles, types, and classes of fuses and circuit breakers. UL now series-tests and issues various combinations of fuses and circuit breakers as submitted by different manufacturers. These UL certifications are preferable to manufacturers' data without third-party certification where available.

Although the proper selection of a fuse to protect a circuit breaker, starter, or cable circuit will generally prevent equipment failure during a fault condition, some apparatus design practices allow damage to bimetals, contacts, and other parts. Unless the combination has been specifically tested and rated as a unit, the application of a fuse of a given interrupting rating in a switch or other fusible device does not confer that rating on the equipment involved. A switch, for example, might not withstand the let-through energy of a current-limiting fuse during certain fault-current conditions. When a combination rating is not available, the rating of the fuse or the device, whichever is less, should be used.

The voltage rating of a fuse should be selected as equal to or higher than the nominal system voltage on which it is used. When applying a high-voltage current-limiting fuse of a given voltage rating on a circuit of a lower voltage rating, consideration should be given to the magnitude and effect of overvoltages that will be induced due to the zero current forcing action of the fuse during the interruption of high-magnitude fault currents. A low-voltage fuse of any labeled voltage rating will always perform satisfactorily on lower service voltages. This is not a problem at 600 V or less.

The current rating of a fuse should be selected so that it clears only on a fault or an overload and not on current inrush. Ambient temperatures and types of enclosures affect fuse performance and should be considered. Fuse manufacturers should be requested to supply correction factors for unusual ambient temperatures.

5.4 Performance Limitations

5.4.1 Load Current and Voltage Wave Shape. The published operating characteristics of all protective relays and trip devices are based on an essentially pure sinusoidal wave shape of current and voltage. Many industrial loads are of such a nature as to produce harmonics in the system current and voltage. This condition is aggravated by the presence of any distribution equipment in the system with nonlinear electrical characteristics. As a result, it is important to understand the nature of the expected system current and voltage as well as the effect that wave-shape distortion might have on the protective devices being applied.

5.4.2 Instrument Transformers [36]. If a protective relay is to operate predictably and reliably, it must receive information that accurately represents conditions that exist on the power system from the circuit instrument transformers. Since current and potential transformers become significantly nonlinear devices under certain conditions, they may not produce an output precisely representative of power system conditions either in wave shape or magnitude. The exact extent of any distortion is a function of the input signal level and transformer burden (total connected impedance) as well as the actual design (accuracy class) of the instrument transformer being applied. Potential transformer performance characteristics are classified by ANSI/IEEE C57.13-1978 [16]. This same standard provides separate classifications for current transformers with regard to burden capability and accuracy for both metering and relaying service. The burden and output requirements of all instrument transformers should be carefully checked against their rating for any relay application to verify that proper relay operation will result. In some cases where a larger burden is encountered or the expected level of fault current is higher than that encompassed by the ANSI/IEEE C57.13-1978 [16] rating structure, it may be necessary to obtain the current transformer saturation curve from the manufacturer in order to analytically establish acceptable performance.

5.5 Principles of Protective Relay Application [38], [40], [50]. (See also ANSI/IEEE C37.91-1985 [13].) Fault-protection relaying can be classified into two groups: primary relaying, which should function first in removing faulted equipment from the system; and backup relaying, which functions only when primary relaying fails.

To illustrate the areas of protection associated with primary relaying, Fig 62 shows the various areas together with circuit breakers that feed each electric element of the system. Note that it is possible to disconnect any piece of faulted equipment by opening one or more circuit breakers. For example, when a fault occurs on the incoming line L_1, the fault is within a specific area of protection (area A) and should be cleared by the primary relays that operate circuit breakers 1 and 2. Likewise a fault on bus 1 is within a specific area of protection, area B, and should be cleared by the primary relaying actuating circuit breakers 2, 3, and 4. If circuit breaker 2 fails to open and the faulted equipment remains connected to the system, the backup protection provided by circuit breaker 1 and its relays must be depended upon to clear the fault.

Figure 62 illustrates the basic principles of primary relaying in which separate areas of protection are established around each system element so that each can be isolated by a separate interrupting device. Any equipment failure occurring within a given area will cause tripping of all circuit breakers supplying power to that area.

Fig 62
One-Line Diagram Illustrating Zones of Protection

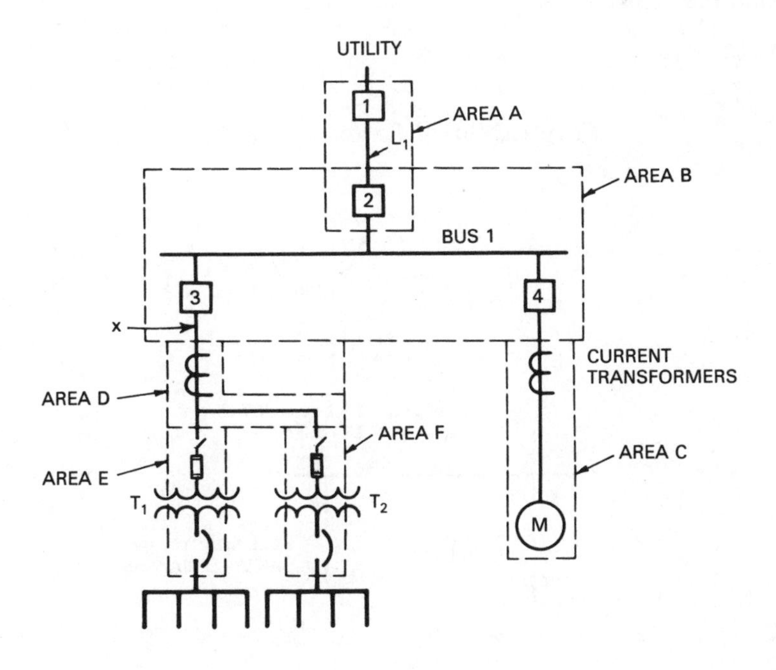

To assure that all faults within a given zone will operate the relays of that zone, the current transformers associated with that zone should be placed on the line side of each circuit breaker so that the circuit breaker itself is a part of two adjacent zones. This is known as *overlapping*. Sometimes it is necessary to locate both sets of current transformers on the same side of the circuit breaker. In radial circuits the consequences of this lack of overlap are not usually very serious. For example, a fault at X on the load side of circuit breaker 3 in Fig 62 could be cleared by the opening of circuit breaker 3 if there were any way to cause it to open circuit breaker 3. Since the fault is between the circuit breaker and the current transformers, the relays of circuit breaker 3 will not see it, and circuit breaker 2 will have to open and consequently interrupt the other load on the bus. When the current transformers are located immediately at the load bushings of the circuit breaker, the amount of circuit exposed to this problem is minimized. The consequences of lack of overlap become more serious in the case of tie circuit breakers between differentially protected buses and bus feeders protected by differential or pilot-wire relaying.

In applying relays to industrial systems, safety, simplicity, reliability, maintenance, and the degree of selectivity required should be considered. Before attempting to design a protective relaying plan, the various elements that make up the distribution system together with the operating requirements should be examined.

5.5.1 Typical Small-Plant Relay Systems [30]. One of the simplest industrial power systems consists of a single service entrance circuit breaker and one distribution transformer stepping the utility's primary distribution voltage down to utilization voltage, as illustrated in Fig 63. There would undoubtedly be several circuits on the secondary side of the transformer, protected by either circuit breakers or combination fused switches.

Fig 63
Typical Small Industrial System

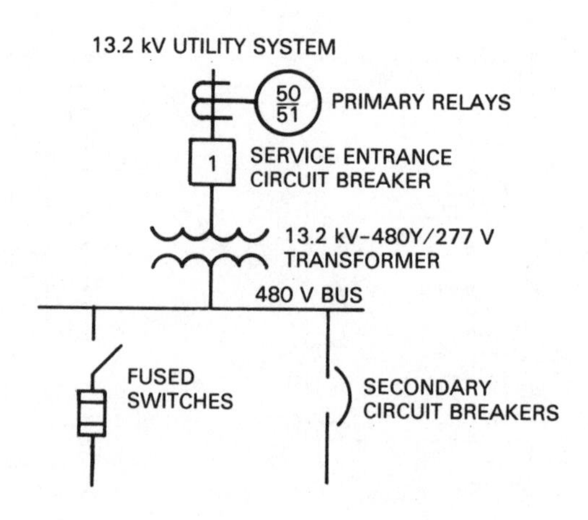

Protection for the feeder circuit between the incoming line and the devices on the transformer secondary would normally consist of conventional overcurrent relays, Device 51. Preferably the relays should have the same time-current characteristics as the relays on the utility system, so that for all values of fault current the local service entrance circuit breaker can be programmed to trip before the utility supply line circuit breaker. The phase relays should also have instantaneous elements, Device 50, to promptly clear high-current faults.

This simple system provides both primary and backup relay protection. For instance, a fault on a secondary feeder should be cleared by the secondary protective device; however, if this device should fail to trip, the primary relays will trip circuit breaker 1.

This simple industrial system can be expanded by tapping the primary feeder and providing fuse protection on the primary of each distribution transformer as shown in Fig 64.

This provides an additional step or area of protection over the simpler system shown in Fig 63. All secondary feeder faults should be cleared by the secondary circuit breakers as before, while faults within the transformer should now be cleared by the transformer primary fuses. The fuses may also act as backup protection for the faults that are not cleared by the secondary feeder protective devices. Primary feeder faults will, as before, be cleared by circuit breaker 1, and it in turn will act as backup protection for the transformer primary fuses.

Fig 64
System of Fig 63 Expanded by Addition of a Transformer and Associated Secondary Circuits

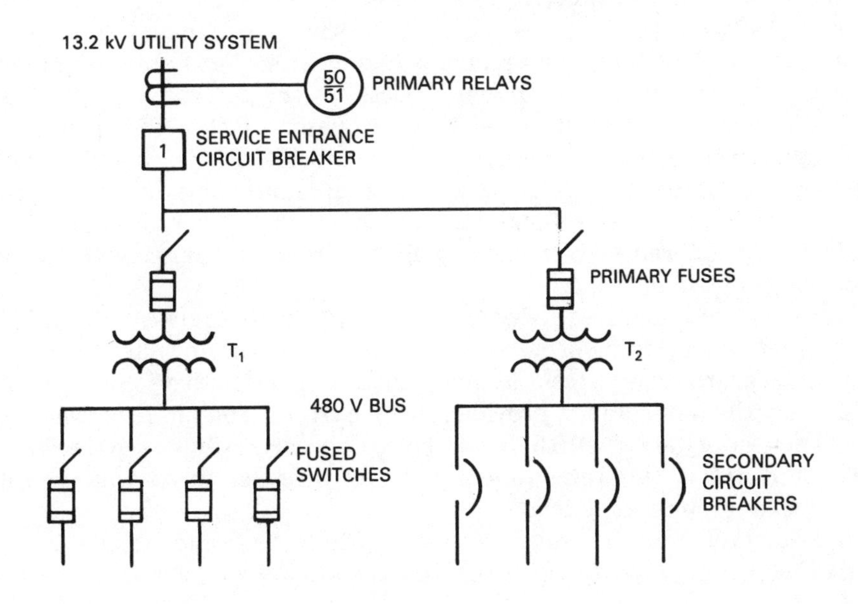

5.5.2 Protective Relaying for a Large Industrial Plant Power System [39], [55], [57], [59]. As an electric system becomes larger, the number of sequential steps of relaying also increases, giving rise to the need for a protective relaying scheme that is inherently selective within each zone of protection. Figure 65 shows the main connections of a large system.

5.5.2.1 Primary Protection. The relay selectivity problem is of great concern to the utilities because their 69 kV supply lines are paralleled and their transformers are connected in parallel with the plant's local generation. The utility company should participate in the selection of relays applied for operation of either incoming circuit breaker in case of a disturbance in the 69 kV bus or transformers. Due to the 69 kV bus tie, a fault in either a bus or a transformer cannot be cleared by the opening of circuit breaker A or B alone but will require the opening of circuit breakers A or B as well as AB and C or D.

Three directionally controlled overcurrent relays (Device 67) should be installed for circuit breakers C and D and connected to trip for current flow toward the respective 69 kV transformer. Directionally controlled overcurrent relays are suggested because their sensitivity is not limited by the magnitude of load current in the normal or nontrip direction. Three overcurrent relays having inverse-time characteristics should be installed at circuit breaker positions A and B as backup protection for faults that may occur on or immediately adjacent to the 69 kV buses. The connection of these overcurrent relays (Device 50/51), shown in Fig 65 as being energized from the output of two current transformers in a summation connection at the incoming 69 kV lines and bus tie, provides the advantage of isolating only the faulted bus section in a shorter time than would be possible if individual circuit breaker relays were used.

The next zones of protection are the 13.8 kV buses 1 and 2. Fault currents are relatively high for any equipment failure on or near the main 13.8 kV buses. For this reason a differential protective relay scheme (Device 87B) is recommended for each bus. Differential relaying is instantaneous in operation and is inherently selective within itself. Without such relaying, high-current bus faults should be cleared by proper operation of overcurrent devices on the several sources. This usually results in long-time clearing since the overcurrent devices have pickup and time settings determined by other than bus fault considerations. General practice is to use separate current transformers with the same ratio and output characteristics for the differential relay scheme. A multicontact auxiliary relay (Device 86B) is used with the differential relays to trip all the circuit breakers connected to the bus whenever a bus fault occurs.

To realize maximum sensitivity, the time-delay ground relays (Device 51N) at the 69–13.8 kV source transformers are connected to the output of current transformers measuring the current in the neutral connection to ground. The 87TN relay is differentially connected to provide sensitive tripping on faults between the transformer secondary and the 13.8 kV main circuit breaker. Unlike the time-delay relays 51N-1 and 51N-2, this relay does not have to be set to coordinate with other downstream ground-fault relays.

Superior protection for the cable tie between buses 2 and 3 is provided by pilot-wire differential relays (Device 87L). In addition to being instantaneous in operation, pilot-wire schemes are inherently selective within themselves and require only two pilot wires if the proper relays are used. Backup protection provided by overcurrent relays should be installed at both ends of the tie line. Nondirectional relays can be applied at circuit breaker M, but at circuit breaker N directional relays are more advantageous since the 10 MVA generator represents a fault source at bus 3.

Separate current transformers are used for the pilot-wire differential relaying to provide reliability and flexibility in the application of other protective devices.

The 9000 hp 13.8 kV synchronous motor is provided with a reactor-type reduced-voltage starting arrangement using metalclad switchgear. Overload protection is provided by a thermal relay (Device 49) whose sensor is a resistance temperature detector (rtd) imbedded in the stator windings. This relay can be used to either trip or alarm. Internal fault protection is provided by the differential relay scheme (Device 87M). Backup fault protection and locked-rotor protection is provided by an overcurrent relay (Device 51/50) applied in all three phases. Undervoltage and reverse phase rotation protection is provided by the voltage-sensitive relay (Device 47) connected to the main bus potential transformers.

Ground-fault protection is provided by the instantaneous zero-sequence current relay (Device 50GS). The current-balance relay (Device 46) protects the motor against damage from excessive rotor heating caused by single phasing or another unbalanced voltage condition. The motor rotor starting winding can be damaged by excessive current due to loss of excitation or suddenly applied loads, which cause the motor to pull out of step. Rotor damage could also result from excessive time for the motor to reach synchronous speed and lock into step. To protect against damage from these causes, loss of excitation (Device 40), pull-out (Device 56PO), and incomplete sequence (Device 48) relays should be provided.

In addition to the protection against (1) internal faults, (2) sustained overloads, (3) undervoltage, and (4) voltage surges (see chapter 4) generators must be protected from overheating caused by external unbalanced low-magnitude faults and from being driven as motors (antimotoring) when the prime mover can be damaged by such operation. The backup overcurrent protection should be capable of detecting an external fault-current condition that corresponds to the minimum level of generator contribution. For the 10 MVA generator connected to bus 3, internal fault detection is provided by the percentage differential relay (Device 87G). The ground-fault current is limited by the 400 A resistor in the generator neutral, and ground-fault detection is afforded by the overcurrent relay (Device 51NG). Loss-of-excitation protection is provided by Device 40, external unbalanced overcurrent (negative-sequence) protection by Device 46, antimotoring protection by Device 32, and backup overcurrent protection by Device 51V/50.

It is good practice for transformers of the size shown on the incoming service, where a circuit breaker is used on both the primary and secondary sides, to install percentage differential relays and inverse characteristic overcurrent relays for backup protection. To prevent operation of the differential relays on magnetizing inrush current when energizing the transformer, the large proportion of currents at harmonic multiples of the line frequency contained in the magnetizing inrush current are filtered out and passed through the restraint winding so that the

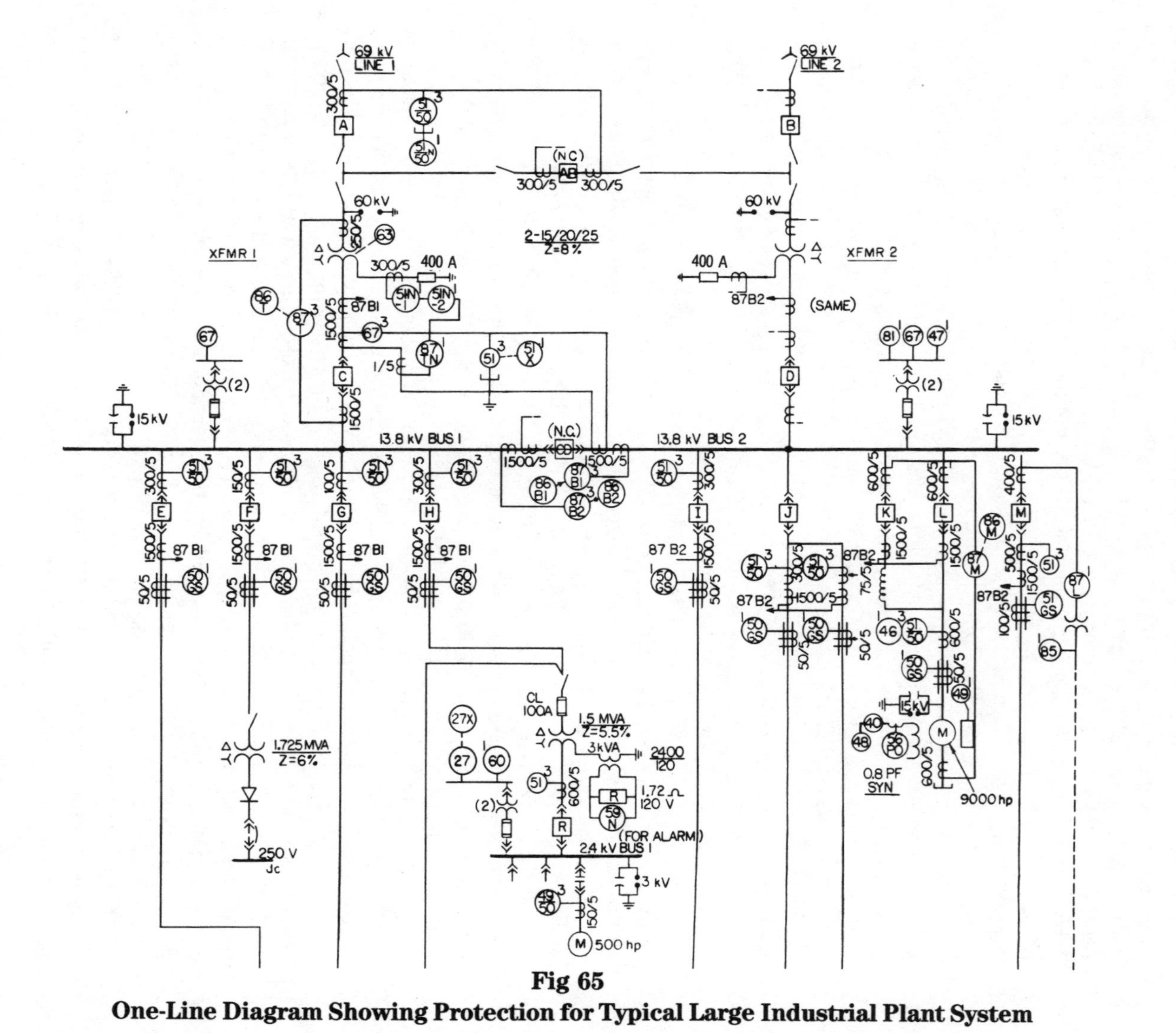

Fig 65
One-Line Diagram Showing Protection for Typical Large Industrial Plant System

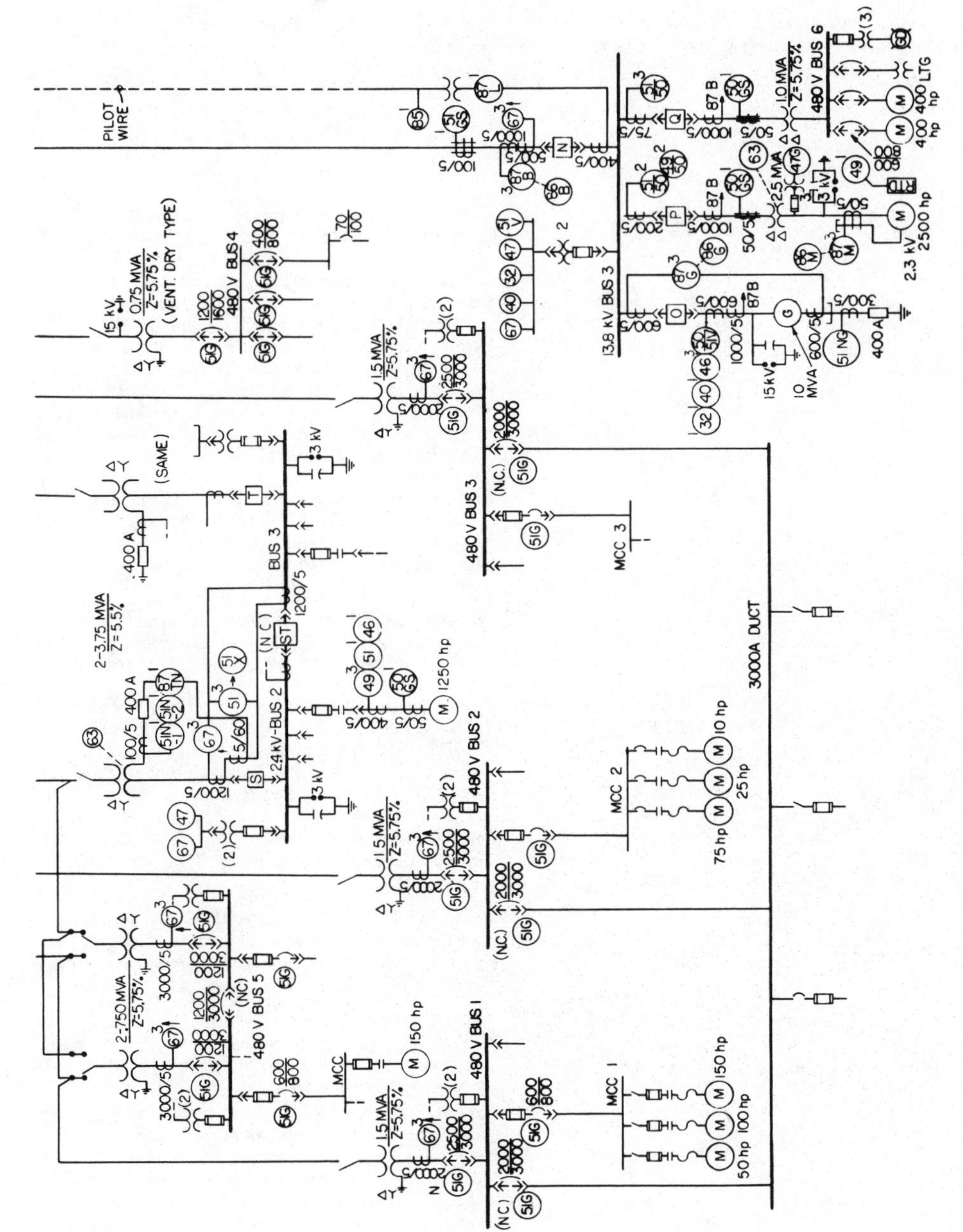

Fig 65 *(continued)*

Protective Device Legend for Fig 65

Location	Device	Description
69 kV supply lines	51/50 51N/50N	Summation phase and ground overcurrent protection for 69 kV bus and backup for transformer differential relaying. Trips circuit-breakers A, AB, and C through auxiliary relay 86T.
15 MVA main transformers	51N-1 51N-2	Backup ground-fault protection for transformer secondary 13.8 kV bus and feeder circuits. 51N-1 trips tie circuit breaker CD; 51N-2 (after time interval) trips circuit breaker.
	63	Sudden pressure relay. Trips circuit breakers A, AB, and C through auxiliary relay 86T.
	67	Directional phase overcurrent as backup to transformer differential. Trips circuit breakers A, AB, and C through auxiliary relay 86T.
	87TN	Sensitive differential protection for ground faults in transformer secondary. Trips circuit breakers A, AB, and C through auxiliary relay 86T.
	87T	Transformer differential protection. Trips circuit breakers A, AB, and C through auxiliary relay 86T.
	86T	Auxiliary trip and lockout relay.
13.8 kV buses 1 and 2	51	Summation phase overcurrent protection as backup for 13.8 kV bus and feeder faults. Trips circuit breakers C and CD through auxiliary relay 51X.
	87B1 87B2	Bus differential. 87B1 trips circuit breakers C, CD, E, F, G, and H through auxiliary relay 86B1. 87B2 trips circuit breakers D, CD, I, J, K, L, and M through auxiliary relay 86B2.
	86B1 86B2	Auxiliary trip and lockout relay.
	81	Underfrequency protection. Initiates load shedding by tripping preselected feeder circuits.
13.8 kV feeders E, F, G, H, I, and J	51/50	Time and instantaneous phase-fault protection. Trips individual feeder circuit breaker.
	50GS	Instantaneous ground-fault protection. Trips individual feeder circuit breaker.
13.8 kV motor control circuit breakers K and L	40 48 56PO	Loss of excitation, incomplete sequence checking and pullout protective relays. Trips circuit breakers K and L.
	46	Current balance relay for single-phase protection. Trips circuit breakers K and L.
	47	Polyphase undervoltage and phase reversal protection. Trips circuit breakers K and L through auxiliary relay 86M.
	49	Overload protection using stator resistance temperature detector. Trips circuit breakers K and L through auxiliary relay 86M.
	50GS	Instantaneous ground-fault protection. Trips circuit breakers K and L through auxiliary relay 86M.
	51/50	Phase overcurrent protection and locked rotor protection. Trips circuit breakers K and L through auxiliary relay 86M.

Protective Device Legend for Fig 65 *(Continued)*

Location	Device	Description
13.8 kV motor control circuit breakers K and L (cont'd)	87M	Motor differential protection. Trips circuit breakers K and L through auxiliary relay 86M.
	86M	Auxiliary trip and lockout relay.
13.8 kV tie line circuit breaker M	51GS	Sensitive ground-fault protection as backup to pilot-wire relaying and backup for bus and feeder ground faults at 13.8 kV bus 3. Trips circuit breaker M.
	51	Phase overcurrent protection as backup to pilot-wire relaying and backup for bus and feeder faults at 13.8 kV buses. Trips circuit breaker M.
	87L	Line differential protection for phase and ground faults using pilot wire. Trips circuit breaker M.
	85	Pilot wire monitoring relay to alarm for open, short-circuited, or grounded pilot wire.
3.75 MVA transformer and 2.4 kV buses 2 and 3	51	Summation phase overcurrent protection for 2.4 kV bus faults and backup protection for feeder faults. Trips circuit breakers S and ST through auxiliary relay 51X.
	51N-1 51N-2	Ground-fault protection for transformer secondary and bus and backup for feeder ground faults. 51N-1 trips tie circuit breaker ST; 51N-2 (after time interval) trips circuit breaker S.
	63	Sudden pressure relay on transformer. Trips circuit breakers S and H.
	67	Directional phase overcurrent protection for transformer faults and 13.8 kV line faults. Trips circuit breaker S.
	87TN	Sensitive differential protection for ground faults in transformer secondary. Trips circuit breakers S and H.
	46	Current balance relay for single-phase protection. Trips contactor.
	47	Polyphase undervoltage and phase sequence protection. Trips circuit breakers S and ST through auxiliary relay 51X.
	49	Replica-type thermal overload protection. Trips contactor.
	50GS	Instantaneous ground-fault protection. Trips contactor.
	51	Overcurrent relay for motor locked rotor protection. Trips contactor.
1.5 MVA transformer and 2.4 kV bus 1	49/50	Replica-type thermal overload protection including instantaneous element for short-circuit protection. Trips contactor.
	51	Phase overcurrent protection. Trips circuit breaker R.
	59N	Sensitive voltage detection of ground faults for high-resistance grounded system. Initiates alarm signal.
	27	Single-phase undervoltage protection. Trips motor contactor through auxiliary relay 27X.
	60	Negative sequence voltage relay detects single phasing of source. De-energizes undervoltage relay 27.

Protective Device Legend for Fig 65 *(Continued)*

Location	Device	Description
13.8 kV tie line, circuit breaker N at 13.8 kV bus 3	67	Directional phase overcurrent protection as backup to pilot wire and backup for bus and feeder faults at 13.8 kV buses. Trips circuit breaker N.
	87L	Line differential protection for phase and ground faults using pilot wire. Trips circuit breaker N.
	85	Pilot-wire monitoring relay to initiate alarm for open, short-circuited, or grounded pilot wire.
	51G	Sensitive ground-fault protection as backup to pilot-wire relaying and backup for bus and feeder faults at 13.8 kV buses. Trips circuit breaker N.
13.8 kV bus 3	87B	Bus differential. Trips circuit breakers N, O, P, and Q through auxiliary relay 86B.
	86B	Auxiliary trip and lockout relay.
10 MVA generator	32	Reverse power or antimotoring protection. Trips circuit breaker O.
	40	Loss of excitation protection. Alarm and subsequent trip of circuit breaker O.
	46	Negative-sequence overcurrent protection for generator due to external unbalanced fault. Trips circuit breaker O.
	51V	Backup overcurrent protection for external three-phase faults. Trips circuit breaker O.
	51NG	Ground-fault protection for generator and backup for differential relays and feeder ground relays. Trips circuit breaker O and field circuit breaker through 86G.
	87G	Generator differential protection. Trips circuit breaker O and field circuit breaker through 86G.
	86G	Auxiliary trip and lockout relay.
2.5 MVA transformer and motor at 13.8 kV bus 3	47	Polyphase undervoltage and phase sequence protection. Trips circuit breaker P.
	49	Temperature overload protection using stator resistance temperature detectors. Initiates alarm.
	49/50	Replica-type thermal overload protection including instantaneous element for short-circuit protection. Trips circuit breaker P.
	50GS	Instantaneous ground-fault protection for 13.8 kV transformer primary. Trips circuit breaker P.
	51/50	Phase overcurrent and locked rotor protection. Trips circuit breaker P.
	63	Sudden pressure relay, transformer mounted. Trips circuit breaker P.
	87M	Motor differential relay. Trips circuit breaker P through auxiliary relay 86M.
	86M	Auxiliary trip and lockout relay.
1.0 MVA transformer at 13.8 kV bus 3	50GS	Instantaneous ground-fault protection. Trips circuit breaker Q.
	51/50	Time and instantaneous phase-fault protection. Trips circuit breaker Q.
480 V transformer secondary circuit breakers	67	Directional phase overcurrent protection for transformer faults and 13.8 kV line faults. Trips 480 V circuit breaker.

current unbalance required to trip is made much greater during the excitation transient than during normal operation.

5.5.2.2 Medium-Voltage Protection. The medium-voltage (2.4 kV) substations shown in Fig 65 are designed primarily for the purpose of serving the medium- and large-size motors. Buses 2 and 3 fed by the 3750 kVA transformers are connected together by a normally closed tie circuit breaker, which is relayed in combination with each main circuit breaker by means of a partial differential or totalizing relaying scheme (Device 51). The current transformers are connected with the proper polarity so that the relay sees only the total current into its bus zone and does not see any current that circulates into a bus zone through either main and leaves through the tie. The relay backs up the feeder circuit breaker relaying connected to its respective bus or operates on bus faults to trip the tie and appropriate main circuit breakers simultaneously, thereby saving one step of relaying time over what is required when the tie and main circuit breakers are operated by separate relays. One possible disadvantage to this scheme occurs when a directional relay or main circuit breaker malfunctions for a transformer fault or when a bus feeder circuit breaker fails to properly clear a downstream fault. The next device in the system that can clear is the opposite primary feeder circuit breaker. If this occurs, a total loss of service to the substation will result. As a result, additional overcurrent relays (Device 51) are sometimes added to the tie circuit breaker on systems where the possibility of this occurrence cannot be tolerated, although they are not shown on the system in Fig 65. These relays can be set so as not to extend any other relay operating time, while providing the necessary backup protection to afford proper circuit isolation for faults upstream from either main circuit breaker.

The source ground relaying for the double-ended 2.4 kV primary unit substation is similar to that described for the 13.8 kV transformer secondary. The single-ended 1500 kVA 2.4 kV primary unit substation illustrates a method for high-resistance grounding utilizing an isolation transformer in the neutral circuit. This scheme limits the magnitude of ground current to a safe level while permitting the use of a lower voltage rated resistor stack. The remainder of the 2.4 kV relaying shown in Fig 65 is, in one form or another, provided for protection of the motor loads.

The application of a combination motor and transformer as shown connected to the 13.8 kV bus 3 is referred to as the unit method. This is done in order to take advantage of the lower cost of the motor and the transformer at 2.4 kV, as compared to the motor alone at 13.8 kV. Motor internal fault protection is provided by instantaneous overcurrent relays, arranged to provide differential protection (Device 87M), by the use of zero-sequence (doughnut-type) current transformers located either at the motor terminals or, preferably, in the starter. The latter current transformer location will also afford protection to the cable feeder. Three current transformers and three relays are applied in this form of differential protection. Thermal overload protection is provided by Device 49 using an rtd as the temperature sensor. Surge protection is provided by the surge arrester and capacitor located at the motor terminals, while undervoltage and reverse-phase-rotation protection is provided by Device 47 connected to the bus potential transformers. The sudden pressure relay (Device 63) is used for detection of transformer internal faults. Branch circuit phase- and ground-fault protection is provided by Devices 51/50 and 50GS, respectively.

The 500 hp induction motor served from the 2.4 kV bus 1 is provided with a nonfused class E contactor. The maximum fault duty on this 2.4 kV bus is well within the 50 000 kVA interrupting rating of the contactor, and therefore fuses are not required. Motor overload protection is furnished by the replica-type thermal relay (Device 49) with the instantaneous overcurrent element (Device 50) applied for phase-fault protection. Separate relaying for motor locked-rotor protection is normally not justified on motors of this size. Undervoltage and single-phasing protection is provided for this and the other motors connected to this bus by Device 27, an undervoltage relay, and by Device 60, a negative-sequence voltage relay connected to the bus potential transformers. Due to the essential function of the motors applied on this bus, a high-resistance grounding scheme is utilized. A line-to-ground fault produces a maximum of 2 A as limited by the 1.72 Ω resistor applied in the neutral transformer secondary. A voltage is developed across the overvoltage relay (Device 59N), which initiates an alarm signal to alert operating personnel.

The 1250 hp induction motor connected to 2.4 kV bus 2 is provided with a fused class E contactor for switching. The fuse provides protection for high-magnitude faults. Motor overload protection is furnished by a replica-type thermal relay (Device 49). Locked-rotor and circuit protection for currents greater than heavy overloads is furnished by Device 51. Protection against single phasing underload is provided by the current-balance relay (Device 46). Instantaneous ground-fault protection is provided by Device 50GS, which is connected to trip the motor contactor since the ground-fault current is safely limited to 800 A maximum. Undervoltage and reverse-phase rotation protection is provided by Device 47.

5.5.2.3 Low-Voltage Protection. Figure 65 illustrates several different types of 480 V unit substation operating modes. Buses 1, 2, and 3, for example, represent a typical low-voltage industrial spot network system that is often used where the size of the system and its importance to the plant operation require the ultimate in service continuity and voltage stability. Multiple sources operating in parallel and properly relayed provide these features. The circuit breakers are provided with solid-state trip devices as the overcurrent protection means. Ground-fault protection is also indicated and would be supplied either as an optional modification to the trip device on the respective circuit breaker, or as a standard zero-sequence relaying scheme on feeder circuits. For tripping of transformer secondary main circuit breakers and protecting the secondary winding, a relay located in the transformer neutral provides another convenient approach.

Since the trip devices of the three main circuit breakers supplying 480 V buses 1, 2, and 3 would normally be set identically to provide selectivity with the tie circuit breakers feeding the 3000 A bus and the other 480 V feeder circuit breakers for downstream faults, directional relays should be provided on these circuit breakers. This will permit selective operation between all 480 V feeder circuit breakers and the main circuit breaker during reverse current flow conditions for transformer or primary faults. Directional relays might also be applied to each of the service tie circuit breakers feeding the 3000 A bus duct so as to provide selective operation between these interrupters for transformer secondary bus faults.

To protect the 800 A frame size feeder circuit breakers from the high level of available fault current at secondary buses 1, 2, 3, and 5, current-limiting fuses should be applied in combination with each circuit breaker. Since the tie circuit breaker at bus 5 is normally closed, the main circuit breakers are also provided with directional relays to ensure selective operation between mains for upstream faults.

The unit substation feeding 480 V bus 4 is a conventional radial arrangement and, except for the addition of ground-fault protection, the circuit breakers shown are equipped with standard trip devices. Bus 6 is fed from a delta-connected transformer and is provided with a ground-fault detection system with both a visible and an audible signal. The small low-current frame-size circuit breakers at this bus have standard trip devices only and do not require the assistance of current-limiting fuses as a result of the lower fault duty on the load side of the 1000 kVA transformer.

5.5.3 Relaying for an Industrial Plant with Local Generation [35], [40], [46], [B14]. When additional power is required in a plant that has been generating all its power, and a parallel-operated tie with a utility system is adopted, the entire fault-protection problem should be reviewed together with circuit breaker interrupting capacities and system component withstand capabilities. In Fig 66 the following assumptions are made:

(1) All circuit breakers in the industrial plant are capable of interrupting the increased short-circuit current.

(2) Each plant feeder circuit breaker is equipped with inverse-time or very-inverse-time overcurrent relays with instantaneous units.

(3) Each of the generators is protected by differential relays and also has external fault backup protection in the form of generator overcurrent relays with voltage-restraint or voltage-controlled overcurrent relays as well as negative-sequence current relays for protection against excessive internal heating for line-to-line faults.

(4) The utility company end of the tie line will be automatically reclosed through synchronizing relays following a tripout.

(5) The utility system neutral is solidly grounded and the neutrals of one or both plant generators will be grounded through resistors.

(6) The plant generators are of insufficient capacity to handle the entire plant load; therefore, no power is to be fed back into the utility system under any condition.

Protection at the utility end of the tie line might consist of three distance relays or time overcurrent relays without instantaneous units. If the distance relays were used, they would be set to operate instantaneously for faults in the tie line up to 10% of the distance from the plant, and with time delay for faults beyond that point in order to allow one step of instantaneous relaying in the plant on heavy faults. If time overcurrent relays were used, they would be set to coordinate with the time delay and instantaneous relays at the plant. At the industrial plant end of the tie at

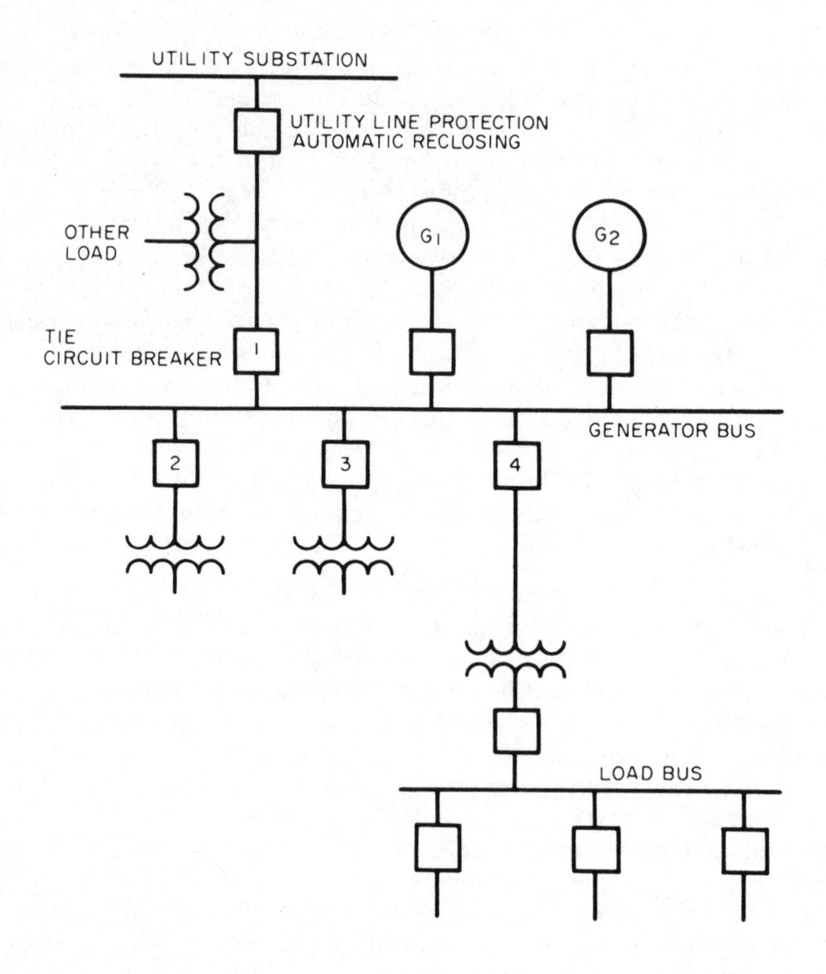

Fig 66
Industrial Plant System with Local Generation

circuit breaker 1, there should be a set of directional overcurrent relays for faults on the tie line, or reverse power relaying to detect and trip for energy flow to other loads on the utility system should the utility circuit breaker open, or both.

The directional overcurrent relays are designed for optimum performance during fault conditions. The tap and time dial should be set to ensure operation within the short-circuit capability of the plant generation, and also to be selective to the extent possible with other fault-clearing devices on the utility system.

The reverse power or power directional relay is designed to provide maximum sensitivity for flow of energy into the utility system where coordination with the utility protective devices is not a requisite of proper performance. A sensitive tap setting can be used, although some time delay is required to prevent nuisance tripping that may occur from load swings during synchronizing.

Due to this time delay a reverse power relay trip of circuit breaker 1 alone may be too slow to prevent generator overload in the event of loss of the utility power source. Further, the amount of power flowing out to the other utility loads may not at all times be sufficient to ensure relay pickup. A complete loss of the plant load can only be prevented by early detection of generator frequency decay to immediately trip not only circuit breaker 1, but also sufficient nonessential plant load so that the remaining load is within the generation capability. An underfrequency relay to initiate the automatic load shedding action is considered essential protection for this system. For larger systems two or more underfrequency relays may be set to operate at successively lower frequencies. The nonessential loads could thereby be tripped off in steps, depending on the load demand on the system.

The proposed relay protection for a tie line between a utility system and an industrial plant with local generation should be thoroughly discussed with the utility to ensure that the interests of each are fully protected. Automatic reclosing of the utility circuit breaker with little or no delay following a tripout is usually normal on overhead lines serving more than one customer. To protect against the possibility of the two systems being out of synchronism at the time of reclosure, the incoming line circuit breaker 1 can be transfer tripped when the utility circuit breaker trips. The synchro-check relaying at the utility end will receive a dead-line signal and allow the automatic reclosing cycle to be completed. Reconnection of the plant system with the utility supply can then be accomplished by normal synchronizing procedures.

Generator external-fault protective relays, usually of the voltage-restraint or voltage-controlled overcurrent type, and negative-sequence current relays provide primary protection in case of bus faults and backup protection for feeder or tie line faults. These generator relays will also operate as backup protection to the differential relays in the event of internal generator faults, provided there are other sources of power to feed fault current into the generator.

5.6 Protection Requirements. The primary purpose of a coordination study is to determine satisfactory ratings and settings for the electric system protective devices. The protective devices should be chosen so that pickup currents and operating times are short but sufficient to override system transient overloads such as inrush currents experienced when energizing transformers or starting motors. Further, the devices should be coordinated so that the circuit interrupter closest to the fault opens before other devices.

Determining the ratings and settings for protective devices requires familiarity with the NEC [22] requirements for the protection of cables, motors, and transformers, and with ANSI/IEEE C57.12.00-1980 [15] for transformer magnetizing inrush current and transformer thermal and magnetic stress damage limits.

5.6.1 Transformers

5.6.1.1 Maximum Overcurrent Protection. The NEC [22], Article 450-3, specifies the maximum overcurrent level at which the transformer protective devices may be set. If there is no secondary protection, transformers with primaries rated for more than 600 V require either a primary circuit breaker or fuse that will operate at no more than 300% or 250% of transformer full-load current, respectively. Better protection will be realized with breaker settings or fuse ratings lower than these NEC [22] maximum levels; the actual value depends on the nature of the specific load involved. When both primary and secondary protective devices are provided, the maximum protective levels depend on the transformer impedance and secondary voltage. These maximum levels of protection are shown in Table 23.

On the other hand, transformers with primaries rated 600 V or less require primary protection rated at 125% of full-load current when no secondary protection is present, and 250% as the maximum rating of the primary feeder overcurrent device when secondary protection is set at no more than 125% of transformer rating. Certain exceptions to these requirements for smaller sized transformers, detailed in NEC [22], Article 450-3, are intended to permit the application of protective devices having standard ratings normally available. The permissible fuse rating is generally lower than the circuit breaker setting due to the difference in the circuit opening characteristics in the overload region.

5.6.1.2 Transformers Withstand Limits. For years, transformer primary protective devices were required to clear a bolted secondary short circuit within time specified limits. These time limits define the transformer withstand capability and are based on the impedance of the transformer as follows:

Impedance (percent)	Current (time base value)	Time (seconds)
4	25	2
5	20	3
6	16.6	4
7 and above	14.3 or less	5

At levels of current in excess of about 400–600% of full load, the transformer withstand characteristic can be conservatively approximated by a constant I^2t (heating) plot, which is represented by a straight line of -2 slope extending to and terminating at the appropriate short-circuit withstand point.

Table 23
Maximum Overcurrent Protection (in Percent)

	Transformers with Primary and Secondary Protection				
	Primary Over 600 V		Secondary Over 600 V		Secondary 600 V or Below
Transformer Rated Impedance	Circuit Breaker Setting	Fuse Rating	Circuit Breaker Setting	Fuse Rating	Circuit Breaker Setting or Fuse Rating
No more than 6%	600	300	300	250	250
More than 6% but no more than 10%	400	300	250	225	250

It has been widely recognized that damage to transformers from through faults is the result of mechanical and thermal effects. The former, in fact, have gained increased recognition as a major factor in transformer failures. Accordingly, a new standard is available that significantly revises this familiar *ANSI withstand point*: ANSI/IEEE C57.109-1985 [19], which relates to liquid-filled transformers; a companion document is currently being prepared for dry-type transformers. A complete discussion of this subject is given in ANSI/IEEE Std 242-1986 [20], Chapter 10, and in the Appendix of ANSI/IEEE C37.91-1985 [13].

The following discussion briefly reviews the through-fault protection guidelines for Category II transformers (501-1667 kVA single-phase, 501-5000 kVA three-phase) and Category III transformers (1668-10 000 kVA single-phase, 5001-30 000 kVA three-phase). The through-fault protection curves take into consideration the fact that transformer damage due to mechanical effects is cumulative, and the number of through-faults to which a transformer can be exposed is different, depending on the transformer application.

Accordingly, two through-fault protection curves have been established for both Category II, Fig 67, and Category III, Fig 68, transformers. One curve is for those applications where faults occur frequently, typically more than 10 in a transformer lifetime; and the second is for infrequently occurring faults, typically not more than 10. Where secondary-side conductors are enclosed in conduit, busway, or otherwise isolated, as found in industrial, institutional, and commercial systems, the incidence of faults is extremely low and the infrequent fault curve may be used to determine the settings of main secondary devices, primary devices, or both. In contrast, transformers with secondary-side overhead lines have a relatively high exposure to through-faults and the use of reclosing-type protective devices may subject the transformer to repeated current surges from each fault. In these cases the frequent fault withstand curve should be used.

Another consideration is a relative shift in the damage point that occurs in delta-wye transformers with the wye on the low side and the neutral point grounded. A secondary single-phase-to-ground fault of one per unit value (using the three-phase fault values as a base) will produce a fault current of one per unit in the primary delta winding, but results in only 0.58 per unit current in the line to the delta winding that contains the protective device.

Therefore, a second damage characteristic, corresponding to that provided by ANSI/IEEE C57.109-1985 [17] and derated for a wye-wound solidly grounded neutral secondary, should be plotted at 0.58 per unit of the normal characteristic.

5.6.1.3 Other Protection Considerations. In selecting the settings or ratings of the primary protective device, the following items should be known and considered.

(1) Voltage rating of the system

(2) Rated load and inrush current of the transformer

(3) Short-circuit load of the supply system in kilovolt-amperes

(4) Type of load, whether steady, fluctuating, or subject to heavy motor, welding, furnace, or other starting surges

(5) Coordination with other protective devices

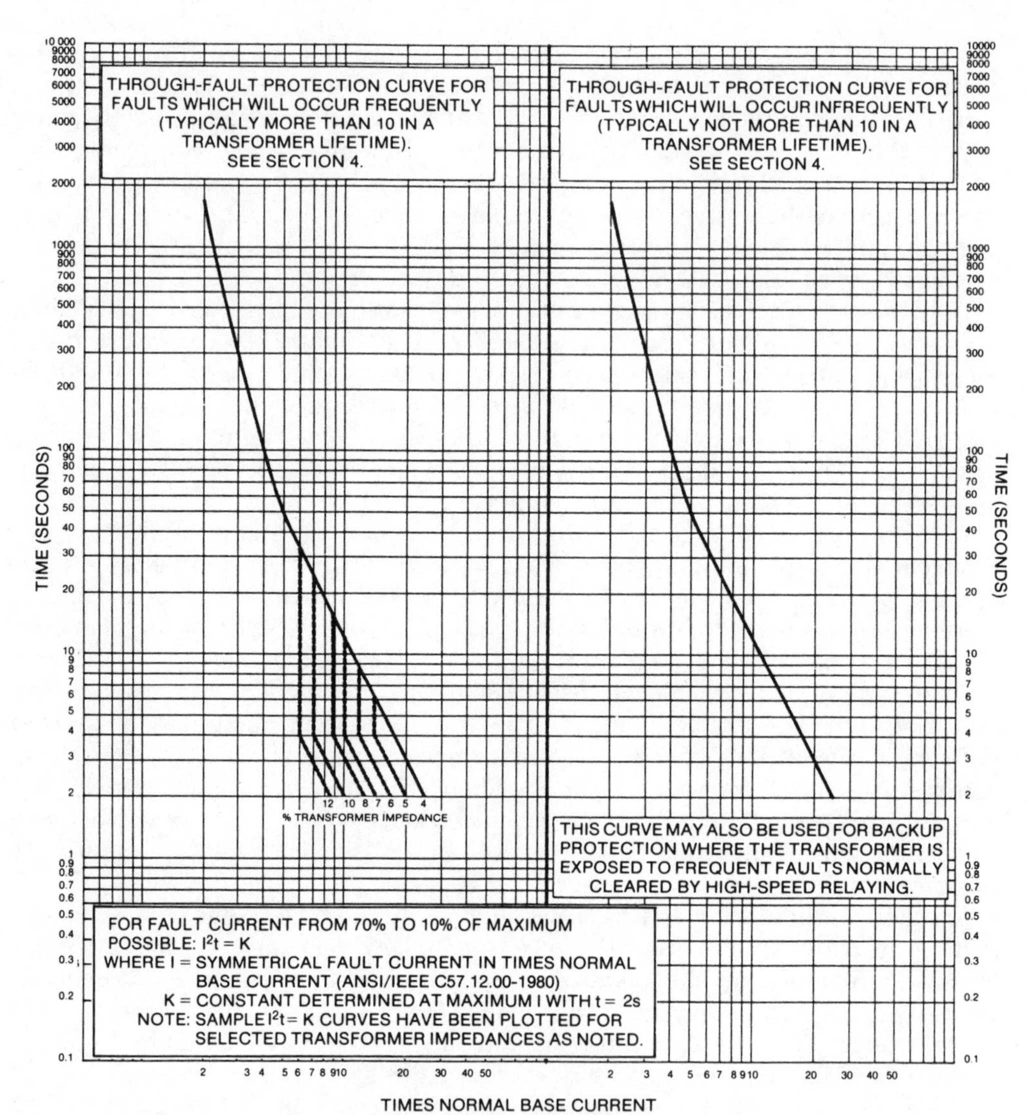

From ANSI/IEEE C57.109-1985 [17].

Fig 67
Category II Transformers

Relays, when used in combination with power circuit breakers for protection of a transformer primary circuit, should have a time-current curve shape similar to that of the first downstream device. Pickup of the time-delay element may typically be 150–200% of the transformer primary full-load current rating. The instantaneous pickup setting should be set at 150–160% of equivalent maximum secondary three-phase symmetrical short-circuit current to allow for the dc component of fault current during the first half-cycle. The setting should also permit the magnet-

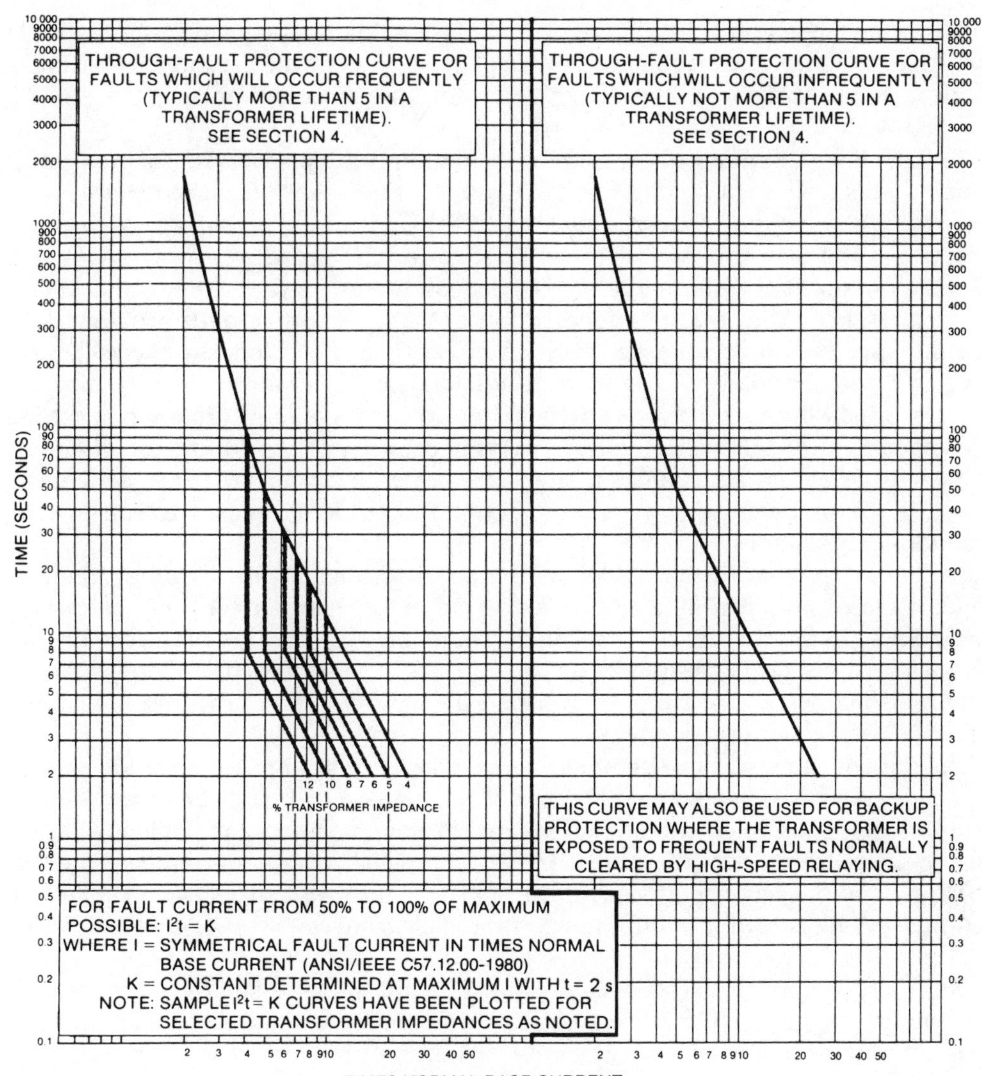

From ANSI/IEEE C57.109-1985 [17].

Fig 68
Category III Transformers

izing inrush current to flow. In general the transformer inrush current is approximately 8–12 times the transformer full-load current for a maximum period of 0.1 s. This point should be plotted on the time–current curve, and it should fall below the transformer primary protection device curve. If there is more than one transformer connected to this feeder, the pickup of the time-delay element should not exceed 600% full-load current of the smallest transformer, assuming that the transformers have secondary protection and an impedance of 6% or less. When used in the transformer secondary circuit, the pickup of the time-delay element should also be between 150 and 200% full-load current of the transformer secondary rating. A typical circuit configuration is illustrated by the one-line diagram insert in Fig 73 in 5.7.2.3.

5.6.2 Feeder Conductors. Restrictions that apply are provided in the NEC [22]. Protection of feeders or conductors rated 600 V or less shall be in accordance with their current-carrying capacity as given in NEC [22] tables, except where the load includes motors. In this case it is permissible for the protective device to be set higher than the continuous capability of the conductor (to permit coordination on faults or starting the largest connected motor while the other loads are operating at full capacity), since running overload protection is provided by the collective action of the overload devices in the individual load circuits. Where protective devices rated 800 A or less are applied that do not have adjustable settings that correspond to the allowable current-carrying capacity of the conductor, the next higher device rating may be used. There are other exceptions that are allowed in the NEC [22], Article 240-3, such as capacitor and welder circuits and transformer secondary conductors.

Feeders rated more than 600 V are required to have short-circuit protection, which may be provided by a fuse rated at no more than 300% of the conductor ampacity or by a circuit breaker set to trip at no more than 600% of the conductor ampacity. Although not required by the NEC [22], improved protection of these circuits is possible when running overload protection is also provided in accordance with the conductor ampacity.

The flow of short-circuit current in an electric system imposes mechanical and thermal stresses on cable as well as circuit breakers, fuses, and the other electric components. Consequently, to avoid severe permanent damage to cable insulation during the interval of short-circuit current flow, feeder conductor damage characteristics should be coordinated with the short-circuit protective device. The feeder conductor damage curve should fall above the clearing-time curve of its protective device.

This damage curve represents a constant I^2t limit for the insulated conductor. It is dependent upon the maximum temperature that the insulation can be permitted to reach during a transient short-circuit condition without incurring severe permanent damage. Recommended short-circuit temperature limits, which vary according to the insulation type, are published by cable manufacturers. For any particular magnitude of current, the time required to reach the temperature limit can be determined from one of the following equations.

For copper conductors

$$\left(\frac{I}{A}\right)^2 t = 0.0297 \log_{10} \left(\frac{T_2 + 234}{T_1 + 234}\right)$$

For aluminum conductors

$$\left(\frac{I}{A}\right)^2 t = 0.0125 \log_{10} \left(\frac{T_2 + 228}{T_1 + 228}\right)$$

where

I = rms current in amperes
t = time in seconds
A = conductor cross-sectional area in circular mils
T_1 = initial conductor temperature in °C
T_2 = final conductor temperature in °C (short-circuit temperature limit)

If the initial and short-circuit temperatures are known, these equations can be used to construct a conductor damage curve which is valid for time intervals up to approximately 10 s. Since the initial temperature depends upon the cable loading and ambient conditions, and therefore cannot usually be determined accurately, it is common to conservatively assume that the initial temperature is equal to the rated maximum continuous temperature of the conductor.

5.6.3 Motors

5.6.3.1 Large Alternating-Current Rotating Apparatus. (See ANSI/IEEE Std C37.96-1976 [14].) The protection of an ac induction motor is a function of its type, size, speed, voltage rating, application, location, and type of service. In addition, a motor may be classified as being in essential or nonessential service, depending upon the effect of the motor being shut down on the operation of the process or plant. Although the discussion earlier in this chapter on the different types of protective devices indirectly touches on some of the problems associated with protecting motors, it is worthwhile to examine such an important subject from the standpoint of the machine itself.

Unscheduled motor shutdowns may be caused by

(1) Internal faults
(2) Subtained overloads and locked rotor
(3) Undervoltage
(4) Phase unbalance or reversal
(5) Voltage surges
(6) Reclosure and transfer switch operations

The ideal relay scheme for an induction motor must provide protection against all these hazards. In the following text the relaying approach to protect against each of these problems will be discussed in general terms. Later in the chapter, several specific applications will be discussed in detail. For a complete discussion on motor protection, see Chapter 9 of ANSI/IEEE Std 242-1986 [20].

(1) *Internal Faults*. Internal fault protection for induction motors can be obtained by either overcurrent relays, but preferably percentage differential relays, as described in 5.3.5.1. When the supply source is grounded, separate and more sensitive ground-fault protection can be provided using the relaying schemes described in 5.3.7.1 and 5.3.7.2. The preferred solution is to use the zero-sequence approach for ground-fault relaying where all three phase leads are passed through the single window-type current transformer. This eliminates false tripping due to unequal current-transformer saturation and allows the use of a fast, sensitive ground-fault relay setting.

(2) *Sustained Overloads and Locked Rotor*. Conventional overcurrent relays do not provide suitable protection against sustained overloads because they will *over-*

protect the motor if set to pickup for the normal overloads encountered. That is, the relay will not allow full use of the thermal capability of the motor and in many cases will not provide sufficient time delay to permit complete starting. This is shown in Fig 69, where for most conditions less than locked-rotor current there is too much margin between the motor thermal capability curve and the relay operating time characteristic. With a higher pickup and appropriate time-dial setting as shown, the overcurrent relay will provide excellent locked-rotor and short-circuit protection of the motor.

Thermal relays, on the other hand, will give adequate protection for light and medium overloads, allowing loading of the motor close to its thermal capability. In general, however, thermal relays will not give adequate protection for heavy overloads, locked rotor, and short circuits. Therefore, in most cases, both types of relays should be used to provide optimum protection for overloads, locked rotor, and short circuits and allow maximum use of the motor capability. In this way the characteristics of the protection can be closely *shaped* to the motor thermal damage curve.

There are two common types of thermal relays available for motor protection as discussed in 5.3.14 and 5.3.16. One operates in response to resistance temperature

Fig 69
Motor and Protective Relay Characteristics

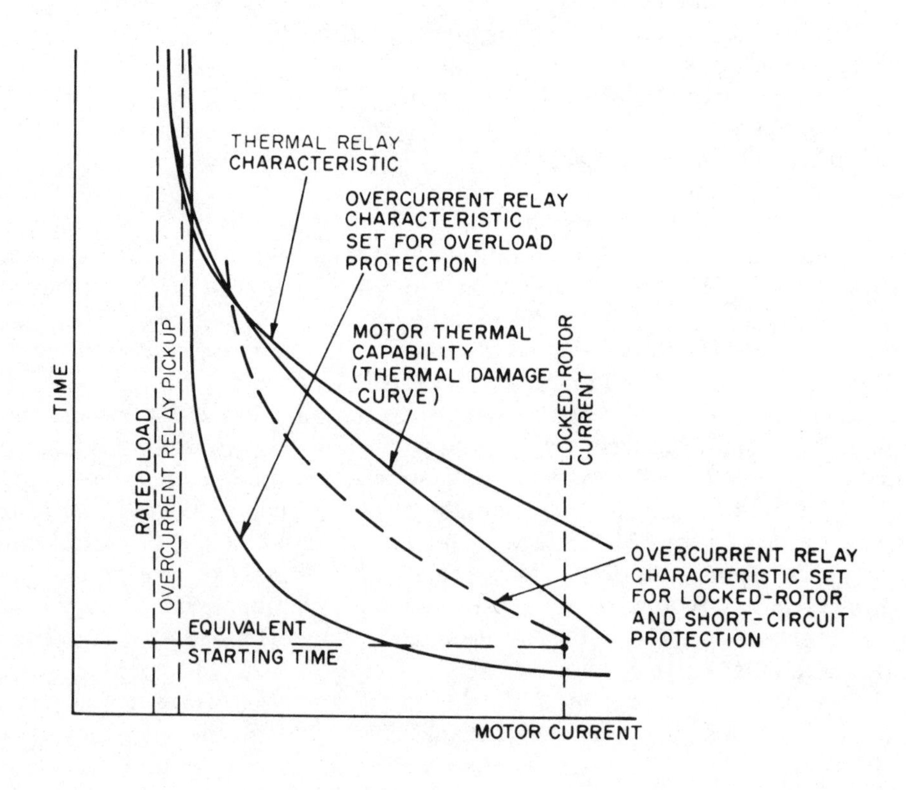

detectors embedded in the machine windings, and the other operates in response to motor current. The latter type normally has adjustable pickup and trip characteristics to compensate for the motor service factor as well as the ambient temperature differences between motor and relay.

Frequently, medium-voltage motors are protected by a contactor with thermal overload relays applied in combination with current-limiting fuses, which are intended to open the circuit for high-fault currents. In addition to matching the overload protection to the motor thermal capabilities for such applications, it is equally important to select the fuses so that they protect the contactor by opening faster on currents in excess of the contactor interrupting rating. Likewise, the overload relays must prevent fuse blowing by tripping before the fuse clears on currents within the contactor capabilities.

(3) *Undervoltage.* Low voltages can prevent motors from coming up to normal operating speed or they can result in overload conditions. Although thermal overload relays will detect an overload resulting from undervoltage, large motors and medium-voltage motors should have separate undervoltage protection. An induction-type undervoltage relay is usually provided to prevent starting when the voltage is unacceptably low, and to prevent operation on momentary voltage dips.

Single-phasing under load or three-phase dissymmetry can cause overheating within the motor at a rate above that perceived by the thermal overload relays, even in the phase with the largest current. If the machine is only partially loaded, the overload relays may never detect that damage is occurring.

(4) *Phase Unbalance or Reversal.* When starting from rest, single phase (one line open) will prevent starting, while reverse-phase rotation can have immediate disastrous results on the motor or the driven equipment. In all cases where such conditions are likely to exist, a phase-failure and reverse-phase relay should be applied.

If not properly protected, three-phase motors are vulnerable to damage when loss of voltage occurs on one phase. There are numerous causes of such loss of voltage, and these can occur anywhere in the distribution system. The chief problems resulting from single-phasing of three-phase motors is overheating, which can cause reduction of life expectancy or complete failure.

The modern practice of applying three overload devices on three-phase motors assures detection of single-phasing in most cases. When operating at normal load, loss of voltage on one phase causes an abnormal current in the remaining phases, which is sensed by the protective devices. However, under some conditions of light loading, three-phase motors can overheat when single-phased, without being detected by the overload protective devices. Even when operating near rated horsepower under single-phase conditions, the motor can be damaged prior to response by conventional protective devices.

Negative-sequence voltage or current-balance relays should be considered for protection of motors above 1000 hp against these conditions.

(5) *Voltage Surges.* Voltage surges are transient overvoltages caused by switching or lightning strokes. They are characterized by a steep wave front. Surge protection equipment consists of a protective capacitor and arrester that should be connected as close to the motor terminals as possible. Further application cri-

teria for treating this problem are discussed in Chapter 4 of this book.

(6) *Reclosure and Transfer Switch Operations* [42]. Under normal operating conditions, the self-generated voltage of an ac motor lags the bus voltage by a few electrical degrees in induction motors and by 25-35 electrical degrees in synchronous motors. The operation of a recloser on the utility power supply or the transfer to an alternate source will cause the power to be interrupted for a fraction of a second or longer. When power is removed from a motor, the terminal voltage does not collapse suddenly, but decays in accordance with the open-circuit machine time constant (time for self-generated voltage to decay to 37% of rated bus voltage). The load with its inherent inertia acts as a prime mover that attempts to keep the rotor turning. The frequency or phase relationship of the motor self-generated voltage no longer follows the bus voltage by a fixed torque angle, but starts to separate farther from it (out-of-phase in electrical degrees) as the motor decelerates.

If the motor is reconnected to the bus voltage with its self-generated voltage at a high level and severely out-of-phase, dangerous stresses that are both mechanical and electrical are placed on the motor. In addition to possible damage to the motor, excessive torque may also damage the motor coupling. Furthermore, the excessive current drawn by the motor may trip the overcurrent protective device.

A check should be made to determine that re-energization occurs at a point where the motor and load will not be subjected to excessive forces. Protection against this problem can be provided by certain types of frequency relays that operate as a function of the rate of change of frequency to remove the motor from the line. An alternate method is to prevent re-energization until the residual voltage has decayed to a safe value.

Automatic transfer switching means can be provided with accessory controls that disconnect motors prior to transfer and reconnect them after transfer and when the residual voltage has been substantially reduced. Another method is to provide in-phase monitors within the transfer controls that prevent transfer until the motor bus voltage and the source are nearly synchronized.

5.6.3.2 Small Motors. The specific requirements for the protection of small induction motors are specified in Article 430 of the NEC [22]. Each motor branch circuit must be provided with a disconnect means, branch circuit protection, and a motor running overcurrent protective device. Two examples are shown in Fig 70.

The branch circuit disconnect and protective means are generally combined in one device such as a molded-case circuit breaker or a fused disconnect switch. The motor running overcurrent protection is provided by overload relays. The motor is energized and de-energized by a controller. This unit may be operated either manually or electrically (magnetic type). The overload relays open the motor controller to provide motor running overcurrent protection. It is not uncommon to have the motor controller included in the same enclosure as the motor branch circuit disconnect and overcurrent protection device. The complete unit is called a combination motor starter, providing motor branch circuit disconnect and overcurrent protection along with motor control and running overcurrent protection.

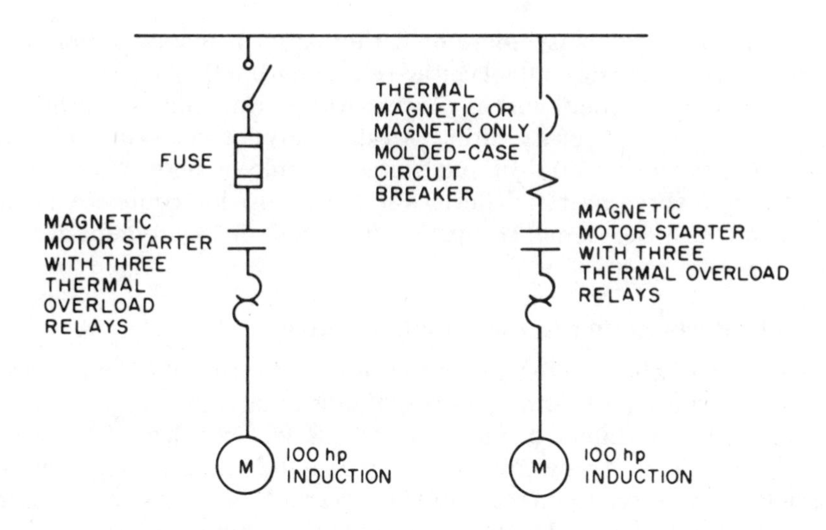

Fig 70
Motor Protection Acceptable to the NEC [22]

The motor branch circuit overcurrent device must allow the motor to start (without opening on motor inrush current), and it must open for short circuits.

A combination disconnect and overcurrent protective device must safely open the circuit under a short-circuit condition and disconnect the circuit with the motor running at full load or with locked rotor. The short-circuit overcurrent protective device must be capable of interrupting the circuit under the maximum available short circuit and in so doing protect the branch circuit. The switch should be quick-make-quick-break, horsepower-rated, and capable of being closed in on a fault of the magnitude available at its application point without damage. The switch must safely withstand the I^2t and peak let-through current of the fuses without realizing an immediate failure or change in operating characteristics, which could lead to problems during normal operation some time later.

For running overcurrent protection it is necessary to select the proper thermal unit for the overload relay. All manufacturers' tables of thermal units are based on the operation of the motor and controller in the same ambient temperature of 40 °C or less. To apply these devices properly the following must be determined:

(1) motor full-load and locked-rotor current from motor nameplate
(2) motor service factor from nameplate
(3) ambient temperature for motor
(4) ambient temperature for controller
(5) motor starting time with load connected

With this information and following the manufacturer's recommendation, an adjusted motor full-load current can be determined to select the proper overload relay thermal unit from the manufacturer's table. Then it must be verified that the trip characteristic will permit starting.

Undervoltage protection is inherent in the use of a magnetic controller and three-wire control since the control voltage is taken from the line or primary side of the controller. Most magnetic motor controllers will drop out when the operating coil voltage drops to 65% of its rating. All units do not have the same drop-out characteristics, so the actual drop-out voltage should be determined by test. For many motors the three overload devices may not provide complete single-phase protection, in which case it can be furnished as a special equipment modification.

5.7 Use and Interpretation of Coordination Curves

5.7.1 Need and Value. Determining the settings for the overcurrent protection on a power system can be a formidable task that is often said to require as much art as technical skill. Continuity of plant electric service requires that interrupting equipment operate selectively. This normally demands slower operation or longer opening delays (for a given current) of the interrupters successively closer to the power source during faults. The necessity for maximum safety to personnel and electric equipment, on the other hand, calls for the fastest possible isolation of faulted circuits.

The coordination curve plot provides a graphical means of displaying the competing objectives of selectivity and protection. This method of analysis is useful when designing the protection for a new power system, when analyzing protection and coordination conditions in an existing system, or as a valuable maintenance reference when checking the calibration of protective devices. The coordination curves provide a permanent record of the time-current operating relationship of the entire protection system.

Actual plotting of the curves on a single sheet of graph paper using a common current scale is essential because rarely do all the fault protective devices involved have time-current curves of the same shape, and it is difficult to visualize the relationship of the many different shapes of curves. A scale corresponding to the currents expected at the lowest voltage level works best. For example, fault-current protective devices on both sides of a 2400-480 V transformer should be plotted on the 480 V current scale. To plot 2400 V device time-current curves on the 480 V scale, first determine the desired time and current settings in the usual manner on the basis of current expected on the 2400 V circuit. Then plot time directly since that scale is unchanged, but multiply the 2400 V currents by 5 (ratio of 2400/480) to obtain equivalent current at 480 V before plotting on the 480 V scale.

Usually the coordination plot is made on log-log graph paper with current as the abscissa (horizontal axis) and time as the ordinate (vertical axis). A choice of the most suitable current and time settings is made for the device to provide the best possible protection and safety to personnel and electric equipment and also to function selectively with other protective devices to disconnect the faulted equipment with as little disturbance as possible to the rest of the system.

5.7.2 Device Performance. The manufacturers of protective devices publish time-current operating characteristic curves and other performance data for all devices used in a protection system. The individual time-current characteristics of thermal overload relays, fuses, solid-state and series trip devices, and relays are transposed onto a common plot for selecting coordinated settings. Typical relay curves are shown in Fig 52. Relay time-current curves normally begin at multiples of 1.5 times minimum closing current or pickup setting, since their performance cannot be accurately predicted below that value. However, curves showing the approximate expected time-current performance of lower values can usually be obtained from the manufacturer, if required. The relay time-current curves define the operation of the relay alone and do not include any circuit breaker interrupting time.

5.7.2.1 Time-Delay Relays in Series. The time-current curves of direct-acting time-delay trip devices, fuses, and time-delay thermal devices include the necessary allowance for overtravel, manufacturing tolerances, etc. The time-current characteristics of relays are represented by families of single-line curves to which tolerance bands must be added. This tolerance band is based on the fact that the second relay in a chain continues to see fault current until the circuit breaker associated with the first relay has opened and the arc extinguished. This is nominally 5-8 cycles for the circuit breakers commonly used in industrial systems, although the actual contact parting time will be 3-5 cycles. After the first circuit breaker has opened the circuit and de-energized the second relay, the contacts of an induction disk relay will continue to coast for 0.1 s (standard inverse-time relay) due to the inertia of the induction disk to which the movable contact is attached.

A minimum total time margin of 0.4 s at maximum fault current is sufficient to afford satisfactory selectivity between inverse-time relays. This margin allows for the 0.08 s circuit breaker opening time (5 cycles), 0.1 s overtravel, and a safety factor of 0.22 s to cover manufacturing variations and inaccuracies in positioning of the time dial or lever when setting the relay. If the relays are more accurately set to a specific curve by instrument, this margin can be safely reduced to approximately 0.30 s. Faster circuit breaker operation and solid-state relays or applications with electromechanical relays that result in less disk overtravel will permit further reduction of this margin (to approximtely 0.2-0.25 s).

When two induction relays in series are set to coordinate at the maximum available fault current, they will be selective on lower values of current if they have the same shape of time-current curves, and the minimum operating current setting (pickup) of the slower relay is higher than that of the faster relay. If the pickup setting is lower, the operating-time curves of the two relays will cross each other at some low value of fault current, and the *slow* relay will beat the *fast* one for all currents below that value.

Plotting the time-current curves on a single graph with a common current for all the relays and other devices in series is illustrated in Fig 71, which reveals two conflicts that might otherwise have escaped notice. In this case the power supply was sufficient to maintain a 250 000 kVA short-circuit duty without appreciable decrement, and there was no fault current contribution from synchronous equipment connected to the 2.4 kV system. On this basis the maximum 2.4 kV system symmetrical fault current is 20 000 A and the maximum 13.8 kV system symmetrical fault current is 10 460 A (60 000 A on a 2.4 kV base). It was also assumed that the end relay in the chain (D) was to be set at a minimum of 0.5 s.

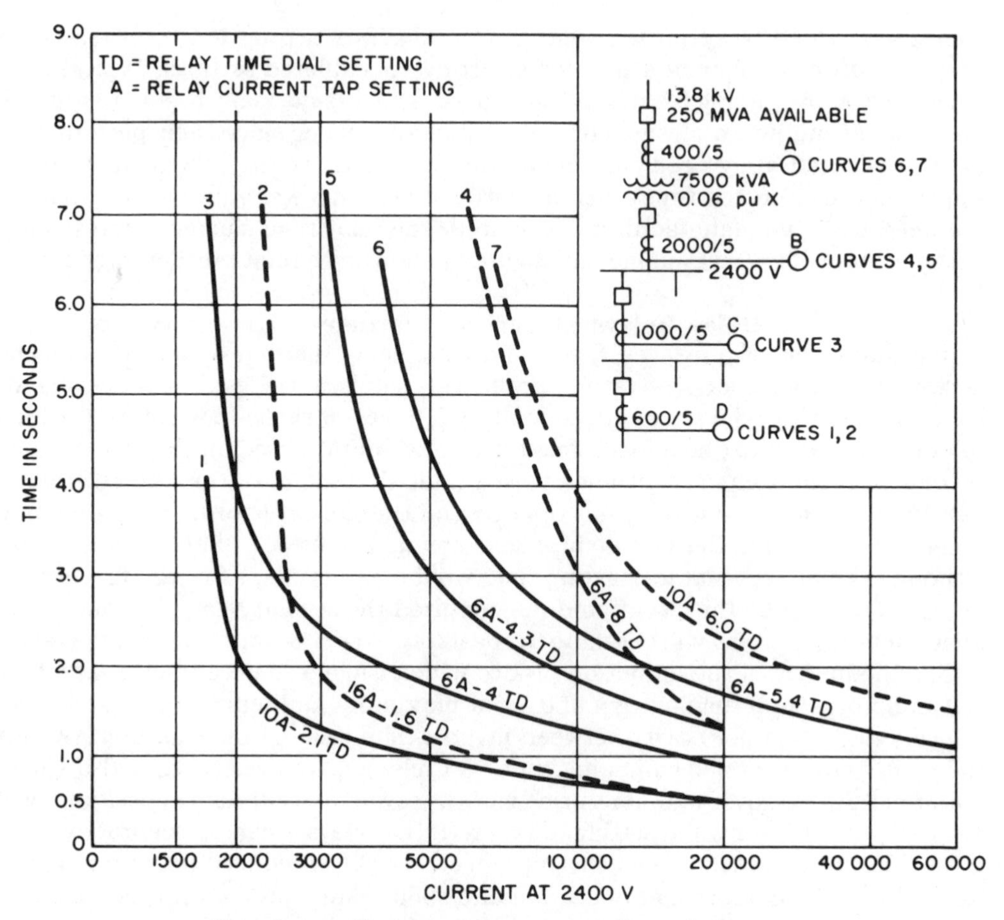

Very Inverse Characteristic	Curve 4
Moderately Inverse Characteristic	Curves 1—3, 5—7

NOTE: The curves are plotted on semilog coordinates for illustrative purposes only. Normally log-log coordinates are used. Also, in actual practice tap settings above 8A are seldom required.

Fig 71
Selecting Time-Current Curves and Relay Tap Settings for an Industrial Plant Distribution System

The three sets of 2.4 kV relays were coordinated by selecting time-current settings that would make their operating times 0.4 s apart at the maximum current of 20 000 A. Then the single set of relays on the 13.8 kV system was coordinated with those on the 2.4 kV system at the same value of fault current (3480 A at 13.8 kV) because this is the current that the relays at A would see during a fault on the 2.4 kV side of the transformer. Coordination was accomplished by selecting time-current settings that would give 0.4 s delay between relays A and B for a 20 000 A fault on the 2.4 kV system.

Following the foregoing procedure could possibly give satisfactory results without actually plotting the curves if the basic rules for coordination relays were observed, that is, (1) relays with the same shape curves were used in series with each other and (2) relays farthest from the source of power had current settings below that of the relays ahead of them.

Unfortunately it is easy to overlook these basic rules, and because of this the effort required to plot the curves proves worthwhile.

As shown in Fig 71, time-current settings on relay D that would give either curve 1 or curve 2 would satisfy the requirement that it take a minimum of 0.5 s at 20 000 A. The curve 2 setting, however, is slower and less sensitive than that represented by curve 1 throughout most of its range. Furthermore, curve 2 crosses curve 3 which represents the desired setting for relay C, thereby requiring desensitization of C to make it selective with D. Thus the curve 2 setting on relay D would greatly reduce the short-circuit protection provided by either relay C or D.

Curves 4, 5, 6, and 7 of Fig 71 illustrate what would happen if relay B had a very-inverse-time characteristic instead of an inverse-time curve as the others do. Curve 4 meets the requirement that it be 0.4 s slower than curve 3 representing relay C, when both are operating at 20 000 A. Also, curve 6 satisfies the requirement that relay A be 0.4 s slower than B when A is operating at the equivalent of the 20 000 A 2.4 kV system short-circuit current. If the curves had not been plotted, there would be reason to believe that the contemplated settings for relays A and B as represented by curves 6 and 4 would be satisfactory. Actually, the very-inverse-time characteristic of relay B causes its curve to cross that of relay A at a high level of fault current, so that the tripping sequence of the circuit breakers would be reversed. For this particular circuit it would not be too serious, since tripping either circuit breaker would shut down the whole circuit, but it would still nullify the effectiveness of the relays in giving indication as to where the trouble was. If the very-inverse-time relay were retained at B, relay A's setting would have to be desensitized and increased in time (curve 7) to be selective with B. This would result in greater damage during a short circuit, so an inverse-time relay should be substituted for the very-inverse-time relay at B, making it possible to set B to give performance as shown by curve 5. This would mean that A and B could both be more sensitive and faster, and consequently would give better protection to the system.

If the very-inverse-time relay were used at B, the backup protection that it could provide would be very poor due to the big *gap* in the pickup currents of relays B and C, as shown by curves 3 and 4.

When choosing between two combinations of current-tap and time-dial settings, either of which will give a desired operating time at maximum fault current, the combination with the lower current and higher time-dial setting is usually preferable, because the relay will be more sensitive and faster on low values of fault current. Suppose that an operating time of 0.5 s is desired with a relay connected to 1000/5 A current transformers in a circuit with an available symmetrical fault current of 20 000 A. Relays with 6 A tap and 2.1 time-dial setting, or 10 A tap and 1.7 time-dial setting will both give the desired time. But in case of a fault involving only 3000 A, the relay with the 6 A setting would operate in 1.25 s compared with 2 s for the 10 A combination. If the current is still further reduced to 2000 A, the first relay will still operate in 2.1 s, but the second one will be very slow, since operation is uncertain when the current is only 1.0 times relay pickup.

A special problem arises when attempting to coordinate overcurrent devices on opposite sides of a delta-wye-connected transformer. For line-to-line faults occurring and measured on the wye side, the available fault current through the transformer will be reduced to approximately 87% ($\sqrt{3}/2 \cdot 100$) of the available three-phase fault current. The highest of the unbalanced phase currents on the delta side, however, will be 100% of the value experienced for a balanced three-phase fault, and the overcurrent device in this phase will operate faster relative to the protection on the wye side. The effect that this change in relative operating characteristics has on coordination for line-to-line faults can be examined graphically by shifting the normal plot of the delta-side protective device by the ratio of 0.87/1.0. This technique will be illustrated by an example in 5.8.

5.7.2.2 Instantaneous Relays. When two circuit breakers in series both have instantaneous overcurrent relays, their selectivity is dependent solely on their current settings. Therefore the relays must be set so that the one nearest the source will not trip when maximum asymmetrical fault current flows through the other circuit breaker. This requires sufficient impedance in the circuit between the two circuit breakers to cause faults beyond both circuit breakers to receive less current than faults near the source circuit breaker, so that the relays of the source circuit breaker can determine the fault location. If this differential is insufficient, selective operation is impossible with instantaneous overcurrent relays and the opening of both circuit breakers on through faults must be tolerated.

Usually the impedance of a transformer is sufficient to achieve selectivity between an instantaneous relay on a primary feeder and the instantaneous trip coil of a low-voltage secondary circuit breaker. Also the impedance of open transmission lines may be sufficient to provide the necessary differential in short-circuit-current magnitude to permit the use of instantaneous relays at both ends.

Generally instantaneous relays at opposite ends of in-plant cable systems cannot be coordinated because the circuit impedance is too low to provide the necessary current differential.

Applications involving both phase overcurrent and residually connected ground relays should be reviewed carefully to determine that steady-state and transient error currents are below the instantaneous pickup setting of the relay. The instantaneous element in a residually connected scheme may not be able to be set at all. Some solid-state instantaneous relays are designed to have inverse operating characteristics to allow a lower pickup setting without tripping during the transient portion of the error current.

Instantaneous attachments are generally furnished on all time-delay overcurrent relays on switchgear equipment so that they will be interchangeable, but they should be employed for tripping only when applicable. The fact that a relay setting study reveals that some of the instantaneous relays must be made inoperative should not be interpreted as a sign of a poorly designed protective system.

5.7.2.3 Low-Voltage Circuit Breakers. The time–current characteristics for typical low-voltage power circuit breakers with solid-state trip devices are represented by bands of curves shown in Fig 72. The maximum and minimum operating

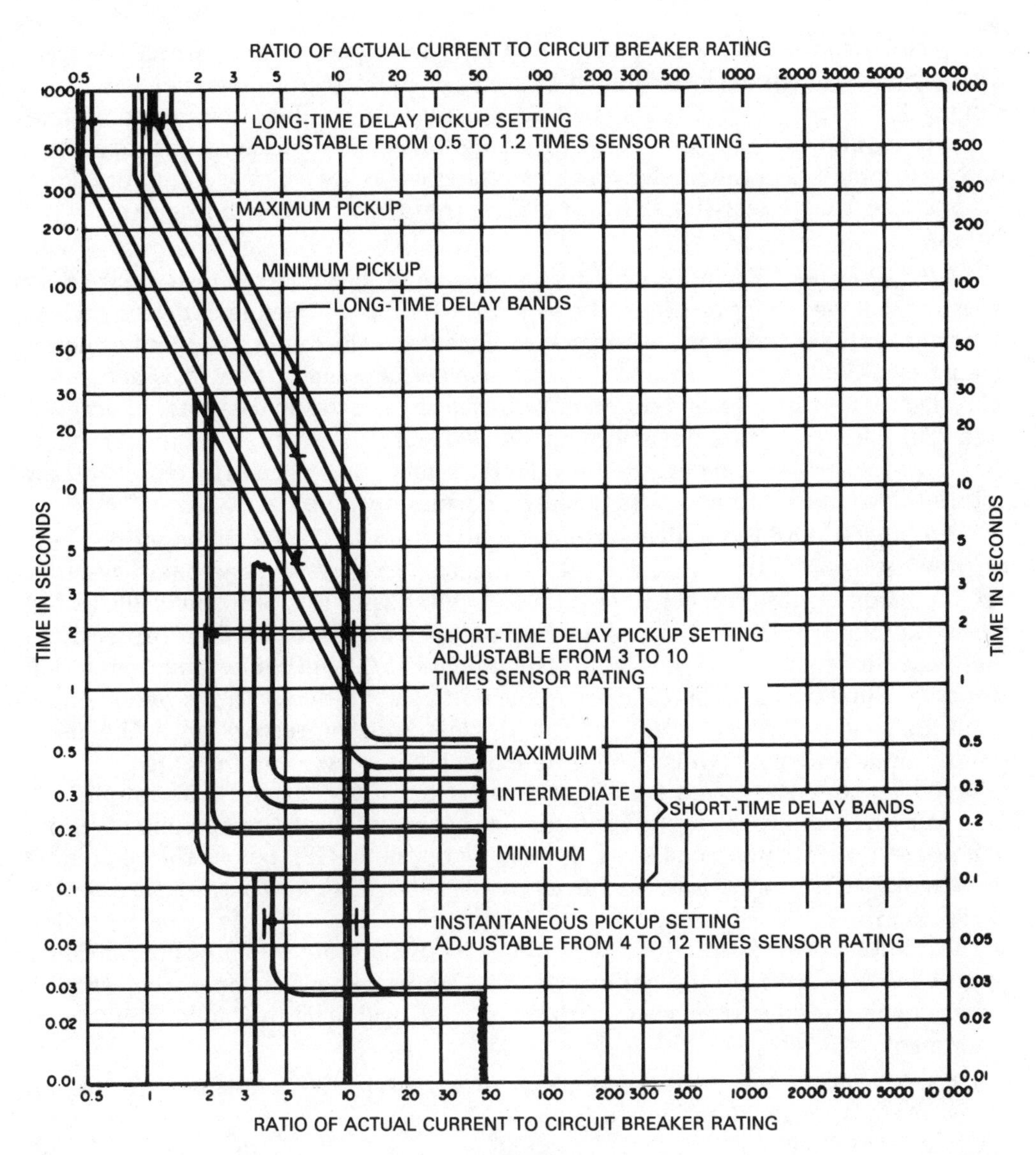

Fig 72
Adjustability Limits of Low-Voltage Power Circuit Breaker Trip Devices

time curves define the operating characteristic of the trip device. The very narrow bandwidth of the trip characteristic is achieved by the use of high-quality, industrial-grade, solid-state components and permits several breakers to be closely coordinated without excessively high current or time-delay settings. Long-time delay, short-time delay, and instantaneous and ground-fault characteristics are available as required and all are individually adjustable in both current and time delay, either by means of discrete tap settings or a continuously adjustable setting.

Figure 73 shows the relative operating characteristics of two circuit breakers with solid-state trip devices applied in series and illustrates how coordination is achieved between circuit breakers having different combinations of long-time, short-time, and instantaneous trip elements. Since all tolerances and operating times are included in the published characteristics for low-voltage circuit breakers, to establish that selectivity exists requires only that the plotted curves do not intersect.

5.7.2.4 Fuses. Typical power fuse performance curves are shown in Fig 74. As with low-voltage circuit breakers, the total time for circuit opening is described by a band that represents the manufacturing tolerance of the fuse, the upper boundary being the maximum operating time. Total fuse performance, however, is defined by a broader band with a lower boundary representing the thermal damage characteristic. The clearing characteristic of a downstream interrupter should be coordinated with the thermal damage characteristic of an upstream fuse to prevent any deterioration of its rating or change in its normal opening delay.

5.7.2.5 Ground-Fault Protection. As discussed in 5.2.4, care must be exercised when using a conventional one-line diagram to analyze three-phase systems. An important application consideration associated with this problem is illustrated when examining the selectivity of a system employing a ground-fault protector in the main service as shown in Fig 75. The analytical flaw becomes apparent when standard time-current curves are employed in combination with the one-line diagram to evaluate coordination of the downstream protectors with the main ground-fault protector (GFP). The simple overlay of curves can indicate complete selectivity and suggest that satisfactory isolation of the circuit is accomplished. However, the change in circuit conditions that occurs after the opening of (only) one phase protector can cause a response by the main GFP before the remaining branch or feeder protectors clear the circuit, resulting in a loss of the entire electrical service.

The addition of sensitive GFP devices to each downstream feeder or branch could improve the coordination significantly. The improvement in operation should, however, be considered in view of the cost and availability of adequately rated equipment.

5.7.3 Preparing for the Coordination Study [57]. The following information be required for a coordination study:

(1) A system one-line diagram showing the complete system details including all protective devices and associated equipment

(2) Schematic diagrams showing protective device tripping functions

(3) A short-circuit analysis providing the maximum and minimum values of short-circuit current that are expected to flow through each protective device whose performance is to be studied under varying operating conditions

(4) Normal loads for each circuit and the anticipated maximum and minimum operating loads and special operating requirements

(5) Machine and equipment impedances and all other pertinent data necessary to establish protective device settings and to evaluate the performance of associated equipment such as current and potential transformer ratios and accuracies

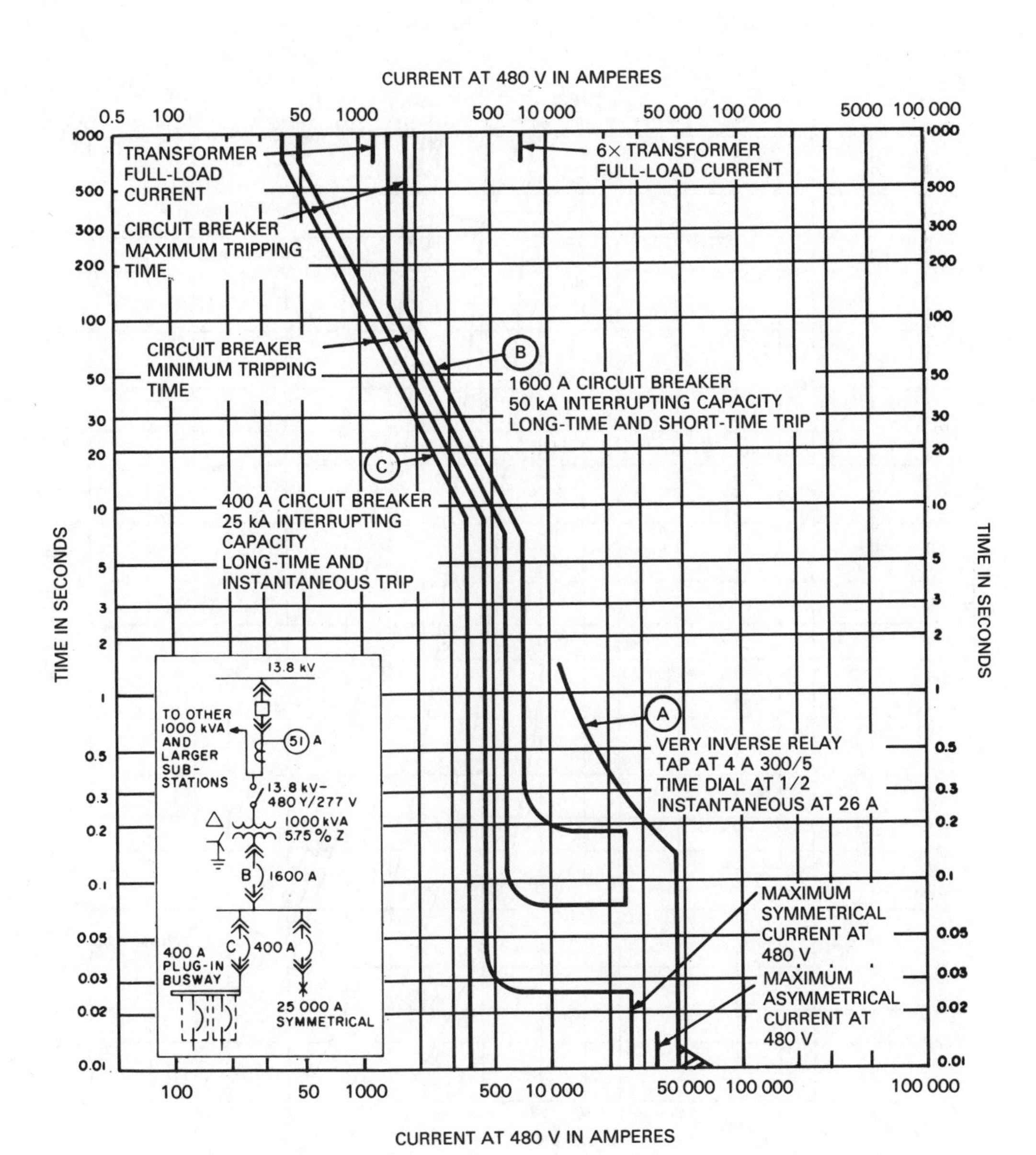

Fig 73
Selective Tripping Time-Current Characteristic Curves
(Low-Voltage Power Circuit Breakers on Secondary Unit Substations)

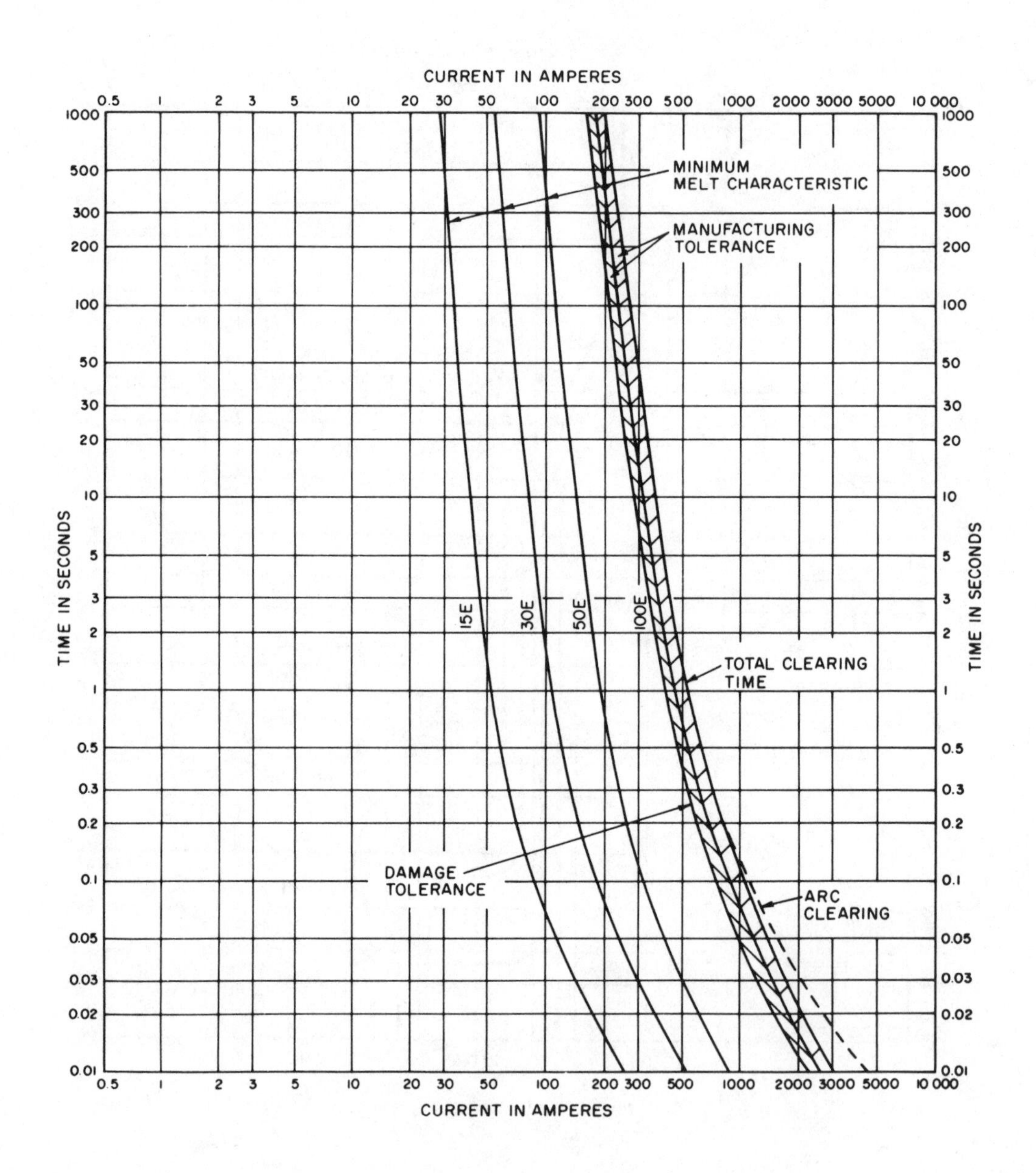

Fig 74
Typical Time-Current Characteristic Curves of Fuses

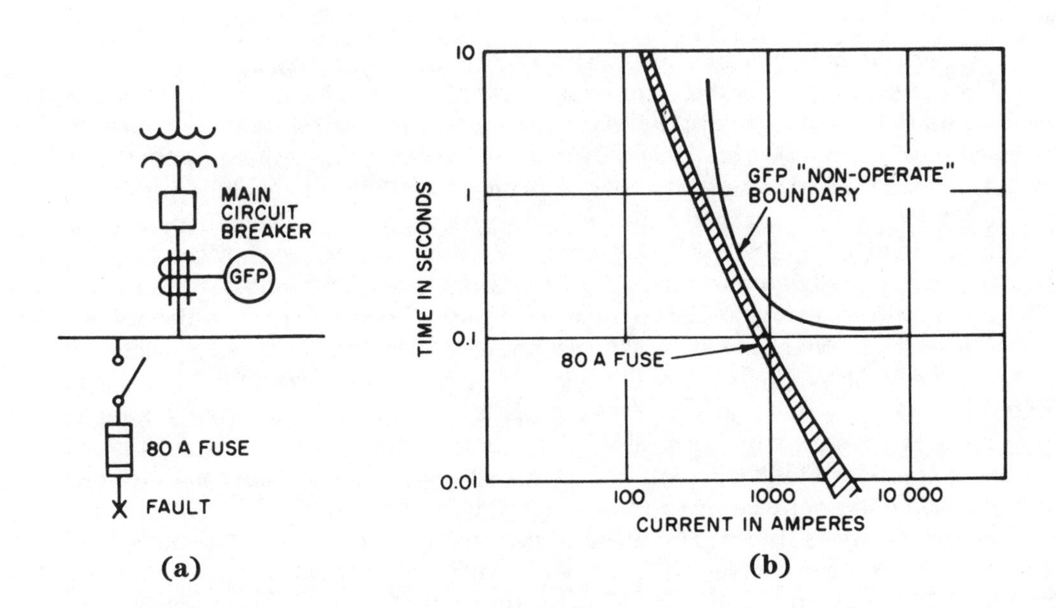

Fig 75
One-Line Diagram and Time-Current Coordination Curve Misrepresenting Proper Fault Clearing

(6) All special requirements of the power company intertie, including the time-current characteristic curve of the utility protection immediately upstream from the system

(7) Manufacturers' instruction bulletins, time-current characteristic curves, and interrupting ratings of all electric protective devices in the power system

(8) NEC [22] or other governing code requirements as a reference

5.8 Specific Examples—Applying the Fundamentals. To illustrate some of the many factors that should be considered and the problems that arise when applying the information and principles provided in the previous sections of this chapter to an actual industrial power system, the completed coordination curves (Figs 76–85) for the system shown in Fig 65 will be discussed in detail. Strong emphasis has been placed on the prime objectives of equipment protection and selective interrupter performance. Relays that are unresponsive to system overcurrents and have no time-current characteristics are not shown on the graphical coordination plots. The selection of settings for these devices is beyond the scope of this text but can be readily determined by referring to the manufacturer's instruction material covering the relays in question.

The examples given are only intended to be illustrations. Each system encountered in practice should be analyzed in detail since effective protective device selection and coordination must apply to a specific situation and not a general case.

5.8.1 Setting and Coordination of the 13.8 kV System Relaying

5.8.1.1 Primary Feeders Supplying Transformer. The overcurrent relays applied on the 13.8 kV circuits that energize load center distribution transformers provide the dual function of primary protection for phase and ground faults occurring on the 13.8 kV cable and transformer primary winding, and backup protection for faults normally cleared by the secondary devices. Since backup protection requires selective tripping with the secondary main circuit breaker, the primary protection is usually compromised to the extent necessary to obtain selectivity. This compromise can be minimized by selecting a relay characteristic which follows the time-current characteristic of the secondary device as closely as possible.

As shown in Fig 83, the overcurrent relays (Device 51/50) chosen for feeders E, G, and J have an extremely inverse characteristic, and the settings provide a curve that ensures selective tripping with the secondary circuit breaker over most of the range of secondary fault current. The upper edge of the secondary main circuit breaker trip curve represents total clearing time, and a margin of 0.2 s between it and the relay curve at the maximum secondary fault current level is recommended. An intersection or crossover with the secondary circuit breaker occurs at a low region of fault current, as shown in Fig 83, for relay G, and this is regarded as an acceptable compromise in order that the transformer would be fully protected within its withstand limits. The transformer withstand limits are plotted on the curve sheet as three-phase fault limitations and also as the equivalent current (0.58 per unit) appearing in the primary protective device for secondary line-to-ground faults when the secondary neutral is solidly grounded. The same degree of protection on secondary line-to-ground faults is not provided for the transformers supplied by feeders E and J, since the associated relaying is set higher to accommodate the other connected loads. Installation of separate protection ahead of these transformers having a trip characteristic no higher than that shown for relay G could correct the problem. Clearing could be accomplished either by a separate interrupter ahead of each transformer (not shown) or by transfer tripping of circuit breakers E and J.

The relay pickup or tap value setting is based on three considerations:

(1) to enable the feeder and transformer to carry its rated capability plus any expected emergency overloading

(2) to provide for selectivity with the transformer secondary circuit breaker

(3) to provide protection for the transformer and cable within the limitations set forth in the NEC [22], Articles 450-3 and 240-100.

The adjusted trip characteristic for relay G demonstrates the relative performance of the primary and secondary protective devices for a secondary line-to-line fault. As compared to a three-phase fault, there is a slightly larger range of possible fault currents over which coordination between the primary relay and secondary circuit breaker is compromised. Had the secondary overcurrent device been a relay with the same characteristic as the primary relay, complete coordination could have been realized, provided a sufficient clearance was maintained between the curves at a current of approximately 25 000 A (28 700 · 0.87) at 480 V.

The instantaneous overcurrent element (Device 50), employed in conjunction with the time element (Device 51), is set to be nonresponsive to the maximum

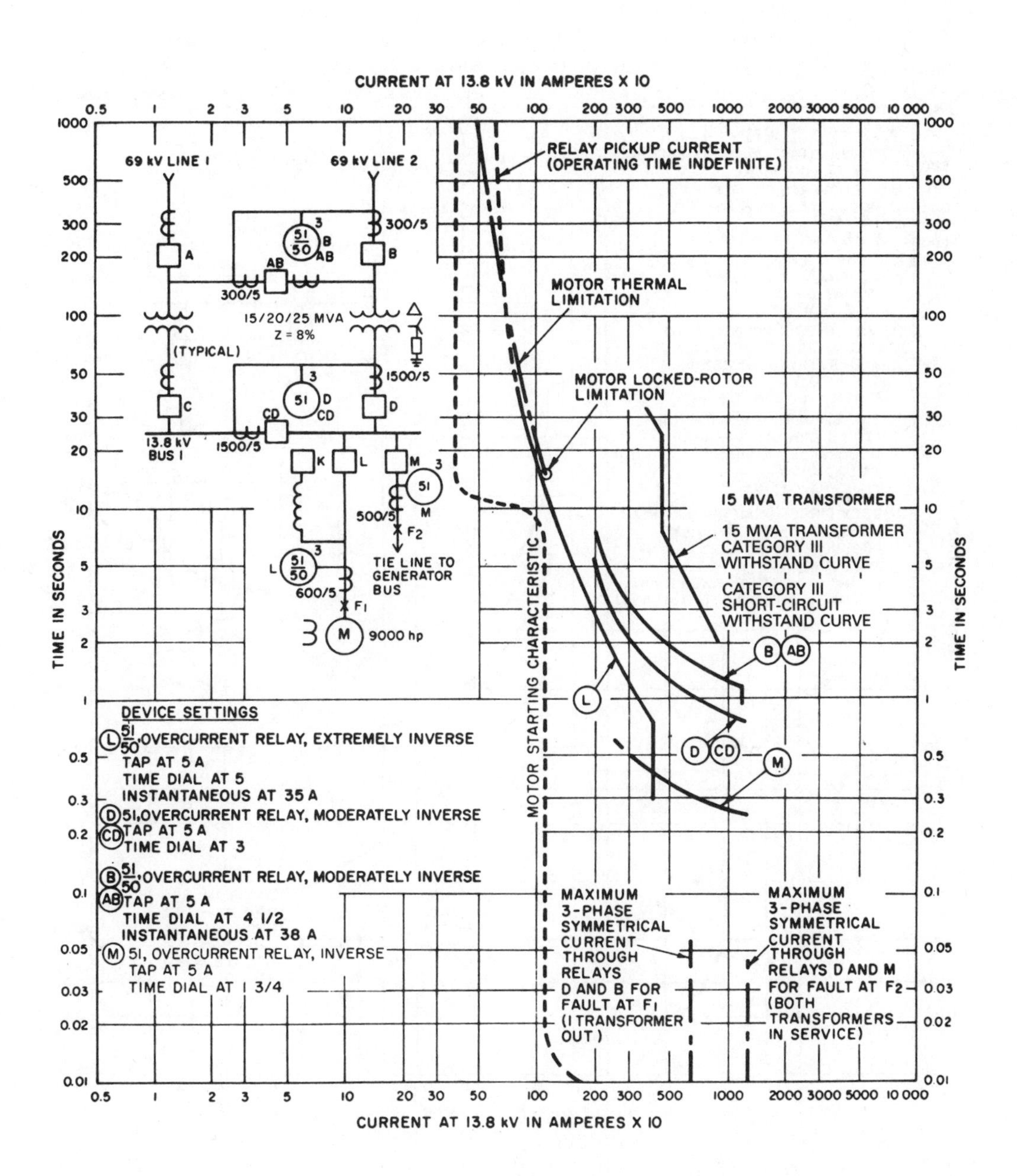

Fig 76
Phase-Relay Time-Current Characteristic Curves for 13.8 kV Feeders L and M and Incoming Line Circuits

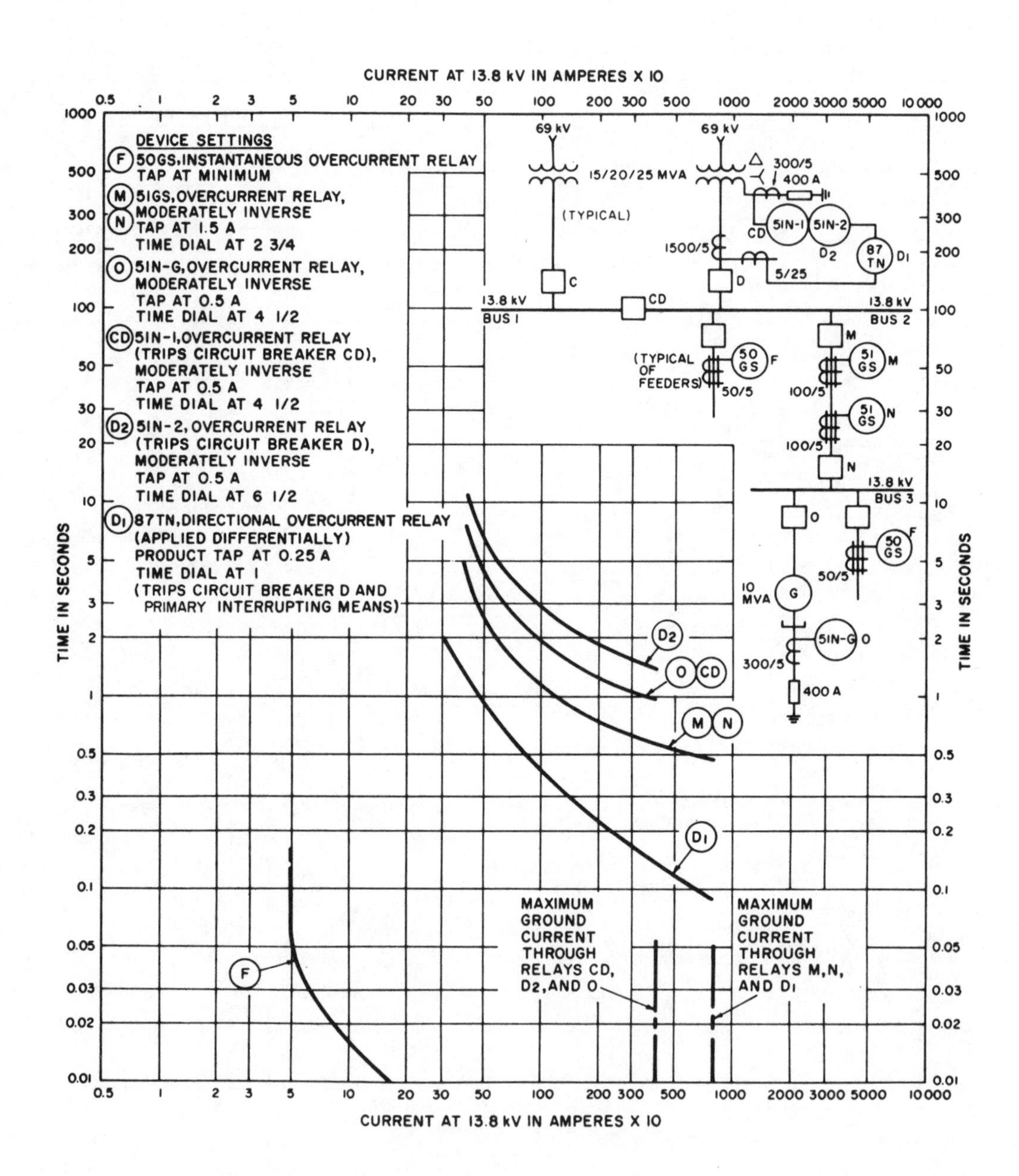

Fig 77
Ground-Relay Time–Current Characteristic Curves for 13.8 kV Source and Feeder Circuits

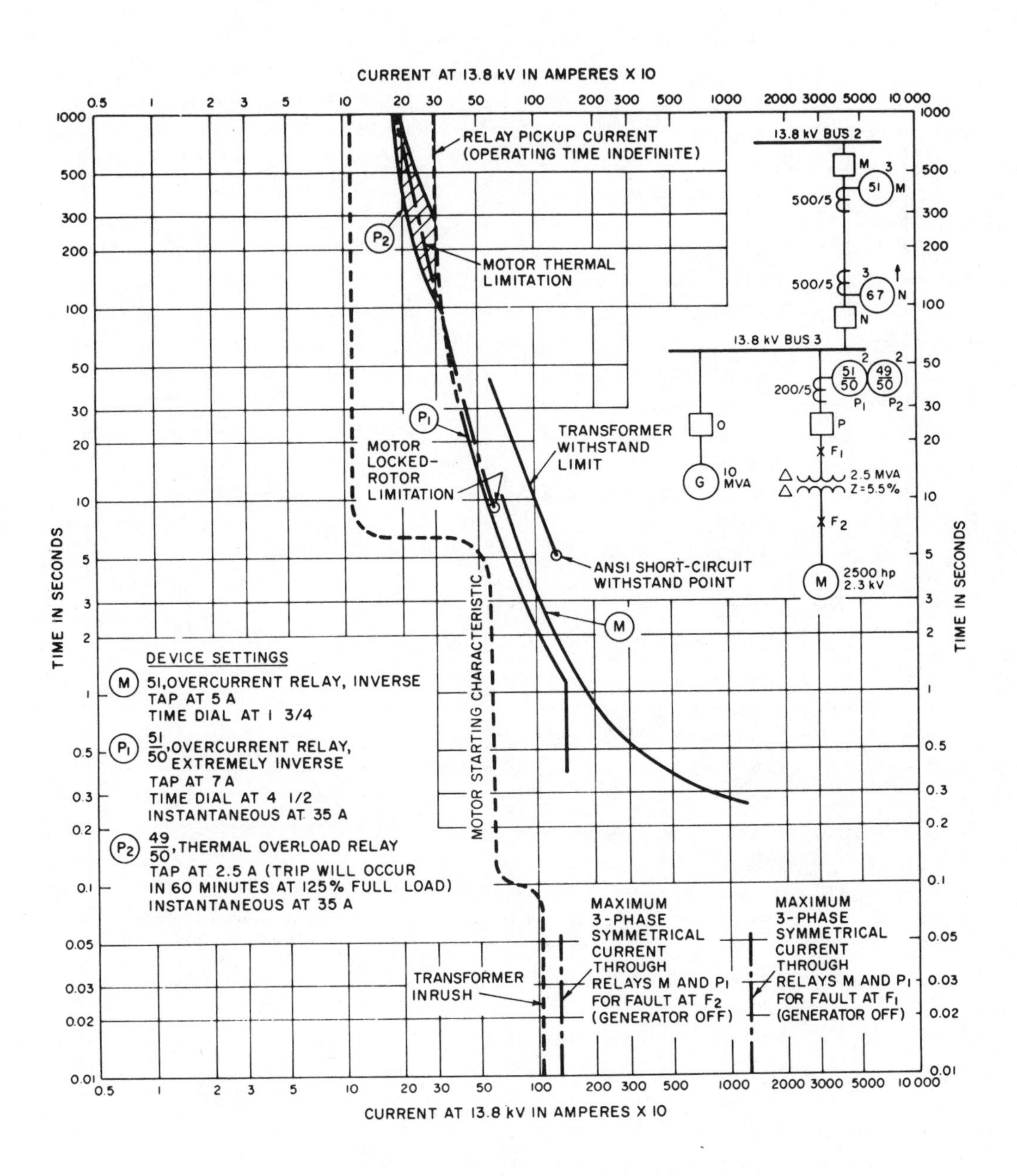

Fig 78
Phase-Relay Time-Current Characteristic Curves for Feeder Relay at 13.8 kV Bus 3

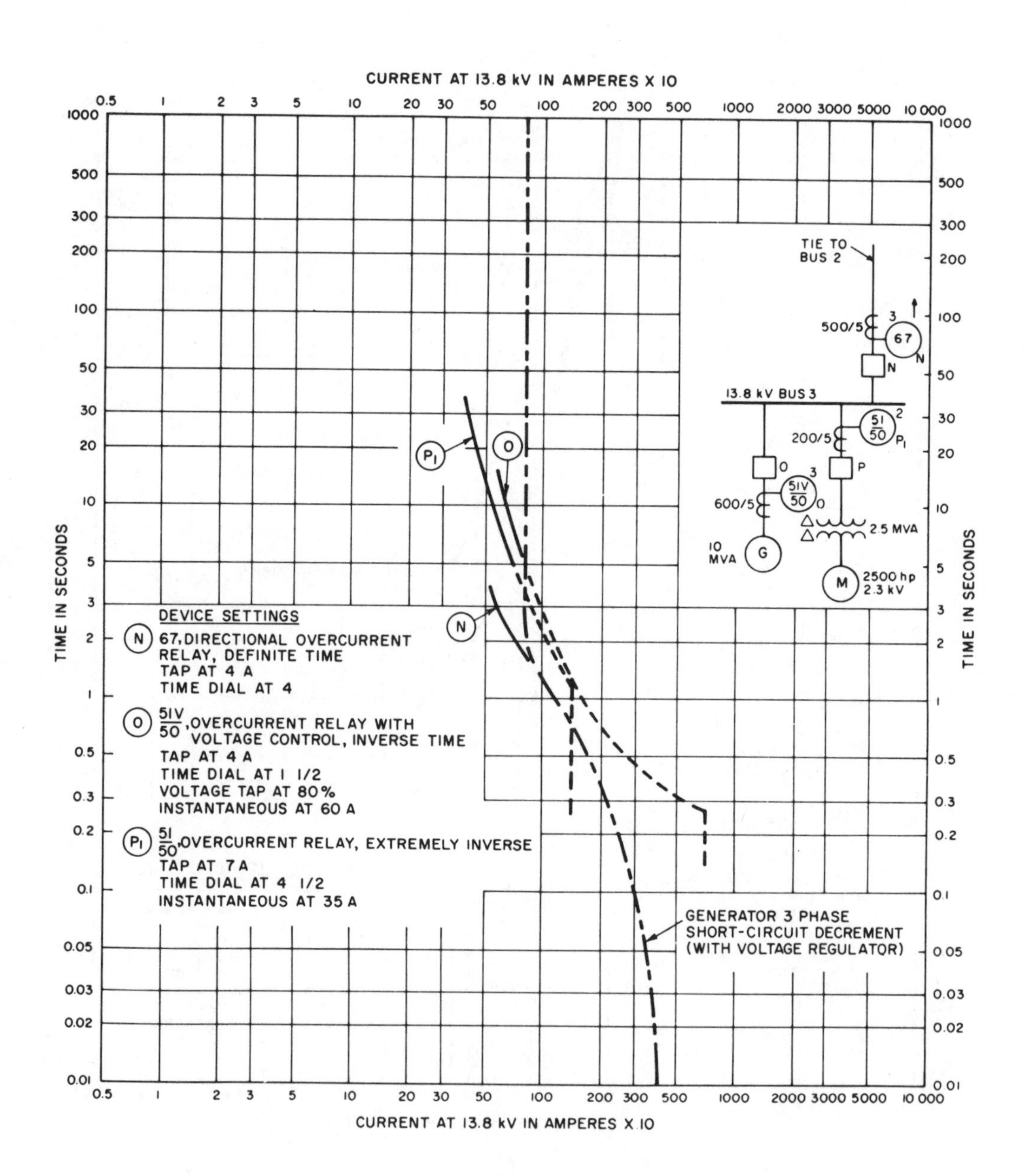

Fig 79
Phase-Relay Time-Current Characteristic Curves for Generator Relay at 13.8 kV Bus 3

asymmetrical rms fault current that it will see for a transformer secondary three-phase fault. The symmetrical value of transformer let-through current is calculated and an asymmetry multiplying factor applied as determined from the *X/R* ratio of the impedance to the point of fault. In addition, a 10% safety margin is added to this calculated setting. When the relayed feeder is energizing more than one transformer, the magnetizing inrush of the transformer group may be the limiting factor for the instantaneous trip setting.

In Fig 81 the overcurrent relays (Device 51/50) for feeders H and I are set to operate selectively with the totalizing relay at the maximum expected 2.4 kV fault current with about a 0.4 s delay between curves. The instantaneous element is set to pick up above the 2.4 kV system asymmetrical fault availability. The pickup setting and characteristics of the extremely inverse relay afford excellent protection of the transformer by staying below the damage curve at all current levels.

For feeders serving relatively small transformers, such as the 750 kVA transformer energized by the bifurcated feeder from circuit breaker J, the short-time thermal withstand capability of the selected cable size should be checked. The full-load rating of this transformer is 31.4 A at 13.8 kV, and a No 8 three-conductor cable of approximately 45 A capacity may have been selected as adequate. A plot of the thermal withstand limit, as shown in Fig 85, reveals that the No 8 cable could be damaged over its length for a 13.8 kV fault exceeding about 3000 A (90 000 A at 480 V). To prevent this possibility, a No 1 size cable should be selected.

Sensitive and prompt clearing of ground faults is possible on all 13.8 kV feeders with the application of the zero-sequence-type current transformer surrounding all three-phase conductors and the associated instantaneous current relay (Device 50GS). Ground-fault sensitivity on the order of 4–10 A is achieved with this combination, depending on the type of relay used. No coordination requirement exists with downstream devices, since these feeder circuits energize transformers with delta-connected primary windings, and ground faults on the secondary side do not produce zero-sequence current in the primary side feeder circuit. The ground relaying is shown in Fig 77.

5.8.1.2 13.8 kV Motor Protection. Figure 78 shows the degree of overload protection provided for the 2500 hp 2.3 kV motor energized through the 2500 kVA transformer from 13.8 kV bus 3. The relaying as applied should protect both the transformer and the motor.

The replica-type thermal relay (Device 49/50) has a tap setting which will trip the circuit breaker when the motor load current is sustained at 125% of rating for a period of 60 min. This pickup setting complies with the NEC [22], Article 430-32, since the machine has a 1.15 service factor. The thermal relay operating characteristic is represented as a band where the lower limit signifies the operating time when the overload occurs after a period of 100% load, and the upper limit signifies the operating time when the overload occurs following zero loading.

The setting for the time element of the phase overcurrent relays (Device 51/50) is determined by the normal starting-time and starting-current requirement for the motor and its locked-rotor thermal limitation. If the permissible locked-rotor time is greater than the required accelerating time, as in the example shown, the overcurrent relay can be set for locked-rotor protection. The pickup or tap setting is usually on the order of 50% of locked-rotor current, and a time lever setting is best determined by several trial starts under actual conditions. For some motor designs, the allowable locked-rotor time may be less than the required accelerating time, and for such conditions an overcurrent relay supervised by a zero-speed switch may be required for locked-rotor protection.

The pickup setting of the instantaneous element of the thermal and phase overcurrent relays is determined by the transformer magnetizing inrush current. Although the magnitude of the inrush current is shown plotted at its approximate minimum possible level of 10 times full load, an actual relay setting of 12–14 times transformer full-load rating should normally be adequate, but can be increased if pickup occurs during trial starts.

The Device 51/50 overcurrent relays also provide primary phase-fault protection for the feeder cable and transformer, and for this reason two relays are applied. The instantaneous ground sensor relay (Device 50GS), set at a minimum tap, completes the protection for this circuit.

Figure 76 illustrates the overcurrent relaying selected for the 9000 hp 13.2 kV synchronous motor. As for the 2500 hp motor, the extremely inverse characteristic is preferred for Device 51/50, for locked-rotor protection, and for cable and motor fault backup protection. Although backup overload protection is also provided by the 160% pickup setting of the time element of Device 51/50, its trip characteristic crosses over the motor thermal damage curve and does not afford complete protection in the light overload region. The stator winding temperature relay (Device 49), whose operating characteristics are not customarily plotted on the time-current curves and which is set to trip rather than alarm, will protect the machine in the region where Device 50/51 does not. The time dial setting falls within the limits of allowable locked-rotor time and required motor-starting time. Since the motor-starting inrush current is limited by the starting reactor, the setting for the instantaneous element of Device 51/50 is based on the asymmetrical current which may be contributed by the motor to a fault on an adjacent circuit. This current is calculated from the subtransient reactance of the machine, and 1.6 and 1.1 multiplying factors are applied to allow for asymmetry and a safety margin.

5.8.1.3 Generator Protection. The 10 MVA generator connected to the 13.8 kV bus 3 has a phase-overcurrent relay with voltage control (Device 51V) applied as backup protection for three-phase faults occurring on the 13.8 kV bus, or on feeder circuits connected to the bus, including the generator circuit. The voltage control or voltage restraint feature of the device permits moderate overloads of the machine without tripping. The instantaneous element for this relay is set above the generator contribution including dc offset to back up the differential relays for faults into the machine from the system. Additional protection for phase-to-phase and phase-to-ground faults is provided by the negative-sequence relay (Device 46) and the ground relay (Device 51G).

Figure 79 illustrates the coordination requirements for the circuits connected to the 13.8 kV bus 3. The generator output under external fault conditions is plotted as a dashed line. The directional overcurrent relay (Device 67), applied at circuit breaker N as backup to the pilot-wire relaying, has a pickup setting that permits full loading of the generator over the tie line. A time lever setting is used that provides

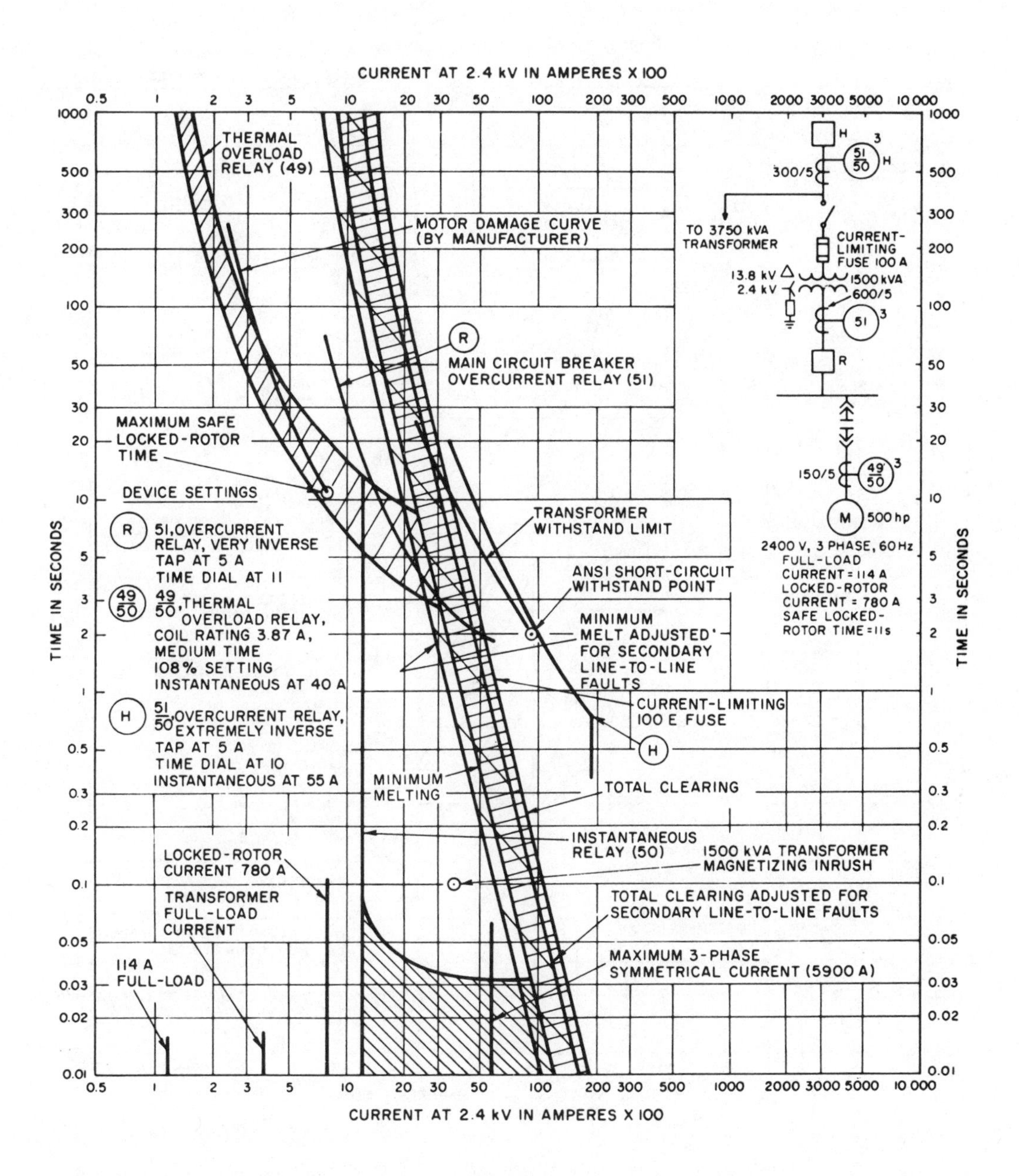

Fig 80
Phase-Relay Time-Current Characteristic Curves for 2.4 kV Bus 1 Coordination

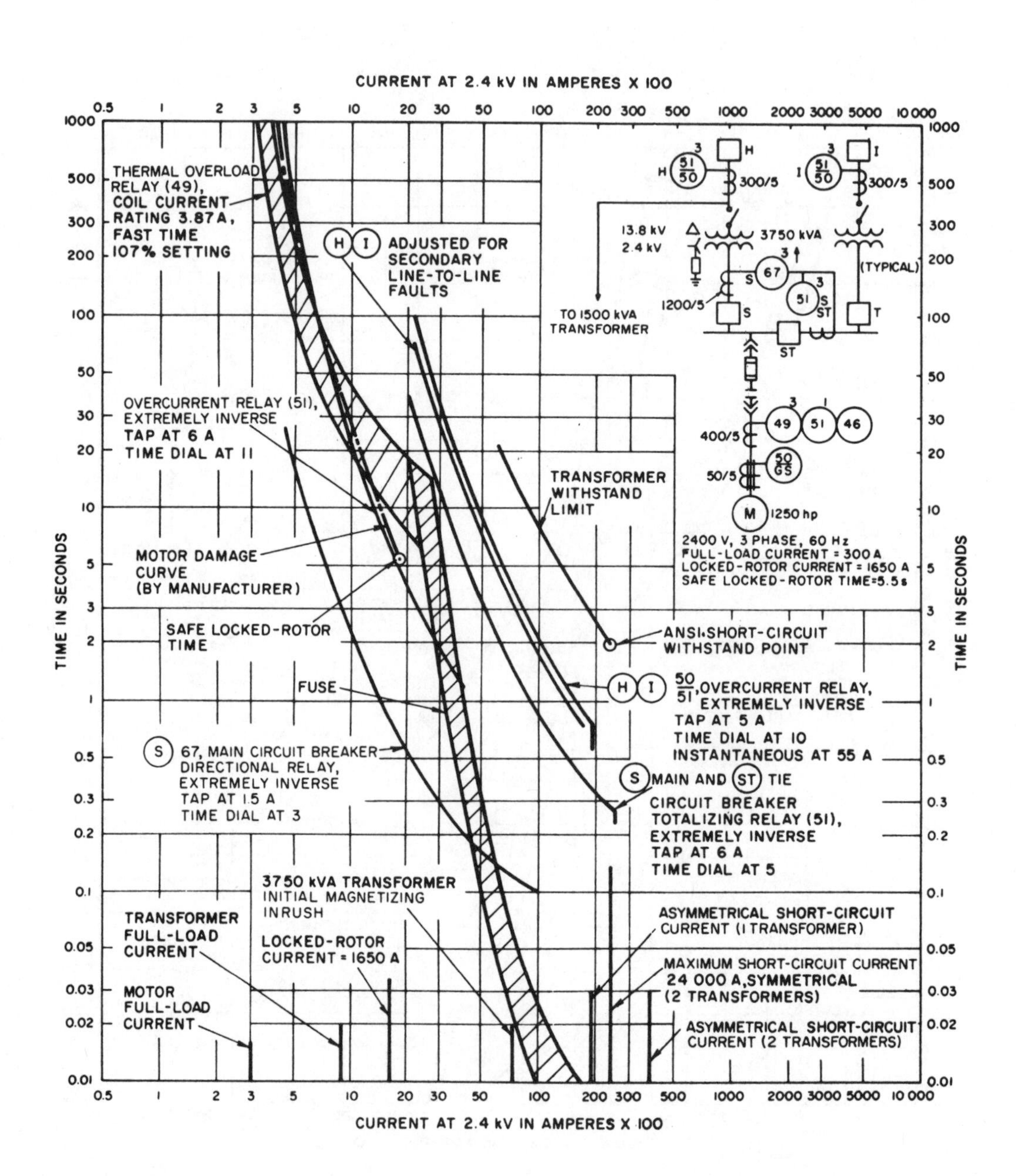

Fig 81
Phase-Relay Time-Current Characteristic Curves for 2.4 kV Bus 2 Coordination

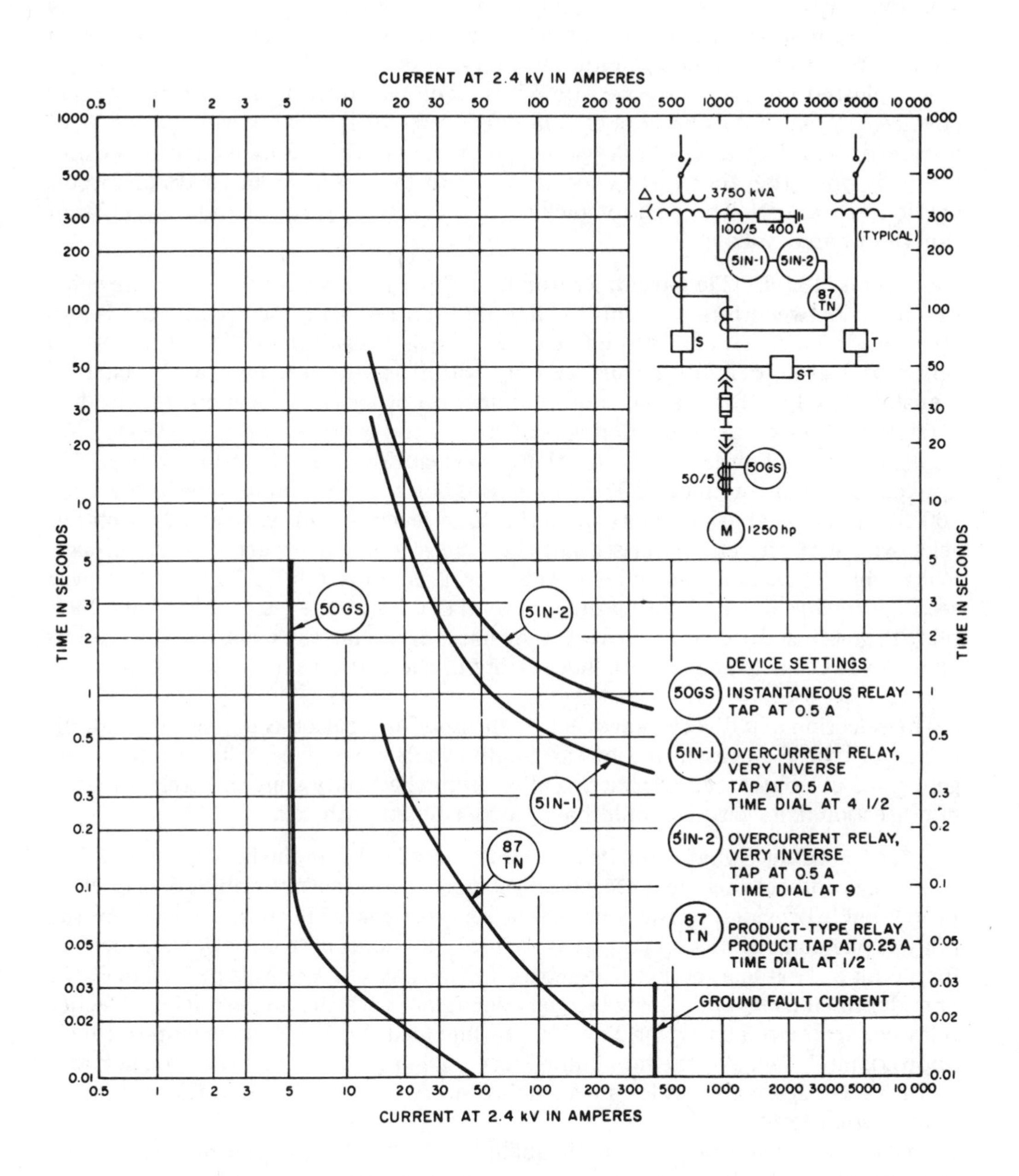

Fig 82
Ground-Relay Time-Current Characteristic Curves for 2.4 kV Buses 2 and 3

selectivity with the 13.8 kV feeder relays of buses 1 and 2 to the extent allowable by the generator short-circuit current, which is too low in comparison to the system contribution to permit coordination in every case.

The plotted inverse characteristic for the voltage-controlled overcurrent relay (Device 51V) at circuit breaker O is in effect only when the bus voltage is 80% of normal or less. This level of voltage can be expected for 13.8 kV feeder faults, and the relay operating time has to coordinate with the overcurrent relays on circuit breakers P_1 and N. The current pickup or tap setting is approximately 115% of generator rated output.

5.8.1.4 13.8 kV Tie Circuit Protection. The primary protection for the cable tie circuit between buses 2 and 3 is line differential, using pilot-wire type relays (Device 87L) at each end of the line. This relay is instantaneous and sensitive to phase and ground faults occurring only within the area zoned by the current transformers. For this reason, coordination with other relaying is not required.

Backup phase-fault protection is provided by the overcurrent relay (Device 51) applied at circuit breaker M, and the directional overcurrent relay (Device 67) applied at circuit breaker N. The tap setting for relay M would be selected near 100% of circuit cable ampacity, and the time lever setting is selected to obtain selectivity with the characteristics of the longest delay overcurrent relay it *overlooks*, which is relay P_1 on feeder P. This setting is plotted in Fig 78. The time lever setting provides a coordinating time interval of 0.6 s at the current level at which relay P_1 goes instantaneous. This is somewhat longer than the usual 0.4 s interval to allow for a possibly required higher setting of the instantaneous trip element on relay P_1.

The selection of a directional overcurrent relay at location N in place of a nondirectional type is necessitated by the limited fault-current contribution from the generator as contrasted to that supplied from the utility source. If relay N were nondirectional, its setting would have to coordinate with relay P_1.

5.8.1.5 Main Substation Protection. The 13.8 kV main buses 1 and 2 are primarily protected by bus differential relaying (Devices 87B1 and 87B2). Backup protection is provided by the overcurrent relays (Device 51) connected so that the relay applied on one bus section sees the total contribution from both transformers for a fault on that bus section. The setting for the Device 51 relay is plotted in Fig 76 and identified as relay D. Relay D must coordinate with the longest delayed feeder relay connected to buses 1 or 2 and the tie line feeder relay M. Its pickup setting is approximately 140% of the maximum force-cooled rating of one transformer, and its time lever setting provides a 0.4 s delay interval with relay M at the maximum fault current level.

The main transformers are individually protected by transformer differential relaying (Device 87T), and as backup protection overcurrent relays (Device 51/50) are applied at the 69 kV level and connected also in a summation arrangement. This relay is identified as relay B in Fig 76, and its tap setting is also at 140% of the maximum rating of each transformer. The time lever setting provides a suitable delay interval with relay D at the maximum simultaneous fault-current value that can be seen by both relays. The instantaneous element supplied with relay B is set above the maximum asymmetrical current that can be seen by relay B for a 13.8 kV fault, which occurs with one transformer out of service.

The relay settings now established at the 69 kV main substation entrance must be reviewed with the power company to ensure that their upstream protective devices will be compatible. In some cases it may be necessary to compromise selectivity to some degree or to establish settings with shorter coordinating intervals in order to meet the maximum clearing times permitted by the utility. Also, the setting of Device 67 that looks out into the utility system should be discussed with the utility to assure compatibility with their system operating procedures.

5.8.1.6 13.8 kV Ground-Fault Protection. Each of the three wye-connected neutrals in the 13.8 kV system are connected to ground through a 19.9 Ω resistor that limits the ground-fault current available from any transformer to 400 A. Depending on the number of transformers in service, a range of 400 A minimum to 1200 A maximum is therefore available for ground relay detection. The sensitivity of the relays applied and their associated current transformers provide a detection capability less than 10% of the 400 A minimum available.

The ground overcurrent relay settings are plotted in Fig 77. All transformer feeders and the 13.8 kV motor feeder are protected with instantaneous current relays energized by zero-sequence-type current transformers of 50/5 ratio. In terms of primary current their pickup sensitivity will be on the order of 5–10 A.

The tie-line ground relays at circuit breaker locations M and N are necessarily time delayed and have identical settings since selective tripping between the two is not important. Their time lever setting provides coordination with the main transformer neutral differential relay (Device 87TN), designated as relay D_1, for faults in the zone that include the transformer secondary and the line side of the main 13.8 kV circuit breakers.

The next level for selective tripping is relay O, the ground relay in the generator neutral. Its setting coordinates with 0.4 s delay with relays M and N at the maximum 400 A level. Likewise relay CD in the transformer neutral is only required to coordinate with relays M and N. Its setting, therefore, can be the same as that selected for relay O.

Relay D_2, also in the transformer neutral, must be delayed 0.4 s beyond relay CD at the 400 A ground-fault level. Relay CD trips the bus tie circuit breaker and thus establishes the location of the fault as being on one side or the other of the bus tie. Whichever transformer is still energizing the faulted bus section will then be tripped off by relay D_2.

5.8.2 Setting and Coordination of the 2.4 kV System Relaying

5.8.2.1 Phase Protection. Figure 81 is the plot of the phase protection for 2.4 kV bus 3 that serves motor loads including the 1250 hp induction motor, representing the largest connected machine. The motor thermal damage curve, which must serve as the starting point for properly designing the protection for any machine, has been plotted as shown. The motor thermal overload relay (Device 49) satisfactorily matches the machine damage characteristics on overloads up to approximately 200% of full-load rating and has been set to protect the motor against sustained overloads. Beyond this point, the extremely inverse time relay (Device 51) matches the motor-damage curve better than Devices 49 and 50 and provides protection for currents up to locked rotor by cutting under the maximum safe locked-rotor time. The 2.4 kV fuse is present to protect the contactor by interrupting heavy fault currents and is sized to withstand locked-rotor current at 10% overvoltage.

The main and tie circuit breaker summation overcurrent relays have been set to coordinate with the motor protection and permit normal expected bus loading. Also, a sufficient delay in relay operating time has been provided to allow the contactor (or overload relay) to selectively clear moderate faults on the same motor circuit should the fault occur on a phase or phases that would escape detection by the single overcurrent relay (Device 51), or even on the same phase should the relay fail to operate.

The current-balance relay (Device 46) provided for single-phase protection of the motor has no time-current operating characteristic as such that would affect the relay coordination on either balanced or unbalanced overcurrent condition. It has sufficient built-in delay to permit other relays such as ground relays to operate first when required. A plot of its performance therefore is not relevant and does not appear in the coordination curve. For best protection, Device 46 should be set at maximum sensitivity provided nuisance tripping does not result.

The time-delay element of the directional overcurrent relay (Device 67) is set for maximum speed and sensitivity to provide the best protection. The relay should have sufficient delay (0.1 s) for reverse flow of motor contribution current into a primary fault. If an instantaneous element was used for improved protection, it must be set to pick up above either this current level or the secondary magnetizing inrush current of the transformer for installations where secondary switching of transformer banks with *dead* primary may occur.

Note from Fig 80, however, that since the setting of the relaying Device 50/51 for circuit breaker H is dictated by the coordination requirements of the 3750 kVA transformer, the 1500 kVA transformer is not adequately protected. This is evident by the fact that the relay curve falls above the transformer damage curve. A 100E primary fuse has been applied to fill this protection void and appears to do so since its clearing time curve falls to the left of the transformer short-circuit withstand point. Also, it provides sufficient operating delay to withstand expected transformer loading continuously as well as transformer magnetizing inrush for the required 0.1 s. However, the fuse does not completely protect the transformer on low-magnitude (arcing) faults due to the crossover of the fuse-interrupting curve and the transformer-damage curve. If such a failure should occur between the transformer and the secondary main circuit breaker R, some amount of transformer damage may be expected, as was discussed in 5.8.1.1. Improved protection would be provided by separate relaying at the transformer primary terminals, which would operate to transfer trip the primary circuit breaker H.

The protection illustrated in Fig 80 for 2.4 kV bus 1 serving the 500 hp motor provides a different approach to locked-rotor protection than that described for the 1250 hp motor. Again, the thermal overload relay has been set to permit continuous operation of the motor at rated current and to provide protection for sustained small overloads. There is, however, no Device 51 to provide protection for heavy overloads or locked rotor. The thermal damage curve intersects the maximum operating time of the overload relay at approximately 300% of full-load current, and beyond this point there is no protection. The condition gets worse with increased current up to locked-rotor level. The overload relay will eventually operate, and if its operating time should fall close to the minimum operating time, it may protect the motor. This condition is normally considered economically justifiable on small or noncritical machines. Nevertheless, some motor damage and reduction in expected life will result if a prolonged high overcurrent condition exists. The instantaneous element of the relay is set to pick up above the motor locked-rotor current (including dc component) to avoid nuisance tripping on starting. The contactor in this example is capable of interrupting the available fault current so that current-limiting fuses are not required in combination with it.

Protection against motor damage from single-phase operation such as would occur following interruption by one transformer primary fuse is provided by the negative-sequence voltage relay (Device 60). The main circuit breaker overcurrent relay (Device 51) has been set to provide transformer overload protection and also to coordinate with the downstream motor protection and the upstream transformer primary fuse. Coordination with the fuse is doubtful for currents above about 3000 A due to the difference in the fuse and relay characteristics.

5.8.2.2 Ground-Fault Protection [44], [45]. (See also ANSI/IEEE Std 242-1986 [20]. Figure 82 illustrates the coordination of the ground-fault protection on 2.4 kV bus 2. All the feeder circuit breaker ground relays (Device 50GS) operate from zero-sequence-type current transformers and are set to trip instantaneously with maximum sensitivity.

The time-delay product type (Device 87TN) detects ground faults only between the transformer and the main circuit breaker and functions to trip the appropriate primary feeder circuit breaker and secondary main circuit breaker. Since it is not necessary to coordinate this relay with the feeder ground relays, it is set on the minimum time lever for fastest possible operation.

The relay 51N-1 must coordinate with relays 50GS and 87TN to trip the 2.4 kV tie circuit breaker ST for bus faults to ground or as backup to the 2.4 kV feeder circuit breaker ground relaying. Relay 51N-2 must coordinate with Devices 50GS, 87TN, and 51N-1. This is the final relay to operate on bus faults and must wait for the tie circuit breaker to open and then isolate the fault on one bus section or the other by tripping the appropriate main circuit breaker.

The ground fault protection for 2.4 kV bus 1 is not plotted since it is a high-resistance grounded system with no tripping. The voltage relay (Device 59N) senses the presence of a ground fault on the system, which is evidenced by a current flow and voltage drop through the resistor R, and operates an alarm.

5.8.3 Setting and Coordination of the 480 V System Protective Equipment. Although there are several types of 480 V systems commonly used in industrial plants, only the *fully selective–fully rated* system will be described since the other systems (such as partially selective systems) are similar but compromise either a degree of equipment protection or selectivity for reduced cost.

5.8.3.1 Phase Overcurrent Protection

(1) *480 V Radial System.* 480 V buses 4 and 6 each represent a small distribution system feeding motors or other loads such as lighting or heating. For optimum protection and selectivity, the following should be considered:

(a) The motor control center (MCC) feeder circuit breaker series trip devices must provide overload and short-circuit protection for its feeder cables, and should also be selective with the downstream fuses or molded-case circuit breakers for all values of fault current up to the maximum available at the MCC bus. This is accomplished by using long-time and short-time series trip devices on the feeder circuit breaker. The *minimum* time delay band setting on the short-time characteristic easily coordinates with the total-clearing characteristic of the molded-case circuit breaker. Overload protection is provided by setting the long-time element to pickup at the lesser of the feeder cable ampacity or approximately 125% of the MCC load.

It is possible to achieve selectivity between a molded-case circuit breaker and a feeder breaker equipped with elecromechanical trip devices having long-time and instantaneous characteristics, provided the maximum let-through energy I^2t of the molded-case circuit breaker is less than the unlatching I^2t of the feeder circuit breaker trip device during the interruption of fault current. Generally this will be true if the molded-case circuit breaker is 75 A or less, although the equipment manufacturer should be consulted to be sure. It would be difficult, if not impossible, to obtain selectivity between the molded-case circuit breaker and the instantaneous tripping charcteristic of a solid-state trip device. The solid-state trip devices are sensitive to rms current and are activated by a voltage signal that is proportional to the instantaneous current magnitude. In contrast, the molded-case circuit breaker requires a finite amount of let-through energy (I^2t) to open the breaker and clear the fault.

(b) The main secondary circuit breaker series trip devices must provide overload protection for the load center distribution transformer and short-circuit protection for the 480 V bus and feeder circuit breakers and must also be selective with the feeder circuit breaker series trip devices. This can be accomplished by using long-time and short-time trip devices in the main secondary circuit breaker with the long-time element set to pick up at the maximum permissible short-time overload capacity of the transformer.

Figure 85 shows the phase protection for 480 V bus 4 using standard molded-case circuit breakers and solid-state trip devices.

(2) *480 V Network System.* 480 V buses 1, 2, and 3 comprise a typical spot-network system. For optimum protection and coordination the following factors should be considered:

(a) Power must not flow out of the network into the supply system as a result of faults in the supply circuits. This is accomplished by using directional overcurrent relays (Device 67) operating from current transformers in the secondary connections from each supply transformer. These relays must be set to trip the associated 480 V supply circuit breaker when current flows into the supply system before the other two supply circuit breakers open as a result of the operation of their non-directional protective devices. If the directional relays are equipped with instantaneous elements capable of directional discrimination (some do not have this capability), the instantaneous elements would be set to pick up above the momentary 480 V induction motor contribution to primary faults so as to avoid an unnecessary opening of the main circuit breakers for trouble on a remote circuit.

(b) Faults on buses 1, 2, or 3 or on feeders fed from these buses must be cleared either by the feeder circuit breakers or by the associated 480 V supply and service bus tie circuit breakers before all service bus tie circuit breakers open. This can sometimes be accomplished by careful selection and setting of the feeder protective devices and the supply and tie circuit breaker trip devices. In most cases, however, special relays will be required to obtain the necessary tripping characteristics for either the tie or the supply circuit breakers. Figure 83 shows that coordination between bus 1, 2, and 3 service ties as well as between a service tie and the 480 V feeder circuit breakers is difficult to obtain using directional relays alone. Instantaneous bus differential relays should be considered as the alternate means for solving the problem.

(c) Certain modes of operation could result in the 13.8 kV tie circuit breaker CD being opened. In such cases a fault on any primary circuit would cause high currents to circulate through this substation as power is transferred between the separated primary systems. If the time dial and instantaneous settings for the directional relays were selected so as to provide coordination between the 480 V main circuit breakers and any 13.8 kV feeder circuit breakers not serving this substation for remote primary faults, a loss of coordination between the 480 V main supply circuit breakers would result.

To ensure selective operation of all the protective devices during the normal mode of operation with the 13.8 kV bus tie circuit breaker CD closed, loss of coordination must be accepted for certain types of faults should this circuit breaker ever be opened. With a fault on a remote primary feeder such as the circuit fed by circuit breaker F, the directional relays on the 480 V main supply circuit breakers of buses 1 and 2 will trip before the relaying on primary feeder F would selectively isolate the fault. A similar situation arises for any remote primary feeder fault with the 13.8 kV tie circuit breaker open. Since this is not the normal mode of operation, it is not considered to be a serious compromise.

(d) Faults on the 3000 A busway or on service feeders fed from this busway must be cleared either by the feeder circuit breaker protective devices or by the service bus tie circuit breakers before all supply circuit breakers open and cause a complete 480 V system outage. This is accomplished by using selective trip devices (long-time–short-time delay) on the tie and supply circuit breakers.

Figure 83 shows that selective phase protection has been achieved (no overlapping of trip characteristics) for 480 V buses 1, 2, and 3 using fused feeder circuit breakers with electromechanical series trip devices, and provided that the tie and supply circuit breakers are equipped with directional relays or differential relays. Similar characteristics could be obtained using a solid-state trip device having long-time and short-time trips.

From the curves it is not obvious that bus 1, 2, or 3 feeder circuit breaker 800 A fuses coordinate with the dual-element 200 A motor-starter fuses. This is determined by examining the I^2t let-through charcteristics of the two fuses. To coordinate satisfactorily, the total clearing I^2t of the 200 A fuse must be less than the damage I^2t of the 800 A fuse. In some cases where coordination is difficult, several fuse ratings (even within the same class) must be evaluated to establish the most suitable design.

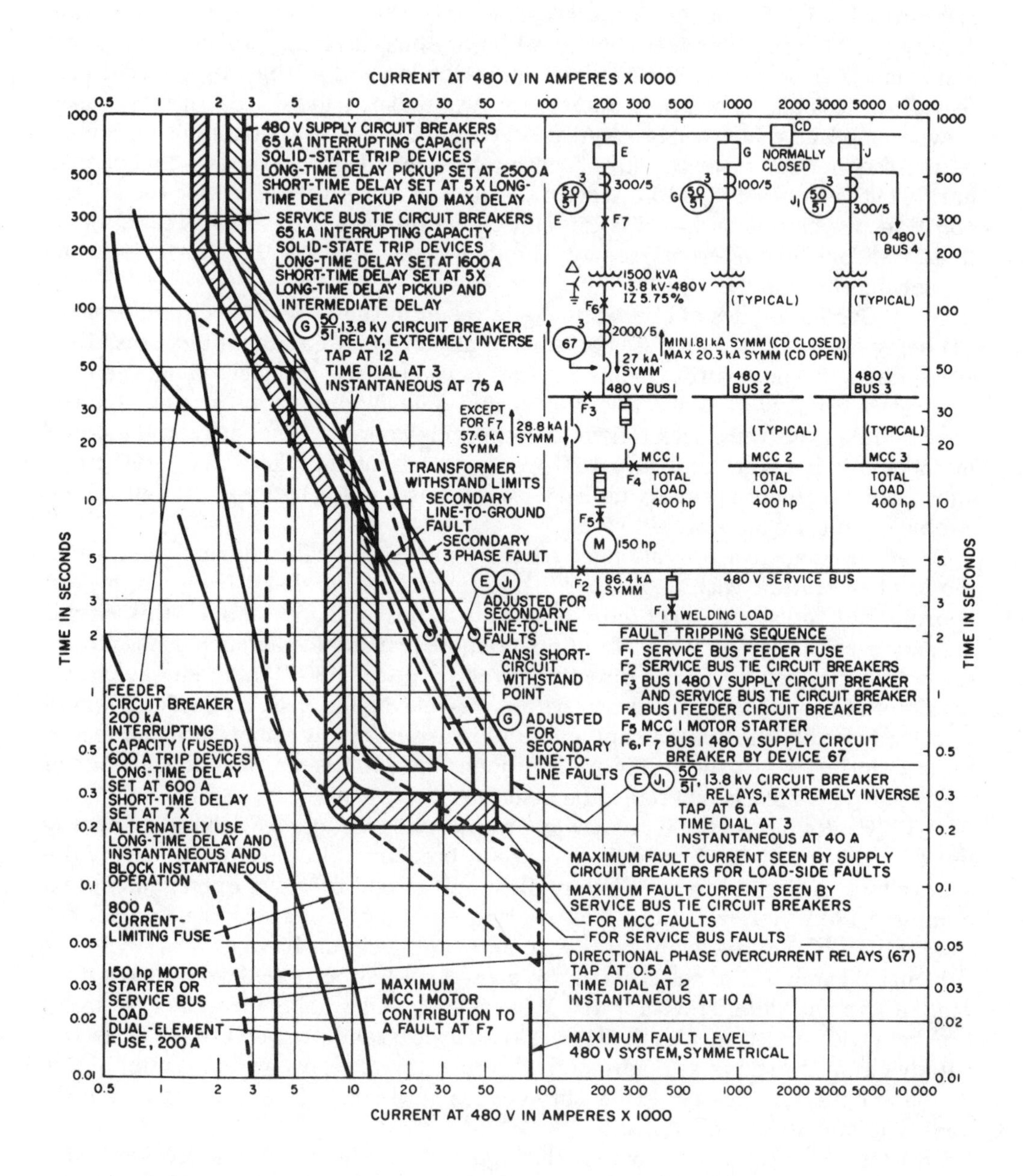

Fig 83
Phase-Protection Time-Current Characteristic Curves for 13.8 kV Feeder E, G, and J and 480 V Bus 1, 2, and 3 Network Coordination

(3) *480 V Double-Ended Secondary Unit Substation with Normally Closed Secondary Tie.* 480 V bus 5 is essentially a variation of the 480 V spot network (buses 1, 2, and 3) previously discussed. The basic protection considerations also apply for this example. The following factors should be considered for optimum protection and coordination:

(a) Directional relays should be applied to the main secondary circuit breakers as described in the discussion for 480 V buses 1, 2, and 3 to selectively isolate primary system faults.

(b) Faults on either bus section or on feeders fed from the bus must be cleared by the feeder circuit breaker protective devices or by the bus tie circuit breaker and the associated 480 V supply circuit breaker before the other 480 V supply circuit breaker operates and causes a complete 480 V outage. This is accomplished by using long-time-short-time delay selective trip devices in the tie and supply circuit breakers.

Figure 83, although specifically representing buses 1, 2, and 3, illustrates the delay sequence that would be used for selecting coordinated device settings for bus 5 as well.

As in the case of 480 V buses 1, 2, and 3, the I^2t characteristics of the motor starter and feeder circuit breaker fuses should be evaluated to ensure coordination between fuses.

5.8.3.2 Ground-Fault Protection. The 480 V spot network in Fig 65 is solidly grounded. The coordination curves for the ground-fault protection for this portion of the system will, therefore, illustrate the principles that apply to obtaining selective performance of the circuit breakers during ground faults. With solid grounding, the maximum line-to-ground fault short-circuit current is virtually equal to the three-phase value, but because of the fault and ground-return impedance much lower magnitude currents flow. Sensitive relaying should therefore be used to ensure that ground-fault currents that are too low to be tripped instantaneously or even detected by the standard trip devices or solid-state series trip devices are safely cleared.

Common methods employed to provide this protection are as follows:

(1) Relays fed from residually connected current transformers in feeder circuits, tie circuit breakers, and main supply circuit breakers

(2) Zero-sequence (doughnut) current transformers enclosing all phase conductors in feeder circuits rather than residually connected current transformers; the connections are otherwise the same as described in [33]

(3) Solid-state ground trip devices integral with feeder, tie, and main supply circuit breakers

(4) Overcurrent relays connected to the secondary of a current transformer sensing the current in the distribution transformer neutral (not illustrated in this diagram)

Very special care is required when applying ground relays to four-wire secondary selective systems where the neutral circuit is not switched if selectivity is to be achieved between main and bus tie circuit breakers (see ANSI/IEEE Std 142-1982 [19]). The principles involved in selecting coordinated ground-relay settings are the same as those described for phase protective devices. Figure 84 shows selective

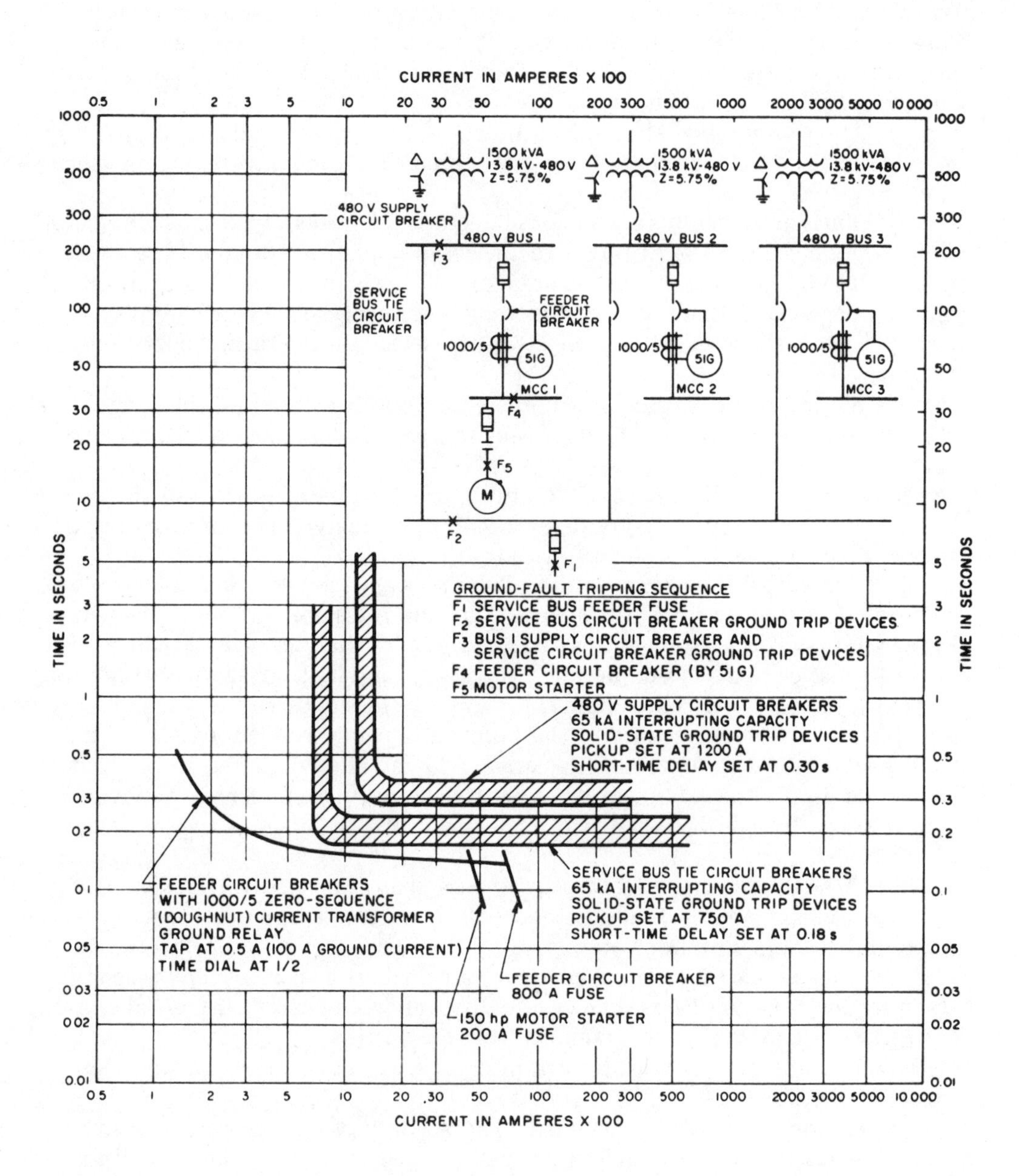

Fig 84
Ground-Relay Time-Current Characteristic Curves for 480 V Bus 1, 2, and 3 Network

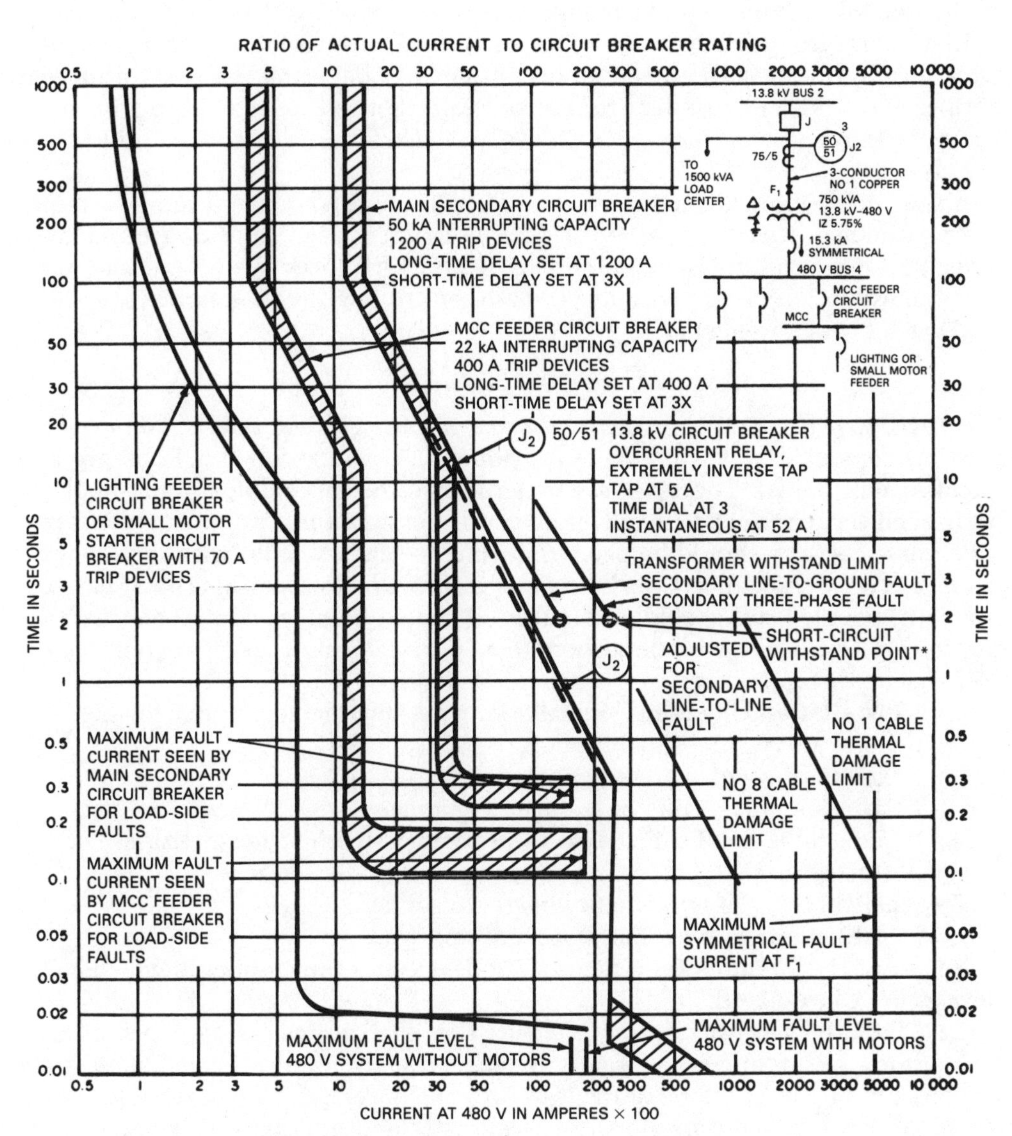

*As defined in ANSI C57.12.00-1973, which has been superseded by ANSI/IEEE C57.12.00-1980 [15].

Fig 85
Phase Protection Time-Current Characteristic Curves for 13.8 kV Feeder J and 480 V Bus 4 Coordination

ground relaying for bus 1, 2, and 3 spot-network systems using a definite time ground relay fed from a window-type current transformer around the feeder cables to motor control centers, and solid-state ground trip devices in the service bus tie and main 480 V supply circuit breakers.

In a solidly grounded system, the zero-sequence current transformer must be of such a ratio that it will provide an output current low enough to be within relay ratings and sufficiently undistorted to accomplish accurate relaying during maximum ground-fault conditions. For the system illustrated, a ratio of 1000/5 has been selected.

A time-delay relay is suggested for the feeder circuit breakers to allow motor-starter and feeder circuit breaker fuses to interrupt high-magnitude ground faults. Selective fault isolation for faults on bus 1, 2, or 3 requires bus differential relays, since coordination cannot be achieved between ground directional relays on the tie and service main breakers.

5.9 Testing [31], [37]. In order to secure the full benefit that a well-designed protective installation is capable of providing, the installation should be properly installed and tested. The tests are exacting and often complex, and should be performed very carefully to avoid endangering persons and equipment. Where possible, these services should be procured from specialists. Otherwise the following guide may be of assistance. Checking a new installation to ascertain that it has been correctly installed, connected, and adjusted should be more comprehensive than the testing that should be done periodically to determine deterioration, loss of adjustment, or tampering.

5.9.1 Installation Checking. Installation checking includes the following:

(1) General survey and diagramming

(2) Preliminary checking of equipment

(3) Calibration and setting of installed equipment

(4) Implementation of safeguards to prevent permanent magnetization of current transformers

(5) Final checking of equipment going into service

5.9.1.1 General Survey and Diagramming

(1) Study the intended function of each device and the manner in which all the devices are coordinated.

(2) Check the wiring diagrams to ensure that each device is so connected that it will perform its intended function. If no diagrams have been provided, make them or obtain them, for it will be difficult to do a safe intelligent job of testing without them. Preserve the diagrams for future reference and update them when changes or additions are made.

(3) Compare the diagrams with the actual connections and, when differences are found, determine whether the error is in the diagram or in the wiring and correct it.

5.9.1.2 Preliminary Checking of Equipment

(1) Inspect equipment for damage or maladjustment caused by shipment or installation.

(2) Verify that all protective relays, auxiliary relays, trip coils, trip circuit seal-in and target coils, fuses, and instrument transformers are the proper types and range.

(3) Remove wedges, ties, and blocks installed by the manufacturer to prevent damage during shipment.

(4) Make electrical continuity checks of all current, potential, and control circuits, referring constantly to the diagrams.

(5) Remove short-circuiting links from current transformers after checking that secondary circuits are complete.

(6) Make ratio and polarity checks of current and potential transformers where these somewhat time-consuming tests are considered necessary.

(7) Make tests of the insulation of all relays, wiring, instrument transformer secondaries, and instruments.

(8) Make provision for future testing of the equipment in conveniently small units. The installation of a relatively small number of test switches, test terminals, and test links before the equipment goes into service may make it possible to test the various elements of the protective installation one by one after they are in service, with a minimum of disturbance to production. The omission of such devices may require major parts of the plant to be shut down for testing. Where current transformers or wiring are shared between the relays of a single position and differential or other relays that are common to groups of positons, see that safe means are provided for separating each position's relays, current transformers, and wiring from the group being tested.

5.9.1.3 Calibration and Setting of Installed Equipment

(1) *Direct-Trip Circuit Breakers*. Low-voltage air circuit breakers are often tripped directly by the current through them without the interposition of current transformers and relays. Electromechanical trip devices are usually set at the factory and to check them in the field requires a test source capable of supplying trip currents. If they cannot be tested, at least verify that the marked instantaneous and time-delay settings are as required for coordination with other circuit breakers and fuses.

Solid-state trip devices have adjustments that are easily set and tested at the job site using a small compact test set designed for that purpose by the breaker manufacturer. The adjustments are usually factory set on their minimum settings; it is advisable, therefore, to set and calibrate the devices to their specified values.

(2) *Relay-Operated Circuit Breakers*. The relays should be checked in accordance with the manufacturer's instructions and with the general guide below in which initial and maintenance checks are compared. However, the actual performance of the relays in service depends on the behavior of the instrument transformers that supply them with current and potential; and these in turn are influenced by the magnitude of their secondary burdens. It is therefore advisable to plan the testing in such a way as to obtain information about the performance of the relays, wiring, and transformer together as a unit, as well as separately.

Figure 86 shows one phase of a typical current-transformer circuit, indicating four different positions at which test current may be applied. The first three positions cause current to flow either toward the relay only, toward the current transformer

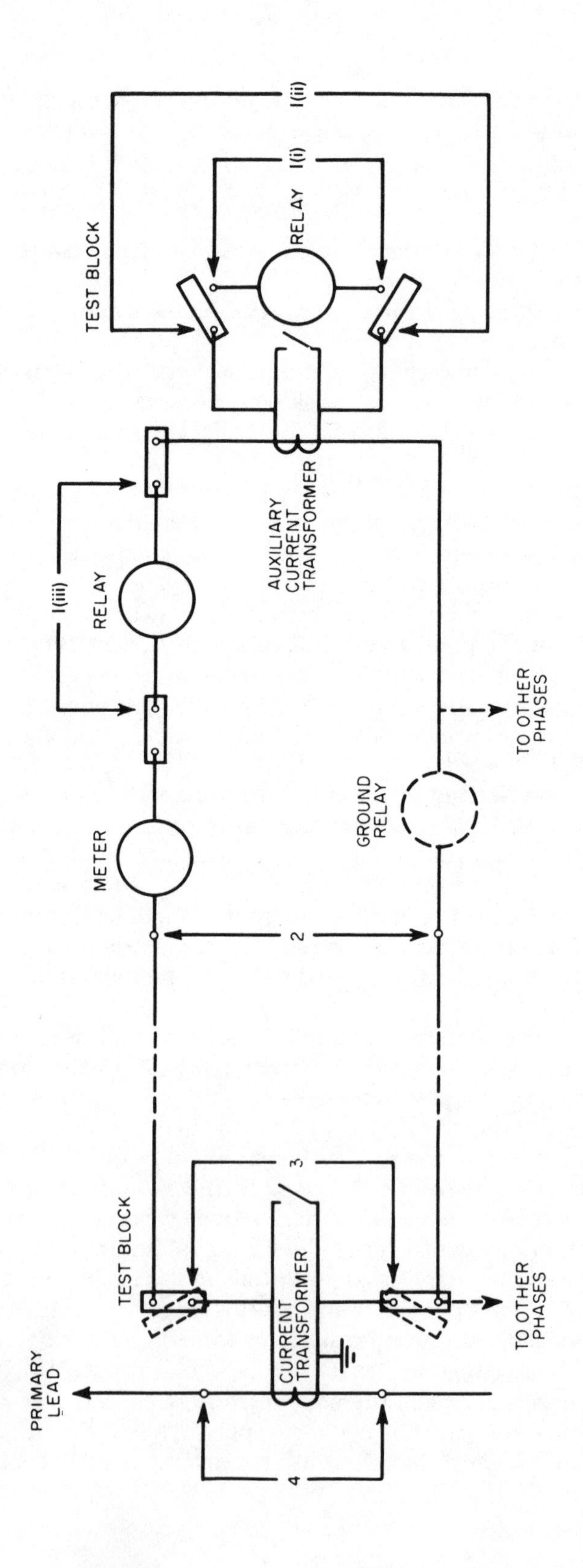

Fig 86
Typical Current-Transformer Circuit

only, or toward both in parallel. The test from position 2 or 3 toward the current transformer only is a secondary impedance or excitation test and should include at least three points on the current-transformer saturation curve, with one at or slightly above the knee. One three-phase set of current transformers may differ widely in impedance from another set, yet each may be satisfactory for its own function if the values are consistent within the set.

(a) *Position 1. Test current applied at the individual relay location.* At this point it is possible to make three measurements, each of which is useful under certain conditions. In order to show position 1 in all three of its variations, an auxiliary current transformer has been added. These auxiliary current transformers are often used in multiple differentials and other complex schemes but are rarely employed in simple circuits like that shown in Fig 86. In testing they are treated the same as any other current transformers.

(i) The relay is disconnected, checked, and calibrated separately as an instrument.

Relays should be checked with current and an accurate timing device when they are placed in service to make sure that they have not been damaged in shipment and that they operate in the desired time shown by the coordination curves. If a relay does not give the desired operating time for a given current on the predetermined time dial setting, the desired time can usually be obtained by adjustment of the time dial. Other adjustments should not be attempted unless the adjuster is quite familiar with relay design and performance or has specific manufacturer's instructions.

(ii) With the relay disconnected and the main current-transformer primary effectively open, the test current is applied to the remainder of the secondary circuit. The current drawn should be low until the voltage is raised to the point where a main or auxiliary current transformer begins to saturate. This test checks for defects in the secondary circuit, except in the disconnected relay, but including current-transformer excitation current, open and short circuits, cross connections to other phases, etc.

If this test discloses appreciable differences in the test voltage required to produce a given value of current in the various phases, find out why. The cause may be an open or short circuit in the secondary wiring, a defective current transformer, or a legitimate unbalance of secondary burden, caused, for example, by single-phase metering or by the omission of a relay on one phase. It is not unusual to find that in the current-transformer common return lead of the three phases the burden, including relays, is much greater than in the phase leads. If it appears to be excessive, the tests should be extended at least to position 2.

(iii) The test current is applied to the relay current terminals, with the secondary wiring to the current transformers and other equipment normally connected. The relay is subjected to the same tests as in position 1 (i), including timing, any difference in the results obtained being of course due to the fraction of current used in the excitation of current transformers.

This test is of particular value, not only because it provides a measure of the extent to which current-transformer performance affects relay pickup and timing, but also because it is the basis of much maintenance testing. If the values obtained in this position during later maintenance tests are substantially the same as those in the installation tests, the entire layout may be assumed unchanged. Only if the tests in positions 1(iii) and 1(ii) show unexplained unbalances, definitely noticeable saturation, or questionable residual burdens, are more extended tests necessary.

(b) *Position 2. Test current applied at switchboard terminals of current-transformer leads.* The test current is applied to an entire phase group of relays, meters, auxiliary current transformers, etc. Since the main current transformers remain in shunt with the burden, their effect on relay performance is included. This is a convenient and fairly effective means of determining whether special relay calibrations are required.

In testing ground relays from position 2, both phase- and ground-relay burdens will be included, which is the condition that will exist in actual operation to clear a ground fault. The phase relays will sometimes be called upon to operate on phase-to-phase faults (test current applied between two current-transformer phase leads) and sometimes on three-phase faults (test current applied between one current-transformer phase lead at a time and the neutral current level, with the neutral burden jumpered out). If there is any significant difference in readings, data should be recorded for both connections.

The connections at position 2 may be opened to test the current transformers without burden other than their leads to the panel. This is particularly advantageous in metal-clad installations, where positions 3 and 4 are inaccessible or difficult to reach.

(c) *Position 3. Test current applied at current-transformer secondary terminals.* The test current is applied across the secondary terminals of the current transformer, or across the secondary leads in the proximity of the current transformer, with all meters, relays, and other burden normally connected and the primary open. The testing is the same as position 2, and the results are the same except that the secondary leads are included with the burden in the same manner as in normal service and that all devices can be readily identified with their respective current transformers. If leads were not positively identified, this is important. The current transformer can be tested alone from this position.

(d) *Position 4. Primary current check.* This is the best method of checking the performance of current transformers and relays together, since all burdens are included along with their normal effect on the saturation characteristics of the current transformer. Unfortunately, it is usually difficult to make the necessary high-current connections to the primary circuit and the equipment required for the high-current test source is unwieldy, so the primary current check is limited to special applications. When a primary current check is made, both ratio and polarity of the current transformers should be determined.

5.9.1.4 Implementation of Safeguards to Prevent Permanent Magnetization of Current Transformers. If during any of the above procedures the test current in the secondary winding of a current transformer is abruptly interrupted, the current transformer core iron may become permanently magnetized by residual flux to an extent determined by the current-transformer turns ratio, the hysteresis characteristics of the core steel, and the magnitude of test current interrupted in the current-transformer winding. Unremoved, this residual magnetism will significantly affect the accuracy of the current transformers when they are placed back into service and may cause the connected relays to nuisance trip or otherwise operate unpredictably.

This misbehavior can be avoided by using a continuously variable current supply for any tests involving current transformers in the connected circuits and instructing the operator to gradually reduce the test current from the test value to zero before opening the circuit to the test power source. As an alternative, the tests can be performed with all current-transformer secondaries short-circuited, and the test procedures modified accordingly.

5.9.1.5 Final Checking of Equipment Going into Service. Once the usual high-potential and phasing checks have been completed, the equipment is energized at normal potential for final check. Instrument-transformer cases should have been grounded with conductors of adequate size, and the secondary wiring grounded either solidly or, if necessary, through well-designed spark gaps.

With proper grounds in place, suitable test switches, jacks, links, etc, installed, and with adequately insulated test leads, many users manipulate connections and make tests with the equipment energized. However, one has to make certain that all the necessary auxiliary testing devices are present, and that every step of the testing procedure has been planned and closely examined in advance to guard against unforeseen hazards.

(1) *Current-Transformer Secondary-Circuit Checks*. All changes of connection, insertion, and removal of meters, etc, should be made in such a manner that the secondary circuits of energized current transformers are not opened, even momentarily. An energized current transformer with an open secondary acts as a stepup transformer with a ratio equal to the turns ratio, and thus dangerously high voltages are generated. All current-transformer secondary circuits should therefore be provided with a test block that requires the current-transformer secondary terminals to be short-circuited before the secondary circuit can be disconnected.

(a) *Null checks*. If current is found where there should be none, a defect is indicated. However, a null check is inconclusive and should be supplemented by an additional check that would detect a false null caused by an open or short circuit.

For differential circuits in which the operating coil normally has no current, check that there is none. This, in conjunction with check (b) or (c), verifies that the current-transformer ratios are correctly balanced and the polarities and phases correctly related.

A zero or negligible current reading in the neutral or common return lead of a three-phase set of current transformers under balanced-load conditions indicates ratio balance and like polarity of the three secondaries.

(b) *Inspection of active relay circuits*. Use an ammeter, voltmeter, wattmeter, or phase-angle meter to check for the proper values and polarities of voltage and current in the various relay circuits. Check the contact positions of directional element and voltage relays, and compare with those expected in view of load conditions.

(c) *Relay operation checks with diverted load currents*. Whenever any relay in service is tested in a manner that may cause it to operate, the consequences of circuit breaker tripping must be considered. If a circuit breaker operation is not permissible, the trip circuit of the relay being checked must be opened.

(i) *Differential relays*. After determining that current is in the individual current-transformer circuits but none in the operating coil circuit, a current may be caused to flow in the operating coil by temporarily short-circuiting and disconnecting all but one of the current transformers.

The current from the remaining current transformer will check not only the operating coil circuit, but also the current transformer and its leads. This should be done with each current transformer circuit in turn if they have not been verified by other tests.

(ii) *Neutral or residual current relays*. Short-circuit and disconnect the current transformer leads of all but one phase. The remaining phase will supply current to the relay.

(2) *Potential-Transformer Secondary Circuit Checks*. Measure the potentials applied to all relay potential coils. If any are inadequate, investigate. Look for blown fuses, short circuits, excessive burdens, and improperly adjusted potential devices.

Check the potential transformer and phase to which each relay potential terminal is connected by removing one potential fuse at a time (at the potential-transformer secondary terminals) and noting the effect on the voltage applied to the relay.

(a) *Ground-fault potential relays and elements*. Voltage relays, or relay elements, that are connected in one corner of a potential-transformer secondary broken-delta connection have no voltage across them in the absence of a ground fault. Such a fault can be simulated as follows:

(i) De-energize the equipment so that it is safe to work on.

(ii) Disconnect the phase lead from one potential-transformer primary terminal and fasten this lead where it can safely be re-energized.

(iii) Short-circuit the vacated primary terminal to its neutral terminal.

(iv) Re-energize the equipment, read voltages, and observe relay operation.

(v) Return connection to normal and re-energize the equipment.

(b) *Ground-fault directional relays*. These have polarizing windings, either a current winding energized from a current transformer in the equipment neutral-to-ground connection or potential windings energized from one corner of a broken-delta-connected set of potential transformers, the same as the preceding type, and their operating current windings energized from the common (neutral or residual) connection of a set of current transformers. Check them with diverted load currents as follows:

(i) Determine the direction of power flow.

(ii) Alter one phase of the potential transformer primary as described in (a) (ii) and (iii).

(iii) Short-circuit and disconnect the current-transformer leads of this same phase. This should cause the ground-fault directional relay to indicate a direction of power flow that is the reverse of that actually existing in the line, that is, if the power flows toward the bus, the relay contacts should close to trip.

(iv) Restore the current-transformer leads and remove their short circuit. Then short-circuit and remove the current-transformer leads of the other phases.

This should cause the relay to indicate a direction of power flow that is the same as that actually in the line.

(v) Restore all connections to normal.

(3) *Stage System Test at Normal or Reduced System Voltage*. This method causes no difficulties with automatic throw-over devices and is the best method of testing them. Nearly all other system tests require setting up staged faults, which are the last resort in testing. The faults are applied to the system at carefully chosen times and places, under controlled and backup-protected conditions, and the action of relays and other equipment is recorded and analyzed. Such tests are seldom used, and can only be justified

(a) if the scheme is intricate, new, or unfamiliar

(b) if the wiring is complicated or inaccessible

(c) if the relay response characteristics are believed to be so critical that the use of normal or diverted load currents would introduce intolerable phase-angle errors

(d) if the scheme has shown otherwise unexplainable misbehavior, or

(e) if the power system is so complex that performance of protective devices cannot be accurately predicted.

A staged fault test should be approved only when no other method of testing will suffice. The plan should be scrutinized from every conceivable safety consideration, and all parties who could possibly be affected should be notified.

5.9.2 Maintenance and Periodic Testing. Protective equipment should be checked periodically to assure that its operation and coordination have not been impaired by exposure to dusty, smoky, oily, or corrosive atmospheres, by mechanical vibration or shock, by excessive temperature, by tampering, or in any other manner. The consequences of neglecting proper testing may at first seem slight but they are cumulative, and such neglect can lead to an increasing loss of protection.

5.9.2.1 Maintenance Benefits. If a plant is interconnected with a utility, the utility will own or control the protective equipment adjacent to one or both sides of the tie point and will make its own initial and periodic tests necessary to protect equipment and the service to other customers. The acceptance of responsibility for the proper maintenance of other equipment may be covered by contracts and correspondence between the utility and the industrial power user, but frequently it is left to custom and practice. However, it is best that the role of the utility and the plant be clearly established in writing so that each party clearly recognizes its responsibilities to the other.

A regular maintenance program is required to minimize the number of unscheduled plant shutdowns. In some plants an hour of down time can be more costly than the expense of a thorough maintenance program. An effective maintenance program involves inspecting, testing, and maintenance of electric protective devices and all associated equipment at regular intervals.

5.9.2.2 Recommended Maintenance Frequency and Procedures. The time interval between tests of protective equipment is a variable, depending upon such factors as cleanliness of atmosphere and surroundings, average operating temperatures, freedom from vibration and shock, quality of operating personnel, etc. The best interval in any given type of installation should be determined by experience, but will lie somewhere between a minimum of six months and a maximum of three years (with supplementary operating checks at intermediate periods). If no deterioration is found on several successive checks, the testing is probably more frequent than necessary, while if equipment is found in an inoperative condition or out of adjustment, the testing has been delayed too long.

Regularly scheduled testing should be supplemented with special tests made at any time there is reason for suspecting that the protective equipment may have been damaged. An intelligent and loyal operating force can assist in noting and reporting signs of abnormal performance. Where such assistance is reliable, periodic tests may be scheduled less frequently. Standard test values must be on file, and although all tests may not be recorded, details of any defects found must be recorded. The records must cover not only relays, but all protective equipment for special as well as periodic tests.

In planning a maintenance program for a plant shutdown, the following preparatory work should be done:

(1) Select the date and time when equipment may be de-energized. In operations where summer vacations or other annual shutdowns occur, these are the best times for maintenance work. Maintenance programs may also be carried out on night shifts and over weekends.

(2) Update the system single-line diagram.

(3) Update the short-circuit analysis and protective-device coordination study using the latest values of short-circuit duty available from the power company's system.

(4) Since maintenance work is generally done on premium time, and shutdown periods must be held to a minimum, the necessary materials, tools, and test sets must be assembled in advance and prepared for use. These include relay test set, circuit breaker test set, any necessary lighting and power during the shutdown, and cleaning aids consisting of clean, white, lint-free cloth, solvents, vacuum cleaners (with insulated attachments), and dry compressed air. Equipment required for personnel safety will include an energized-phase detector, an energized-phase-detector tester, three-phase ground straps, rubber floor mats, safety glasses and shoes, and rubber gloves. Personnel must be made safety conscious by being required to wear nonconducting hard hats, keep sleeves rolled down, and remove watches, rings, metals, and other conducting articles.

5.9.3 Scope of Testing. Testing of protective equipment is generally considered as relay testing, but it goes far beyond mere verification of the relay calibration. The greater number of irregularities found do not involve the relays as instruments. Testing should be planned to check, as far as possible, the entire system from instrument transformers to circuit breaker operation.

During shutdown the maintenance program should include but not necessarily be limited to the following work:

(1) Remove, inspect, and electrically test the protective relays. Testing consists of passing current through the relay coil and recording the time required for contact closing to verify proper calibration and operation.

(2) Remove from service, inspect, and electrically test the electromechanical-trip low-voltage circuit breakers to ascertain that their tripping characteristics comply

with the manufacturer's time-current characteristic curves or the coordination study. Testing consists of passing current through each pole and recording the current and time necessary to trip the circuit breaker.

A trip test kit, which applies a current signal to the trip device simulating the output of the circuit breaker current transformers during overload or fault conditions, should be used to test static-trip low-voltage circuit breakers to verify trip calibration. At least one test in each phase using the same procedure as recommended for electromechanical-trip circuit breakers should be performed to verify actual operation.

(3) Inspect and electrically test high- and medium-voltage circuit breakers if provisions are available. Check them mechanically, and lubricate at specified points. If slow closing is possible, determine if the arcing contacts make before the main contacts do and open after the main contacts open.

(4) Check medium-voltage fuses for correct application, size, continuity, and signs of overheating.

(5) Check the key interlocking system between the medium-voltage and low-voltage components to determine if it is operating correctly to provide the safety features required.

(6) Open medium- and high-voltage cubicles and visually inspect for general conditions and for loose bus and cable connections, cable terminations, bus insulator cracks, switchgear and conduit grounding connections, control wiring, etc.

(7) Check the grounding system for all switchgear and unit substations.

A specific procedure must be established for the maintenance work during the shutdown. First, electrically test relays and circuit breakers with the unit substations energized, and then de-energize the switchgear and unit substations and perform the necessary inspection and maintenance. With this procedure the test set can be connected directly to the substation. Make connections with heavy-duty alligator clips or bolt-type clamps to the load side studs of a cubicle that contains a spare circuit breaker.

5.9.3.1 Instrument Transformers and Wiring. Inspect equipment visually for obvious defects such as broken studs, loosened nuts, damaged insulation, etc.

The indication of normal potential at the relay by lamp or voltmeter is considered adequate verification for potential transformers and circuits. If the combined relay and current-transformer check made at installation is repeated and substantially the same results are obtained, this is sufficient proof that there is no short circuit in the current transformer or its leads. A check under load with a low-burden ammeter in series or shunt with the relay will establish proof of continuity. More elaborate checking as outlined in 5.10.1 is required if there has been any change in the equipment or wiring, or if a change in setting materially alters the current or potential-transformer burdens. If there is evidence of improper performance, the equipment must be completely de-energized, the protective ground connections removed, and the insulation of the current, potential, and control wiring tested.

5.9.3.2 Relays. Protective relays may be temporarily removed for inspection and test. The relay is very accurate and dependable when operated in the proper environment and under the right conditions and should require little maintenance. Whenever removed from its case, it should be protected from dust, moisture, and excessive shock. Before testing any relay, its mode of operation and its relationship to the system should be known. Since the metallic enclosure forms a path for stray eddy currents that can affect relay performance, overcurrent relays should be tested while installed in a relay case. Manufacturers' instructions for specific relays provide useful information concerning connections, adjustments, repairs, timing data, and so on.

(1) Visually inspect the relay by removing the cover and checking for deposits of dust, dirt, or other foreign matter. Disable the circuit breaker trip circuit by removing the relay from service. Since the majority of relays operate from the secondaries of current transformers, removing the relay must be done very carefully because if the secondary circuit is opened, a very high voltage will result.

(2) Remove all foreign material such as dust or filings in the relay using a small hand air syringe or dry compressed air of 15 lb/in^2 or lower.

(3) Make sure all connections are tight.

(4) Remove any rust or filings from disk or magnet poles with a magnet cleaner or brush.

(5) Hold relay up to the light to make sure the disk has proper clearance between magnet poles.

(6) Inspect the relay for the presence of moisture.

(7) Check the condition of bearings and pivots by rotating the disk manually to close the contacts and observe that operation is smooth. Special oil is required for jewel-bearing lubrication.

(8) Clean pitted or burned contacts with contact burnisher or fine file.

(9) Check for damaged coils, resistors, or wiring, and for damaged or maladjusted indicator targets or holding devices.

The following list compares initial and periodic tests for some of the more common types of relays.

(1) *Differential Relays*

(a) *Initial.* Verify the published differential characteristics by checking at minimum operating current and at several other values.

(b) *Periodic.* Check minimum operating current with current in each restraint coil. Check operation and time at a higher value with one restraint coil.

(2) *Undervoltage Relays*

(a) *Initial.* Check and adjust pickup and dropout values. Record time, at least on drop from normal to zero voltage.

(b) *Periodic.* Check dropout voltage and time.

(3) *Overvoltage Relays*

(a) *Initial.* Check and adjust pickup and dropout values. Check time characteristics, if specified, and operational functions.

(b) *Periodic.* Check pickup. Check operation according to desired functions.

(4) *Directional Elements*

(a) *Initial.* On energizing the current winding or the potential winding alone, there should be no more than a slow drift toward closing. Using high current and low voltage, determine the minimum power in volt-amperes required for effective closing. Set test-feed for high (60 A or more) current and no voltage and apply momentarily; the contacts should not close. In applying time settings to a directional overcurrent relay, arrange the test to include the time of both the directional and the overcurrent elements.

NOTE: On current-polarized directional units and on current-product-type directional relays, the current coils can usually be connected in series for ease in testing and calibrating.

(b) *Periodic*. Test for action of current or voltage alone as above. Check overall time, including the associated element. Check directional element time separately only if the selectivity margins are so close as to require it.

(5) *Instantaneous Overcurrent Relays and Elements*. In checking the pickup current of these relays, raise the current gradually in order to avoid transient effects. Also, some plunger-type relays exhibit a marked increase of burden impedance when the gap of the magnetic circuit closes in operating, so that it may be difficult to obtain certain values of current on test. This effect can be made less troublesome by the use of relatively high values of test voltage, including a resistance in series with the test source. If a relay heats excessively during current adjustment, make a preliminary adjustment of the current with the relay bypassed.

(a) *Initial*. Check pickup current and adjust if required. Dropout is important in a few applications. If a plunger-type relay has trip coils or other protective devices in series with the relay coil, check from position 2, 3, or 4 to be sure that the pickup relay does not make the total current-transformer burden too great for the operation of these other devices.

(b) *Periodic*. Check minimum pickup current and functional response.

(6) *Induction Time-Overcurrent Relays and Elements*. These are particularly susceptible to variation in both pickup current and timing if the waveform of the current used for testing is distorted. Such distortion can arise from nonlinear magnetic effects either in the relay itself or in a test transformer. It can be minimized by using a relatively high-voltage test source (several times the voltage required across the relay winding itself), absorbing the excess in a rheostat connected in series with the relay. Some relays also change characteristics with heating, which may arise from ambient temperatures, load currents, or test currents.

(a) *Initial*. Check the pickup current for the tap setting selected. Verify the timing characteristic by checking two points on the relay operating curve, one at a minimum value or four times pickup and the second at a larger current. Make all these tests both with the relay isolated and with the current transformers and other burden included and, if there is any significant difference, record both sets of data for future use. If the relay does not operate according to its published characteristics, adjust following the manufacturer's instructions and repeat the test.

(b) *Periodic*. Check pickup current and timing, preferably in a combined test.

A very important part of the periodic maintenance is keeping adequate records for future use in detecting changes. Figure 87 illustrates a typical inspection and test form for a time-delay overcurrent relay with instantaneous attachment. It also serves as a record for model, type, and manufacturer with *as found* and *as left* calibration settings.

5.9.3.3 Low-Voltage Circuit Breakers. Manufacturers' literature should be consulted prior to inspection and testing. Visual and mechanical inspection should include the following:

PLANT ENGINEERING SURVEY
OVERCURRENT RELAY INSPECTION AND TEST RESULTS

Plant Location ______________________________ Date ______________

Relay Location ____

Relay Description ____

Relay Type ____

Relay Manufacturer ____

Current-Transformer Ratio ___ Accuracy ___ Potential-Transformer Ratio ___ Accuracy ___
Burden: ________ Burden: ________

TEST RESULTS

Settings			Relay Tests					
Tap (amperes)	Time Dial	Instantaneous	Pickup (amperes)	Tap Value × 4 (amperes)	Time (seconds)	Instantaneous Pickup (amperes)	Target (dc amperes)	
								Specified
								As Found
								As Left
			For Future Tests					Date

Seal-in holds at (dc amperes) ______________

Comments:

Fig 87
Typical Relay Inspection and Test Form

(1) Test the interlock feature that prevents plug-in or drawout while the circuit breaker is closed to assure positive action.

(2) Trip the circuit breaker and draw it out from its enclosure.

(3) Lubricate the tracks, jack screws, or other mechanism present in the cubicle for drawing the circuit breaker in and out using the lubricant recommended by the manufacturer.

(4) Cover the opening to the cubicle if a spare circuit breaker is not available.

(5) Clean the circuit breaker.

(6) Examine the primary fingers on drawout contacts on the circuit breaker for evidence of overheating or embrittlement cracks. Check springs for tension. Clean and service contact surfaces as required. Apply a thin film of contact lubricant.

(7) Remove arc chutes; examine for cracks, dirt, splashed copper, and broken or cracked snuffers. Clean.

(8) Inspect and clean main and arcing contacts in accordance with the manufacturer's instructions. Note that commercial products are available to resilver these contacts.

(9) Open and close circuit breaker several times to determine that operation is smooth and there is no binding.

(10) If provisions for slow closing are available, manually slow-close contacts and observe that all three phases make at the same time and that arcing contacts (if fitted) make first on closing and break last on opening. Make necessary adjustments.

(11) With the circuit breaker contacts open, insert a piece of paper backed with carbon paper between the fixed and moving contacts. Close the circuit breaker, then manually trip it. Remove paper and examine the *contact print* to make sure that both moving and fixed contacts are mating properly. Clean the contacts. Check proper operation of counter and flag.

(12) Tighten all screwed and bolted connections except pivot joints.

(13) Manually operate the armatures from all trip devices including undervoltage and shunt trip devices, strike the trip bar, and trip the circuit breaker.

(14) Lubricate mechanical joints and the racking device using lubricant recommended. Keep lubricants away from electrical contact surfaces.

(15) Examine all overload trip devices to verify that they are of the same type, have the same rating, and have proper settings.

Some of the common causes of malfunction of circuit breakers are as follows:

(1) Improper main contact pressure resulting in local overheating and burning of contact surfaces. In extreme cases, contacts may actually be welded together through this local overheating. Deterioration of insulation may result. The dimensional change or wear of parts may vary the wipe or overtravel of contacts resulting in improper contact pressure.

(2) Improper fit of circuit breaker primary fingers or drawout contacts on the bus tabs. Local overheating can destroy the tension in the primary finger springs.

(3) Cracked or dirty arc chutes can cause a phase-to-phase or phase-to-ground flashover when the circuit breaker is operated under load.

(4) Stiff or frozen mechanical joints in either the circuit breaker mechanism or trip device mechanism may cause delayed operation or failure to operate.

(5) Vibration may shift the trip bar so that the armature of the overload device fails to strike properly to release the latch to trip the circuit breaker.

(6) Worn parts or loss of oil in the overload time delay dashpot.

(7) Contaminated oil in the overload device dashpot freezing the armature, resulting in failure to trip.

(8) Improper calibration or setting of the overload trip device.

Improper delay in opening automatically under overload or fault is most dangerous to safety. Normal operational procedures and careful maintenance inspections will reveal most of the other conditions that may remove protection from the circuit. Unless overload tests are run, there is no way that improper delay can be detected until the device operates improperly.

Testing the direct-acting trip devices of low-voltage circuit breakers is becoming a widely accepted practice. The following procedures are used for electrical testing of circuit breakers.

(1) The connection of the circuit breaker drawout contacts to the stabs of the test set should be tight. This may present problems especially with the large size circuit breakers. Hand-operated hydraulic platform trucks and chain hoist suspension systems have been used with some success to raise the circuit breaker to the proper height.

(2) The circuit breaker is tested one phase at a time.

(3) Trip devices must be allowed to fully reset between tests.

(4) Each of the trip features on each pole should be individually tested—long-time, short-time, and instantaneous—for proper action, operating current, and time to trip. Test current and trip times need to be compared to the coordination study or manufacturers' time-current characteristic curves, and the trip mechanism should be adjusted if necessary to the proper setting.

(5) Several tests are recommended at each setting.

Test current must not be passed through the fuses when fused circuit breakers are tested. Testing must be arranged so that neither the test set nor a circuit breaker is overheated during the tests. Records should be kept of all tests and results. Figure 88 shows a typical inspection and test form for a circuit breaker with or without fuses.

On circuit breakers with solid-state trip devices, the trip devices can be easily calibrated and breaker tripping verified using a portable test set. In many cases, the test set furnished by the manufacturer can be quickly connected to the trip device by means of test jacks installed for that purpose. Both current pickup and time delay settings can be precisely calibrated. The test set cannot, however, test the transformation ratio of the current sensor and, therefore, the full-current test used for testing breakers with electromechanical trips must be used when this is desired.

Complete electrical test and maintenance for circuit breakers should be performed every two years and the circuit breakers should be exercised by tripping and closing several times every year. After the relays and circuit breakers have been tested electrically, the switchgear and substations must be de-energized, inspected, and maintained as necessary.

Records should be kept on the general condition of the apparatus for comparison at subsequent inspection periods. Figure 89 illustrates an inspection checklist for a

PLANT ENGINEERING SURVEY
LOW-VOLTAGE POWER CIRCUIT BREAKER INSPECTION AND TEST RESULTS

Plant Location ______________________ Date ____________

Circuit Breaker Location ______________________

Circuit Description ______________________

Circuit Breaker Manufacturer ______________________ Type ____________

Manufacturer's Serial Number or Shop Order Number ______________________

Trip Device Manufacturer ______________________ Type ____________

Trip Coil Rating ____________ ____ Long-Time Delay Range ____________

Short-Time Delay Range ____________

Instantaneous Range ____________

Cable Size ______________________

Fuse Manufacturer and Catalog Number ______________________

TEST RESULTS

Trip Unit		As Found Settings per Phase			As Left Settings per Phase		
		A	B	C	A	B	C
Long-Time Delay (amperes)							
Short-Time Delay (amperes)							
Instantaneous (amperes)							
Test Results	Long-Time Delay Current (amperes)						
	Time (seconds)						
	Short-Time Delay Current (amperes)						
	Time (seconds)						
	Instantaneous Current (amperes)						

Comments:

Fig 88
Typical Low-Voltage Power Circuit Breaker Inspection and Test Form

SUBSTATION # . . .

Yes No

A. *High-Voltage Section*

1. Concerning the two-position, three-pole gang-operated air interrupter switches for each feeder,
   a. Do the switches operate freely and with the proper action if "quick-make"—"quick-break"? ☐ ☐
   b. Do the three blades make contact at the same time? ☐ ☐
   c. Are the arcing contacts intact? ☐ ☐
   d. Are there insulating barriers between the poles? ☐ ☐
2. Has the circuit phase rotation been checked for the two feeders? ☐ ☐
3. What are the nameplate data for the three current-limiting fuses on the transformer primary?
4. Does the key interlocking system work correctly? ☐ ☐
5. Are the stress cones and terminations made correctly? ☐ ☐
6. Are the phase buses insulated completely? ☐ ☐
7. Are bus connections tight? ☐ ☐
8. Have the insulators been checked for cracks, cleanliness, and tracking? ☐ ☐
9. Have the system ground connections been made to the ground bus? ☐ ☐
10. Have the steel plates been provided to prevent access to the high-voltage compartment? ☐ ☐
11. Has a mimic bus been painted on the switchgear? ☐ ☐
12. Has any discoloration of current-carrying metal parts occurred? ☐ ☐

B. *Transformer Section*

1. What are the transformer nameplate data?
2. What are the insulation resistances of the primary and secondary windings?
   a. Primary to ground _____________ megohms
   b. Secondary to ground _____________ megohms
   c. Primary to secondary _____________ megohms
3. What are the resistances of the delta-connected primary windings?
   a. Phase A to phase B _____________ ohms
   b. Phase A to phase C _____________ ohms
   c. Phase B to phase C _____________ ohms
4. Has the transformer been checked for leaking conditions? ☐ ☐
5. Have the provisions been made for four 2½% taps in the high-voltage windings (two above and two below rated primary voltage)? ☐ ☐
   Tap changer is set at position _____________
6. Have the connections at the transformer bushings for both the high side and low side been checked for tightness and good contact surfaces? ☐ ☐
7. Are the connections to the transformer primary and secondary bushings made through flexible connectors? ☐ ☐
8. Have the transformer bushings been checked for cracks? ☐ ☐
9. Has the neutral of the transformer secondary been brought out through a bushing and has it been connected to the neutral bus? ☐ ☐

Fig 89
Typical Unit Substation Inspection Checklist

	Yes	No
B. *Transformer Section* (Cont'd)		
10. Has the neutral bus been connected to the grounding pad at the secondary end of the transformer by means of a removable bus-bar link?	☐	☐
11. Have the transformer accessory devices been checked?	☐	☐
12. Has the phase designation been maintained in connections from the transformer secondary bushing to the low-voltage bus? (Front to back, top to bottom, and left to right shall be phase A, phase B, and phase C, respectively)	☐	☐
13. Has the gas absorber been installed completely?	☐	☐
14. Have provisions been made to install auxiliary fan cooling with all necessary control devices wired?	☐	☐
C. *Low-Voltage Section*		
1. Have buses been provided for between the transformer secondary bushings and the transformer secondary main circuit breaker?	☐	☐
2. Are the three-phase main bus bars continuous and rated for ______________ amperes?	☐	☐
3. Is the neutral bus continuous and rated for ______________ amperes?	☐	☐
4. Is the ground bus		
a. Continuous?	☐	☐
b. Connected to the system ground?	☐	☐
c. Solidly bolted to the steel framework?	☐	☐
5. Is the bus phase designation from front to back, top to bottom and left to right phase A, phase B, and phase C, respectively, when viewed from the front of the switchgear?	☐	☐
6. Has the circuit phase rotation been checked in the bus-tie cubicle?	☐	☐
7. Are all bus connections tight?	☐	☐
8. Has any discoloration of current-carrying metal parts occurred?	☐	☐
9. Is it possible to draw a circuit breaker in or out while it is in the closed position?	☐	☐
10. If aluminum cable is used,		
a. Have the outer strands of the conductor been scored or nicked causing oxidation at the point of connection?	☐	☐
b. Are the cable terminator lugs designed for use on aluminum cable?	☐	☐
11. Have primary current-limiting fuses been provided for potential transformers?	☐	☐
12. Do the ammeter and voltmeter read correctly?	☐	☐
13. Are the ammeter and voltmeter test blocks associated with the main secondary circuit breaker wired correctly?	☐	☐
14. Are there transit barriers or protective boots over primary studs for each space cubicle?	☐	☐
15. Are the cable connections made directly to the buses only for metering and/or control?	☐	☐
16. Are all feeder conduits grounded within the substation?	☐	☐
17. Are feeder cables adequately supported?	☐	☐
18. Have feeder nameplates been installed on the switchgear?	☐	☐
19. In the fire pump feeder circuit,		
a. Is the circuit breaker overcurrent protective device set so as not to open the circuit under stalled rotor current or other motor starting conditions of the fire pump motor under maximum plant load?	☐	☐
b. Does the circuit breaker overcurrent protective device (fuses are not recommended) have overcurrent setting for short-circuit protection only (instantaneous trip without long- and short-time delay)?	☐	☐

Fig 89 *(continued)*

unit substation that has a primary selective design. It is divided into a high-voltage section, a transformer section, and a low-voltage section. A similar checklist should be tailored to other items of equipment.

5.9.3.4 Control Wiring and Operation. Periodic testing of protective equipment must ensure that the operation of any tripping relay will result in the circuit breaker being tripped. After all terminals and exposed portions of the trip circuit wiring and the condition and adjustment of any circuit breaker auxiliary switches in the trip circuit have been visually checked, the relay trip contacts should be manually closed to simulate an actual trip operation. Where there are too many relays to trip the circuit breaker from each, the trip circuit can normally be tested through at least one relay that is connected with the others to a common point on the opening control circuit wire. A record should be kept of each relay from which the circuit breaker was tripped, so that all relays may in turn be covered during successive tests.

5.9.3.5 Completing the Job. Before re-energizing the protective system, a visual inspection by at least two qualified persons should be made to ascertain that all temporary ground connections are removed, and that all tools, rags, and other cleaning aids have been removed from the interior of switchgear and unit substations.

5.9.4 Summary—Requirements for Protective Equipment Testing

(1) Adequate test sources.

(2) Adequate test switches or similar means of isolating equipment that will permit testing equipment both individually and in functional groups.

(3) Control and metering equipment for the applied currents and voltages. Some types of relays require special equipment, which is specified in the manufacturers' instruction books.

(4) A suitable means for timing trip circuit closures and openings and for timing circuit breaker closing and opening times.

(5) Complete records of all connections, electrical constants, settings, test values, operating performance, and failures or weaknesses found on test.

(6) Trained personnel.

(7) Written test procedures that are comprehensive and easy to follow.

5.10 References. This standard shall be used in conjunction with the following publications:

[1] ANSI C37.6-1971, American National Standard Schedules of Preferred Ratings for AC High-Voltage Circuit Breakers Rated on a Total Current Basis.

[2] ANSI C37.46-1981, American National Standard Specifications for Power Fuses and Fuse Disconnecting Switches.

[3] ANSI C37.47-1981, American National Standard Specifications for Distribution Fuse Disconnecting Switches, Fuse Supports, and Current-Limiting Fuses.

[4] ANSI C50.10-1977, American National Standard General Requirements for Synchronous Machines.

[5] ANSI C50.13-1977, American National Standard Requirements for Cylindrical Rotor Synchronous Generators.

[6] ANSI C62.2-1981, American National Standard Guide for Application of Valve Type Lightning Arresters for Alternating-Current Systems.

[7] ANSI C97.1-1972, American National Standard Low-Voltage Cartridge Fuses 600 Volts or Less.

[8] ANSI/IEEE C37.2-1979, IEEE Standard Electrical Power System Device Function Numbers.

[9] ANSI/IEEE C37.13-1981, IEEE Standard Low-Voltage AC Power Circuit Breakers Used in Enclosures.

[10] ANSI/IEEE C37.40-1981, IEEE Standard Service Conditions and Definitions for High-Voltage Fuses, Distribution Enclosed Single-Pole Air Switches, Fuse Disconnecting Switches, and Accessories.

[11] ANSI/IEEE C37.41-1981, IEEE Standard Design Tests for High-Voltage Fuses, Distribution Enclosed Single-Pole Air Switches, Fuse Disconnecting Switches, and Accessories.

[12] ANSI/IEEE C37.90-1978, IEEE Standard Relays and Relay Systems Associated with Electric Power Apparatus.

[13] ANSI/IEEE C37.91-1985, IEEE Guide for Protective Relay Applications to Power Transformers.

[14] ANSI/IEEE C37.96-1976, IEEE Guide for AC Motor Protection.

[15] ANSI/IEEE C57.12.00-1980, IEEE Standard General Requirements for Liquid-Immersed Distribution, Power, and Regulating Transformers.

[16] ANSI/IEEE C57.13-1978, IEEE Standard Requirements for Instrument Transformers.

[17] ANSI/IEEE C57.109-1985, IEEE Guide for Transformer Through-Fault-Current Duration.

[18] ANSI/IEEE Std 100-1984, IEEE Standard Dictionary of Electrical and Electronics Terms.

[19] ANSI/IEEE Std 142-1982, IEEE Recommended Practice for Grounding of Industrial and Commercial Power Systems.

[20] ANSI/IEEE Std 242-1986, IEEE Recommended Practice for Protection of Coordination of Industrial and Commercial Power Systems.

[21] ANSI/NEMA MG1-1978, Motors and Generators.

[22] ANSI/NFPA 70-1987, National Electrical Code.

[23] ANSI/NFPA 70B-1983, Recommended Practice for Electrical Equipment Maintenance.

[24] ANSI/UL 198B-1982, Safety Standard for Class H Fuses.[20]

[25] ANSI/UL 198C-1981, Safety Standard for High-Interrupting Capacity Fuses, Current-Limiting Types.

[26] ANSI/UL 198D-1982, Safety Standard for Class K Fuses.

[27] ANSI/UL 198E-1982, Safety Standard for Class R Fuses.

[28] ANSI/UL 198F-1982, Safety Standard for Plug Fuses.

[29] ANSI/UL 198G-1981, Safety Standard for Fuses for Supplementary Overcurrent Protection.

[30] AIEE Committee Report. Bibliography of Industrial System Coordination and Protection Literature. *IEEE Transactions on Applications and Industry*, vol 82, Mar 1963, pp 1-2.

[31] AIEE Committee Report. A Survey of Relay Test Methods. *AIEE Transactions (Power Apparatus and Systems)*, pt III, vol 75, Jun 1956, pp 254-260.

[32] ANANIAN, L. G. and COLVIN, F. L. The Industrial Electrical Engineer's Responsibilities and How They Reflect on Management. *IEEE Transactions on Industry and General Applications*, vol IGA-7, Mar/Apr 1971, pp 169-177.

[33] *Application and Protection of Pilot-Wire Circuits for Protective Relaying.* AIEE Paper S-117, 1960.

[34] BAKER, D. S., Generator Backup Overcurrent Protection, *IEEE Transactions on Industry Applications*, vol IA-18, Nov/Dec 1982, pp 632-640.

[35] BARNES, H. C., MURRAY, C. S., and VERRALL, V. E. Relay Protection Practices in Steam Power Stations. *AIEE Transactions (Power Apparatus and Systems)*, pt III, vol 77, Feb 1959, pp 1360-1367.

[36] BEEMAN, D. L., Ed. *Industrial Power Systems Handbook*. New York: McGraw-Hill, 1955.

[37] BOURBONNAIS, T. L., II. The Coordination and Testing of Protective Relays in Industrial Plants. *AIEE Transactions (Power Apparatus and Systems)*, pt III, vol 78, Apr 1959, pp 1-10.

[38] BRIGHTMAN, F. P. More About Setting Industrial Relays. *AIEE Transactions (Power Apparatus and Systems)*, pt III, vol 73, Apr 1954, pp 397-406.

[39] BRIGHTMAN, F. P. Selecting AC Overcurrent Protective Device Settings for Industrial Plants. *AIEE Transactions (Applications and Industry)*, pt II, vol 71, Sept 1952, pp 203-211.

[20] ANSI/UL publications can be obtained from the Sales Department, American National Standards Institute, 1430 Broadway, New York, NY 10018, or from Publication Stock, Underwriters Laboratories, Inc, 333 Pfingsten Rd, Northbrook, IL 60020.

[40] BRIGHTMAN, F. P. Short-Circuit Protection for Industrial Plant Generators. General Electric Company, Bull GER-779, 1953.

[41] CAMERON, F. L. The Coordination of High-Voltage Current-Limiting Fuses. *Proceedings, 1964 1st Annual Conference on Industrial and Commercial Power Systems*, IEEE T-163, pp 100-108.

[42] CASTENSCHIOLD, R. Solutions to Industrial and Commercial Needs Using Multiple Utility Services and Emergency Generator Sets. *IEEE Transactions on Industry Applications*, vol IA-10, Mar/Apr 1974, pp 205-208.

[43] CONRAD, R. R. and DALASTA, D. A. New Ground Fault Protective System for Electrical Distribution Circuits. *IEEE Transactions on Industry and General Applications*, vol IGA-3, May/Jun 1967, pp 217-227.

[44] DALZIEL, C. F. and STEINBACK, E. W. Underfrequency Protection of Power Systems for System Relief: Load Shedding—System Splitting. *AIEE Transactions (Power Apparatus and Systems)*, pt III, vol 78, Dec 1959, pp 1227-1238.

[45] DE WITT, J. S., Ed. *Ground Fault Protection Handbook*. Salem, MA: Electrical Protection, Inc.

[46] FAWCETT, D. V. The Tie Between a Utility and an Industrial when the Industrial Has Generation. *AIEE Transactions (Industry and Applications)*, pt II, vol 77, Jul 1958, pp 136-143.

[47] FITZGERALD, E. M. and STEWART, V. N. High-Capacity Current-Limiting Fuses Today. *AIEE Transactions (Power Apparatus and Systems)*, pt III, vol 78, Oct 1959, pp 937-947.

[48] GOFF, L. E., ROOK, M. J., and POWELL, L. J. Pilot Wire Relay Applications in Industrial Plants Utilizing a New Static Pilot Wire Relay, *IEEE Industry Applications Society*, 1979 Conference Record, 79CH1484-5IA, pp 1195-1203.

[49] HARDER, E. L., KLEMMER, E. H., SONNEMANN, W. K., and WENTZ, E. C. Linear Couplers for Bus Protection. *AIEE Transactions*, vol 61, May 1942, pp 241-248.

[50] HIGGINS, T. D. and PEACH, N. First Principles of System Coordination and Protection—A Proposed First Chapter for a Publication on Preferred Practice. *IEEE Industry and General Applications Group*, 1966 Conference Record 34C-36, pp 245-252.

[51] IEEE Committee Report. Coordination of Lightning Arresters and Current-Limiting Fuses. *IEEE Transactions on Power Apparatus and Systems*, vol PAS-91, May/Jun 1972, pp 1075-1078.

[52] JACOBS, P. C., Jr. Current-Limiting Fuses: Their Characteristics and Applications. *AIEE Transactions (Power Apparatus and Systems)*, pt III, vol 75, Oct 1956, pp 988-993.

[53] KAUFMANN, R. H. Nature and Causes of Overvoltages in Industrial Power Systems. *Iron and Steel Engineer*, Feb 1952.

[54] LARNER, R. A. and GRUESEN, K. R. Fuse Protection of High-Voltage Power Transformers. *AIEE Transactions (Power Apparatus and Systems)*, pt III, vol 78, Oct 1959, pp 864-878.

[55] LATHROP, C. M. and SCHLECKSER, C. E. Protective Relaying on Industrial Power Systems. *AIEE Transactions*, pt II, vol 70, 1951, pp 1341-1345.

[56] Mason, C. R. *The Art and Science of Protective Relaying*. New York: Wiley, 1956.

[57] McFADDEN, R. H. Power-System Analysis: What It Can Do for Industrial Plants. *IEEE Transactions on Industry and General Applications*, vol IGA-7, Mar/Apr 1971, pp 181-188.

[58] SHIELDS, F. J. The Problem of Arcing Faults in Low-Voltage Power Distribution Systems. *IEEE Transactions on Industry and General Applications*, vol IGA-3, Jan/Feb 1967, pp 15-25.

[59] WEDDENDORF, W. A. Evidence of Need for Improved Coordination and Protection of Industrial Power Systems. *IEEE Transactions on Industry and General Applications*, vol IGA-1, Nov/Dec 1965, pp 393-396.

5.11 Bibliography

[B1] ANSI/IEEE C37.2-1979, IEEE Standard Electrical Power System Device Function Numbers.

[B2] *Applied Protective Relaying*. Newark, NJ: Westinghouse Electric Corporation, Relay Instrument Division, B-7235-E, 1976.

[B3] CONCORDIA, C. and PETERSON, H. A. Arcing Faults in Power Systems. *AIEE Transactions*, vol 60, 1941, pp 340-346.

[B4] *Electrical Transmission and Distribution Reference Book*. East Pittsburgh, PA: Westinghouse Electric Corporation, 1965.

[B5] *Engineering Dependable Protection for an Electrical Distribution System, Part III—Component Protection for Electrical Systems*. McGraw-Edison Company, Bussman Manufacturing Division, St Louis, MO, 1975.

[B6] FAWCETT, D. V. How to Select Overcurrent Relay Characteristics. *IEEE Transactions on Applications and Industry*, vol 82, May 1963, pp 94-104.

[B7] GILBERT, M. M. and BELL, R. N. Directional Relays Provide Differential-Type Protection on Large Industrial Plant Power System. *AIEE Transactions (Applications and Industry)*, pt II, vol 74, Sept 1955, pp 220-227.

[B8] HOFFMANN, D. C. and REIFSCHNEIDER, P. J. The System Application of Low-Voltage Power Circuit Breakers with Solid-State Trip Devices. *Industrial and Commercial Power Systems and Electric Space Heating Joint Technical Conference*, 1967 Conference Record, IEEE 34C-55, pp 29-44.

[B9] IEEE Committee Report. Bibliography of Relay Literature 1967-1969. *IEEE Transactions on Power Apparatus and Systems*. Sept/Oct 1971, pp 1952-1988.

[B10] IEEE Committee Report. Protection Fundamentals for Low-Voltage Electrical Distribution Systems in Commercial Buildings. IEEE JH 2112-1, 1974.

[B11] KELLY, A. R. Allowing for Decrement and Fault Voltage in Industrial Relaying. *IEEE Transactions on Industry and General Applications*, vol IGA-1, Mar/Apr 1965, pp 130-139.

[B12] KELLY, A. R. Relay Response to Motor Residual Voltage During Automatic Transfers. *AIEE Transactions (Applications and Industry)*, pt II, vol 74, Sept 1955, pp 245-252.

[B13] KIMBARK, E. W. *Power System Stability*, vol 2. New York: Wiley, 1950.

[B14] POTOCHNEY, G. J. and POWELL, L. J. Application of Protective Relays on a Large Industrial-Utility Tie with Industrial Cogeneration. *IEEE Transactions on Industry Applications*, vol IA-19, May/Jun 1983, pp 461-469.

[B15] WARRINGTON, A. R. van C. *Protective Relays: Their Theory and Practice*, vol 1. New York: Wiley, 1968.

[B16] SHOTT, H. S. and PETERSON, H. A. Criteria for Neutral Stability of Wye-Grounded-Primary Broken-Delta-Secondary Transformer Circuits. *AIEE Transactions*, vol 60, Nov 1941, pp 997-1002.

[B17] STACEY, E. M. and SELCHAU-HANSEN, P. V. SCR Drives—AC Line Disturbance, Isolation and Short-Circuit Protections. *IEEE Transactions on Industry Applications*, vol IA-10, Jan/Feb 1974, pp 88-105.

[B18] *Protective Relays—Application Guide*. Stafford, England: QEC Measurements Ltd, 1975.

6. Fault Calculations

6.1 Introduction. Even the best designed electric systems occasionally experience short circuits resulting in abnormally high currents. Overcurrent protective devices such as circuit breakers and fuses should isolate faults at a given location safely, with minimal circuit and equipment damage and minimal disruption of the plant's operation. Other parts of the system such as cables, bus ducts, and disconnecting switches shall be able to withstand the mechanical and thermal stresses resulting from maximum flow of fault current through them. The magnitudes of fault currents are usually estimated by calculation, and equipment is selected using the calculation results.

The current flow during a fault at any point in a system is limited by the impedance of circuits and equipment from the source or sources to the point of fault and is not directly related to the load on the system. However, additions to the system that increase its capacity to handle a growing load, while not affecting the normal load at some existing parts of the system, may drastically increase the fault currents. Whether an existing system is expanded or a new system is installed, available fault currents should be determined for proper application of overcurrent protective devices.

Calculated maximum fault currents are nearly always required. In some cases, the minimum sustained values are also needed to check the sensitivity requirements of the current-responsive protective devices.

This chapter has three purposes: first, to present some fundamental considerations of fault calculations; second, to illustrate some commonly used methods of making fault calculations with typical examples; and third, to furnish typical data that can be used in making fault calculations.

The size and complexity of many modern industrial systems may make longhand fault calculations impractically time consuming. Computers are generally used for major fault studies. Whether or not computers are available, a knowledge of the nature of fault currents and calculating procedures is essential to conduct such studies.

6.2 Sources of Fault Current. Power frequency currents that flow during a fault come from rotating electric machinery. Power capacitors can also produce extremely high transient fault or switching currents, but usually of short duration and of natural frequency much higher than power frequency. These are discussed in Chapter 8. Rotating machinery in industrial-plant fault calculations may be analyzed in four categories:

(1) Synchronous generators
(2) Synchronous motors and condensers
(3) Induction machines
(4) Electric utility systems

The current from each rotating machinery source is limited by the impedance of the machine and the impedance between the machine and the fault. The impedance of a rotating machine is not a simple value, but is complex and variable with time.

6.2.1 Synchronous Generators. If a short circuit is applied to the terminals of a synchronous generator, the short-circuit current starts out at a high value and decays to a steady-state value some time after the inception of the short circuit. Since a synchronous generator continues to be driven by its prime mover and to have its field externally excited, the steady-state value of fault current will persist unless interrupted by some switching means. To represent this characteristic, one can use an equivalent circuit consisting of a constant driving voltage in series with an impedance that varies with time (Fig 90). This varying impedance consists primarily of reactance.

For purposes of fault-current calculations, industry standards have established three specific names for values of this variable reactance, called subtransient reactance, transient reactance, and synchronous reactance.

Figure 90
Equivalent Circuit for Generators and Motors
E = (Driving Voltage, X Varies with Time)

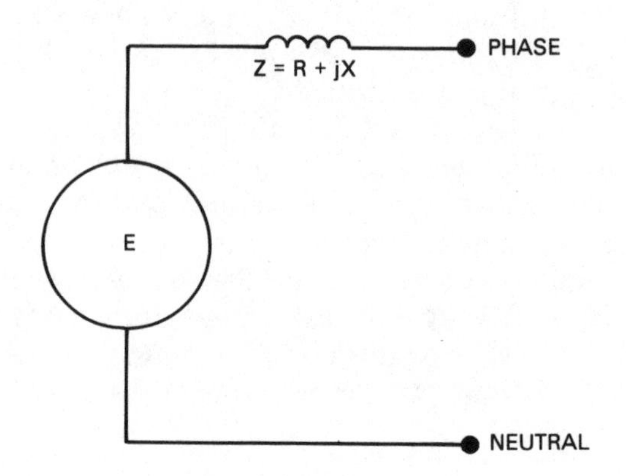

X_d'' = Subtransient reactance; determines current during first cycle after fault occurs. In about 0.1 s reactance increases to

X_d' = Transient reactance; assumed to determine current after several cycles at 60 Hz. In about ½–2 s reactance increases to

X_d = Synchronous reactance; this is the value that determines the current flow after a steady-state condition is reached.

Because most fault-interrupting devices such as circuit breakers and fuses operate well before steady-state conditions are reached, generator synchronous reactance is seldom used in calculating fault currents for application of these devices.

Synchronous generator data available from some manufacturers includes two values for direct axis reactance—for example, subtransient reactance X_{dv}'' and X_{di}''. The X_{dv}'' value should be used for short-circuit calculations.

6.2.2 Synchronous Motors and Condensers. Synchronous motors supply current to a fault much as synchronous generators do. When a fault causes system voltage to drop, the synchronous motor receives less power from the system for rotating its load. At the same time, the internal voltage causes current to flow to the system fault. The inertia of the motor and its load acts as a prime mover and, with field excitation maintained, the motor acts as a generator to supply fault current. This fault current diminishes as the magnetic field in the machine decays.

The generator equivalent circuit is used for synchronous motors. Again, a constant driving voltage and the same three reactances X_d'', X_d', and X_d are used to establish values of current at three points in time.

Synchronous condensers are treated in the same manner as synchronous motors.

6.2.3 Induction Machines. A squirrel-cage induction motor will contribute fault current to a circuit fault. This is generated by inertia driving the motor in the presence of a field flux produced by induction from the stator rather than from a dc field winding. Since this flux decays on loss of source voltage caused by a fault at the motor terminals, the current contribution of an induction motor to a terminal fault reduces and disappears completely after a few cycles. Because field excitation is not maintained, there is no steady-state value of fault current as for synchronous machines.

Again, the same equivalent circuit is used, but the values of transient and synchronous reactance approach infinity. As a consequence, induction motors are assigned only a subtransient value of reactance X_d''. This value is about equal to the locked-rotor reactance.

For fault calculations, an induction generator can be treated the same as an induction motor. Wound-rotor induction motors normally operating with their rotor rings short-circuited will contribute fault current in the same manner as a squirrel-cage induction motor. Occasionally, large wound-rotor motors operated with some external resistance maintained in their rotor circuits may have sufficiently low short-circuit time constants that their fault contribution is not significant and may be neglected. A specific investigation should be made to determine whether to neglect the contribution from a wound-rotor motor.

6.2.4 Electric Utility Systems. The remote generators of the electric utility system are a source of short-circuit current, often delivered through a supply transformer. The generator equivalent circuit can be used to represent the utility system. The utility generators are usually remote from the industrial plant. The current contributed to a fault in the remote plant appears to be merely a small increase in load current to the very large central station generators, and this current contribution tends to remain constant. Therefore, the electric utility system is usually represented at the plant by a single-valued equivalent impedance referred to the point of connection.

6.3 Fundamentals of Fault-Current Calculations. Ohm's law, $I = E/Z$, is the basic relationship used in determining fault current, where I is the fault current, E is the driving voltage of the source, and Z is the impedance from source to fault including the impedance of the source.

Most industrial systems have multiple sources supplying current to a short circuit since each motor can contribute. One step in fault calculation is the simplification of the multiple-source system to the condition where the basic relationship applies.

6.3.1 Purpose of Calculations. System and equipment complexity and the lack of accurate parameters make precise calculations of short-circuit currents exceedingly difficult, but extreme precision is unnecessary. The calculations described provide reasonable accuracy for the maximum and minimum limits of short-circuit currents. These satisfy the usual reasons for making calculations.

The maximum calculated short-circuit current values are used for selecting interrupting devices of adequate short-circuit rating, to check the ability of components of the system to withstand mechanical and thermal stresses, and to determine the time-current coordination of protective relays. The minimum values are used to establish the required sensitivity of protective relays. Minimum short-circuit values are sometimes estimated as fractions of the maximum values. If so, it is only necessary to calculate the maximum values of fault current.

6.3.2 Type of Fault. In an industrial system, the three-phase fault condition is frequently the only one considered, since this type of fault generally results in maximum current.

Line-to-line fault currents are approximately 87% of three-phase fault currents. Line-to-ground fault currents can range in utility systems from a few percent to possibly 125% of the three-phase value, but in industrial systems line-to-ground fault currents of more than the three-phase value are rare.

Assuming a three-phase fault condition also simplifies calculations. The system including the fault remains symmetrical about the neutral point, whether or not the neutral point is grounded and regardless of wye or delta transformer connections. The balance three-phase current can be calculated using a single-phase circuit which has only line-to-neutral voltage and impedance.

In calculating the maximum current, it is assumed that the fault is a zero-impedance (bolted) fault with no current-limiting effect due to the fault itself. It should be recognized, however, that actual faults often involve arcing, and variable arc impedance can reduce low-voltage fault current magnitudes appreciably.

In low-voltage systems, the minimum values of fault current are sometimes calculated from known effects of arcing. Analytical studies indicate that the arcing-fault currents, in per unit of bolted-fault values, may be typically as low as

(1) 0.89 at 480 V and 0.12 at 208 V for three-phase arcing

(2) 0.74 at 480 V and 0.02 at 208 V for line-to-line single-phase arcing

(3) 0.38 at 277 V and 0.01 at 120 V for line-to-neutral single-phase arcing

6.3.3 Basic Equivalent Circuit. The basic equation finds the current of a simple circuit having one voltage source and one impedance. In the basic equation, the voltage E represents a single overall system-driving voltage, which replaces the array of individual unequal generated voltages acting within separate rotating machines. This voltage is equal to the prefault voltage at the point of fault connection. The impedance Z is a network reduction of the impedances representing all significant power-system elements.

This equivalent circuit of the power system is a valid circuit transformation in accordance with Thevenin's theorem. It permits a determination of short-circuit current corresponding to the values of system impedances used.

Ordinarily, the prefault voltage is taken as the system nominal voltage at the point of fault because this is close to the maximum operating voltage under fully loaded system conditions, and therefore the short-circuit currents will approach maximum. Higher than nominal voltage might be used in an unusual case when full load system voltage is observed to be above nominal.

The single-phase representation of a three-phase balanced system uses per-phase impedances and the line-to-neutral system driving voltage. Line-to-neutral voltage is line-to-line voltage divided by $\sqrt{3}$. Calculations may use impedances in ohms and voltages in volts, or both in per unit. Per-unit calculations simplify short-circuit studies for industrial systems that involve voltages of several levels. When using the per-unit system, the driving voltage is equal to 1.0 per unit if voltage bases are equal to system nominal voltages.

The major elements of impedance must always be included in a short-circuit calculation. These are impedances of transformers, busways, cables, conductors, and rotating machines. There are other circuit impedances, such as those associated with circuit breakers, wound or bar-type current transformers, bus structures, and bus connections, that are usually small enough to be neglected in short-circuit calculations, because the accuracy of the calculation is not generally affected. Omitting them provides slightly conservative (higher) short-circuit currents. However, on low-voltage systems, and particularly at 208 V, there are cases where their inclusion can significantly reduce the calculated short-circuit current.

Also, the usual practice is to disregard the presence of static loads (such as lighting and electric heating) in the network, despite the fact that their associated impedance is actually connected in shunt with other network branches. This approach is considered valid since usually the relatively high power factor static-load impedances are large and approximately 90 degrees out of phase compared to the other highly reactive parallel branches of the network.

In ac circuits, the impedance Z is the vector sum of resistance R and reactance X. It is always acceptable to calculate short-circuit currents using vector impedances in the equivalent circuit. For many short-circuit current-magnitude calculations at medium or high voltage, and for a few at low voltage, it is sufficiently accurate, conservative, and simpler to ignore resistances and use reactances only, since the reactances are usually much larger.

However, for many lower-voltage calculations, resistance should not be ignored because the calculated currents would be overconservative.

Resistance data are definitely needed for calculations of X/R ratios when applying high- and medium-voltage circuit breakers, but they need to be kept separate from the reactance.

6.4 Restraints of Simplified Calculations. The short-circuit calculations described in this chapter are a simple E/Z evaluation of extensive electric power-system networks. Before describing the step-by-step procedures in making these calculations, it is appropriate to review some of the restraints imposed by the simplification.

6.4.1 Impedance Elements. When an ac electric-power circuit contains resistance R, inductance L, and capacitance C, such as the series connection shown in Fig 91, the expression relating current to voltage includes the terms shown in Fig 91. A solution for the current magnitude requires the solution of a differential equation.

If two important restraints are applied to this series circuit, the following simple equation using vector impedances ($X_L = \omega L$ and $X_C = 1/\omega C$) is valid:

$$E = I\left[R + j\left(\omega L - \frac{1}{\omega C}\right)\right]$$

These restraints are that, first, the electric driving force be a sine wave and, second, the impedance coefficients R, L, and C be constants. Unfortunately, in short-circuit calculations these restraints may be invalidated. A major reason for this is switching transients.

6.4.2 Switching Transients. The vector-impedance analytical tool recognizes only the steady-state sine-wave electrical quantities and does not include the

Fig 91
Series RLC Circuit

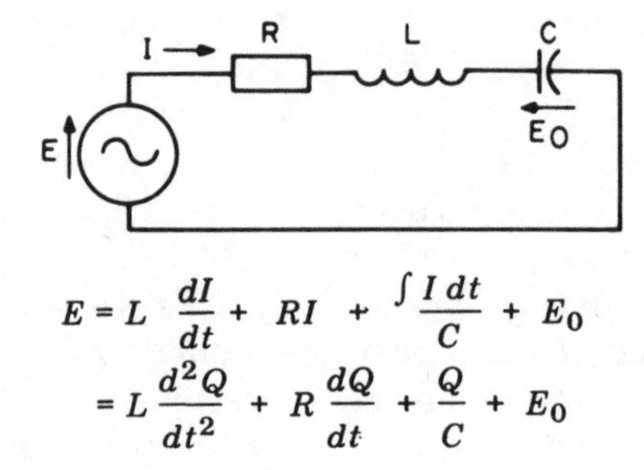

$$E = L\frac{dI}{dt} + RI + \frac{\int I\,dt}{C} + E_0$$

$$= L\frac{d^2Q}{dt^2} + R\frac{dQ}{dt} + \frac{Q}{C} + E_0$$

effects of abrupt switching. Fortunately, the effects of switching transients can be analyzed separately and added. (An independent solution can be obtained from a solution of the formal differential equations.)

In the case of resistance R (Fig 92), the closure of switch SW causes the current to immediately assume the value that would exist in the steady state. No transient adder is needed.

In the case of inductance L (Fig 93), an understanding of the switching transient can best be acquired using the expression

$$E = L \frac{dI}{dt}$$

$$\frac{dI}{dt} = \frac{E}{L}$$

This expression tells us that the application of a driving voltage to an inductance will create a time rate of change in the current magnitude. The slope of the current-time curve in the inductance will be equal to the quantity E/L.

The steady-state current curve is displayed at the right-hand side of the graph of Fig 93. It lags the voltage wave by 90 degrees and is rising at the maximum rate in the positive direction when the voltage is at the maximum positive value. It holds at a fixed value when the driving voltage is zero. This curve is projected back to the time of circuit switching (dashed curve). Note that at the instant the switch is closed, the steady-state current would have been at a negative value of about 90% of crest value. Since the switch was previously open, the true circuit current must be zero. After closing the switch, the current wave will display the same slope as the steady-state wave. This is the solid-line current curve beginning at the instant of switch closing. Note that the difference between this curve and the steady state is a positive dc component of the same magnitude that the steady-state wave would have had at the instant of switch closing, in the negative direction. Thus the switching transient takes the form of a dc component whose value may be anything between zero and the steady-state crest value, depending on the angle of closing.

Fig 92
Switching Transient *R*

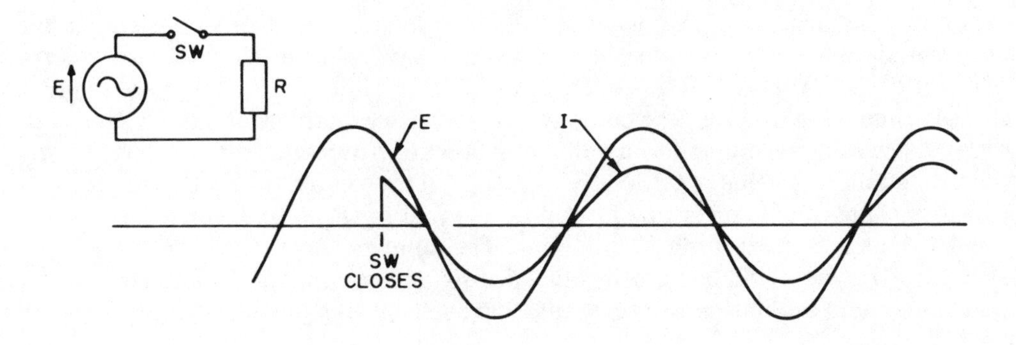

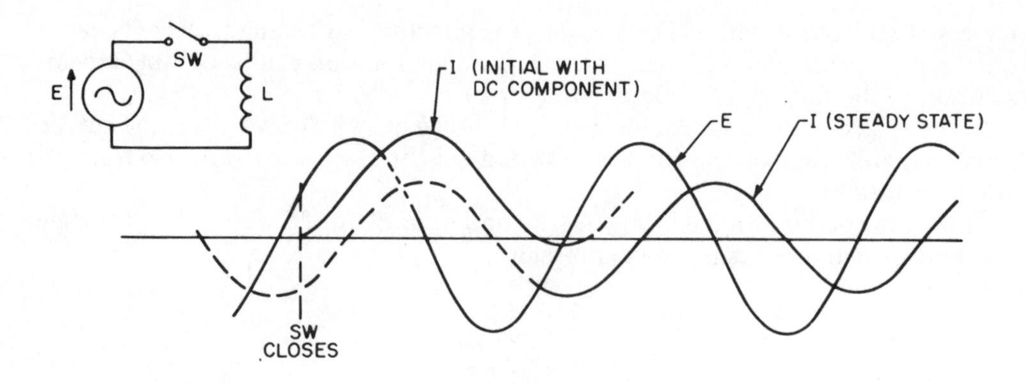

Fig 93
Switching Transient *L*

If the circuit contained no resistance, the current would continue forever in the displaced form. The presence of resistance causes the dc component to be dissipated exponentially. The complete expression for the current would take the form

$$I = \frac{E}{j\omega L} \sin \omega t + I_{dc}\, e^{-Rt/L}$$

The presence of dc components may introduce unique problems in selective coordination between some types of overcurrent devices. It is particularly important to bear in mind that these transitory currents are not disclosed by the vector-impedance circuit solution, but must be introduced artificially by the analyst or by the guide rules he follows.

6.4.3 Decrement Factor. The value at any time of a decaying quantity, expressed in per unit of its initial magnitude, is the decrement factor for that time. Refer to Fig 94 for decrement factors of an exponential decay. The significance of the decrement factor can be understood better if the exponential is expressed in terms of the time constant. If, as indicated in Fig 94, the exponent is expressed as $-t/t'$ with the time variable t in the numerator and the rest combined as a single constant t' (called the time constant) in the denominator, the transitory quantity begins its decay at a rate that would cause it to vanish in one time constant. The exponential character of the decay results in a remnant of 36.8% remaining after an elapsed time equal to one time constant. Any value of the transitory term selected at, say, time t will be reduced to 0.368 of that value after a subsequent elapsed time equal to one time constant. A transitory quantity of magnitude 1.0 at time zero would be reduced to a value of 0.368 after an elapsed time equal to one time constant, to a value of 0.135 after an elapsed time equal to two time constants, and to a value of 0.05 after an elapsed time equal to three time constants.

6.4.4 Multiple Switching Transients. The analyst usually assumes that the switching transient will occur only once during one excursion of short-circuit current flow. An examination of representative oscillograms of short-circuit currents

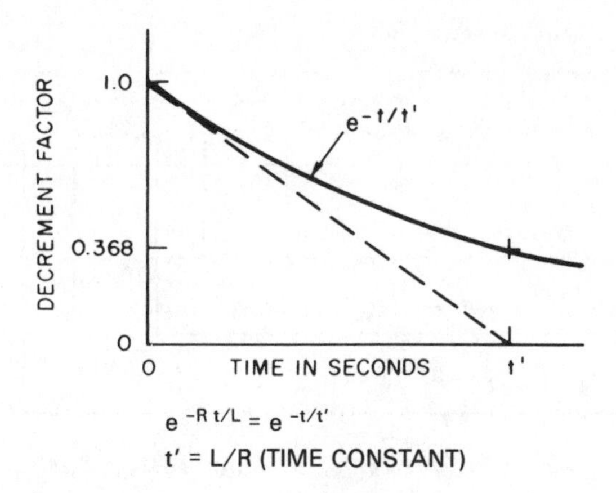

Fig 94
Decrement Factor

will often display repeated instances of momentary current interruptions. At times, an entire half-cycle of current will be missing. In other cases, especially in low-voltage circuits, there may be a whole series of chops and jumps in the current pattern. A switching interrupter, especially when switching a capacitive circuit, may be observed to restrike two or perhaps three times before complete interruption is completed. The restrike generally occurs when the voltage across the switching contacts is high. It is entirely possible that switching transients, both simple dc and ac transitory oscillations, may be reinserted in the circuit current a number of times during a single incident of short-circuit current flow. The analyst should remain mindful of possible trouble.

6.4.5 Practical Impedance Network Synthesis. One approach to an adequate procedure for computing the phase A current of a three-phase system is indicated in Fig 95. For each physical conducting circuit, the voltage drop is represented as the sum of the self-impedance drop in the circuit and the complete array of mutually coupled voltage drops caused by current flow in other coupled circuits. The procedure is complex even in those instances where the current in both the neutral and ground conductors is zero.

The simplified analytical approach to this problem assumes balanced symmetrical loading of a symmetrical polyphase system. With a symmetrical system operating with a symmetrical loading, the effects of all mutual couplings are similarly balanced. What is happening in phase A in the way of self- and mutually coupled voltages is also taking place in phase B with exactly the same pattern, except displaced 120 degrees, and it is also taking place in phase C with the same pattern, except displaced another 120 degrees. The key to the simplification is the fact that the ratio of the total voltage drop in one phase circuit to the current in that phase circuit is the same in all three phases of the system. Thus it appears that each phase possesses a firm impedance value common with the other phases. This unique impedance quantity is identified as the single-phase line-to-neutral imped-

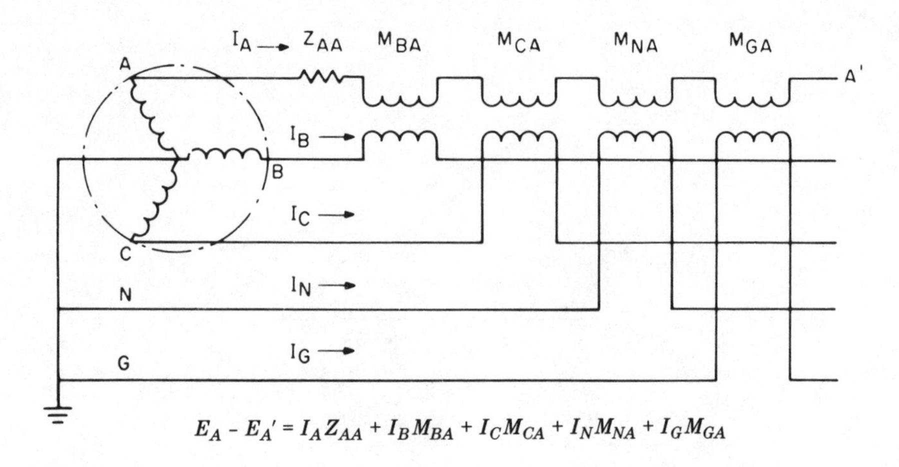

$$E_A - E_{A'} = I_A Z_{AA} + I_B M_{BA} + I_C M_{CA} + I_N M_{NA} + I_G M_{GA}$$

Fig 95
Three-Phase, Four-Wire Circuit, Unbalanced Loading

ance value. Any one line-to-neutral single-phase segment of the system may be sliced out for the analysis, since all are operating with the same load pattern.

The impedance diagram of the simplified concept appears in Fig 96. The need to deal with mutual coupling has vanished. Since each phase circuit presents identically the same information, it is common to show only a single phase segment of the system in a one-line diagram as illustrated simply by Fig 90. The expressions below the sketch in Fig 96 contain some unfamiliar terms. Their meaning will be discussed in succeeding paragraphs.

One restraint associated with this simple analytical method is that all phases of the system share symmetrical loading. While a three-phase short circuit would satisfy this restraint, some short-circuit problems that must be solved are not balanced. For these unbalanced short-circuit problems, we use the concept of symmetrical components. This concept discloses that any conceivable condition of unbalanced loading can be correctly synthesized by the use of appropriate magnitudes and phasing of several systems of symmetrical loading. In a three-phase system, with a normal phase separation of 120 degrees, there are just three possible symmetrical loading patterns. These can be quickly identified with the aid of Fig 97. Loadings of the three-phase windings A, B, and C must follow each other in sequence, separated by some multiple of 120 degrees. In Fig 97(a) they follow each other with a 120 degree separation, in Fig 97(b) with a 240 degree separation, and in Fig 97(c) with a 360 degree separation. Note that separation angles of any other multiples of 120 degrees will duplicate one of the three already shown. These loading patterns satisfy the restraints demanded by the analytical method to be used.

Note that Fig 97(a), identified as the positive sequence, represents the normal balanced operating mode. Thus there are only two sequence networks that differ

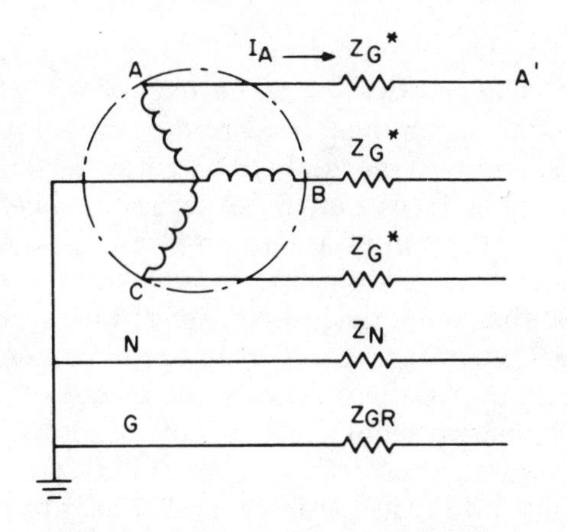

Impedance identity for each symmetrical pattern:

Positive sequence	Z_{G1}
Negative sequence	Z_{G2}
Zero sequence	$Z_{G0} + 3Z_{GR}$*

*Based on zero current in conductor N.

$$E_A - E_A' = I_{A1} Z_{G1} + I_{A2} Z_{G2} + I_{A0} (Z_{G0} + 3Z_{GR})$$

Fig 96
Three-Phase, Four-Wire Circuit, Balanced Symmetrical Loading

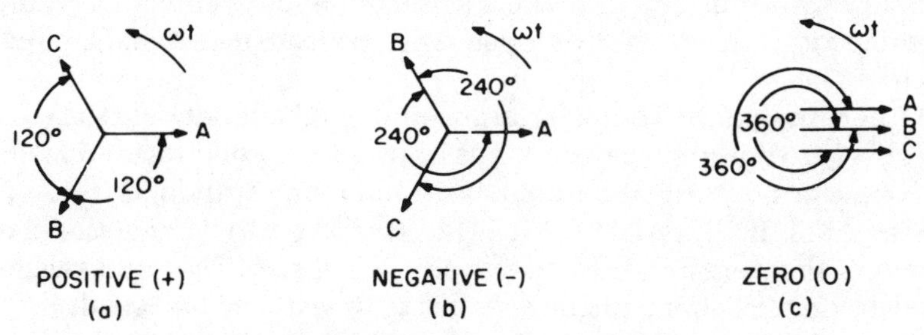

The individual phase quantities A, B, and C follow each other with an angular spacing of

Positive sequence	120°
Negative sequence	240°
Zero sequence	360°

Fig 97
Three-Phase Symmetrical Load Patterns Applicable to a Three-Phase System

from the normal. Figure 97(b), called the negative sequence, identifies a loading pattern very similar to the positive sequence, except that the electrical quantities come up with the opposite sequence. A current of this pattern flowing in a motor stator winding would create a normal-speed rotating field, but with backward rotation. The pattern of Fig 97(c), called the zero sequence, represents the case in which the equal currents in each phase are in phase. Each phase current reaches its maximum in the same direction at the same instant.

It is understandable that machine interwinding mutual coupling and other mutual coupling effects will be different in the different sequence systems. Hence it is likely that the per-phase impedance of the negative- and zero-sequence systems will differ from that of the positive sequence. Currents of zero sequence, being in phase, do not add up to zero at the end terminal as do both the positive- and negative-sequence currents. They add arithmetically and return to the source via an additional circuit conductor. The zero-sequence voltage drop of this return conductor is accounted for in the zero-sequence impedance value. With this understanding of the three symmetrical loading patterns, the significance of the notes below the sketch in Fig 96 becomes clear.

The simplifications in analytical procedures accomplished by the per-phase line-to-neutral balanced system concepts carry with them some important restraints.

(1) The electric power system components shall be of symmetrical design pattern.

(2) The electric loading imposed on the system shall be balanced and symmetrical.

Wherever these restraints are violated, it is necessary to construct substantially hybrid network interconnections that bridge the zones of unbalanced conditions. In the field of short-circuit current calculations, the necessary hybrid interconnections of the sequence networks to accommodate the various unbalanced fault connections can be found in a variety of published references. It is harder to find the necessary hybrid interconnections to accommodate a lack of symmetry in the circuit geometry, as needed for an open-delta transformer bank, an open-line conductor, etc.

6.4.6 Other Analytical Tools. A large number of valid network theorems can be used effectively to simplify certain kinds of problems encountered in short-circuit analysis. These are described and illustrated in many standard texts on ac circuit analysis; see ANSI/IEEE C37.13-1981 [4][21] (Chapter 8). Of exceptional importance are Thevenin's theorem and the superposition theorem. Thevenin's theorem allows an extensive complex single-phase network to be reduced to a single driving voltage in series with a single impedance, referred to the particular bus under study. The superposition theorem allows the local effect of a remote voltage change in one source machine to be evaluated by impressing the magnitude of the voltage change, at its point of origin, on the complete impedance network; the current reading in an individual circuit branch is treated as an adder to the prior current magnitude in

[21] The numbers in brackets correspond to those in the references at the end of this chapter; when preceded by B, they correspond to the bibliography at the end of this chapter.

that branch. These analytical tools, like the others, have specific restraints that must be observed to obtain valid results.

6.4.7 Respecting the Imposed Restraints. Throughout this discussion, emphasis has been placed on the importance of respecting the restraints imposed by the analytical procedure in order to obtain valid results. Mention has been made of numerous instances in short-circuit analysis where it is necessary to artificially introduce appropriate corrections when analytical restraints have been violated. One remaining area associated with short-circuit analysis involves variable impedance coefficients. When an arc becomes a series component of the circuit impedance, the *R* it represents is not constant. If it is 100 Ω at a current of 1 A, it might be 0.1 Ω at a current of 1000 A. During each half-cycle of current flow, the arc resistance might traverse this range. It is difficult to determine a proper value to insert in the 60 Hz network. Correctly setting this value of *R* does not compensate for the violation of the restraint that demands that *R* be a constant. The variation in *R* lessens the impedance to high-magnitude current, which results in a wave shape of current that is much more peaked than a sine wave. The current now contains harmonic terms. Since they result from a violation of analytical restraints, they will not appear in the calculated results. Their character and magnitude can be determined by other means and the result artificially introduced into the solution for fault current. A similar type of nonlinearity may be encountered in electromagnetic elements in which iron plays a part in setting the value of *L*. If the ferric parts are subject to large excursions of magnetic density, the value of *L* may be found to drop substantially when the flux density is driven into the saturation region. As with variable *R*, the effect of this restraint violation will result in the appearance of harmonic components in the true circuit current.

6.4.8 Conclusions. The purpose of this review of fundamentals is to obtain a better understanding of the basic complexities involved in ac system short-circuit calculations. In dealing with the day-to-day practical problems, the analyst should adopt the following goals:

(1) Select the optimum location and type of fault to satisfy the purpose of the calculation.

(2) Establish the simplest electric-current model of the problem that will both accomplish this purpose and minimize the complexity of the solution.

(3) Recognize the presence of system conditions that violate the restraints imposed by the analytical methods in use.

(4) Artificially inject corrections in computed results to compensate if these conditions are large enough to be significant.

Some conclusions of the preceding section apply to the simplified procedures of this chapter. A balanced three-phase fault has been assumed, and a simple equivalent circuit has been described. The current E/Z calculated with the equivalent circuit is an alternating symmetrical rms current, because E is the rms voltage. Within specific constraints to be discussed, this symmetrical current may be directly compared with equipment ratings, capabilities, or performance characteristics that are expressed as symmetrical rms currents.

The preceding analysis of inductive circuit switching transients indicates that simplified procedures should recognize and account for asymmetry as a system condition. The correction to compensate for asymmetry considers the asymmetrical short-circuit current wave to be composed of two components. One is the ac symmetrical component E/Z. The other is a dc component initially of maximum possible magnitude, equal to the peak of the initial ac symmetrical component, assuming that the fault occurs at the point on the voltage wave where it creates this condition. At any instant after the fault occurs, the total current is equal to the sum of the ac and dc components (Fig 98).

Since resistance is always present in an actual system, the dc component decays to zero as the stored energy it represents is expended in I^2R loss. The decay is assumed to be an exponential, and its time constant is assumed to be proportional to the ratio of reactance to resistance (X/R ratio) of the system from source to fault. As the dc component decays, the current gradually changes from asymmetrical to symmetrical (Fig 98).

Asymmetry is accounted for in simplified calculating procedures by applying multiplying factors to the alternating symmetrical current. The resulting estimate of the asymmetrical rms current is used for comparison with equipment ratings, capabilities, or performance characteristics that are expressed as total (asymmetrical) rms currents.

The alternating symmetrical current may also decay with time, as indicated in the discussion of sources of short-circuit current. Changing the impedance representing the machine properly accounts for ac decay of the current to a short circuit

Fig 98
Typical System Fault Current

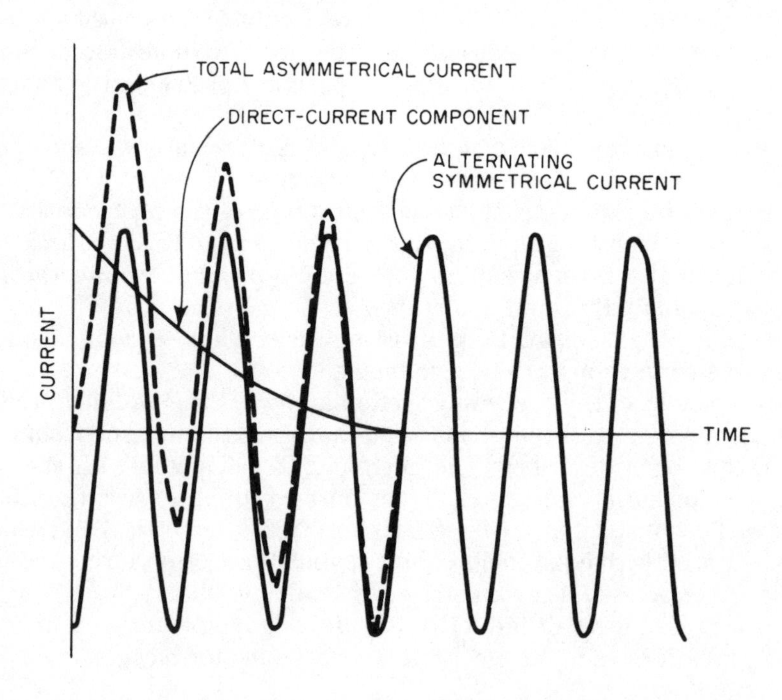

at rotating machine terminals. The same impedance changes are assumed to be applicable when representing rotating machines in extensive power systems.

6.5 Detailed Procedure. A significant part of the preparation for a short-circuit current calculation is establishing the impedance of each circuit element and converting impedances to be consistent with each other for combination in series and parallel. Sources of impedance values for circuit elements are nameplates, handbooks, manufacturers' catalogs, tables included in this chapter, and direct contact with the manufacturer.

Two established consistent forms for expressing impedances are ohms and per unit (per unit differs from percent by a factor of 100). Individual equipment impedances are often given in percent, which makes comparisons easy, but percent impedances are rarely used without conversion in system calculations. In this chapter, the per-unit form of impedance is used because it is more convenient than the ohmic form when the system contains several voltage levels. Impedances expressed as per unit on a defined base can be combined directly, regardless of how many voltage levels exist from source to fault. To obtain this convenience, the base voltage at each voltage level must be related according to the turns ratios of the interconnecting transformers.

In the per-unit system, there are four base quantities: base apparent power in volt-amperes, base voltage, base current, and base impedance. The relationship of base, per-unit, and actual quantities is as follows:

$$\text{per-unit quantity (voltage, current, etc)} = \frac{\text{actual quantity}}{\text{base quantity}}$$

Usually a convenient value is selected for base apparent power in volt-amperes, and a base voltage at one level is selected to match the transformer rated voltage at that level. Base voltages at other levels are established by transformer turns ratios. Base current and base impedance at each level are then obtained by standard relationships. The following formulas apply to three-phase systems, where the base voltage is the line-to-line voltage in volts or kilovolts and the base apparent power is the three-phase apparent power in kilovolt-amperes or megavolt-amperes:

$$\text{base current (amperes)} = \frac{\text{base kVA (1000)}}{\sqrt{3}\ \text{(base volts)}} = \frac{\text{base kVA}}{\sqrt{3}\ \text{(base kV)}}$$

$$= \frac{\text{base MVA } (10^6)}{\sqrt{3}\ \text{(base volts)}} = \frac{\text{base MVA (1000)}}{\sqrt{3}\ \text{(base kV)}}$$

$$\text{base impedance (ohms)} = \frac{\text{base volts}}{\sqrt{3}\ \text{(base amperes)}} = \frac{(\text{base volts})^2}{\text{base kVA (1000)}}$$

$$= \frac{(\text{base kV})^2\ (1000)}{\text{base kVA}} = \frac{(\text{base kV})^2}{\text{base MVA}}$$

Impedances of individual power system elements are usually obtained in forms that require conversion to the related bases for a per-unit calculation. Cable impedances are generally expressed in ohms. Converting to per unit using the indicated relationships leads to the following simplified formulas, where the per-unit impedance is Z_{pu}:

$$Z_{pu} = \frac{\text{actual impedance in ohms (base MVA)}}{(\text{base kV})^2}$$

$$= \frac{\text{actual impedance in ohms (base kVA)}}{(\text{base kV})^2\ (1000)}$$

Transformer impedances are in percent of self-cooled transformer ratings in kilovolt-amperes and are converted using

$$Z_{pu} = \frac{\text{percent impedance (base kVA)}}{\text{kVA rating (100)}}$$

$$= \frac{\text{percent impedance (10) (base MVA)}}{\text{kVA rating}}$$

Motor reactance may be obtained from tables providing per-unit reactances on element ratings in kilovolt-amperes and are converted using

$$X_{pu} = \frac{\text{per-unit reactance (base kVA)}}{\text{kVA rating}}$$

The procedure for calculating industrial system short-circuit currents consists of the following steps:

(1) Prepare system diagrams
(2) Collect and convert impedance data
(3) Combine impedances
(4) Calculate short-circuit current

Each step will be discussed in further detail.

6.5.1 Step 1—Prepare System Diagrams. A one-line diagram of the system should be prepared to show all sources of short-circuit current and all significant circuit elements. Figure 99, used for a subsequent example, is a one-line diagram of a hypothetical industrial system.

Impedance information may be entered on the one-line diagram after initial data collection and after conversion. Sometimes it is desirable to prepare a separate diagram showing only the impedances after conversion. If the original circuit is complex and several steps of simplification are required, each may be recorded on additional impedance diagrams as the calculation progresses.

The impedance diagram might show reactances only, or it might show both reactances and resistances if a vector calculation is to be made. For calculation of a system X/R ratio, as described later for high-voltage circuit-breaker duties, a resistance diagram showing only the resistances of all circuit elements shall be prepared.

6.5.2 Step 2—Collect and Convert Impedance Data. Impedance data should be collected for important elements and converted to per unit on bases selected for the study (see Note at end of this chapter.)

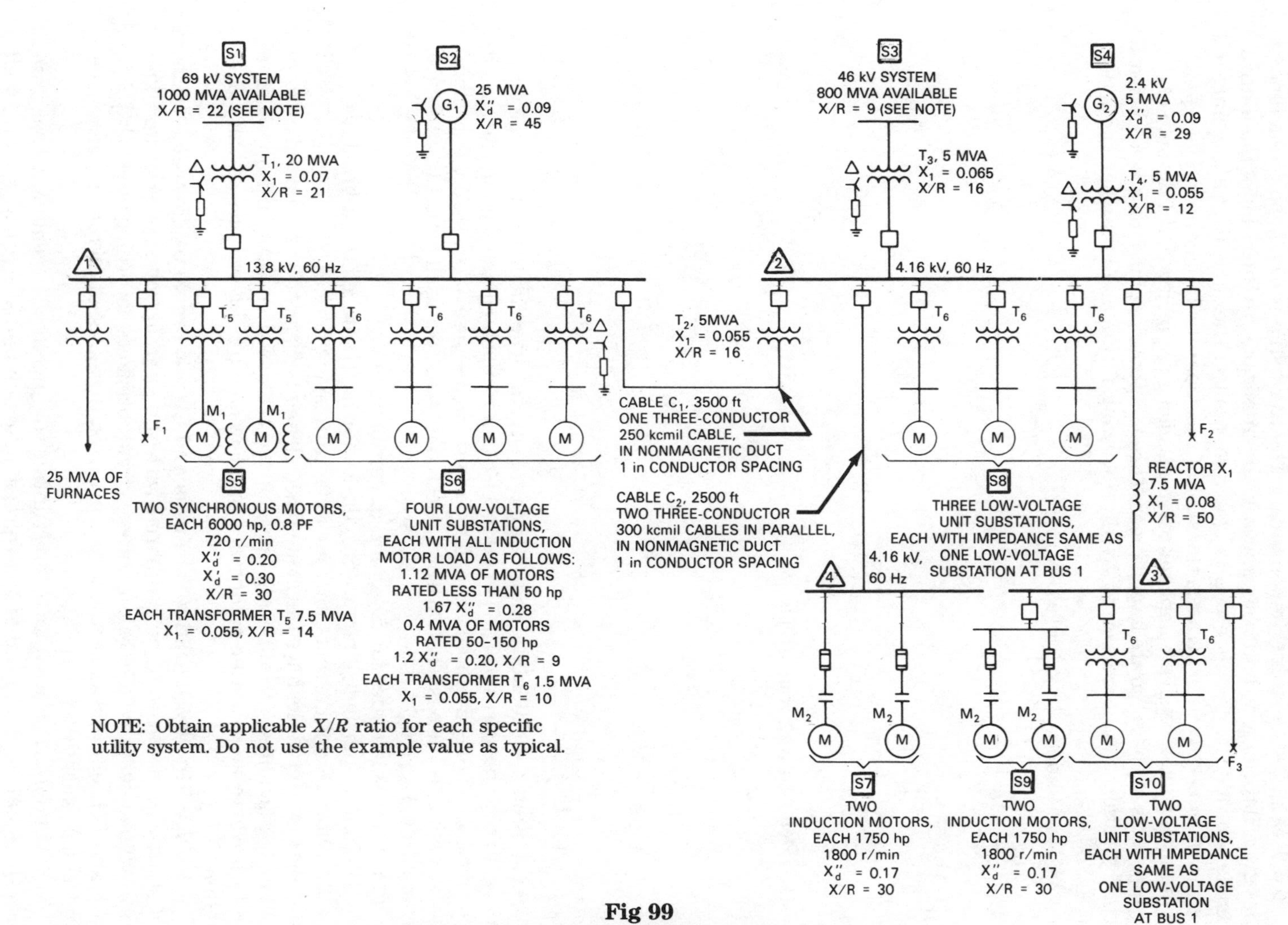

NOTE: Obtain applicable X/R ratio for each specific utility system. Do not use the example value as typical.

Fig 99
One-Line Diagram of Industrial System Example

6.5.3 Step 3—Combine Impedances. The third step is to combine reactances or vector impedances, and resistances where applicable, to the point of fault into a single equivalent impedance, reactance, or resistance. The equivalent impedance of separate impedances in series is the sum of the separate impedances. The equivalent impedance of separate impedances in parallel is the reciprocal of the sum of the reciprocals of the separate impedances. Three impedances forming a wye or delta configuration can be converted by the following formulas for further reduction (Fig 100).

(1) Wye to delta [Fig 100(a)]:

$$A = \frac{bc}{a} + b + c$$

$$B = \frac{ac}{b} + a + c$$

$$C = \frac{ab}{c} + a + b$$

(2) Delta to wye [Fig 100(b)]:

$$a = \frac{B \cdot C}{A + B + C}$$

$$b = \frac{A \cdot C}{A + B + C}$$

$$c = \frac{A \cdot B}{A + B + C}$$

6.5.4 Step 4—Calculate Short-Circuit Current. The final step is to calculate the short-circuit current. The impedances of rotating machines used in the circuit to calculate short-circuit currents depend on the purpose of the study. This chapter examines three basic networks of selected impedances used for the results most commonly desired:

(1) First-cycle duties for fuses and circuit breakers
(2) Contact-parting (interrupting) duties for high-voltage circuit breakers
(3) Short-circuit currents for time-delayed relaying devices

The three networks have the same basic elements except for the impedances of rotating machines. The differing impedances are based on standard application guides where interrupting equipment applications are the purpose of the calculation:

6.5.4.1 First-Cycle Duties for Fuses and Circuit Breakers. For calculations of short-circuit duties to be compared with the interrupting ratings of low- and high-voltage fuses or of only low-voltage circuit breakers according to [1], [4], [5], [6], and [7], subtransient impedances are used to represent all rotating machines in the equivalent network.

The standards for low-voltage interrupting equipment allow a modified subtransient reactance for a group of low-voltage induction and synchronous motors fed

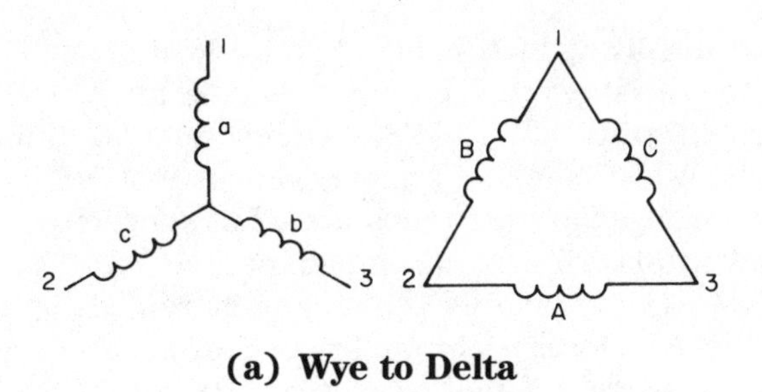

(a) Wye to Delta

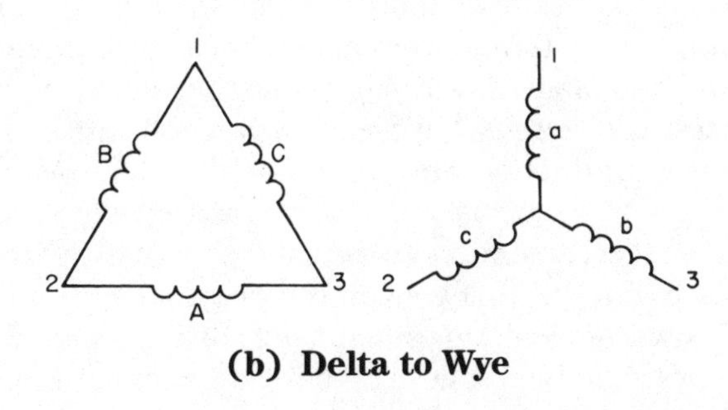

(b) Delta to Wye

Fig 100
Wye and Delta Configurations

from a low-voltage substation. If the total of motor horsepower ratings at 480 or 600 V is approximately equal to the transformer self-cooled rating in kilovolt-amperes, a per-unit reactance of 0.25 on the transformer self-cooled rating may be used as a single impedance to represent the group of motors.

High-voltage short-circuit duties calculated with these impedances are used when applying high-voltage fuses and when finding high-voltage system available short-circuit duties for use as factors in subsequent low-voltage calculations.

For calculations of short-circuit duties to be compared with only high-voltage circuit-breaker closing and latching capabilities (post-1964 rating basis) or momentary ratings (pre-1964 rating basis) according to reference standards ANSI/IEEE C37.010-1979 [2] and ANSI/IEEE C37.5-1979 [3] multiplying factors shown in the first-cycle column of Table 24 are applied to rotating machine reactances (or impedances). For motors, this approximates the effect of motor short-circuit current-contribution ac decay.

The preceding description indicates that different treatment of induction motors might uneconomically necessitate two first-cycle calculations for comprehensive industrial system short-circuit studies covering both low and high (including medium) voltages, if procedures of applicable standards are followed without interpretation. The high-voltage circuit breaker application procedure described in ANSI/IEEE C37.010-1979 [2] and ANSI/IEEE C37.5-1979 [3] defines three induction motor size groups, recommends omitting the group of motors each less than

50 hp, and applies multiplying factors of 1.2 or 1.0 to subtransient impedances of motors in the groups of larger and larger sizes. The low-voltage circuit breaker application guide, ANSI/IEEE C37.13-1981 [4], recommends impedances based on subtransient and allows estimates of typical symmetrical first-cycle contributions from connected low-voltage motors to substation bus short circuits at 4 times rated current (the equivalent of 0.25 per unit impedance).

The *4 times rated current* short-circuit contribution estimate is determined approximately in the low-voltage circuit breaker application guide, ANSI/IEEE C37.13-1981 [4], by assuming a typical connected group having 75% induction motors at 3.6 times rated current and 25% synchronous motors at 4.8 times rated. Other typical group assumptions could be made; for example, many groups now have larger low-voltage induction motors instead of synchronous motors, but these larger motors also have higher and longer-lasting short-circuit contributions. Accordingly, a *4 times rated current* approximation continues to be accepted practice when the load is all induction motors of unspecified sizes.

To simplify comprehensive industrial system calculations, a single combination first-cycle network is recommended to replace the two different networks just described. It is based on the following interpretation of referenced standards. Because the initial symmetrical rms magnitude of the current contributed to a terminal short circuit might be 6 times rated for a typical induction motor, using a *4.8 times rated current* first-cycle estimate for the large low-voltage induction motors (described as *all others, 50 hp and above* in Table 24) is effectively the same as multiplying subtransient impedance by approximately 1.2. For this motor group, there is reasonable correspondence of low- and high-voltage procedures. For smaller induction motors (*all smaller than 50 hp* in Table 24) a conservative estimate is the *3.6 times rated current* (equivalent of 0.28 per unit impedance) first-cycle assumption of low-voltage standards, and this is effectively the same as multiplying subtransient impedance by 1.67.

With this interpretation as a basis, the following induction motor treatment is recommended to obtain a single combination first-cycle short-circuit calculation for multivoltage industrial systems:

Table 24
Rotating-Machine Reactance (or Impedance) Multipliers

Type of Rotating Machine	First-Cycle Network	Interrupting Network
All turbine generators; all hydrogenerators with amortisseur windings, all condensers	$1.0\ X_d''$	$1.0\ X_d''$
Hydrogenerators without amortisseur windings	$0.75\ X_d'$	$0.75\ X_d'$
All synchronous motors	$1.0\ X_d''$	$1.5\ X_d''$
Induction motors		
Above 1000 hp at 1800 r/min or less	$1.0\ X_d''$	$1.5\ X_d''$
Above 250 hp at 3600 r/min	$1.0\ X_d''$	$1.5\ X_d''$
All others, 50 hp and above	$1.2\ X_d''$	$3.0\ X_d''$
All smaller than 50 hp	Neglect	Neglect

From ANSI/IEEE C37.010-1979 [2] and ANSI/IEEE C37.5-1979 [3].

(a) Include connected motors, each less than 50 hp, using either a 1.67 multiplying factor for subtransient impedances, if available, or an estimated first-cycle impedance of 0.28 based on motor rating.

(b) Include larger motors using the impedance multiplying factors of Table 24. Most low-voltage motors 50 hp and larger are in the 1.2 times subtransient reactance group. An appropriate estimate for this group is first-cycle impedance of 0.20 per unit based on motor rating.

The last two lines of Table 24 are replaced by Table 25 for the recommended combination network.

The single combination first-cycle network adds conservatism to both low-voltage and high-voltage standard calculations. It increases calculated first-cycle short-circuit currents at high voltage by the contributions from small induction motors and at low voltage, when many motors are 50 hp or larger, by the increased contribution of larger low-voltage induction motors.

Once the first-cycle network has been established and its impedances are converted and reduced to a single equivalent per unit impedance Z_{pu} (or reactance X_{pu}) for each fault point of interest, the symmetrical short-circuit current duty is calculated by dividing the per unit prefault operating voltage E_{pu} by Z_{pu} (or X_{pu}) and multiplying by base current:

$$I_{sc\ sym} = \frac{E_{pu}}{Z_{pu}} \cdot I_{base}$$

where $I_{sc\ sym}$ is a three-phase symmetrical zero-fault-impedance (bolted) first-cycle short-circuit rms current.

The calculated short-circuit current results for low-voltage buses are now directly applicable for comparison with low-voltage circuit breaker, fuse and other equipment short circuit ratings or capabilities expressed as symmetrical rms currents.

When the equipment rating or capability is expressed as a total (asymmetrical) rms current, the calculated symmetrical short-circuit current duty is multiplied by a multiplying factor found in the applicable standard to obtain the appropriate first-cycle total (asymmetrical) rms current for comparison.

Table 25
Combined Network Rotating Machine Reactance (or Impedance) Multipliers (Changes to Table 24 for Comprehensive Multivoltage System Calculations)

Type of Rotating Machine	First-Cycle Network	Interrupting Network
Induction Motors		
All others, 50 hp and above	$1.2\ X_d''$ *	$3.0\ X_d''$ ‡
All smaller than 50 hp	$1.67\ X_d''$ †	Neglect

* or estimate the first-cycle network X = 0.20 per unit based on motor rating
† or estimate the first-cycle network X = 0.28 per unit based on motor rating
‡ or estimate the interrupting network X = 0.50 per unit based on motor rating

Closing and latching capabilities (or momentary ratings) of high-voltage circuit breakers are total (asymmetrical) rms currents. The appropriate calculated first-cycle duty for comparison is obtained using the 1.6 multiplier specified in ANSI/IEEE C37.010-1979 [2] and ANSI/IEEE C37.5-1979 [3] and the fault point reactance X_{pu} (or impedance Z_{pu}) obtained by network reduction:

$$I_{sc\ tot} = \frac{E_{pu}}{X_{pu}} \cdot 1.6 \cdot I_{base}$$

where $I_{sc\ tot}$ is a three-phase total (asymmetrical) zero-fault impedance (bolted) first-cycle short-circuit rms current.

6.5.4.2 Contact-Parting (Interrupting) Duties for High-Voltage (Above 1 kV, Including Medium-Voltage) Circuit Breakers. First considered are the duties for comparison with interrupting ratings of circuit breakers rated on the pre-1964 total rms current rating basis. The procedures of ANSI/IEEE C37.5-1979 [3]apply regardless of circuit breaker age.

The multiplying factors for reactances of rotating machines in the network are obtained from the *Interrupting* columns of Tables 24 and 25.

For these interrupting duty calculations, the resistance (R) network is also necessary. In the resistance network each rotating machine resistance value must be multiplied by the same factor as its corresponding reactance multiplier from Table 24.

Reduce the reactance network to a single equivalent reactance X_{pu} and reduce the resistance network to a single equivalent resistance R_{pu}. Determine the X/R ratio by dividing X_{pu} by R_{pu}; determine E_{pu}, the prefault operating voltage; and determine E/X by dividing E_{pu} by X_{pu}.

Select the multiplying factor for E/X correction from curves of Figs 101 and 102. To use the curves, it is necessary to know the circuit breaker's contact parting time as well as the proximity of generators to the fault point (local or remote). Local generator multiplying factors apply only when generators that are predominant contributors to short-circuit currents are located in close electrical proximity to the fault as defined in the caption of Fig 101 (and Fig 103).

Minimum contact parting times are usually used and are defined in Table 26.

Table 26
Definition of Minimum Contact-Parting Time for AC High-Voltage Circuit Breakers (ANSI/IEEE C37.010-1979 [2] and ANSI/IEEE C37.5-1979 [3])

Rated Interrupting Time, Cycles at 60 Hz	Minimum Contact-Parting Time, Cycles at 60 Hz
8	4
5	3
3	2
2	1.5

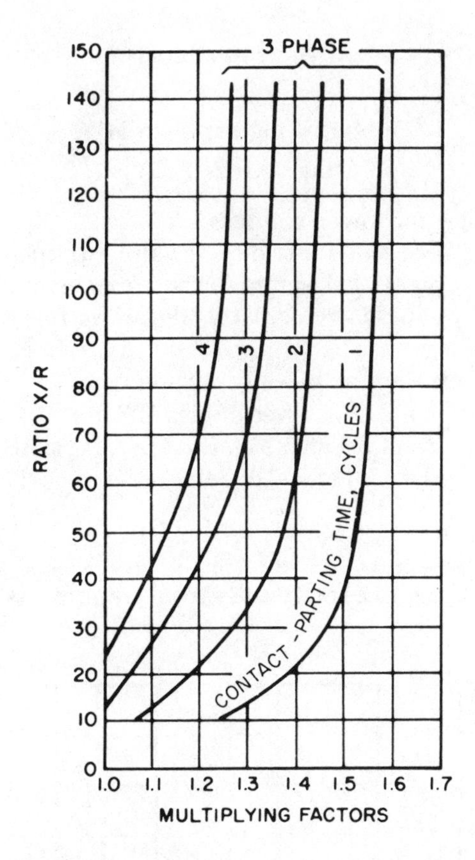

NOTE: Fed predominantly from generators through no more than one transformation or with external reactance in series that is less than 1.5 times generator subtransient reactance (local) (ANSI/IEEE C37.5-1979 [3])

Fig 101
Multiplying Factors (Total Current Rating Basis) for Three-Phase Faults

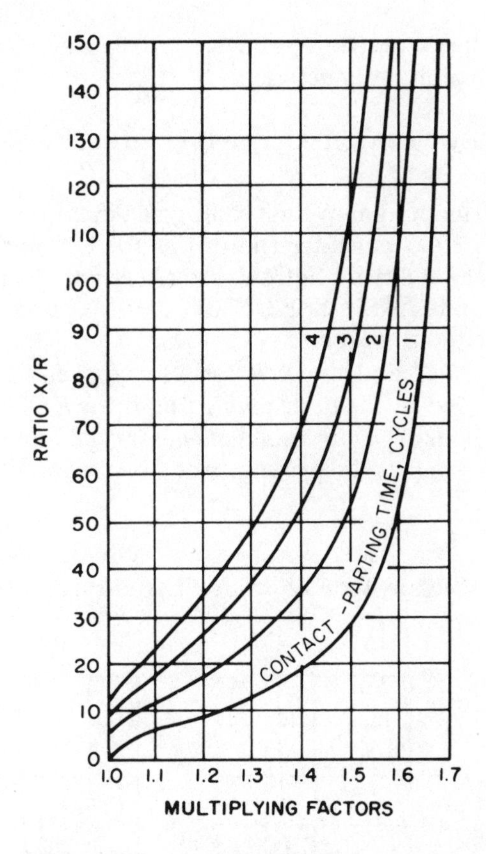

NOTE: Fed predominantly through two or more transformations or with external reactance in series equal to or above 1.5 times generator subtransient reactance (remote) (ANSI/IEEE C37.5-1979 [3])

Fig 102
Multiplying Factors (Total Current Rating Basis) for Three-Phase and Line-to-Ground Faults

Multiply E_{pu}/X_{pu} by the multiplying factor and the base current:

$$\frac{E_{pu}}{X_{pu}} \cdot \text{multiplying factor} \cdot I_{base}$$

This is the calculated total rms current interrupting duty to be compared to the circuit breaker's interrupting capability. For older circuit breakers with three-phase interrupting ratings in megavolt-amperes, the short-circuit current capability in kA is found by dividing the rating in megavolt-amperes by $\sqrt{3}$ and by the operating voltage in kV when the voltage is between the rated maximum and minimum limits.

$$\text{asymmetrical interrupting capability in kA} = \frac{\text{interrupting rating in MVA}}{\text{operating voltage in kV} \cdot \sqrt{3}}$$

The minimum-limit voltage calculation applies for lower voltages.

Next consider the duties for comparison with the short-circuit (interrupting) capabilities of circuit breakers rated on the post-1964 symmetrical rms current basis. ANSI/IEEE C37.010-1979 [2] procedures apply to calculating duties for these circuit breakers.

E/X and the X/R ratio for a given fault point are as already calculated.

Select the multiplying factor for E/X correction from curves of Figs 103 and 104. To use the curves, it is necessary to know the circuit breaker's contact parting time as well as the proximity of generators to the fault point (local or remote), as before.

Fig 103
Multiplying Factors for Three-Phase Faults Fed Predominantly from Generators

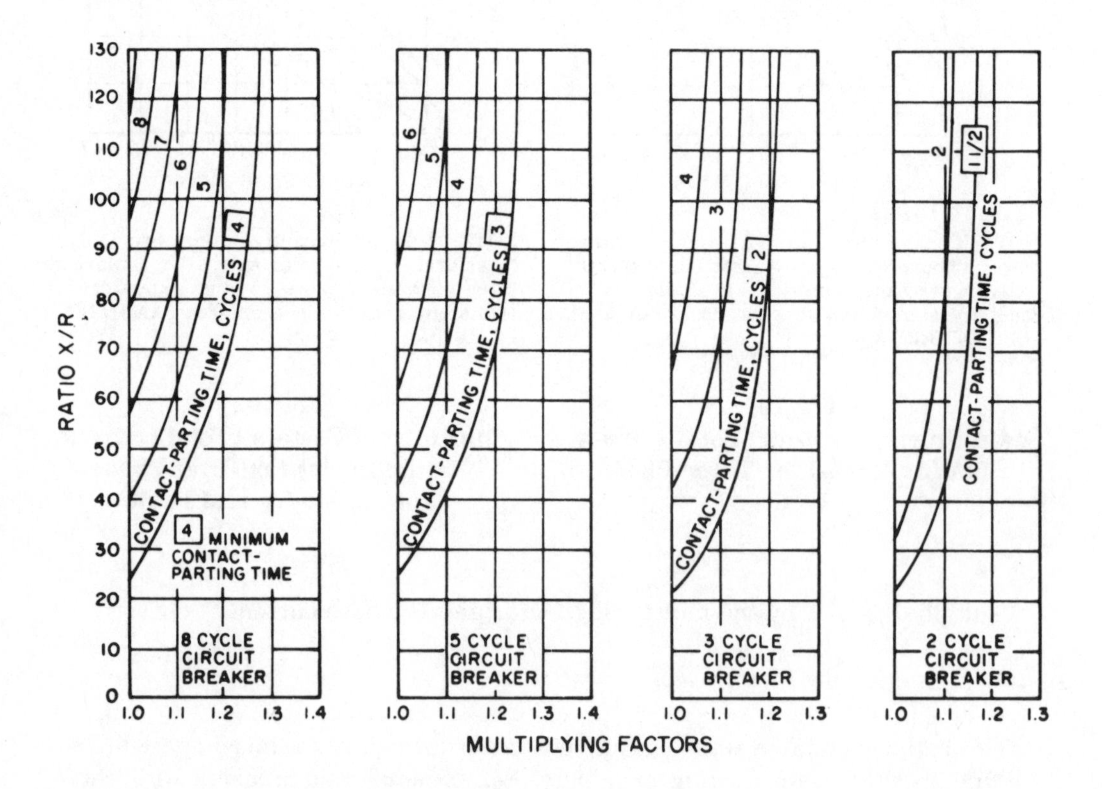

NOTE: Through no more than one transformation or with external reactance in series that is less than 1.5 times generator subtransient reactance (local) (ANSI/IEEE C37.010-1979 [2])

Multiply E_{pu}/X_{pu} by the multiplying factor and the base current:

$$\frac{E_{pu}}{X_{pu}} \cdot \text{multiplying factor} \cdot I_{base}$$

The result is an interrupting duty to be compared with the rated symmetrical interrupting capability of a circuit breaker, but it is a symmetrical interrupting duty only if the multiplying factor for E/X is 1.0. The rated symmetrical interrupting capability of the circuit breaker is calculated as follows:

$$\text{symmetrical interrupting capability} = \frac{(\text{rated } I_{sc})\ (\text{rated maximum } E)}{\text{operating } E}$$

This calculated current shall not exceed the maximum symmetrical interrupting capability listed for the circuit breaker.

Fig 104
Multiplying Factors for Three-Phase and Line-to-Ground Faults Fed Predominantly from Generators

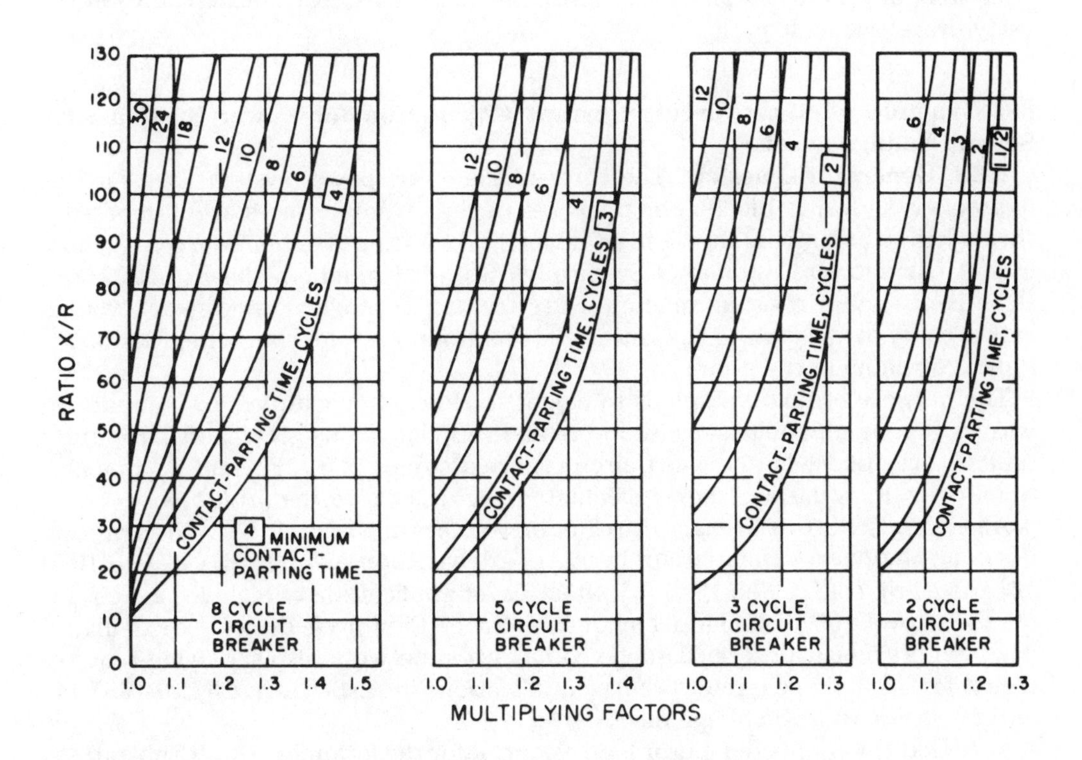

NOTE: Through two or more transformations or with external reactance in series that is equal to or above 1.5 times generator subtransient reactance (remote) (ANSI/IEEE C37.010-1979 [2])

The calculating procedures described for first-cycle and interrupting networks are different in several respects from procedures detailed in earlier editions of this publication that were based on standards now superseded. The differences are intended to account more accurately for contributions to high-voltage interrupting duty from large induction motors, for the exponential decay of the dc component of short-circuit current, and for the ac decay of contributions from nearby generators.

6.5.4.3 Short-Circuit Currents for Time-Delayed Relaying Devices. For the application of instantaneous relays, the value of the first-cycle short-circuit current determined by the first-cycle network should be used. For an application of time-delay relays beyond six cycles, the equivalent system network representation will include only generators and the passive elements such as transformers and cables between them and the fault point. The generators are represented by a transient or a larger impedance related to the magnitude of decaying generator short-circuit current at the specified calculation time. All motor contributions are omitted. Only the generators that contribute fault current through the relay under consideration to the fault point are considered for the relay application. The dc component will have decayed to near zero and is not considered. The short-circuit symmetrical rms current is E_{pu}/X_{pu}, where X_{pu} is derived from the equivalent reactance network consisting of generators and passive equipment (cables, transformers, etc) in the relay protective fault path.

6.6 Example of Short-Circuit Current Calculation for Power System with Several Voltage Levels

6.6.1 General Discussion. The three-phase 60 Hz power system used for this example is shown in Fig 99. For purposes of the example, buses are numbered 1 through 4 with numbers shown in triangles, and rotating machine sources of short-circuit currents are numbered S_1 through S_{10} with numbers shown in squares. Groups of similar rotating machines are treated as single sources, each with a rating equal to the sum of the ratings in the group and the characteristics of the typical machine in the group.

The purpose of the example is to calculate short-circuit duties for comparison with ratings or capabilities of circuit breakers applied at buses 1, 2, and 3. Separate faults (three-phase bolted short circuits) are assumed at F_1, F_2, and F_3, one at a time. When F_1 is the fault being calculated, bus 1 is called the *fault bus*.

All fault buses are at primary distribution voltages of 13.8 or 4.16 kV. Interrupting duty calculations for their circuit breakers are based on ANSI/IEEE C37.010-1979 [2] and ANSI/IEEE C37.5-1979 [3], which cover applications of high-voltage circuit breakers (over 1000 V, including medium voltage). First-cycle duties are calculated with the previously described single combination network also satisfying requirements for low-voltage circuit-breaker applications in ANSI/IEEE C37.13-1981 [4] and for low- and high-voltage fuses.

Note that the connected motor load assumed for the low-voltage unit substations of this example is lower than that observed for many actual substations. Experience has shown that the rated kVA summation for connected motors often greatly exceeds substation transformer kVA. This is a factor to be considered in studies intended to account for future growth.

6.6.2 Utility System Data. In-plant generators operating in parallel with utility system ties are the main sources both at bus 1 and at bus 2. The representation of remote utility generators for plant short-circuit calculations is often based on the utility *available* short-circuit current, or apparent power in MVA, which should be the highest applicable magnitude (future rather than present, for conservative equipment selection, also specifying the X/R ratio) of short-circuit contribution delivered by the utility from all sources outside the plant (not including contributions from in-plant sources). These data are converted to an equivalent impedance. Obtaining original equivalent impedance data from the utility is equally useful.

6.6.3 Per-Unit Calculations and Base Quantities. This example uses per-unit quantities for calculations. The base for all per-unit power quantities throughout the system is 10 MVA (any other value could have been selected). Voltage bases are different for different system voltage levels, but it is necessary for all of them to be related by the turns ratios of interconnecting transformers, as specified in kilovolts at each numbered bus in Fig 99. Any actual quantity is the per-unit magnitude of that quantity multiplied by the applicable base. For example, 1.1 per-unit voltage at bus 1 is actually 1.1 · the 13.8 kV base voltage at bus 1 = 15.18 kV. Per-unit system bases and actual quantities have identical physical relationships. For example, in three-phase systems the relationship

$$\text{total MVA} = \sqrt{3}\,(E_{\text{L-L}} \text{ in kV})\,(I_{\text{line}} \text{ in kA})$$

applies both to actual quantities and to bases of per-unit quantities. Other useful base quantities for this example, derived using the 10 MVA base and the base voltages of Fig 99 in the equations of 6.5, are listed as follows:

	Base Line-to-Line Voltage $E_{\text{L-L}}$	
	13.8 kV	4.16 kV
Base line current, kA	0.4184	1.388
Base line-to-neutral impedance, ohms	19.04	1.73

This example calculates balanced per-unit three-phase short-circuit current duties using one of three identical per-unit line-to-neutral positive-sequence circuits, energized by per-unit line-to-neutral voltage. Only line-to-line base voltages are listed, however; for balanced three-phase circuits, line-to-line voltages in per unit of these bases are identical to line-to-neutral voltages in per unit of their corresponding line-to-neutral base voltages.

6.6.4 Impedances Represented by Reactances. The usual calculation of short-circuit duties at voltages over 1000 V involves circuits in which resistance is small with respect to reactance, so computations are simplified by omitting the resistance from the circuit. The slight error introduced makes the solution conservative. This example employs this simplification by using only the reactances of elements when finding the magnitudes of short-circuit duties. However, element resistance data are necessary to determine X/R ratios as described later in this example.

6.6.5 Equivalent Circuit Variations Based on Time and Standards. Calculations of high-voltage circuit breaker short-circuit current duty may make use of several equivalent circuits for the power system, depending on when duties are calculated and on the procedure described in the standard used as a basis.

The circuit used for calculating first-cycle (momentary) short-circuit current duties uses subtransient reactance, sometimes modified as shown in Tables 24 and 25, for all rotating machine sources of short-circuit current. Synchronous machines and large induction motors (over 250 hp at 3600 r/min or 1000 hp at 0–1800 r/min) are represented with unmodified subtransient reactance. Medium induction motors (all others 50 hp and above) have subtransient reactance multiplied by 1.2 (or first-cycle X is estimated at 0.20 per unit). Small induction motors (less than 50 hp each) have subtransient reactance multiplied by 1.67 (or first-cycle X is estimated at 0.28 per unit).

The circuit used for calculating short-circuit (interrupting) duties, at circuit-breaker minimum contact-parting times of 1.5 to 4 cycles after the short circuit starts, retains synchronous generator subtransient reactance unchanged. It also represents synchronous motors and large induction motors with subtransient reactance multiplied by 1.5, as well as medium induction motors with subtransient reactance multiplied by 3.0 (or interrupting X is estimated at 0.50 per unit); it neglects induction motors with less than 50 hp.

Passive element reactances are the same in all equivalent circuits.

Resistances are necessary to find fault point X/R ratios used in short-circuit (interrupting) duty calculations based on ANSI/IEEE C37.010-1979 [2] and ANSI/IEEE C37.5-1979 [3]. The fault point X/R ratio is the fault-point X divided by the fault-point R. A fault-point X is found by reducing the reactance circuit described in preceding paragraphs to a single equivalent X at the fault point. A fault-point R is found by reducing a related resistance-only circuit. This is derived from the reactance circuit by substituting the resistance in place of the reactance of each element, obtaining the resistance value by dividing the element reactance by the element X/R ratio. For motors whose subtransient reactance is increased by a multiplying factor, the same factor must be applied to the resistance in order to preserve the X/R ratio for the motor.

The X/R data for power-system elements of this example, shown in Fig 99, are *medium typical* data obtained in most cases from tables and graphs that are included in the applicable standards and are reproduced in the Note at the end of this chapter.

The *approximately 30 cycle* network often is a minimum-source representation intended to investigate whether minimum short-circuit currents are sufficient to operate current-actuated relays. Minimum-source circuits might apply at night or when production lines are down for any reason. Some of the source circuit breakers may be open and all motor circuits may be off. In-plant generators are represented with transient reactance or a larger reactance related to the magnitude of decaying generator short-circuit current at the desired calculation time, for this example assumed at 1.5 times subtransient reactance in the absence of better information.

6.6.6 Impedance Data and Conversions to Per Unit. Reactances of passive elements, obtained from Fig 99, are listed in Table 27, along with the conversion of each reactance to per unit on the 10 MVA base.

Most of the data given in Fig 99 are per unit, based on the equipment's nameplate rating. Any original percent impedance data were divided by 100 to obtain a per-unit impedance for Fig 99. Conversions are changes of MVA base, multiplied by the ratio of the new MVA base (10 MVA for the example) to the old MVA base (rated MVA). When the equipment's rated voltage is not the same as the base voltage, it is also necessary to make voltage base conversions using the square of the ratio of rated voltage to example base voltage as the multiplier (see 6.5). This is not illustrated in this example.

Physical descriptions of cables are used to establish their reactances in ohms based on data in tables N1.3 and N1.6. Dividing an impedance in ohms by the base impedance in ohms converts it to per unit.

6.6.7 Subtransient Reactances of Rotating Machines, and Reactances for the Circuit to Calculate First-Cycle (Momentary) Short-Circuit Current Duties. Subtransient reactances of rotating machine sources of short-circuit current modified for the combination first-cycle network based on interpretation of reference low- and high-voltage standards—ANSI/IEEE C37.010-1979 [2], ANSI/IEEE C37.5-1979 [3], and ANSI/IEEE C37.13-1981 [4]—are listed in Table 28 together with conversions to per unit on the study base.

Reactances representing the rotating machines of utility systems are found by assuming that the *available* megavolt-amperes are 1.0 per unit of a base equal to itself, and 1.0 per-unit megavolt-amperes corresponds to 1.0 per-unit reactance at 1.0 per-unit voltage.

The circuit development and impedance simplifications are described subsequently.

Table 27
Passive-Element Reactances, in Per Unit, 10 MVA Base

Transformer T_1,	$X = 0.07\ (10/20) = 0.035$ per unit
Transformer T_2,	$X = 0.055\ (10/5) = 0.110$ per unit
Transformer T_3,	$X = 0.065\ (10/5) = 0.130$ per unit
Transformer T_4,	$X = 0.055\ (10/5) = 0.110$ per unit
Transformer T_5,	$X = 0.055\ (10/7.5) = 0.0734$ per unit
Transformer T_6,	$X = 0.055\ (10/1.5) = 0.367$ per unit
Reactor X_1,	$X = 0.08\ (10/7.5) = 0.107$ per unit

Cable C_1, from Tables N1.3 and N1.6 for 250 kcmil at 1 in spacing,

$$X = 0.0922 - 0.0571 = 0.0351\ \Omega/1000\ \text{ft}$$

(There are no reactance corrections as this is three-conductor cable in nonmagnetic duct.)

For 3500 ft of cable, the conversion to per unit on a 10 MVA 13.8 kV base is

$$X = (3500/1000)\ (0.0351/19.04) = 0.0064 \text{ per unit}$$

Cable C_2, 300 kcmil at 1 in spacing,

$$X = 0.0902 - 0.0571 = 0.0331\ \Omega/1000\ \text{ft}$$

For 2500 ft of two cables in parallel at 4.16 kV,

$$X = (2500/1000)\ (1/2)\ (0.0331/1.73) = 0.0239 \text{ per unit}$$

Table 28
Subtransient Reactances of Rotating Machines, Modified for First-Cycle (Momentary) Duty Calculations, in Per Unit, 10 MVA Base

69 kV system, $X = 1.0\ (10/1000) = 0.01$ per unit
Generator 1, $X_d'' = 0.09\ (10/25) = 0.036$ per unit

46 kV system, $X = 1.0\ (10/800) = 0.0125$ per unit
Generator 2, $X_d'' = 0.09\ (10/5) = 0.18$ per unit

Large synchronous motor M_1, using the assumption that the horsepower rating of an 0.8 power factor machine is its kVA rating,
$X_d'' = 0.20\ (10/6) = 0.333$ per unit, each motor

Large induction motor M_2, using the assumption that hp = kVA,
$X_d'' = 0.17\ (10/1.75) = 0.971$ per unit

Low-voltage motor group, 0.4 MVA, from 50 to 150 hp,
first-cycle $X = 1.2\ X_d'' = 0.20\ (10/0.4) = 5.0$ per unit

Low-voltage motor group, 1.12 MVA, less than 50 hp each
first-cycle $X = 1.67\ X_d'' = 0.28\ (10/1.12) = 2.5$ per unit

6.6.8 Reactances and Resistances for the Circuit to Calculate Short-Circuit (Interrupting) Current Duties. Reactances, and resistances derived from them as described previously, are detailed in Table 29.

6.6.9 Reactances for the Circuit to Calculate *Approximately 30-Cycle* Minimum Short-Circuit Currents. Minimum generation for this problem occurs with Generator 1 down, the 46 kV utility system connection open, and all motors disconnected. Reactance details are given in Table 30.

6.6.10 Circuit and Calculation of First-Cycle (Momentary) Short-Circuit Current Duties. The circuit used for calculating the symmetrical alternating currents that establish the first-cycle (momentary) short-circuit duties based on a combination of current circuit breaker and fuse standards is shown in Fig 105(a). Source circuits S_5 through S_{10} have been simplified using the series and parallel combinations indicated in Table 31, based on the per-unit element impedances obtained directly from Tables 27 and 28. The identities of buses and sources are retained in Fig 105(a), even after the individual element impedances from Fig 99 lose identification when reactances are combined.

The connection of an ac source, the voltage magnitude of which is the prefault voltage at the fault bus, between the dotted *common* connection and the fault at the fault bus causes the flow of per-unit alternating short-circuit current that is being calculated.

The reactances of Fig 105(a) are further simplified as shown in Fig 105(b), without losing track of the three fault locations. The reactance simplifications are summarized in Table 32. The table contains columns of reactances and reciprocals. Arrows are used to indicate the calculation of a reciprocal. Sums of reciprocals are used to combine reactances in parallel. A dashed line in the reactance column indicates that reactances above the line have been combined in parallel.

Table 29
Reactances, X/R Ratios, and Resistances for AC High-Voltage Circuit Breaker Contact-Parting Time (Interrupting) Short-Circuit Duties

Transformer T_1, $X/R = 21$, $R = 0.035/21 = 0.001\,667$ per unit
Transformer T_2, $X/R = 16$, $R = 0.110/16 = 0.006\,88$ per unit
Transformer T_3, $X/R = 16$, $R = 0.130/16 = 0.008\,12$ per unit
Transformer T_4, $X/R = 12$, $R = 0.11/12 = 0.009\,16$ per unit
Transformer T_5, $X/R = 14$, $R = 0.0734/14 = 0.005\,24$ per unit
Transformer T_6, $X/R = 10$, $R = 0.0367/10 = 0.0367$ per unit
Reactor X_1, $X/R = 50$, $R = 0.107/50 = 0.002\,14$ per unit

Cable C_1, ac resistance at 50 °C from Table 20 is 0.0487 Ω/1000 ft, correction for 75 °C = 1.087
For 3500 ft of cable converted to per unit on a 10 MVA 13.8 kV base,
$R = (3500/1000)\,(1.087)\,(0.0487/19.04) = 0.009\,72$ per unit

Cable C_2, ac resistance from Table 20 is 0.0407 Ω/1000 ft
For 2500 ft of two cables in parallel on a 10 MVA 4.16 kV base at 75 °C,
$R = (2500/1000)\,(1.087/2)\,(0.0407/1.73) = 0.0320$ per unit

69 kV system, $X/R = 22$, $R = 0.01/22 = 0.000\,445$ per unit
Generator 1, $X/R = 45$, $R = 0.036/45 = 0.0008$ per unit

46 kV system, $X/R = 9$, $R = 0.0125/9 = 0.001\,389$ per unit
Generator 2, $X/R = 29$, $R = 0.18/29 = 0.0062$ per unit

Large synchronous motor M_1, using $X = 1.5\,X_d'' = 1.5\,(0.333) = 0.5$ per unit, $X/R = 30$,
$R = 0.5/30 = 0.016\,67$ per unit
Large induction motor M_2, using $X = 1.5\,X_d'' = 1.5\,(0.971) = 1.457$ per unit, $X/R = 30$,
$R = 1.457/30 = 0.048\,57$ per unit
Low-voltage motor group 50–150 hp, using $X = 3.0\,X_d'' = (3/1.2) \cdot 5.0 = 12.5$ per unit, $X/R = 9$,
$R = 12.5/9 = 1.389$ per unit
Low-voltage motor group below 50 hp is omitted

NOTE: See Tables 27 and 28 for reactances of passive elements, utility systems, and generators.

The final simplification of reactances to obtain one fault-point X for each fault location is detailed in Table 33. The results for the specified fault buses are the last entries in the reactance columns.

Alternating short-circuit currents are calculated from the circuit reactance reductions assuming the prefault voltage is 1.0 per unit, and alternating rms current is, of course, E/X per unit. Multiplying by base current converts to real units. The resulting symmetrical (alternating only) first-cycle (momentary) short-circuit rms currents are

at F_1, $I_{sym} = (1.0/0.016)\,(0.4184) = 26.15$ kA
at F_2, $I_{sym} = (1.0/0.0423)\,(1.388) = 32.81$ kA
at F_3, $I_{sym} = (1.0/0.1048)\,(1.388) = 13.25$ kA

These currents are useful as primary *available* symmetrical short-circuit current data for calculations of short-circuit duties at low-voltage buses of unit substations connected to these medium-voltage buses.

Asymmetrical short-circuit current duties for comparison with ac high-voltage (over 1000 V, including medium-voltage) circuit-breaker closing and latching capabilities (post-1964 rating basis) or momentary ratings (pre-1964 rating basis) are

Table 30
Reactances for *Approximately 30-Cycle* Short-Circuit Currents

Utility system S_1 reactance is unchanged
Generator 2, S_4 is represented with reactance larger than subtransient, assumed at $1.5\ X_d'' = 1.5 \cdot 0.18 = 0.27$ per unit
All other sources, S_2, S_3, S_5–S_{10}, are disconnected

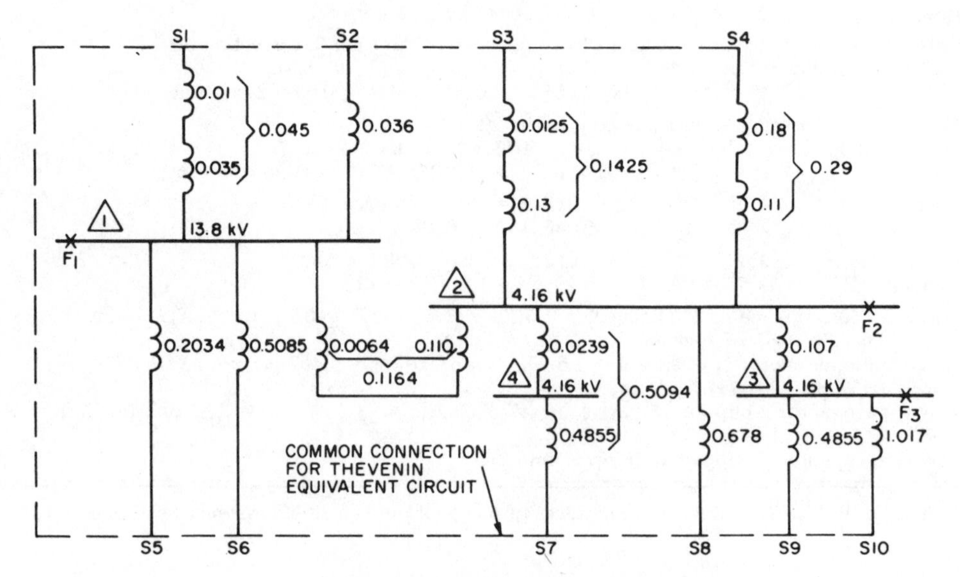

(a) Reactance Diagram

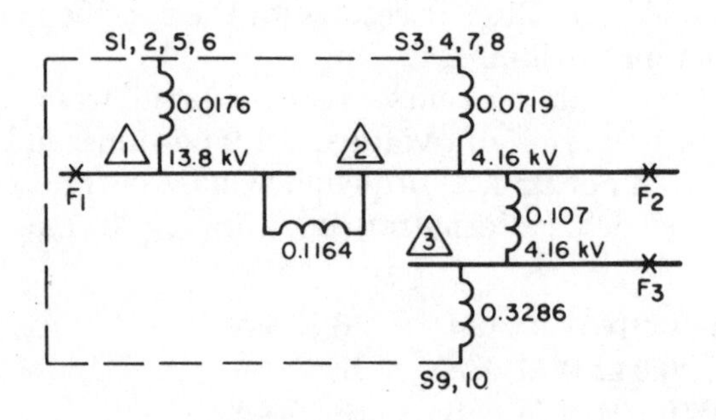

(b) Simplified Reactance Diagram

Fig 105
Circuits of Power System Reactances for Calculation of First-Cycle (Momentary) Short-Circuit Current Duties for Fuses and Low-Voltage Circuit Breakers

Table 31
Reactances for Fig 105(a)

S_5 to bus 1, two circuits in parallel, each with M_1 motor X_d'' and T_5 transformer X,

X_d'' = (1/2) (0.3333 + 0.0734) = 0.2034 per unit

S_6 to bus 1 (after combining all the motors of one substation for an equivalent low-voltage motor X_d'' = 2.5 (5)/(2.5 + 5) = 1.667), four circuits in parallel, each with an equivalent low-voltage motor X_d'' in series with a T_6 transformer,

X_d'' = (1/4) (1.667 + 0.367) = (1/4) (2.034) = 0.5085 per unit

S_7 to bus 4, two M_2 induction motors,

X_d'' = (1/2) (0.971) = 0.4855 per unit

S_8 to bus 2, three circuits, each as for the S_6 to bus 1 calculation,

X_d'' = (1/3) (2.034) = 0.678 per unit

S_9 to bus 3, two M_2 induction motors,

X_d'' = (1/2) (0.971) = 0.4855 per unit

S_{10} to bus 3, two circuits, each as for the S_6 to bus 1 calculation,

X_d'' = (1/2) (2.034) = 1.017 per unit

Table 32
Reactance Combinations for Fig 105(a)

S_1, S_2, S_5, S_6			S_3, S_4, S_7, S_8			S_9, S_{10}		
X		$1/X$	X		$1/X$	X		$1/X$
0.045	→	22.22	0.1425	→	7.02	0.4855	→	2.060
0.036	→	27.78	0.29	→	3.45	1.017	→	0.983
0.2034	→	4.91	0.5094	→	1.96	0.3286	←	3.043
0.5085	→	1.97	0.678	→	1.47			
0.0176	←	56.88	0.0719	←	13.90			

Table 33
Reactance Combinations for Fault-Point X at Each Fault Bus of Fig 105(b)

Fault at F_1			Fault at F_2			Fault at F_3		
X		$1/X$	X		$1/X$	X		$1/X$
0.3286			0.0176			0.1340	→	7.46
0.107			0.1164			0.0179	→	13.90
0.4356	→	2.30	0.1340	→	7.46	0.0468	←	21.36
0.0719	→	13.90	0.107			0.107		
0.0617	←	16.20	0.3286			0.1538	→	6.502
0.1164			0.4356	→	2.30	0.3286	→	3.043
0.1781	→	5.62	0.0719	→	13.90	0.1048	←	9.545
0.0176	→	56.82	0.0423	←	23.66			
0.016	←	62.44						

found using a 1.6 multiplying factor according to ANSI/IEEE C37.010-1979 [2] and ANSI/IEEE C37.5-1979 [3]. These total (asymmetrical) first-cycle (momentary) short-circuit rms currents are

at F_1, I_{tot} = 1.6 (26.15) = 41.8 kA
at F_2, I_{tot} = 1.6 (32.81) = 52.5 kA
at F_3, I_{tot} = 1.6 (13.25) = 21.2 kA

Asymmetrical short-circuit duties are necessary for comparison with total rms current ratings of ac high-voltage (and medium-voltage) fuses, such as those in the fused motor-control equipment connected to buses 3 and 4. These are found using multiplying factors from ANSI C37.41-1981 [5]. The applicable standard for the circuit of Fig 105 suggests a general-case multiplying factor of 1.55, but a *special case* multiplier of 1.2 may be substituted if the voltage is less than 15 kV and if the X/R ratio is less than 4. The circuit of this example will not have X/R ratios as low as 4. The asymmetrical (total) first-cycle (momentary) short-circuit rms current for fuse applications are

at F_1, I_{tot} = 1.55(26.15) = 40.73 kA
at F_2, I_{tot} = 1.55(32.81) = 50.86 kA
at F_3, I_{tot} = 1.55(13.25) = 20.54 kA

6.6.11 Circuit and Calculation of Contact-Parting Time (Interrupting) Short-Circuit Current Duties for High-Voltage Circuit Breakers. In addition to a circuit of power system reactances for calculating alternating currents ($I_{pu} = E/X$), a resistance-only circuit is needed to establish fault-point X/R ratios. Duties are calculated by applying multiplying factors to E/X. The multiplying factors depend on the fault-point X/R and also on other factors defined subsequently.

The circuits used for calculating E/X and fault-point R are shown in Figs 106(a) and 107(a), respectively. The rotating machine reactances for the circuit of Fig 106(a), if changed from subtransient, are shown in Table 29. Table 34 details how these changes affect the Table 31 simplifications of source circuits S_5 through S_{10}. Table 34 also includes resistance simplifications of source circuits for Fig 107(a).

Figures 106(b) and 107(b) show the last steps of reactance and resistance simplifications, respectively, before the several fault-location identities are lost. Tables 35 and 36 detail the reactance and resistance simplifications starting from Figs 106(a) and 107(a), respectively. The final simplifications of reactances and resistances to obtain one fault-point X and one fault-point R for each fault location are detailed in Tables 37 and 38, respectively.

Values of per-unit E/X for each fault bus are readily obtained from Table 37 when E = 1.0 (as for this example); they are the final entries in the $1/X$ columns, opposite the fault-point X entries. Values converted to actual currents are

at F_1, E/X = 59.03(0.4184) = 24.70 kA
at F_2, E/X = 20.75(1.388) = 28.80 kA
at F_3, E/X = 7.841(1.388) = 10.88 kA

Values of X/R for each fault bus are obtained from the fault-point X and R entries of Tables 37 and 38 as follows:

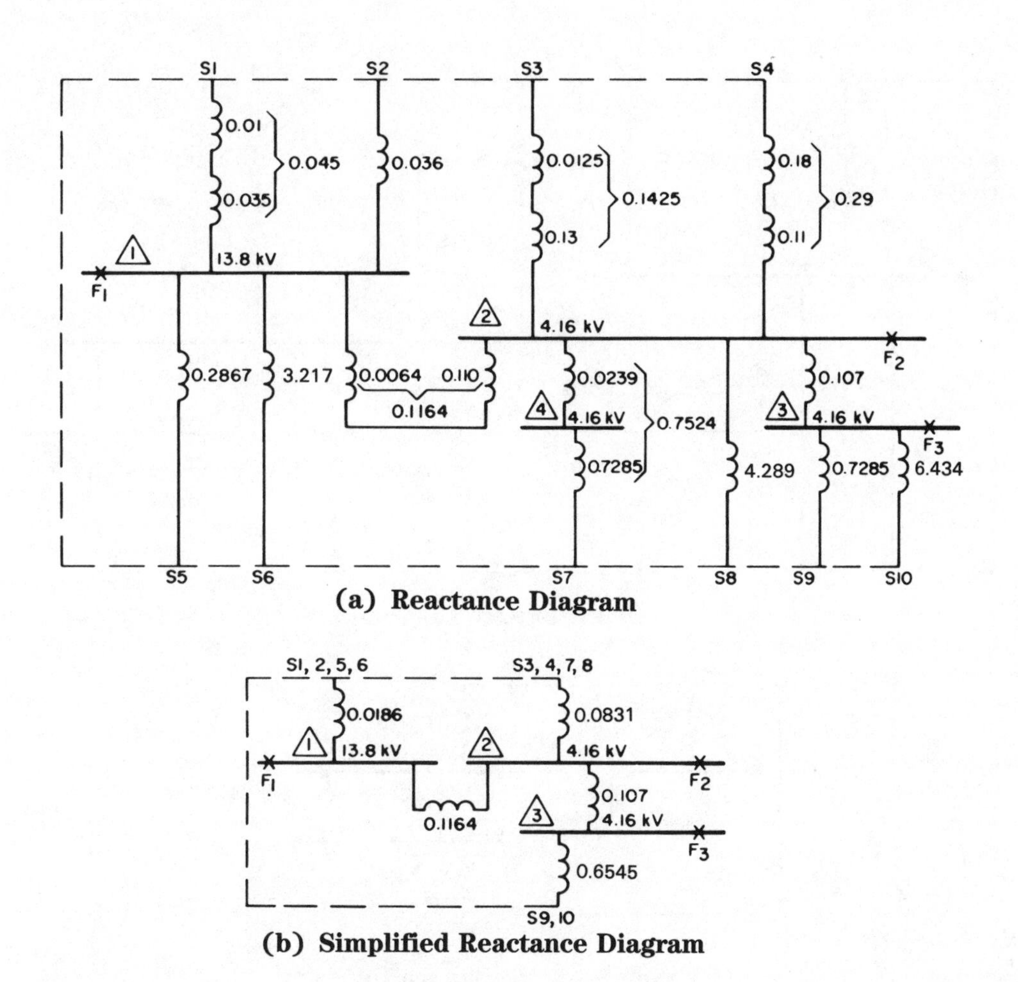

(a) Reactance Diagram

(b) Simplified Reactance Diagram

Fig 106
Circuits of Power System Reactances for Calculation of E/X and Fault-Point X for Contact-Parting Time (Interrupting) Short-Circuit Current Duties for High-Voltage Circuit Breakers

at F_1, X/R = 0.0169/0.000537 = 31.47
at F_2, X/R = 0.0482/0.00348 = 13.85
at F_3, X/R = 0.1275/0.00488 = 26.13

The reference standards contain graphs of multiplying factors that determine calculated short-circuit current duties when applied to E/X values. The proper graph is selected with the following information:

(1) Three-phase or single-phase short-circuit current (three-phase for this example)

(2) Rating basis of the circuit breaker being applied (previous total-current short-circuit ratings or present symmetrical-current short-circuit ratings)

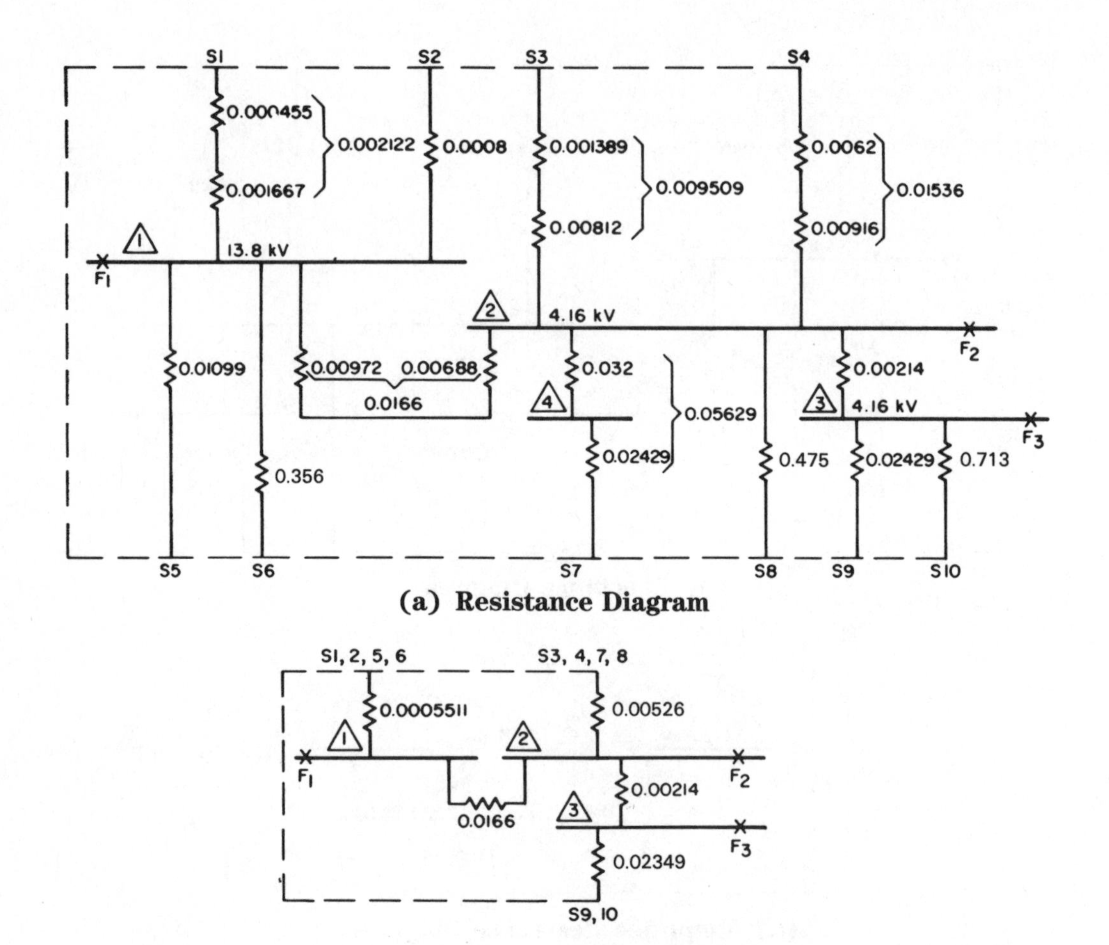

(a) Resistance Diagram

(b) Simplified Resistance Diagram

Fig 107
Circuits of Power System Resistances for Calculation of Fault-Point *R* for Contact-Parting Time (Interrupting) Short-Circuit Current Duties for High-Voltage Circuit Breakers

(3) Rated interrupting time of the circuit breaker being applied
(4) Fault-point X/R ratio
(5) Proximity of generators

The proximity of generators determines the choice between graphs (a) for faults fed predominantly from generators through not more than one transformation or with external impedance in series that is less than 1.5 times generator X_d'' (*local* in this example) and (b) for faults fed predominantly through two or more transformers or with external impedance in series which is equal to or exceeds 1.5 times generator X_d'' (*remote* in this example). The local and remote multiplying factor

Table 34
Reactances for Fig 106(a) and Resistances for Fig 107(a)

S_5 to bus 1, two circuits in parallel, each with 1.5 X_d'' of synchronous motor M_1 and transformer T_5,
X = (1/2) (0.5 + 0.0734) = 0.2867 per unit
R = (1/2) (0.016 67 + 0.005 31) = (1/2) (0.021 98) = 0.010 99 per unit

S_6 to bus 1, four circuits in parallel, motor group and transformer T_6,
X = (1/4) (12.5 + 0.367) = (1/4) (12.867) = 3.217 per unit
R = (1/4) (1.389 + 0.0367) = (1/4) (1.4257) = 0.356 per unit

S_7 to bus 4, two motors M_2,
X = (1/2) (1.457) = 0.7285 per unit
R = (1/2) (0.048 57 + 0.024 29) per unit

S_8 to bus 2, three circuits, each as for the S_6 to bus 1 calculation,
X = (1/3) (12.867) = 4.289 per unit
R = (1/3) (1.4257) = 0.475 per unit

S_9 to bus 3, two motors M_2,
X = (1/2) (1.457) = 0.7285 per unit
R = (1/2) (0.048 57) = 0.024 29 per unit

S_{10} to bus 3, two circuits, each as for the S_6 to bus 1 calculation,
X = (1/2) (12.867) = 6.434 per unit
R = (1/2) (1.4257) = 0.713 per unit

Table 35
Reactance Combinations for Fig 106(a)

S_1, S_2, S_5, S_6			S_3, S_4, S_7, S_8			S_9, S_{10}		
X		$1/X$	X		$1/X$	X		$1/X$
0.045	→	22.22	0.1425	→	7.018	0.7285	→	1.373
0.036	→	27.78	0.29	→	3.448	6.434	→	0.155
0.2867	→	3.49	0.7524	→	1.329	0.6545	←	1.528
3.217	→	0.31	4.289	→	0.233			
0.0186	←	53.80	0.0831	←	12.028			

graphs of ANSI/IEEE C37.010-1979 [2] and ANSI/IEEE C37.5-1979 [3] are Figs 101–104. The local multiplying factors are smaller because they include the effects of generator ac (symmetrical current) decay. Remote multiplying factors are based on no decay of the remote generator ac (symmetrical current) up to circuit breaker contact-parting time. Utility contributions are considered to be from remote generators in most industrial system duty calculations.

For many systems having only remote sources and no in-plant generators, it is clear that the remote multiplying factor is the only choice. For the few systems that have in-plant generator primary power sources, both multiplying factors may be necessary, as explained subsequently.

In this example, short-circuit duties are calculated for total-current (TOT) short-circuit rated (previous basis) circuit breakers with 8-cycle and 5-cycle rated interrupting times (TOT 8 and TOT 5) and symmetrical-current (SYM) short-circuit rated (present basis) circuit breakers with 5-cycle rated interrupting times (SYM 5).

Table 36
Resistance Combinations for Fig 107(a)

S_1, S_2, S_5, S_6			S_3, S_4, S_7, S_8			S_9, S_{10}		
R		1/*R*	*R*		1/*R*	*R*		1/*R*
0.002 122	→	471.3	0.009 509	→	105.2	0.024 29	→	41.17
0.000 8	→	1250.0	0.015 36	→	65.10	0.713	→	1.403
0.010 99	→	90.99	0.056 29	→	17.77	0.023 49	←	42.57
0.356	→	2.81	0.475	→	2.11			
0.000 551 1	←	1815	0.005 26	←	190.3			

Table 37
Reactance Combinations for Fault-Point *X* at Each Fault Bus of Fig 106(b)

Fault at F_1			Fault at F_2			Fault at F_3		
X		1/*X*	*X*		1/*X*	*X*		1/*X*
0.6545			0.0186			0.135	→	7.407
0.107			0.1164			0.0831	→	12.03
0.7615	→	1.313	0.135	→	7.407	0.0514	←	19.44
0.0831	→	12.03	0.7615	→	1.313	0.107		
0.0750	←	13.34	0.0831	→	12.03	0.1584	→	6.313
0.1164			0.0482	←	20.75	0.6545	→	1.528
0.1914	→	5.225				0.1275	←	7.841
0.0186	→	53.80						
0.0169	←	59.03						

Table 38
Reactance Combinations for Fault-Point *R* at Each Fault Bus of Fig 107(b)

Fault at F_1			Fault at F_2			Fault at F_3		
R		1/*R*	*R*		1/*R*	*R*		1/*R*
0.023 49			0.000 551 1			0.017 15	→	58.31
0.002 14			0.016 6			0.005 26	→	190.3
0.025 63	→	39.02	0.017 15	→	58.31	0.004 02	←	248.6
0.005 26	→	190.3	0.005 26	→	190.3	0.002 14		
0.004 36	←	229.32	0.025 63	→	39.02	0.006 16	→	162.3
0.016 6			0.003 48	←	287.6	0.023 49	→	42.57
0.020 96	→	47.71				0.004 88	←	204.9
0.000 551 1	→	1815						
0.000 537	←	1863						

Multiplying factors obtained from both the local and remote graphs of Figs 101-104 are shown in Table 39 for the other conditions previously established in this example.

In this example, with each of two main buses connected to a utility (remote) source and an in-plant generator (local for nearby faults) source, it is not immediately apparent which multiplying factor applies. One technique that perhaps provides an extra margin of conservatism is to use only the larger remote multiplying factors as described in the next paragraphs. An alternative and also conservative procedure that interpolates between multiplying factors requires additional calculations (see 6.6.12.1).

The calculated short-circuit interrupting duty rms currents for three-phase faults at bus 1, using remote multiplying factors, are

for TOT 8 circuit breakers, 1.19(24.70) = 29.39 kA-T
for TOT 5 circuit breakers, 1.27(24.70) = 31.37 kA-T
for SYM 5 circuit breakers, 1.15(24.70) = 28.41 kA-S

The kA-T designation denotes an rms current duty in kiloamperes to be compared with the total-current short-circuit (interrupting) capability of a total rated circuit breaker. This is a total (asymmetrical) rms current duty.

The kA-S designation denotes an rms current duty in kiloamperes to be compared with the symmetrical-current short-circuit (interrupting) capability of a symmetrical rated circuit breaker. This is a symmetrical rms current duty only if the multiplying factor for E/X is 1.0; otherwise it is neither symmetrical nor asymmetrical, but partway in between.

The F_2 fault calculation for a TOT 5 circuit breaker is not detailed in this example. TOT 8 and SYM 5 duties at bus 2 are already available, since 1.0 multiplying factors apply, as follows:

for TOT 8 circuit breakers, 1.0(28.80) = 28.80 kA-T
for SYM 5 circuit breakers, 1.0(28.80) = 28.80 kA-S

Table 39
Three-Phase Short-Circuit Current
Multiplying Factors for E/X for Example Conditions

Fault Location	Fault-Point X/R Ratio	Circuit Breaker-Type	Multiplying Factor Local	Multiplying Factor Remote
F_1	31.47	TOT 8	1.05	1.19
		TOT 5	1.14	1.27
		SYM 5	1.03	1.15
F_2	13.85	TOT 8	1.0*	1.0*
		TOT 5	1.01	1.06
		SYM 5	1.0*	1.0*
F_3	26.13	TOT 8	1.02	1.14
		TOT 5	1.10	1.21
		SYM 5	1.00	1.10

*Standards indicate that a 1.0 multiplying factor applies without further checking when X/R = 15 or less for SYM circuit breakers of all rated interrupting times and for TOT 8 circuit breakers.

The calculated short-circuit (interrupting) duty rms currents for three-phase faults at bus 3, using remote multiplying factors, are

for TOT 8 circuit breakers, 1.14(10.88) = 12.40 kA-T
for TOT 5 circuit breakers, 1.21(10.88) = 13.16 kA-T
for SYM 5 circuit breakers, 1.10(10.88) = 11.97 kA-S

6.6.12 Circuit Breaker Short-Circuit Capabilities Compared with Calculated Remote Multiplying Factor Short-Circuit Current Duties. Short-circuit ratings, or capabilities derived from them, for circuit breakers that might be applied in the example system are listed in Table 40. The headings of the table also show in parentheses the type of calculated short-circuit duties to be compared with listed equipment capabilities or ratings. The capabilities derived from symmetrical short-circuit ratings using a ratio of rated maximum voltage to operating voltage are computed using the example operating voltages listed in the table.

Circuit breakers for bus 1 application, both TOT 8 and SYM 5 types, having short-circuit ratings or capabilities equal to or greater than the corresponding calculated duties at bus 1, are listed in Table 41 with the calculated duties for comparison. Circuit breakers for bus 2 and 3 applications are listed in Tables 42 and 43, respectively, with short-circuit ratings or capabilities and calculated duties.

6.6.12.1 Contact-Parting Time (Interrupting) Duties for High-Voltage Circuit Breakers Using Weighted Interpolation Between Multiplying Factors. For a system with several sources including in-plant generators that might be classified local or remote depending on fault location, logical calculations make use of both remote and local multiplying factors in a weighting process. The weighting consists of applying the remote multiplying factor to the part of the E/X symmetrical short-circuit current contributed by remote sources and the local multiplying fac-

Table 40
AC High-Voltage Circuit Breaker
Short-Circuit Ratings or Capabilities, in Kiloamperes

		8-Cycle Total-Rated Circuit Breakers		5-Cycle Symmetrical-Rated Circuit Breakers	
Circuit Breaker Nominal Size Identification	Example Maximum System Operating Voltage (kV)	Momentary Rating (Total First-Cycle rms Current)	Interrupting Rating (Total rms Current at 4-Cycle Contact-Parting Time)	Closing and Latching Capability (Total First-Cycle rms Current)	*Short-Circuit* Capability (Symmetrical rms Current at 3-Cycle Contact-Parting Time)
4.16—75	4.16	20	10.5	19	10.1
4.16—250	4.16	60	35	58	33.2
4.16—350	4.16	80	48.6	78	46.9
13.8—500	13.8	40	21	37	19.6
13.8—750	13.8	60	31.5	58	30.4
13.8—1000	13.8	80	42	77	40.2

tor to the remainder of E/X. The application of either a local or a remote multiplying factor to the motor contribution part of E/X is permitted by ANSI/IEEE C37.010-1979 (5.4.1, note 5 of the table) [2]. The remote sources part of E/X includes the contribution of an in-plant generator if it is less than 0.4 times the generator current to a short circuit at its terminals; any larger generator current corresponds to a reactance in series that is less than 1.5 times generator X_d'' and supports the use of a local multiplier for the generator contribution, according to ANSI/IEEE C37.010-1979 [2] and ANSI/IEEE C37.5-1979 [3].

Additional calculations are necessary to find the short-circuit currents contributed by each utility and in-plant generator source to the faults being investigated.

Table 41
Calculated Bus 1 Short-Circuit Duties Compared with Ratings or Capabilities of AC High-Voltage Circuit Breakers

Type of Circuit Breaker	TOT 8 (8-Cycle Total Rated)	SYM 5 (5-Cycle Symmetrical Rated)
First-cycle duty, total rms current	40.8 kA	40.8 kA
Short-circuit (interrupting) duty, rms current	29.4 kA-T	28.4 kA-S
Circuit breaker nominal size	13.8-750	13.8-750
Momentary current rating, or closing and latching capability	60 kA	58 kA
Interrupting rating, or short-circuit current capability	31.5 kA	30.4 kA

Table 42
Calculated Bus 2 Short-Circuit Duties Compared with Ratings or Capabilities of AC High-Voltage Circuit Breakers

Type of Circuit Breaker	TOT 8 (8-Cycle Total Rated)	SYM 5 (5-Cycle Symmetrical Rated)
First-cycle duty, total rms current	49.4 kA	49.4 kA
Short-circuit (interrupting) duty, rms current	28.8 kA-T	28.8 kA-S
Circuit breaker nominal size	4.16-250	4.16-250
Momentary current rating, or closing and latching capability	60 kA	58 kA
Interrupting rating, or short-circuit current capability	35 kA	33.2 kA

The magnitude of an in-plant generator contribution for each fault determines whether it is included with utility sources in the remote part of E/X.

The additional calculation of currents in the source branches of the equivalent circuit during a fault at a specified location is a multistep process not illustrated here (and greatly facilitated by available computer programs). The results of the necessary calculations for this example are given in Table 44. Also shown are remote or local classifications for the in-plant generator contributions to faults at F_1, F_2, and F_3.

Table 43
Calculated Bus 3 Short-Circuit Duties Compared with Ratings or Capabilities of AC High-Voltage Circuit Breakers

Type of Circuit Breaker	TOT 8 (8-Cycle Total Rated)	SYM 5 (5-Cycle Symmetrical Rated)
First-cycle duty, total rms current	19.5 kA	19.5 kA
Short-circuit (interrupting) duty, rms current	12.4 kA-T	12.0 kA-S
Circuit breaker nominal size	4.16-250	4.16-250
Momentary current rating, or closing and latching capability	60 kA	58 kA
Interrupting rating, or short-circuit current capability	35 kA	33.2 kA

Table 44
Current Contributions of Separate Sources (Generators) to E/X Symmetrical Short-Circuit (Interrupting) Duties, with Sources Classified Remote or Local (Currents are in Per Unit on the 10 MVA Base of this Example)

Fault Contributions and Classifications*	Fault at F_1	Fault at F_2	Fault at F_3
Fault Point E/X Symmetrical Short-Circuit Current	59.03	20.75	7.84
S_1—69 kV Utility Contribution	22.22	3.06	0.99
Classification	Remote	Remote	Remote
S_2—25 MVA Generator Contribution	27.78	3.83	1.24
Classification†	Local	Remote	Remote
S_3—46 kV Utility Contribution	2.75	7.02	2.28
Classification	Remote	Remote	Remote
S_4—5 MVA Generator Contribution	1.35	3.45	1.12
Classification‡	Remote	Local	Remote

* Utility is always remote, in-plant generator is remote if contribution is less than 0.4 E/X''
† E/X'' (for three-phase short circuit at terminals) = 27.78 per unit
‡ E/X'' (for three-phase short circuit at terminals) = 5.56 per unit

The weighted interpolation has significance only for the fault at F_1. For the fault at F_2, the local and remote multiplying factors are both 1.0 (for TOT 8 and SYM 5 duties) and interpolation has no effect. For the fault at F_3, since all sources including in-plant generators are classified as remote, the remote multiplying factor applies.

For the fault at F_1, the remote part of $E/X = 22.22 + 2.75 + 1.35 = 26.32$ per unit, and the remainder of $E/X = 59.03 - 26.32 = 32.71$ per unit. The calculated short-circuit interrupting duty rms currents for three-phase faults at bus 1 (F_1), using weighted interpolation of multiplying factors, are

for TOT 8 circuit breakers, 1.19 (26.32) + 1.05 (32.71) = 65.67 per unit or 65.67 (0.4184) = 27.5 kA-T

for SYM 5 circuit breakers, 1.15 (26.32) + 1.03 (32.71) = 64.0 per unit or 64.0 (0.4184) = 26.8 kA-S

Comparison of these results with previously calculated bus 1 results, Table 41, shows the previous use of only remote multiplying factors gives an extra margin of conservatism of 6 or 7% in this example.

6.6.13 Circuit and Calculation of *Approximately 30-Cycle* Minimum Short-Circuit Currents. The circuit used is shown in Fig 108. The rotating machine reactances are shown in Table 30. Table 45 details the reactance simplifications starting from Fig 108(b).

A prefault voltage of 1.0 per unit is assumed, I is calculated at E/X per unit, and the conversion is made to amperes. There is no dc component remaining to cause asymmetry. The resulting symmetrical *approximately 30-cycle* short-circuit currents are

at F_1, $I = (1.0/0.0413)\ (0.4184) = 10.14$ kA
at F_2, $I = (1.0/0.1133)\ (1.388) = 12.25$ kA
at F_3, $I = (1.0/0.2203)\ (1.388) = 6.30$ kA

6.7 Example of Short-Circuit Current Calculation for a System under 1000 V. As in portions of a power system over 1000 V, calculation of short-circuit currents at various locations in a system under 1000 V is essential for proper application of circuit breakers, fuses, buses, and cables. All should withstand momentarily the thermal and magnetic stresses imposed by the maximum possible short-circuit current. In addition, circuit breakers and fuses should safely interrupt these maximum fault currents.

For the three-phase system, the three-phase fault will usually produce the maximum fault current. On a balanced three-phase system, the line-to-line fault current will never exceed 87% of the three-phase value. With a system neutral solidly grounded, the line-to-ground fault current could exceed the three-phase short-circuit current by a small percentage; however, this is apt to occur only when there is little or no motor load and the primary system fault contribution is small.

The calculation of symmetrical short-circuit current duties is normally sufficient for the application of circuit breakers and fuses under 1000 V because they have published symmetrical current interrupting ratings. The ratings are based on the

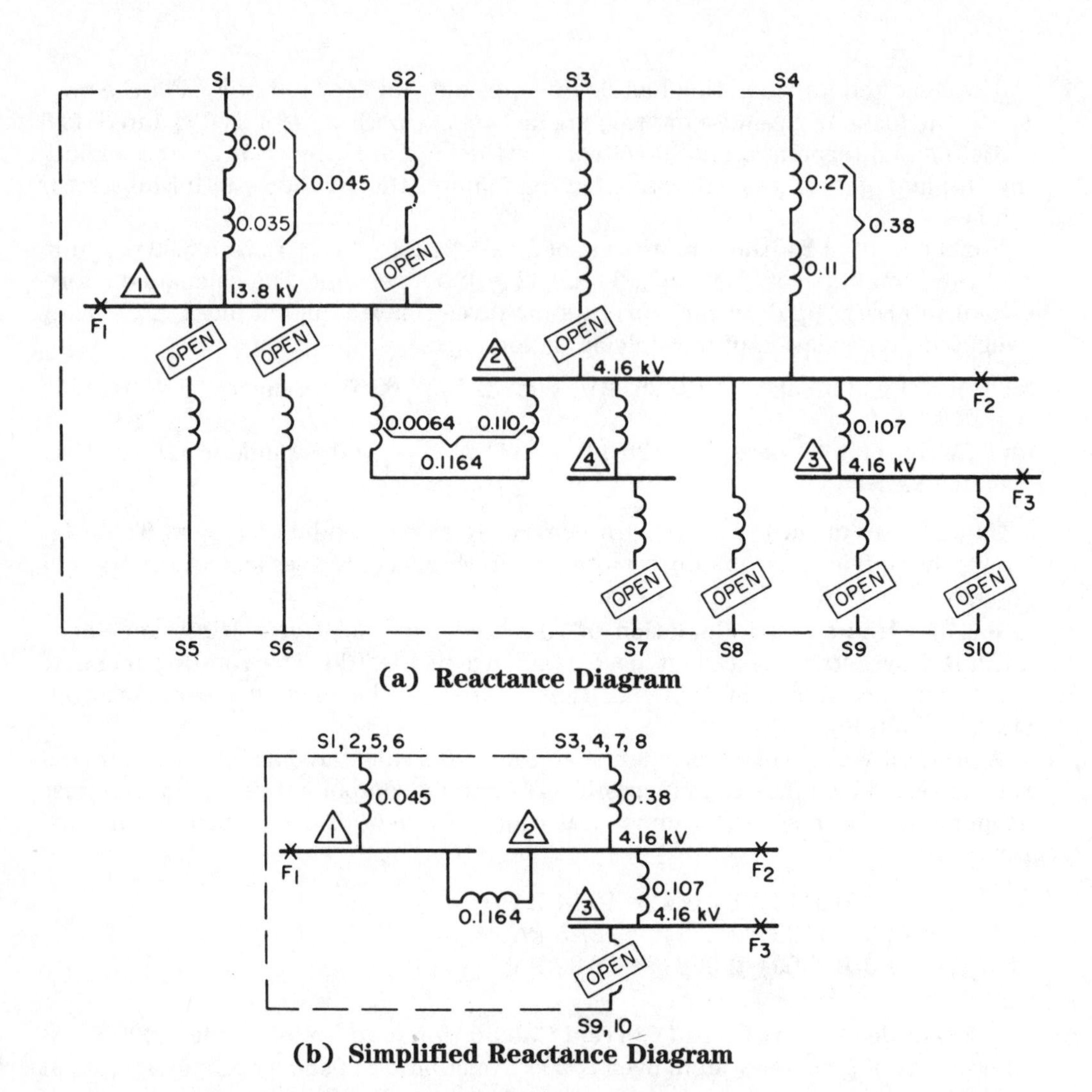

(a) Reactance Diagram

(b) Simplified Reactance Diagram

Fig 108
Circuits of Power System Reactances for Calculation of *Approximately 30-Cycle* Minimum Short-Circuit Currents

Table 45
Reactance Combinations for Fault-Point X at Each Fault Bus of Fig 108(b)

Fault at F_1 X		$1/X$	Fault at F_2 X		$1/X$	Fault at F_3 X
0.38			0.045			0.1133
0.1164			0.1164			0.107
0.4964	→	2.0145	0.1614	→	6.1958	0.2203
0.045	→	22.2222	0.38	→	2.6316	
0.041 26	←	24.2367	0.1133	←	8.8274	

first-cycle symmetrical rms current, determined at ½ cycle after fault inception, and incorporate an asymmetrical capability as necessary for a circuit X/R ratio of 6.6 or less (short-circuit power factor of 15% or greater). A typical system served by a transformer rated 1000 or 1500 kVA will usually have a short-circuit X/R ratio within these limits. For larger or multitransformer systems, it is advisable to check the X/R ratio; if it is greater than 6.6, the circuit breaker or fuse application should be based on asymmetrical current limitations (see ANSI/IEEE C37.13-1981 [4]).

The low-voltage fault calculation procedure differs very little from that used for finding first-cycle short-circuit duties in higher voltage systems. All connected motor ratings are included as fault-contributing sources, and this contribution is based on the subtransient reactance of the machines. The contribution from the primary system should be equivalent to that calculated for its *momentary duty*. Due to the quantity and small ratings of motors usually encountered in low-voltage systems, it is customary to use an assumed typical value for their equivalent reactance in the low-voltage short-circuit network. This typical value is 25% on the individual motor rating or the total rating of a group of motors, both in kilovolt-amperes (see 6.5.4).

The example fault calculation presented here is for a 480 V three-phase system, illustrated by the single-line diagram of Fig 109. The system data shown are typical of those required to perform the calculations.

Bolted three-phase faults F_1 and F_2 are assumed at each of the bus locations, and zero-impedance (bolted) line-to-line faults F_3 and F_4 are assumed at the 120/240 V single-phase locations. Both resistance and reactance components of the circuit element impedances are used in order to illustrate the more precise procedure and to obtain X/R ratios.

Resistances are usually significant in low-voltage short-circuit current calculations. Their effect may be evaluated either by a complex impedance reduction or by separate X and R reductions. The complex reduction leads to the most accurate short-circuit current magnitude. The separate X and R reductions are simpler, conservative, and have the added benefit that they give the best approximation for the X/R ratio at the fault point. They are illustrated by this example.

6.7.1 Step 1—Convert All Element Impedances to Per-Unit Values on a Common Base. The assumed base power is 1000 kVA and the base voltage E_b = 480 V:

$$\text{base current } I_b = \frac{\text{kVA}\,(1000)}{\sqrt{3}\cdot E_b} = \frac{1000\cdot 1000}{\sqrt{3}\cdot 480} = 1202.8\ \text{A}$$

$$\text{base impedance } Z_b = \frac{E_b/\sqrt{3}}{I_b} = \frac{480/\sqrt{3}}{1202.8} = 0.2304\ \Omega$$

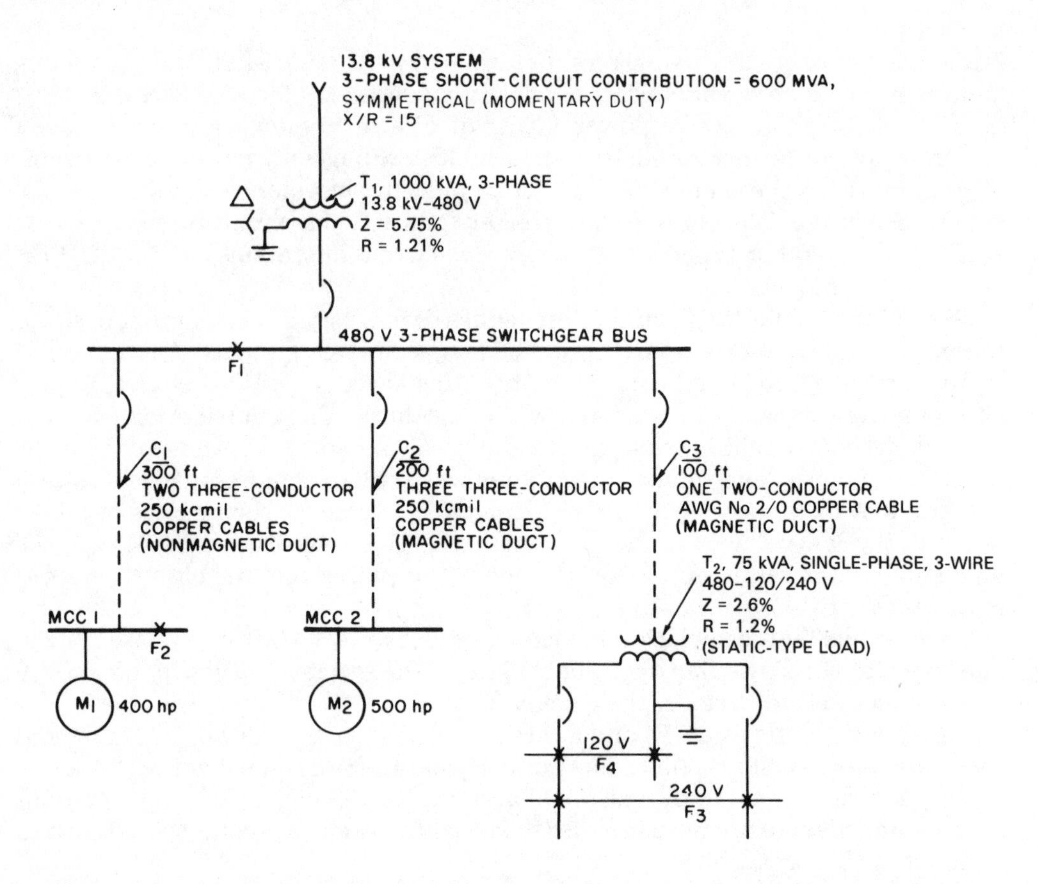

NOTE: The motor horsepower indicated at MCC 1 and 2 represents a lumpted total of small induction three-phase machines ranging in size from 10–150 hp.

Fig 109
Low-Voltage System

(1) *13.8 kV Source Impedance*. The short-circuit current contribution from the 13.8 kV system will usually be expressed in symmetrical megavolt-amperes or amperes, giving a specific X/R ratio. This three-phase fault duty should be the maximum possible available at the primary terminals of the transformer and equivalent to the symmetrical *momentary duty* level. For this example, the 13.8 kV fault duty is 600 MVA or 25 102 A symmetrical at an X/R ratio of 15. The R_s and X_s values comprising the equivalent impedance Z_s can be obtained as follows:

$$Z_s = \frac{\text{base kVA}}{\text{short-circuit kVA}} = \frac{1000}{600\,000} = 0.00166 \text{ per unit}$$

Since $Z_s = \sqrt{(R_s)^2 + (X_s)^2}$ and $X_s/R_s = 15$, the value of $R_s = Z_s/\sqrt{1 + (15)^2} =$ 0.00011 per unit, and the value of $X_s = 15/R_s = 0.00165$ per unit.

(2) *1000 kVA Transformer Impedance.* The transformer manufacturer provides the information that the impedance is 5.75% on the self-cooled base rating of 1000 kVA, and the resistance is 1.21% (R_{T1}). Reactance $X = \sqrt{Z^2 - R^2} = 5.62\%$ (X_{T1}). The per-unit values are

$$R_{T1} = \frac{\text{base kVA}}{\text{transformer kVA}} \cdot \frac{\%R_{T1}}{100} = \frac{1000}{1000} \cdot \frac{1.21}{100} = 0.0121 \text{ per unit}$$

$$X_{T1} = \frac{\text{base kVA}}{\text{transformer kVA}} \cdot \frac{\%X_{T1}}{100} = \frac{1000}{1000} \cdot \frac{5.62}{100} = 0.0562 \text{ per unit}$$

(3) *Cable* C_1 (300 ft of two 250 kcmil three-conductor copper cables in nonmagnetic duct). From published tables, the ac resistance R_{C1} is 0.0541 Ω per conductor per 1000 ft, and the reactance X_{C1} is 0.0330 Ω per conductor per 1000 ft.

For 300 ft of two paralleled conductors,

$$R_{C1} = \frac{0.0541}{2} \cdot \frac{300}{1000} = 0.00812\ \Omega$$

$$X_{C1} = \frac{0.0330}{2} \cdot \frac{300}{1000} = 0.00495\ \Omega$$

Converting impedances to per unit,

$$R_{C1} = \frac{\text{actual ohms}}{\text{base ohms}} = \frac{0.00812}{0.2304} = 0.0352 \text{ per unit}$$

$$X_{C1} = \frac{\text{actual ohms}}{\text{base ohms}} = \frac{0.00495}{0.2304} = 0.0215 \text{ per unit}$$

(4) *Cable* C_2 (200 ft of three 250 kcmil three-conductor copper cables in magnetic duct). From published tables, the ac resistance R_{C2} is 0.0552 Ω per conductor per 1000 ft, and the reactance X_{C2} is 0.0379 Ω per conductor per 1000 ft.

For 200 ft of three parallel conductors,

$$R_{C2} = \frac{0.0552}{3} \cdot \frac{200}{1000} = 0.00368\ \Omega$$

$$X_{C2} = \frac{0.0379}{3} \cdot \frac{200}{1000} = 0.00253\ \Omega$$

Converting impedances to per unit,

$$R_{C2} = \frac{0.00368}{0.2304} = 0.01597 \text{ per unit}$$

$$X_{C2} = \frac{0.00253}{0.2304} = 0.01098 \text{ per unit}$$

(5) *Cable* C_3 (100 ft of one AWG No 2/0 two-conductor copper cable in magnetic duct). From published tables, the ac resistance R_{C3} is 0.102 Ω/1000 ft, and the reactance X_{C3} is 0.0407 Ω/1000 ft.

For 100 ft,

$$R_{C3} = 0.102 \cdot \frac{100}{1000} = 0.0102\ \Omega$$

$$X_{C3} = 0.0407 \cdot \frac{100}{1000} = 0.00407\ \Omega$$

Converting impedances to per unit,

$$R_{C3} = \frac{0.0102}{0.2304} = 0.0443 \text{ per unit}$$

$$X_{C3} = \frac{0.00407}{0.2304} = 0.01766 \text{ per unit}$$

(6) *Motor Contribution*. The running motor loads at motor-control center 1 and 2 buses total 400 hp and 500 hp, respectively. Typical assumptions made for 480 V small motor groups are that 1 hp = 1 kVA, and the average subtransient reactance is 25%. The resistance is 4.167%, based on a typical X/R ratio of 6.

Converting impedances to per unit on the 1000 kVA base,

$$R_{M1} = \frac{\text{base kVA}}{\text{motor kVA}} \cdot \frac{R_{M1}}{100} = \frac{1000}{400} \cdot \frac{4.167}{100} = 0.1042 \text{ per unit}$$

$$X_{M1} = \frac{\text{base kVA}}{\text{motor kVA}} \cdot \frac{X_{M1}}{100} = \frac{1000}{400} \cdot \frac{25}{100} = 0.625 \text{ per unit}$$

$$R_{M2} = \frac{1000}{500} \cdot \frac{4.167}{100} = 0.0833 \text{ per unit}$$

$$X_{M2} = \frac{1000}{500} \cdot \frac{25}{100} = 0.500 \text{ per unit}$$

6.7.2 Step 2 — Draw Separate Resistance and Reactance Diagrams Applicable for Fault Locations F_1 and F_2 (Figs 110 and 111). Since the single-phase 120/240 V system has no fault-contributing sources, it will not be represented in these diagrams.

6.7.3 Step 3 — For Each Fault Location Reduce R and X Networks to Per-Unit Values and Calculate Fault Current. The reduction of the R and X networks at fault location F_1 is shown in Figs 112 and 113. The fault current at F_1 is then calculated as follows. The total impedance Z is

$$Z = \sqrt{R^2 + X^2} = \sqrt{(0.01009)^2 + (0.04811)^2} = 0.04916 \text{ per unit}$$

The total three-phase symmetrical fault current at F_1 is $E/Z \cdot$ base current, that is,

$$\frac{\text{base amperes}}{\text{per unit } Z} = \frac{1202.8}{0.04916} = 24\,470 \text{ A}$$

and the X/R ratio of the system impedance for the fault at F_1 is

$$X/R = \frac{0.04811}{0.01009} = 4.77$$

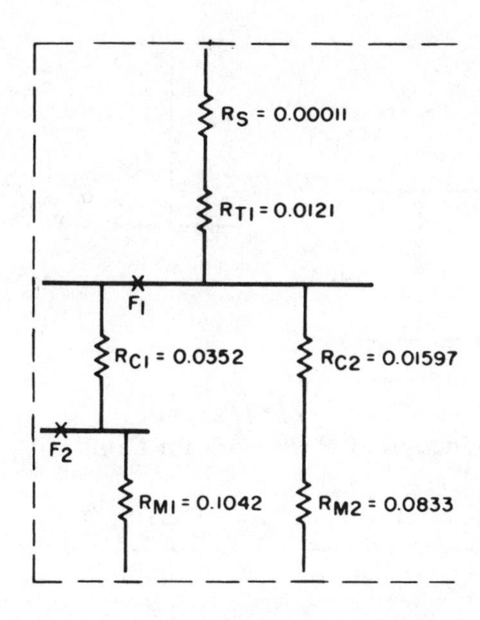

Fig 110
Resistance Network for Faults at F_1 and F_2

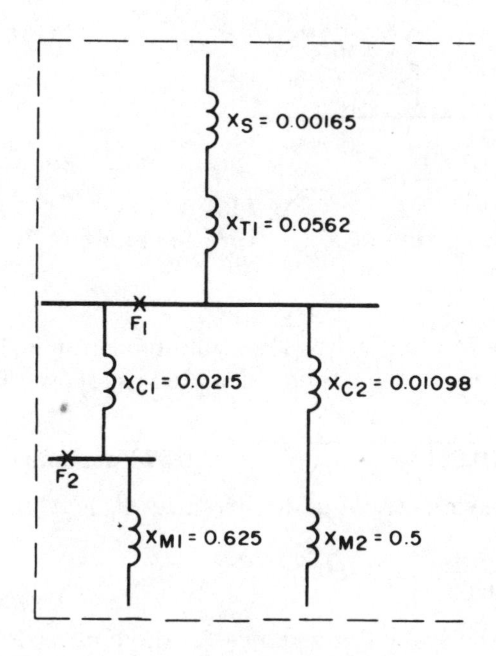

Fig 111
Reactance Network for Faults at F_1 and F_2

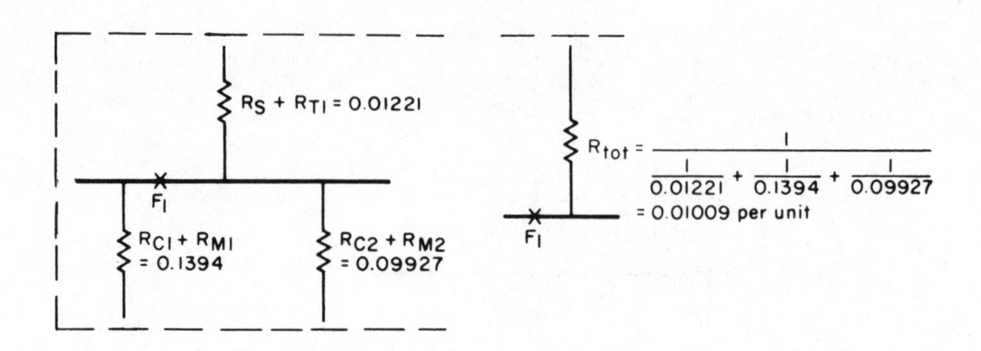

Fig 112
Reduction of *R* Network for Fault at F_1

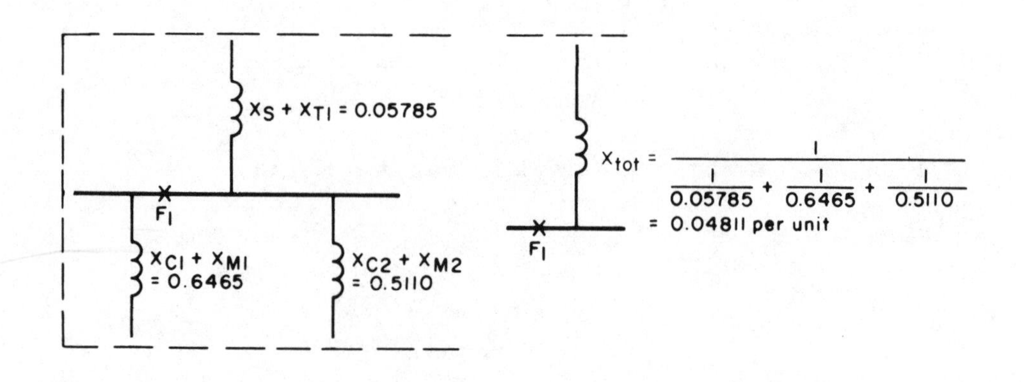

Fig 113
Reduction of *X* Network for Fault at F_1

The reduction of the R and X networks at fault location F_2 is shown in Figs 114 and 115. The fault current at F_2 is then calculated as follows. The total impedance Z is

$$Z = \sqrt{R^2 + X^2} = \sqrt{(0.0319)^2 + (0.0657)^2} = 0.073 \text{ per unit}$$

The total three-phase symmetrical fault current at F_2 is

$$\frac{\text{base amperes}}{\text{per unit } Z} = \frac{1202.8}{0.073} = 16\,480 \text{ A}$$

and the X/R ratio of the system impedance for the fault at F_2 is

$$X/R = \frac{0.0657}{0.0319} = 2.06$$

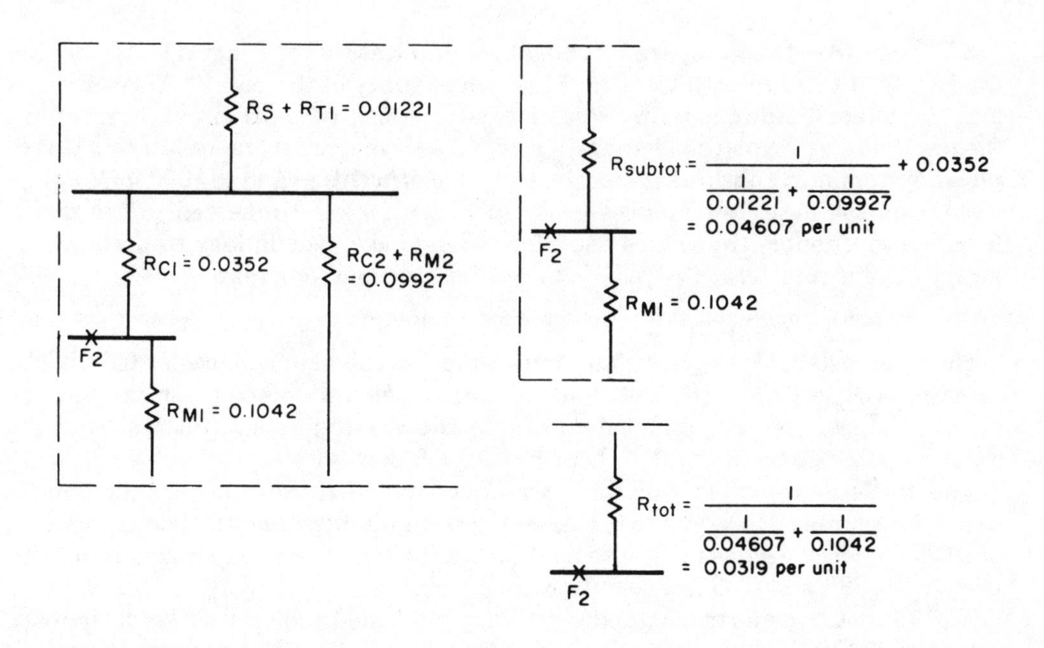

Fig 114
Reduction of *R* Network for Fault at F_2

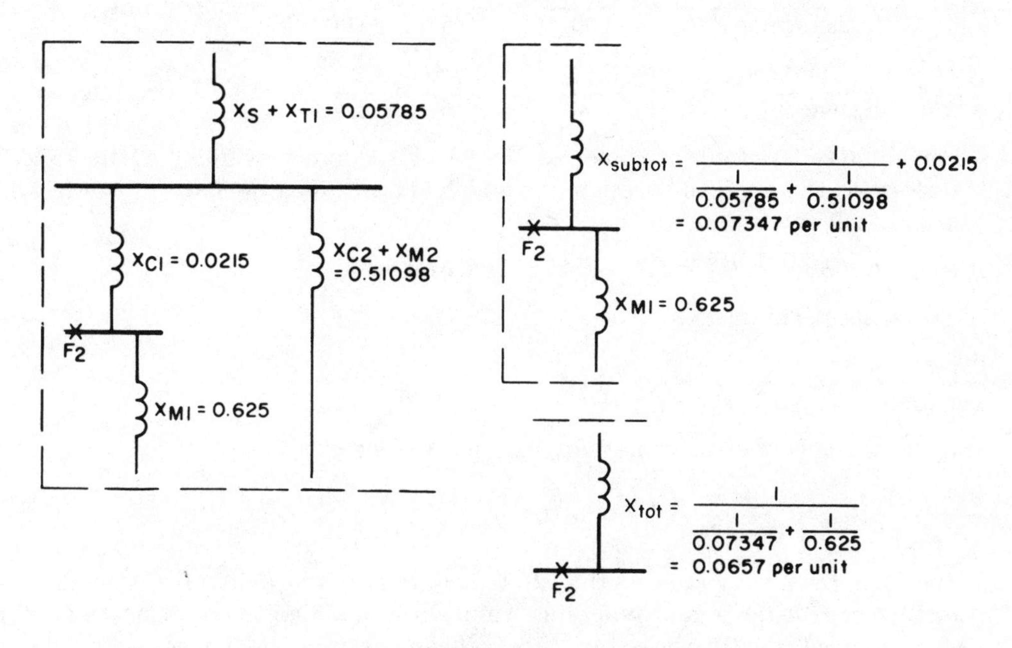

Fig 115
Reduction of *X* Network for Fault at F_2

6.7.4 Step 4—Draw Separate Resistance and Reactance Diagrams Applicable for Faults at the 120/240 V Single-Phase Secondary of the 75 kVA Transformer, and Calculate Fault Currents. Per-unit calculations of short-circuit currents at the low-voltage side of a single-phase transformer connected line-to-line to a three-phase system may continue to use the same base, in this example 1000 kVA, but as a single-phase base. Impedances in the primary system connected to the transformer have double the values used for three-phase calculations to account for both outgoing and return paths of single-phase primary currents.

NOTE: This procedure assumes that the positive- and negative-sequence impedances are equal.

The total system three-phase fault impedance, as calculated above for fault at F_1, consists of R_s = 0.0101 per unit and X_s 0.0481 per unit. Since these are line-to-neutral values, they are doubled to obtain the line-to-line equivalents. Thus R_s becomes 0.0202 per unit and X_s becomes 0.0962 per unit.

The single-phase cable circuit C_3 was determined to have a per-unit line-to-neutral resistance R_{C3} equal to 0.0443 and a per-unit line-to-neutral reactance X_{C3} of 0.01766. These values must also be doubled for the line-to-fault calculation, and become 0.0886 and 0.0353 per unit, respectively.

The 75 kVA transformer impedance, from published tables, is 2.6% on the base rating of 75 kVA and including the full secondary winding. The impedance components are 1.2% resistance R_{T2} and 2.3% reactance X_{T2}.

The per-unit values on the common 1000 kVA base are

$$R_{T2} = \frac{\text{base kVA}}{\text{transformer kVA}} \cdot \frac{\%R_{T2}}{100} = \frac{1000}{75} \cdot \frac{1.2}{100} = 0.16 \text{ per unit}$$

$$X_{T2} = \frac{\text{base kVA}}{\text{transformer kVA}} \cdot \frac{\%X_{T2}}{100} = \frac{1000}{75} \cdot \frac{2.3}{100} = 0.3067 \text{ per unit}$$

For a line-to-line fault at F_3 across the 240 V secondary winding of the 75 kVA transformer, the applicable resistance and reactance diagrams are shown in Figs 116 and 117. The total impedance Z is

$$Z = \sqrt{(0.2688)^2 + (0.4382)^2} = 0.5141 \text{ per unit}$$

the total short-circuit kVA is

$$\frac{\text{base kVA}}{\text{per unit } Z} = \frac{1000}{0.5141} = 1945 \text{ kVA}$$

and the total symmetrical rms short-circuit current is

$$\frac{\text{kVA (1000)}}{E_{\text{L-L}}} = \frac{1945 \cdot 1000}{240} = 8104 \text{ A}$$

For a line-to-line fault across the 120 V secondary of the 75 kVA transformer, the transformer resistance and reactance values are modified to compensate for the half-winding effect. On the same 75 kVA base rating, impedances of one 120 V winding are obtained from those of the 240 V winding using a resistance multiplier of approximately 1.5 and a reactance multiplier of approximately 1.2. These multi-

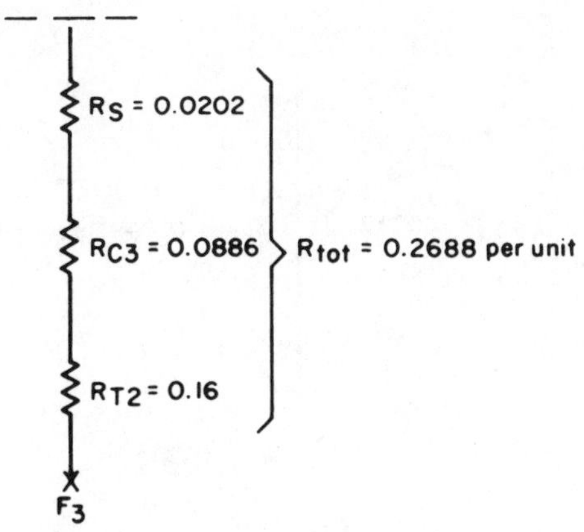

$R_S = 0.0202$
$R_{C3} = 0.0886$ $R_{tot} = 0.2688$ per unit
$R_{T2} = 0.16$
F_3

Fig 116
Resistance Network for Fault at F_3

$X_S = 0.0962$
$X_{C3} = 0.0353$ $X_{tot} = 0.4382$ per unit
$X_{T2} = 0.3067$
F_3

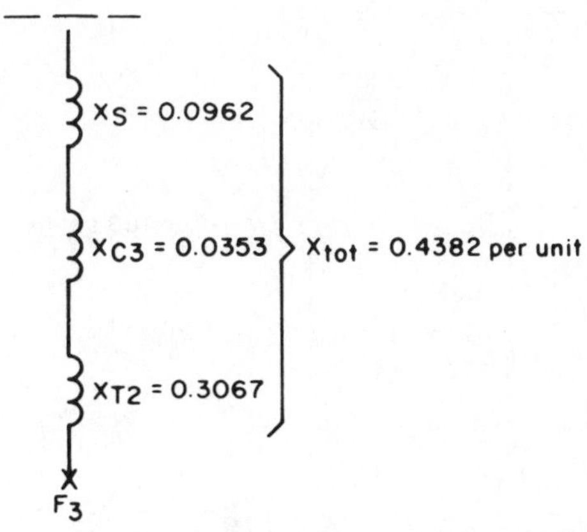

Fig 117
Reactance Network for Fault at F_3

pliers are typical for a single-phase distribution-class transformer. However, for greater accuracy, the transformer manufacturer should be consulted.

For a fault at F_4 the resistance and reactance diagrams are shown in Figs 118 and 119. The total impedance Z is

$$Z = \sqrt{(0.3488)^2 + (0.4995)^2} = 0.6092 \text{ per unit}$$

the total short-circuit kVA is

$$\frac{\text{base kVA}}{\text{per unit } Z} = \frac{1000}{0.6092} = 1642 \text{ kVA}$$

and the total symmetrical rms short-circuit current is

$$\frac{\text{kVA (1000)}}{E_{\text{L-L}}} = \frac{1642 \cdot 1000}{120} = 13\,683 \text{ A}$$

6.8 Calculation of Fault Currents for DC Systems. The calculation of dc fault values is essential in the design and application of distribution and protective apparatus used in dc systems. A knowledge of mechanical stresses imposed by these fault currents is also important in the installation of cables, buses, and their supports.

As in the application of ac protective devices, the magnitude of the available dc short-circuit current is the prime consideration. Since high-speed or semi-high-speed dc protective devices can interrupt the flow of fault current before the maximum value is reached, it is necessary to consider the rate of rise of the fault current, along with the interruption time, in order to determine the maximum current that will actually be obtained. Lower-speed protective devices will generally permit the maximum value to be reached before interruption.

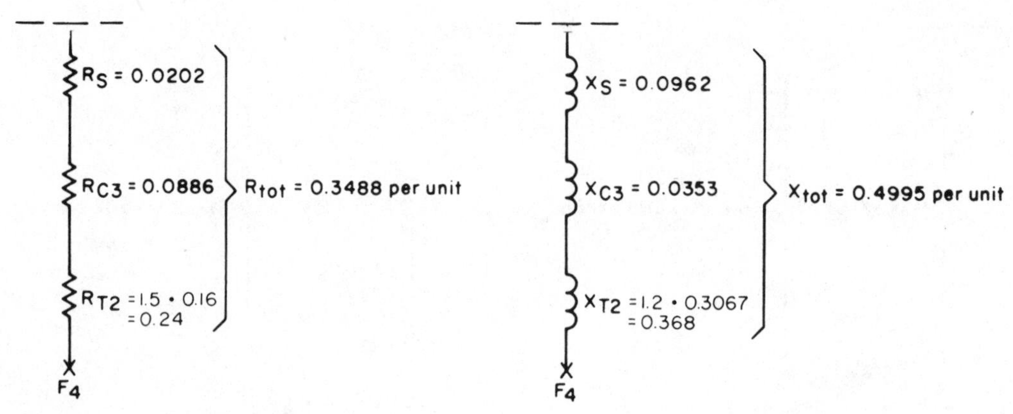

Fig 118
Resistance Network for Fault at F_4

Fig 119
Reactance Network for Fault at F_4

The sources of dc short-circuit currents are
(1) Generators
(2) Synchronous converters
(3) Motors
(4) Electronic rectifiers
(5) Semiconductor rectifiers
(6) Batteries
(7) Electrolytic cells

Simplified procedures for the calculation of dc short-circuit currents are not well established, and therefore this chapter can only provide reference to publications containing helpful information (see [1], [3], [5], [6], and [7]).

6.9 References. This standard shall be used in conjunction with the following publications:

[1] ANSI C97.1-1972, American National Standard for Low-Voltage Cartridge Fuses 600 Volts or Less.

[2] ANSI/IEEE C37.010-1979, IEEE Application Guide for AC High-Voltage Circuit Breakers Rated on a Symmetrical Current Basis.

[3] ANSI/IEEE C37.5-1979, IEEE Guide for Calculation of Fault Currents for Application of AC High-Voltage Circuit Breakers Rated on a Total Current Basis.[22]

[4] ANSI/IEEE C37.13-1981, IEEE Standard for Low-Voltage AC Power Circuit Breakers Used in Enclosures.

[5] ANSI/IEEE C37.41-1981, IEEE Standard Design Tests for High-Voltage Fuses, Distribution Enclosed Single-Pole Air Switches, Fuse Disconnecting Switches, and Accessories.

[22] ANSI/IEEE C37.5-1979 has been withdrawn. Archival copies may be obtained from the Standards Office of the Institute of Electrical and Electronics Engineers, 820 Second Avenue, New York, NY 10017.

[6] NEMA AB1-1975, Molded-Case Circuit Breakers.[23]

[7] NEMA SG3-1981, Low-Voltage Power Circuit Breakers.

[8] AIEE Committee Report. Protection of Electronic Power Converters. *AIEE Transactions*, vol 69, 1950, pp 813-829.

[9] CRITES, W. R. and DARLING, A. G. Short-Circuit Calculating Procedure for DC Systems with Motors and Generators. *AIEE Transactions (Power Apparatus and Systems)*, pt III, vol 73, Aug 1954, pp 816-825.

[10] DORTORT, I. K. Equivalent Machine Constants for Rectifiers. *AIEE Transactions (Communications and Electronics)*, pt I, vol 72, Sept. 1953, pp 435-438.

[11] DORTORT, I. K. Extended Regulation Curves for Six-Phase Double-Way and Double-Wye Rectifiers. *AIEE Transactions (Communications and Electronics)*, pt I, vol 72, May 1953, pp 192-202.

[12] HERSKIND, C. C., SCHMIDT, A., Jr., and RETTIG, C. E. Rectifier Fault Currents—II. *AIEE Transactions*, vol 68, 1949, pp 243-252.

[13] STEVENSON, W. D., Jr. *Elements of Power System Analysis*, New York: McGraw-Hill, 1982.

6.10 Bibliography

[B1] BEEMAN, D. L., Ed. *Industrial Power Systems Handbook*. New York: McGraw-Hill, 1955, chapter 2.

[B2] *Electrical Transmission and Distribution Reference Book*. East Pittsburgh, PA: Westinghouse Electric Corporation, 1964.

[B3] GREENWOOD, A. Basic Transient Analysis for Industrial Power Systems. *Conference Record, 1972 IEEE Industrial and Commercial Power Systems and Electric Space Heating Joint Technical Conference*, IEEE 72CHO600-7-IA, pp 13-20.

[B4] HUENING, W. C., Jr. Interpretation of New American National Standards for Power Circuit Breaker Applications. *IEEE Transactions on Industry and General Applications*, vol IGA-5, no 5, Sep/Oct 1969.

[B5] REED, M. B. *Alternating Current Circuit Theory*, 2nd ed. New York: Harper and Brothers, 1956.

[B6] ST. PIERRE, C. R. *Time-Sharing Computer Programs (DATUMS) for Power System Data Reduction*. Schenectady, NY: General Electric Company, 1973.

[B7] WAGNER, C. F. and EVANS, R. D. *Symmetrical Components*. New York: McGraw-Hill, 1933.

---

[23] NEMA publications can be obtained from the National Electrical Manufacturers Association, 2101 L Street, NW, Washington, DC 20037.

Note
Typical Impedance Data for Short-Circuit Studies
List of Data Tables and Figures

The following tables appear in other chapters:

Table N1.1
Typical Reactance Values for Induction and Synchronous Machines, in Per-Unit of Machine kVA Ratings*

	X''_d	X'_d
Turbine generators†		
2 poles	0.09	0.15
4 poles	0.15	0.23
Salient-pole generators with damper windings†		
12 poles or less	0.16	0.33
14 poles or more	0.21	0.33
Synchronous motors		
6 poles	0.15	0.23
8-14 poles	0.20	0.30
16 poles or more	0.28	0.40
Synchronous condensers†	0.24	0.37
Synchronous converters†		
600 V direct current	0.20	—
250 V direct current	0.33	—
Individual large induction motors, usually above 600 V	0.17	—
Smaller motors, usually 600 V and below	See Tables 24 and 25 in text.	

NOTE: Approximate synchronous motor kVA bases can be found from motor horsepower ratings as follows:
0.8 power factor motor - kVA base = hp rating
1.0 power factor motor - kVA base = 0.8 · hp rating

* Use manufacturer's specified values if available.
† X'_d not normally used in short-circuit calculations.

Table N1.2
Representative Conductor Spacings for Overhead Lines

Nominal System Voltage (volts)	Equivalent Delta Spacing (inches)
120	12
240	12
480	18
600	18
2400	30
4160	30
6900	36
13 800	42
23 000	48
34 500	54
69 000	96
115 000	204

NOTE:
When the cross section indicates conductors are arranged at points of a triangle with spacings A, B, and C between pairs of conductors, the following formula may be used:

$$\text{equivalent delta spacing} = \sqrt[3]{A \cdot B \cdot C}$$

When the conductors are located in one plane and the outside conductors are equally spaced at distance A from the middle conductor, the equivalent is 1.26 times the distance A:

$$\text{equivalent delta spacing} = \sqrt[3]{A \cdot A \cdot 2A} = 1.26\,A$$

Table N1.3
Constants of Copper Conductors for 1 ft Symmetrical Spacing*

Size of Conductor (cmil)	Size of Conductor (AWG No.)	Resistance R at 50 °C, 60 Hz (Ω/conductor/1000 ft)	Reactance X_A at 1 ft Spacing, 60 Hz (Ω/conductor/1000 ft)
1 000 000		0.0130	0.0758
900 000		0.0142	0.0769
800 000		0.0159	0.0782
750 000		0.0168	0.0790
700 000		0.0179	0.0800
600 000		0.0206	0.0818
500 000		0.0246	0.0839
450 000		0.0273	0.0854
400 000		0.0307	0.0867
350 000		0.0348	0.0883
300 000		0.0407	0.0902
250 000		0.0487	0.0922
211 600	4/0	0.0574	0.0953
167 800	3/0	0.0724	0.0981
133 100	2/0	0.0911	0.101
105 500	1/0	0.115	0.103
83 690	1	0.145	0.106
66 370	2	0.181	0.108
52 630	3	0.227	0.111
41 740	4	0.288	0.113
33 100	5	0.362	0.116
26 250	6	0.453	0.121
20 800	7	0.570	0.123
16 510	8	0.720	0.126

NOTE: For a three-phase circuit the total impedance, line to neutral, is

$Z = R + j(X_A + X_B)$

* Use spacing factors of X_B of Tables N1.5 and N1.6 for other spacings.

Table N1.4
Constants of Aluminum Cable, Steel Reinforced (ACSR), for 1 ft Symmetrical Spacing*

Size of Conductor (cmil)	Size of Conductor (AWG No)	Resistance R at 50 °C, 60 Hz (Ω/conductor/1000 ft)	Reactance X_A at 1 ft Spacing, 60 Hz (Ω/conductor/1000 ft)
1 590 000		0.0129	0.0679
1 431 000		0.0144	0.0692
1 272 000		0.0161	0.0704
1 192 500		0.0171	0.0712
1 113 000		0.0183	0.0719
954 000		0.0213	0.0738
795 000		0.0243	0.0744
715 500		0.0273	0.0756
636 000		0.0307	0.0768
556 500		0.0352	0.0786
477 000		0.0371	0.0802
397 500		0.0445	0.0824
336 400		0.0526	0.0843
266 800		0.0662	0.0945
	4/0	0.0835	0.1099
	3/0	0.1052	0.1175
	2/0	0.1330	0.1212
	1/0	0.1674	0.1242
	1	0.2120	0.1259
	2	0.2670	0.1215
	3	0.3370	0.1251
	4	0.4240	0.1240
	5	0.5340	0.1259
	6	0.6740	0.1273

NOTE: For a three-phase circuit the total impedance, line to neutral, is

$Z = R + j\,(X_A + X_B)$

* Use spacing factors of X_B of Tables N1.5 and N1.6 for other spacings.

Table N1.5
60 Hz Reactance Spacing Factor X_B, in Ohms per Conductor per 1000 ft

	Separation (inches)											
(feet)	0	1	2	3	4	5	6	7	8	9	10	11
0	—	−0.0571	−0.0412	−0.0319	−0.0252	−0.0201	−0.0159	−0.0124	−0.0093	−0.0066	−0.0042	−0.0020
1	—	0.0018	0.0035	0.0051	0.0061	0.0080	0.0093	0.0106	0.0117	0.0129	0.0139	0.0149
2	0.0159	0.0169	0.0178	0.0186	0.0195	0.0203	0.0211	0.0218	0.0255	0.0232	0.0239	0.0246
3	0.0252	0.0259	0.0265	0.0271	0.0277	0.0282	0.0288	0.0293	0.0299	0.0304	0.0309	0.0314
4	0.0319	0.0323	0.0328	0.0333	0.0337	0.0341	0.0346	0.0350	0.0354	0.0358	0.0362	0.0366
5	0.0370	0.0374	0.0377	0.0381	0.0385	0.0388	0.0392	0.0395	0.0399	0.0402	0.0405	0.0409
6	0.0412	0.0415	0.0418	0.0421	0.0424	0.0427	0.0430	0.0433	0.0436	0.0439	0.0442	0.0445
7	0.0447	0.0450	0.0453	0.0455	0.0458	0.0460	0.0463	0.0466	0.0468	0.0471	0.0473	0.0476
8	0.0478											

Table N1.6
60 Hz Reactance Spacing Factor X_B, in Ohms per Conductor per 1000 ft

	Separation (quarter inches)			
(inches)	0	1/4	2/4	3/4
0	—	—	−0.072 9	−0.063 6
1	−0.0571	−0.051 9	−0.047 7	−0.044 3
2	−0.0412	−0.038 4	−0.035 9	−0.033 9
3	−0.0319	−0.030 1	−0.028 2	−0.026 7
4	−0.0252	−0.023 8	−0.022 5	−0.021 2
5	−0.0201	−0.017 95	−0.017 95	−0.016 84
6	−0.0159	−0.014 94	−0.013 99	−0.013 23
7	−0.0124	−0.011 52	−0.010 78	−0.010 02
8	−0.0093	−0.008 52	−0.007 94	−0.007 19
9	−0.0066	−0.006 05	−0.005 29	−0.004 74
10	−0.0042	—	—	—
11	−0.0020	—	—	—
12	—	—	—	—

Table N1.7
60 Hz Reactance of Typical Three-Phase Cable Circuits, in Ohms per 1000 ft

	System Voltage				
Cable Size	600 V	2400 V	4160 V	6900 V	13 800 V
4 to 1					
3 single-conductor cables in magnetic conduit	0.0520	0.0620	0.0618	—	—
1 three-conductor cable in magnetic conduit	0.0381	0.0384	0.0384	0.0522	0.0526
1 three-conductor cable in nonmagnetic duct	0.0310	0.0335	0.0335	0.0453	0.0457
1/0 to 4/0					
3 single-conductor cables in magnetic conduit	0.0490	0.0550	0.0550	—	—
1 three-conductor cable in magnetic conduit	0.0360	0.0346	0.0346	0.0448	0.0452
1 three-conductor cable in nonmagnetic duct	0.0290	0.0300	0.0300	0.0386	0.0390
250-750 kcmil					
3 single-conductor cables in magnetic conduit	0.0450	0.0500	0.0500	—	—
1 three-conductor cable in magnetic conduit	0.0325	0.0310	0.0310	0.0378	0.0381
1 three-conductor cable in nonmagnetic duct	0.0270	0.0275	0.0275	0.0332	0.0337

NOTE: These values may also be used for magnetic and nonmagnetic armored cables.

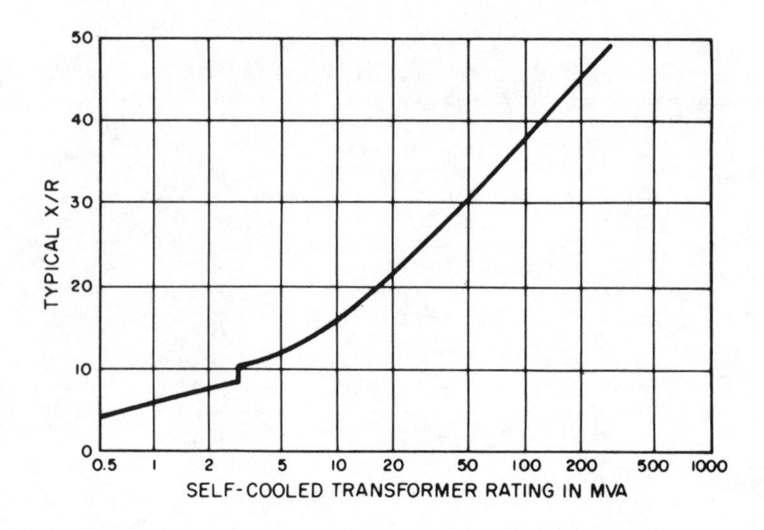

Fig N1.1
X/R Ratio of Transformers (Based on ANSI/IEEE C37.010-1979 [2])

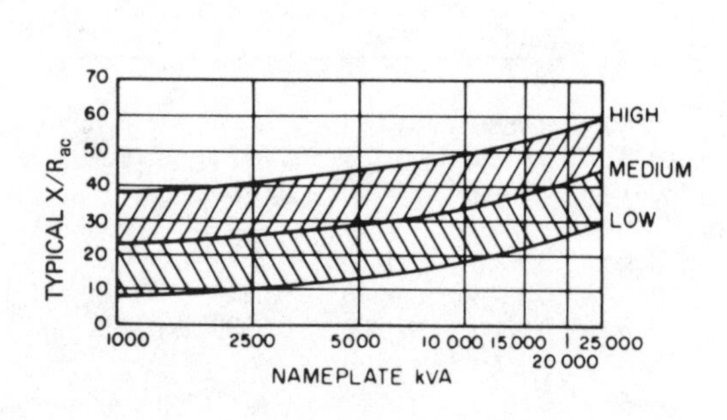

Fig N1.2
X/R Range for Small Generators and Synchronous Motors (Solid Rotor and Salient Pole) (From ANSI/IEEE C37.010-1979 [2])

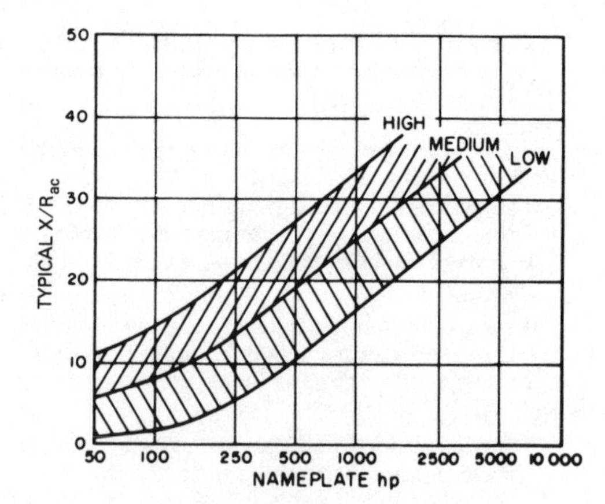

Fig N1.3
X/R Range for Three-Phase Induction Motors (From ANSI/IEEE C37.010-1979 [2])

7. Grounding

7.1 Introduction. All phases of the subject of grounding applicable to the scope of the IEEE Industrial Power Systems Department have been studied and documented in ANSI/IEEE Std 142-1982 [4][23] and that standard constitutes the basic source of technical guidance for this chapter.

Chapter 7, therefore, will simply identify and discuss those facets of the grounding technology that relate to industrial plants. The subject matter will be subdivided into the following five catagories:

(1) System grounding
(2) Equipment grounding
(3) Static and lightning protection grounding
(4) Connection to earth
(5) Grounding resistance measurement

7.2 System Grounding. Alternating-current electric power distribution system grounding is concerned with the nature and location of an intentional electric interconnection between the electric system conductors and ground (earth). The common classifications of grounding to be found in industrial plant ac power distribution systems are

(1) Ungrounded
(2) Resistance grounded
(3) Reactance grounded
(4) Solidly grounded

The nature of electric system grounding may have a striking effect on the magnitude of line-to-ground voltages that must be endured under both steady-state and transient conditions. In ungrounded electric systems that allow severe overvoltage, reduced useful life of insulation can be expected, which may cause insulation

---

[24] The numbers in brackets correspond to those of the references listed at the end of this chapter; when preceded by B, they correspond to the bibliography at the end of this chapter.

failures (circuit faults). In rotating electric machines where insulation space is limited, this conflict between voltage stress and useful life is particularly acute.

In addition to the control of system overvoltages, intentional electric system neutral grounding makes possible sensitive and speedy fault protection based on detection of ground-current flow. Grounded systems in most cases are arranged so that circuit protective devices will remove a faulty circuit from the system regardless of the type of fault. Any contact from phase to ground in the grounded system thus results in immediate isolation of the faulty circuit and its loads. However, the experience of many engineers has been that greater service continuity can be obtained with grounded-neutral than with ungrounded-neutral systems. Furthermore, a very high order of rotating machine ground-fault protection may be acquired by a simple, inexpensive ground overcurrent relay. It is likely that the protective qualities of rotating machine differential protection will be enhanced by grounding the power supply system.

The following practice is recommended for establishing the system grounding connection:

(1) Systems used to supply phase-to-neutral loads that must be solidly grounded are

120/240 V, single-phase, three-wire
208Y/120 V, three-phase, four-wire
480Y/277 V, three-phase, four-wire

(2) Systems that should be resistance grounded are

480 V, three-phase, three-wire
600 V, three-phase, three-wire
2400 V, three-phase, three-wire
4160 V, three-phase, three-wire
6900 V, three-phase, three-wire
13 800 V, three-phase, three-wire wye

7.2.1 Ungrounded Systems. The (so-called) ungrounded system is actually high-reactance capacitively grounded as a result of the capacitance coupling to ground of every energized conductor. The operating advantage sometimes claimed for the ungrounded system stems from the fact that a single line-to-ground fault, if sustained, will not result in an automatic tripout of the circuit. This results merely in the flow of a small charging current to ground. It is generally conceded that this practice introduces potential hazards to other apparatus [8].

There is divided opinion among engineers as to the seriousness of the overvoltage problem on ungrounded systems (600 V and less) and the likelihood of its affecting the electric-service continuity. Many engineers feel that service continuity is improved and insulation failures are reduced by using the grounded system. Others feel that under proper operating conditions the ungrounded system offers an added degree of service continuity not jeopardized by any serious likelihood of dangerous transient overvoltages. A detailed discussion of the factors influencing a choice of the grounded or ungrounded system is given in ANSI/IEEE Std 142-1982 [4], Chapter 1.

For the duration of the ground fault on one conductor, the other two phase conductors throughout the entire metallic system are subjected to 73% overvoltage.

It is, therefore, extremely important to locate the faulty circuit promptly and repair or remove it before the abnormal voltage stresses produce breakdown on other machine windings or circuits. Because of the capacitance coupling to ground, the ungrounded system is subject to dangerous overvoltages (five times normal or more) as a result of an intermittent contact ground fault (arcing ground) or a high inductive reactance connected from one line to ground. As long as no disturbing influences occur on the system, the line-to-ground potentials (even on an ungrounded system) remain steady at about 58% of the line-to-line voltage value. Accumulated operating experience indicates that, in general-purpose industrial power distribution systems, the overvoltage incidents associated with ungrounded operation diminish the useful life of insulation in such a way that electric circuit and machine failures occur more frequently than they do on grounded systems. The advantage of an ungrounded system in not immediately dropping load upon the occurence of a ground fault may be largely destroyed by the practice of ignoring a ground fault and allowing it to remain on the system until a second one occurs causing a power interruption. An adequate detection system together with an organized program for removing ground faults is considered essential for operation of the ungrounded system. These observations are limited to ac systems. Direct-current system operation is not subject to many of the overvoltage hazards present in ac systems.

7.2.2 Resistance-Grounded Systems. Resistance-grounded systems employ an intentional resistance connection between the electric system neutral and ground. This resistance appears in parallel with the system-to-ground capacitive reactance and makes this circuit behave more like a resistor than a capacitor. Resistance-grounded systems can take the forms of

(1) High-resistance grounded systems

(2) Low-resistance grounded systems

In a high-resistance connection ($R \leq X_{co}/3$, where R is the intentional resistance between the electric system neutral and ground, and $X_{co}/3$ is the total system-to-ground capacitive reactance), the overvoltage-producing tendencies of a pure capacitively grounded system will be sufficiently reduced. In a low-resistance grounded system, line-to-ground potentials are rigidly controlled, and sufficient line-to-ground fault current is also available to operate ground-fault relays selectively.

High-resistance grounding provides the same advantages of ungrounded systems yet limits the severe transitory overvoltages associated with ungrounded systems. High-resistance grounded systems do not require immediate clearing of a ground fault since ground-fault current is of a very low magnitude (usually less than 10 A). The ohmic value of the resistance is of the same order (or lower than) as the total system-to-ground capacitive reactance ($X_{co}/3$), that is, the fault current should be at least equal to the system total charging current.

This will limit to moderate value the overvoltages created by an inductive reactance connection from one phase to ground or from an intermittent-contact line-to-ground short circuit. It will not avoid the sustained 73% overvoltage on two phases during the presence of a ground fault on the third phase. Nor will it have much effect on a low-impedance overvoltage source, such as an interconnection with conductors of a higher voltage system, a ground fault on the outer end of an extended winding transformer or step up autotransformer, or a ground fault at the transformer-capacitor junction connection of a series capacitor welder.

Recent investigations recommend that high-resistance grounding should be restricted to 5 kV class systems with charging currents of about 5.5 A or less and should not be attempted on 15 kV systems [B14].

Low-resistance grounding requires a grounding connection of very much lower resistance. The resistance value is tailored to provide a ground-fault current acceptable for relaying purposes. Typical current values used range from 400 A on modern systems using sensitive window current transformer ground-sensor relaying up to perhaps 2000 A in the larger systems using ground-responsive relays connected in current transformer residual circuits. In mobile electric shovel application, much lower levels of ground-fault current (50–25 A) are dictated by the acute shock-hazard considerations.

7.2.3 Reactance-Grounded System. Reactance-grounded systems are not ordinarily employed in industrial power systems. The permissible reduction in available ground-fault current without risk of transitory overvoltages is limited. The criterion for curbing the overvoltages is that the available ground-fault current be at least 25% of the three-phase fault current ($X_0/X_1 \leq 10$, where X_0 is the zero-sequence inductive reactance, and X_1 the positive-sequence inductive reactance of the system). The resulting fault current can be high and present an objectionable degree of arcing damage at the fault, leading to a preference for resistance grounding. Much greater reduction in fault-current value is permissible with resistance grounding without risk of overvoltage.

7.2.4 Solidly Grounded System. Solidly grounded systems exercise the greatest control of overvoltages but result in the highest magnitudes of ground-fault current. These high-magnitude fault currents may introduce new problems and intensify design problems in the equipment grounding system. Solidly grounded systems are used extensively at operating voltages of 600 V and less and at higher voltages than 13.8 kV. At high voltages the cost of grounding equipment is high. Large magnitude ground-fault current generally does not affect electrical equipment. Also large magnitude available ground-fault current is desirable to secure optimum performance of phase-overcurrent trips or interrupters. The low line-to-neutral driving voltage of the supply system (346 V in the 600 V system and 277 V in the 480 V system) lessens the likelihood of dangerous voltage gradients in the ground-return circuits even when higher than normal ground-return impedances are present.

7.2.5 System-Grounding Design Deviations. The intent of the preceding advisory recommendations is to promote broad application of the fewest variety of system-grounding patterns that will satisfy the operational requirements of industrial plant electric power systems in general. Even minor deviations in design practice within a particular variety are to be avoided as much as possible. Nonetheless, it is admitted that the list of recommended patterns is not all inclusive and hence is not to be regarded as mandatory.

Local problems within individual electric power systems will evolve that will justify a deviation from the patterns listed in previous paragraphs. Circumstances can arise that may well justify *solid grounding* with circuit patterns other than those named in these recommendations. For example, in the case where the power supply is obtained from the utility company via feeders from a 4.16Y/2.4 kV solidly grounded substation bus, the user will be justified in adopting that pattern. In such inevitable situations it is imperative that adequate ground-return conductors be provided to minimize the inherent step-and-touch potentials of high ground-fault currents associated with solidly grounded systems. Furthermore, should the user desire to serve 120 V single-phase one-side-grounded load circuits, there could be firm justification for solidly grounding the midpoint of one phase of a 240 V delta system to obtain a 240 V three-phase four-wire delta pattern. Also, the use of a 120 V three-phase delta system for general-purpose power could well justify solid corner-of-the-delta grounding.

In designing the electric power supply system to serve electrically operated excavating machinery, the existence of a greatly accentuated degree of electric-shock-voltage exposure may justify the use of a system-grounding pattern employing a 25 A resistive grounding connection (to establish a 25 A level of available ground-fault current). The achievement of keeping personnel secure from dangerous electric-shock injury, both operators and bystanders, may overshadow the reduction in rotating-machine fault-detection sensitivity, which is therefore sacrificed.

There may be sound justification for the insertion of a reactor in the neutral connection of a generator that is to be connected to a solidly grounded three-phase four-wire distribution system in order to avoid excessive generator-winding current in response to line-to-ground fault on the system. The reactance of the neutral grounding reactor for generator grounding is calculated such that current in any winding does not exceed three-phase short-circuit current and is not less than 25% of three-phase short-circuit current. A minimum short-circuit current of 25% is required to minimize transient overvoltages.

The foregoing examples clearly illustrate the need for design flexibility to tailor the system grounding pattern to cope with the unique and unusual. However, the decision to deviate from the advisory recommendations should be based on a specific engineering evaluation of a need for that deviation (in contrast to a mere desire to do things in a different way).

7.3 Equipment Grounding. Equipment grounding pertains to the system of electric conductors (grounding conductor and ground buses) by which all noncurrent-carrying metallic structures within an industrial plant are interconnected and grounded. The main purposes of equipment grounding are as follows:

(1) To maintain low potential difference between metallic members, ensuring freedom from electric shocks to personnel in the area

(2) To contribute to superior protective performance of the electric system

(3) To avoid fires from volatile materials and the ignition of gases in combustible atmospheres by providing an effective electric conductor system for the flow of ground-fault currents and lightning and static discharges to essentially eliminate arcing and other thermal distress in electrical equipment

All electric conductor housings (for example, metallic conduits, cable trays, junction boxes, etc), equipment enclosures, and motor frames should be interconnected by an equipment grounding conductor system that will satisfy the foregoing requirements. The rules for achieving these objectives are given in ANSI/NFPA 70-1987 [5] (National Electrical Code [NEC]) and ANSI C2-1987 [1] (National Electric Safety Code [NESC]).

When an insulation failure occurs along an electric power circuit, causing an electrical connection between the energized conductor and a metal enclosure, there exists a tendency to raise the enclosure to the same electrical potential that exists on the power conductor. Unless all such enclosures have been grounded in an effective manner, an insulation breakdown will cause dangerous electric potential to appear on the enclosure constituting an electric shock hazard to anyone touching it. The energy released during an arcing ground fault may be sufficient to cause a fire or explosion. Proper setting of ground relays and by intentionally grounding the metallic enclosures in a manner that assures the presense of both adequate ground-fault current capacity and a low value of ground-fault circuit impedance will interrupt the flow of ground-fault current and will thus avoid electric shock and fire hazards [13].

Figures 120 and 121 show typical system and equipment grounding for a three-phase electric system. Solidly grounded, resistance-grounded, and ungrounded systems all have the same equipment grounding requirements. The equipment grounding conductors are connected to provide a low-impedance path for ground-fault current from each metallic enclosure or from equipment to the grounded terminal at the transformer (Figs 120 and 121). The impedance of the complete ground-fault circuit should be low enough to ensure ample flow of ground-fault current for fast operation of the appropriate circuit protective devices, and minimize the potential for stray ground currents on solidly grounded systems. In order to provide a ground-fault current path of low impedance and adequate capacity, either the

Fig 120
Grounding Arrangement for Ground-Fault Protection in Solidly Grounded System, Three-Phase, Three-Wire Circuits

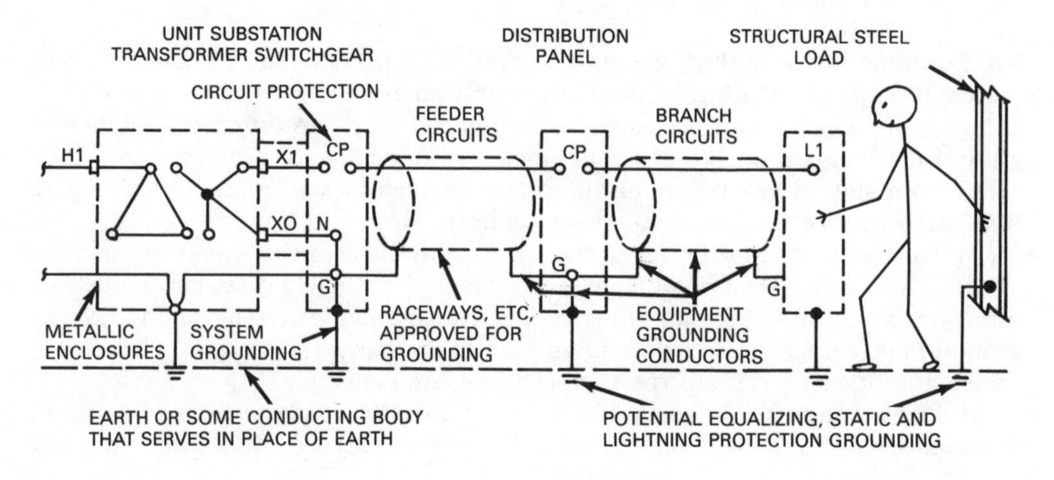

cross-sectional area of the raceway must be large or a parallel grounding conductor must be run inside the raceway (see Fig 121). As shown in Figs 122 and 123, equipment grounding conductors are also required in resistance-grounded and ungrounded systems for personnel shock protection. They must provide paths of sufficient capacity to operate protective devices when phase-to-phase-to-ground faults occur at different locations on a power system.

Economics and operating requirements have resulted in an increasing number of industrial plants owning and operating the stepdown transformer substation interconnecting the industrial plant with the electric utility. In addition to providing the proper equipment grounding in such a substation, step and touch potentials must also be maintained at a safe level. An appropriately designed grounding mat has traditionally served the purposes of providing for both the safety of

Fig 121
Fault-Current Path through Ground-Fault Conductors in Solidly Grounded System, Three-Phase, Three-Wire Circuits

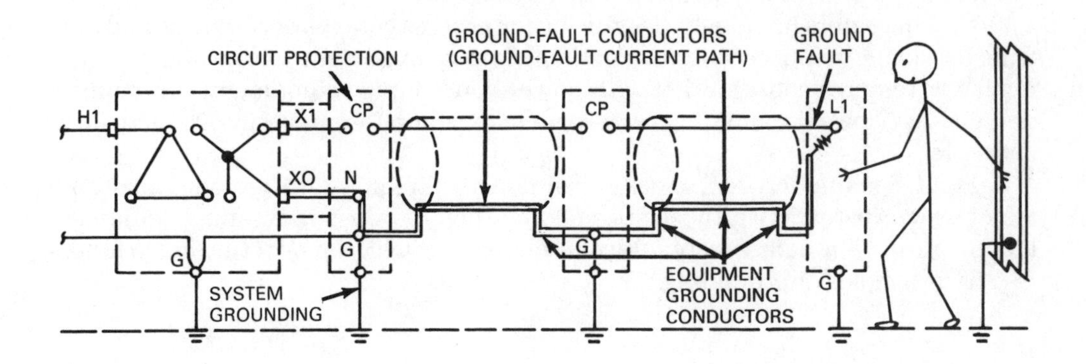

Fig 122
Grounding Arrangement for Ground-Fault Protection in Resistance-Grounded System, Three-Phase, Three-Wire Circuits

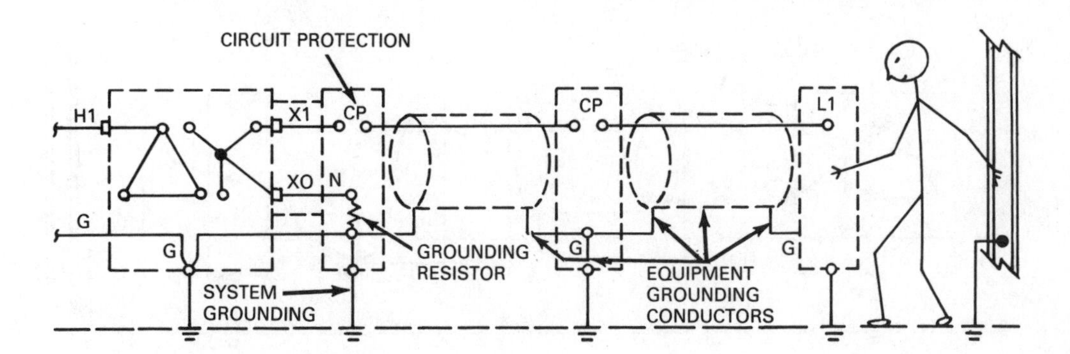

personnel in and near the substation and proper grounding of the substation equipment. Empirical methods have been used extensively in the past for ground mat design due to the great number of calculations required for a perfectly rigorous ground mat analysis. Recently, digital computer programs have added to the accuracy and ease of ground mat design. Persons involved in substation grounding are advised to refer to ANSI/IEEE Std 80-1986 [2] for substation grounding design requirements and detailed calculation procedures.

The ground grid of the stepdown substation is often interconnected with the industrial plant grounding system, either intentionally by a buried ground wire or unintentionally through cable tray, conduit systems, or bus duct enclosures. As a result of this interconnection, the plant grounding system is elevated to the same potential above remote earth as the substation grid during a high-voltage fault in the stepdown substation. Dangerous surface potentials within an industrial plant as well as the substation shall also be prevented. In certain cases, hazardous surface potentials may be eliminated by effectively isolating the substation ground system from the plant ground system; but in most cases, integrating the two grids together and suitably analyzing both systems for step and touch potentials have reduced these potentials to acceptable levels.

Use of computers in industry is growing for process control, accounting, and data transmission. Computer system grounding is very important for optimum performance. It requires coordination with power-conditioning equipment, communication circuits, special grounding requirements of computer logic circuits, and surge arresters.

Computer manufacturers specify grounding techniques for their equipment but may be inconsistent with those specified by some power-conditioning equipment manufacturer. Mutually acceptable solutions can be achieved by returning to fundamental principles of grounding.

Fig 123
Grounding Arrangement for Ground-Fault Protection in Ungrounded System, Three-Phase, Three-Wire Circuits

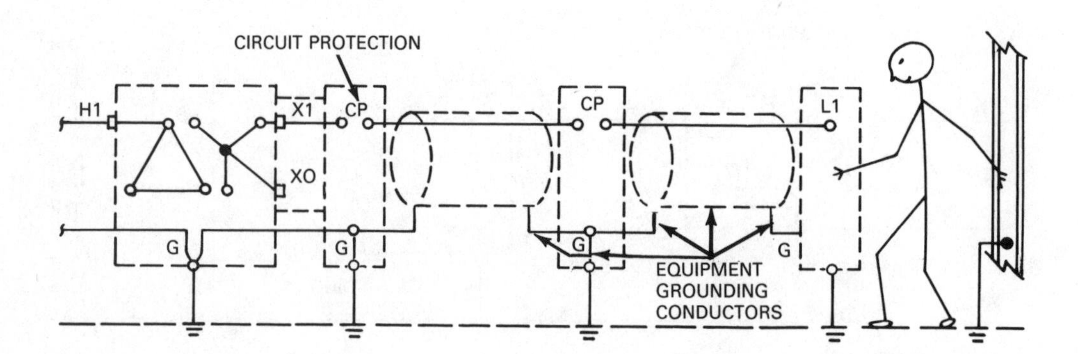

Computer system grounding accomplishes multiple functions, such as safety to operation personnel, a low-impedance fault-return path, and maintenance of the equipotential ground of all units of a computer system.

Connecting the frames of all units of a computer system to a common point should ensure that they stay at the same potential. Connecting that point to the ground should ensure that the equipotential is also ground potential. These objectives are achieved when the units are connected to an ac power source and include a safety equipment ground conductor in each cable or conduit that carries power that comes from a common source [B11].

However, when there is more than one power source, each with its separate ground, this system will create noise currents in the grounding system that are connected to the units of the computer system. In such cases, a *signal reference grid* may be used. This grid may be a large sheet of copper foil installed under the computer or a 2 ft × 2 ft mesh of copper conductors laid out on the subfloor (see Fig 124) to equalize the voltage over a broader frequency range. All computer units should be bonded to this grid in addition to the equipment ground conductor. The signal reference grid is grounded to the same central grounding point as the frames of the system components.

An alternative signal reference grid could be a raised floor metal supporting structure, which is electrically conducting and suitably bonded at all joints. All precautions outlined in [B11] should be followed for the application of a signal reference grid.

Fig 124
Computer Units Connected to Signal Reference Grid and to AC Ground

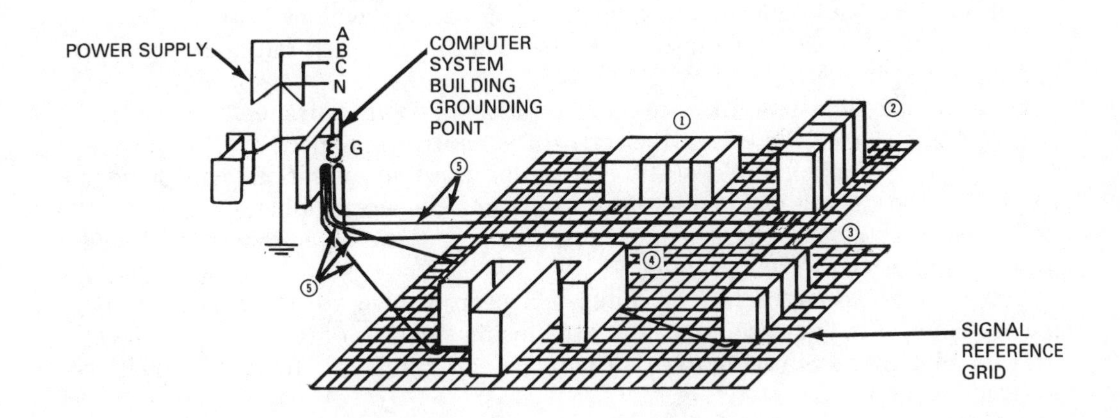

NOTES:
① through ④ are typical computer system modules.
⑤ is the *green wire* safety equipment grounding conductor.
(*Green wire* is connected to the system ground through point G.)

Preferred methods of grounding the following types of equipment are given in detail in ANSI/IEEE Std 142-1982 [4], Chapter 2:

(1) Structures
(2) Outdoor stations
(3) Large generators and motor rooms
(4) Conductor enclosures
(5) Motors
(6) Portable equipment
(7) Surge/lightning protective devices

In many application areas, the applicable national/state electrical codes prescribe such grounding practices as a mandatory requirement.

7.4. Static and Lightning Protection Grounding

7.4.1 Static Grounding. Industrial plants handling solvents, dusty materials, or other flammable products often have a potentially hazardous operating condition because of static charge accumulating on equipment, on materials being handled, or even on operating personnel.

The discharge of a static charge to ground or to other equipment in the presence of flammable or explosive materials is often the cause of fires and explosions, which result in the loss of many lives each year and an accompanying large property loss.

The simple expedient of grounding individual equipment is not always the solution to the problem and is not possible in many processes. Each installation should be studied in order that an adequate method of control may be selected.

The protection of human life is a prime objective in the control of electrostatic charges. In addition to the direct danger to life from explosion or fire created by an electrostatic discharge, there is the possibility of personal injury from being startled by an electric shock, which may, in turn, induce an accident such as a fall from a ladder or platform.

Another objective in controlling static electricity is the avoidance of

(1) Investment losses, buildings, contained equipment, or stored materials
(2) Lost production

The avoidance of losses in this manner represents good insurance.

An additional need for effective electrostatic control may be that of improved product quality. For example, static charges in grinding operations can prevent grinding to the degree of fineness desired in the finished product, or in certain textile operations static charges may cause fibers to stand on end instead of lying flat, resulting in an inferior product.

Material handled by chutes or ducts has been known to accumulate static charges, causing the material to cling to the inside surfaces of the chutes or ducts and thus clog them. Static charges on persons can result in damage to sensitive electronic components.

ANSI/IEEE Std 142-1982 [4], Chapter 3, contains a detailed treatment of the following topics:

(1) Purpose of static grounding
(2) Fundamental causes of static, magnitudes, and conditions required for a static charge to cause ignition

(3) Measurement and detection of static potentials
(4) Hazards in various facilities and mechanisms
(5) Recommended control methods

7.4.2 Lightning Protection Grounding. Lightning protection grounding is concerned with the conduction of current discharges, in the atmosphere, originating in cloud formations, to earth. The function of the lightning grounding system is to convey these lightning discharge currents safely to earth without incurring damaging potential differences across electrical insulation in the industrial power system, without overheating lightning grounding conductors, and without the disruptive breakdown of air between the lightning ground conductors and other metallic members of the structure [7], [11].

Lightning represents a vicious source of overvoltage. It is capable of imparting a potential of one half million volts or more to a stricken object. The current in the direct discharge may be as high as 200 000 A. The rate at which this current builds up may be as much as 10 000 A/μs.

The presence of such high-magnitude fast-rising surge current emphasizes the need for high-discharge capability in surge arresters and low impedance in the connecting leads. For example, should a direct stroke that has made contact with a lightning rod or a mast on an industrial building encounter an inductance of as little as 1 μH with a current buildup rate of 10 000 A/μs, there would result a 10 000 V potential drop across this inductance. Ligntning protection consists of placing suitable air terminals or diverter elements at the top of or around the structure to be protected, and connecting them by an adequate down conductor to the earth itself. A necessary principle is that the adequate down conductor not include any high-resistance or high-reactance portions or connections and should present the least possible impedance to earth. There should be no sharp bends or loops in surge protective grounding circuits. Bend radii should be as long as possible, since sharp bends increase the reactance of the conductor. Reactance is much more critical than resistance, because of the very high frequency of the surge front.

Steel-framed structures, adequately grounded, meet these requirements with the provision of air terminals. In the absence of a steel framework, a down conductor providing at least two paths to earth for a stroke to any air terminal is generally adequate.

All air terminals should be connected by down conductors and should form a two-way path from each air terminal to make connection to the grounding electrode.

Locations of down conductors will depend on the location of air terminals, size of structure being protected, most direct coursing, security against damage or displacement, and location of metallic bodies, water pipes, grounding electrode, and ground conditions. If the structure has metallic columns, these columns will act as down conductors. The air terminals must be interconnected by conductors to make connection with the columns. The average distance between down conductors should not exceed 100 ft (30 m).

Every down conductor must be connected, at its base, to an *earthing* or grounding electrode. This electrode needs to be not more than 2 ft (0.6 m) away from the base of the building and should extend below the building foundation if possible. The length of the grounding conductor is highly important. A horizontal run of, say, 50 ft (15.2 m) to a better electrode (such as a water pipe) is much less effective than a connection to a driven rod alongside the structure itself. Electrodes should make contact with the earth from the surface downward to avoid flashing at the surface. Earth connections should be made at uniform intervals about the structure, avoiding as much as possible the grouping of connections on one side. Properly made connections to earth are an essential feature of a lightning rod system for protection of buildings.

The larger the number of down conductors and grounding electrodes, the lower the voltage developed within the protection system will be, and the better it will perform. This is one of the great advantages of the steel-framed building. Also, at the bottom of each column is a footing, which is a very effective electrode. However, it should be pointed out that internal column footings of large buildings dry up and can become ineffective since they seldom are exposed to ground water.

Interior metal parts of a nonmetal-framed building that are within 6 ft (1.8 m) of a down conductor need to be connected to that down conductor. Otherwise, they may sustain sideflashes from it, incurred because of voltage drop in the lower portion of that down conductor and electrode.

It is highly desirable to keep the stroke current away from buildings and structures involving hazardous liquids, gases, or explosives. Separate diverter protection systems should be used for tanks, tank farms, and explosive manufacture and storage. The diverter element is one or more masts, or one or more effectively grounded elevated wires between masts that are effectively grounded.

Tanks not protected by a diverter system should be well grounded to conduct the current of direct strokes to earth.

Lightning protection of power stations and substations includes the protection of station equipment by means of surge arresters. (Refer to ANSI/IEEE Std 142-1982 [4], Chapter 3.) These arresters should be mounted on, or closely connected to, the frames of the principal equipment that they are protecting, generally a transformer. They may also be mounted on the steel framework of the station or substation where all components are closely interconnected by means of the grounding grid. The surge arrester grounding conductor should be as short and straight as possible and connected to the common station ground bus. The NEC [5] requires that it be not less than AWG No 6 (4.11 mm), but larger sizes may be desirable with larger systems, based on the magnitude of power follow current.

7.5 Connection to Earth

7.5.1 General Discussion. Connections to earth having acceptably low values of impedance are needed to discharge lightning stroke currents, dissipate the released bound charge resulting from nearby strokes, and drain off static voltage accumulations (ANSI/IEEE Std 142-1982 [4], Chapter 4).

The presence of overhead high-voltage transmission circuits may introduce a requirement for a connection to earth to safely pass the ground-fault current that would result from a broken line conductor falling on some part of the building structure.

To a great extent the internal electric distribution system installed within commercial buildings and industrial plants is entirely enclosed in grounded metal. Except for cable tray systems, conductors are enclosed in conduit, metallic armor, or metal raceway. The other electric elements of equipment and machines can be expected to be encased in metal cabinets or metallic machine frames. All of these metallic enclosures and cable trays will be intentionally interconnected and, in turn, will be bonded to other metallic components within the area, such as building structural members, piping systems, messenger cables, etc. Thus the local electric system will be self-contained within its own shell of conducting metal and can be designed to operate adequately and safely without any connection to earth itself. This can be likened to the electric distribution system as installed on an airplane. The airplane structure constitutes an adequate grounding system. No connection to earth is needed to achieve an adequate, safe electric system.

7.5.2 Recommended Acceptable Values. The most elaborate grounding system that can be designed may prove to be inadequate unless the connection of the system to the earth is adequate and has a low resistance [7]. The earth connection is one of the most important parts of the grounding system. It is also the most difficult part to design and to obtain.

The perfect connection to earth would have zero resistance, but this is impossible to obtain. Ground resistances of less than 1 Ω can be obtained, although such a low resistance may not be necessary. Since the resistance required varies inversely with the fault current to ground, the larger the fault current, the lower the resistance must be.

For larger substations and generating stations, the earth resistance should not exceed 1 Ω. For smaller substations and for industrial plants, in general, a resistance of less than 5 Ω should be obtained if practicable. The NEC [5], Article 250, approves the use of a single-made electrode for the system-grounding electrode, if its resistance does not exceed 25 Ω.

7.5.3 Resistivity of Soils. The resistivity of the earth is a prime factor in establishing the resistance of a grounding electrode. The resistivity of soil varies with the depth from the surface, with the moisture and chemical content, and with the soil temperature. For representative values of resistivity for general types of soils and the effects of moisture and temperature, see ANSI/IEEE Std 142-1982 [4], Chapter 4.

7.5.4 Soil Treatment. Soil resistivity may be reduced anywhere from 15–90% by chemical treatment, depending upon the kind and texture of the soil. There are a number of chemicals suitable for this purpose, including sodium chloride, magnesium sulfate, copper sulfate, and calcium chloride. Common salt and magnesium sulfate are most commonly used.

Chemicals are generally applied by placing them in the circular trench around the electrode in such a manner as to prevent direct contact with the electrode. While the effects of treatment will not become apparent for a considerable period, they may be accelerated by saturating the area with water. Also, such treatment is not permanent and must be renewed periodically, depending on the nature of chemical treatment and the characteristics of the soil.

7.5.5 Existing Electrodes. All grounding electrodes fall into one of two categories: those that are an inherent part of the establishment, and those that have been *made* for electrical grounding purposes.

The NEC [5], Article 250, designates underground metal water piping, available on the premises, as part of a preferred grounding electrode system. This preference prevails regardless of length, except that when the effective length of buried pipe is less than 10 ft, it shall be supplemented with an electrode of the type named in Article 250.

For safety grounding and for small distribution systems where the ground currents are of relatively low magnitude, such buried metal-pipe electrodes are usually preferred because they are economical in first cost. However, before reliance can be placed on any electrodes of this group, it is essential that their resistance to earth be measured to ensure that some unforeseen discontinuity has not seriously affected their suitability. The use of plastic pipe in new water systems and of wooden ones in older systems may seriously impair the grounding value of the electrode. Even iron or steel piping may include gaskets that act as insulators. Sometimes small metal (brass) wedges are used to ensure electrical continuity. These wedges should be replaced when repairs are made. Interior piping systems that are likely to become energized must be bonded to the electric system grounding conductor. If the piping system contains a member designed to permit easy removal, a bonding jumper must be installed bridging the removable member.

7.5.6 Concrete-Encased Grounding Electrodes. Concrete below ground level is a good electrical conductor, as good as moderately low-resistivity earth. Consequently metal electrodes encased in such concrete will function as excellent grounding electrodes [11], [14], (see also the NEC [5], Article 250). In areas of poor soil conductivity, the beneficial effects of the concrete encasement are most pronounced.

To create a *made* electrode by encasement of a metal electrode in concrete would probably not be economical, but most industrial establishments employ much concrete-encased metal below grade for other purposes. The reinforcing steel in concrete foundations and footings is a good example. The concrete encasement of steel, in addition to contributing to low-grounding resistance, serves to immunize the steel against corrosive disintegration, such as would take place if the steel were in direct contact with the earth [5]. Even though copper and steel are in contact with each other within the bed of moist concrete, destructive disintegration of the steel member does not take place.

Steel reinforcing bars (re-bars) in foundation piers usually consist of groups of four or more vertical members held by horizontal spacer square *rings* at regular intervals. The vertical members are wired to heavy horizontal members in the spread footing at the base of the pier. Measurements show that such a pier has an electrode resistance of about half the resistance of a simple ground rod driven to the same depth in earth. Electrical connection to the re-bar system is conveniently made by a bar welded to one vertical re-bar and a J-bolt for the column base plate. The J-bolt then becomes the electrode connection. A weld to a re-bar at a point where the bar is in appreciable tension is to be avoided.

Usually such footings appear every 15–20 ft in all directions in industrial buildings. A good rule of thumb for determining the effective overall resistance of the grounding mat is to divide the resistance of one typical footing by half the number of footings around the outside wall of the building. (Inner footings aid little in lowering the overall resistance.)

Copper cable embedded in concrete is similarly benefitted, a fact that may be of particular value under circumstances of high earth resistivity.

The NEC [5], Sections 250-82 and 250-83, recognizes concrete-encased copper and steel as an effective grounding electrode under either the category of an *available* or a *made* electrode.

7.5.7 Made Electrodes. Made electrodes may be subdivided into driven electrodes, buried strips or cables, grids, buried plates, and counterpoises. The type selected will depend upon the type of soil encountered and the available depth. Driven electrodes are generally more satisfactory and economical where bedrock is 10 ft or more below the surface, while grids, buried strips, or cables are preferred for lesser depths. Grids are frequently used for substations or generating stations to provide equipotential areas throughout the entire station where hazards to life and property would justify the higher cost. They also require the least amount of buried material per mho of ground conductance. Buried plates have not been used extensively in recent years because of the higher cost as compared to rods or strips. Also, when used in small numbers, they are the least reliable type of made electrode. The counterpoise is a form of the buried cable electrode, and its use is generally confined to locations having high-resistance soils, such as sand or rock, where other methods are not satisfactory.

A discussion on methods of calculating resistance to earth, current-loading capacity of soils, as well as recommended methods and techniques of constructing connections to earth may be found in ANSI/IEEE Std 142-1982 [4], Chapter 4.

7.5.8 Galvanic Corrosion. There has developed an increased awareness of possible aggravated galvanic corrosion of buried steel members if cross-bonded to buried dissimilar metal, such as copper [10], [12], [15].

The result has been a trend to seek a design of electrical grounding electrode that is, galvanically, neutral with respect to the steel structure. In some cases the grounding electrode design employs steel-exposed metal electrodes with insulated copper cable interconnections [10].

The corrosion of buried steel takes place even in the absence of a cross-connection to buried dissimilar metal. The exposed surface of the buried steel inherently contains bits of dissimilar conducting material, foreign metal fragments, or slag inclusions, which create local galvanic cells and local circulating currents. At spots where current leaves the metal surface, metal ions leave the parent metal and account for destructive corrosion. The cross-bonding to dissimilar metal may aggravate the rate of corrosion, but is not the only cause for the action.

Electrical engineering technology should recognize the problem and seek grounding electrode designs that will produce no observable increase in the rate of corrosive disintegration of nonelectrical buried metal members. An overriding priority dictates that the electrical grounding electrode itself not suffer destruction by galvanic corrosion. Relative economics will be an inevitable factor in the design choice.

The recent increase in the use of plastic pipes for water supply to buildings removes one of the most common sources of complaint. The absence of buried metal piping, however, demands that some other suitable grounding electrode be located or created.

A timely release of new knowledge bearing on this problem is the electrical behavior pattern of concrete-encased metal below grade (see 7.5.6). The relationship to galvanic corrision is

(1) There is generally an extensive array of concrete-encased steel-reinforcing members within the foundations and footings, which collectively account for a huge total surface area, resulting in extremely low-current density values at the steel surface

(2) The concrete-encased re-bars themselves constitute an excellent, permanent, low-resistance earthing connection with little or no economic penalty

(3) Current flow across the steel-concrete boundary does not disintegrate the steel as it would if the steel were in contact with earth

7.6 Ground Resistance Measurement. The ground resistance as defined in ANSI/IEEE Std 142-1982 [4] is "... the ohmic resistance between it and a remote grounding electrode of zero resistance." In other words, ground resistance is the resistance of the soil to the passage of electric current from the electrode into the surrounding earth.

Grounding system resistance, expressed in ohms, should be measured after a system is installed and at periodic intervals thereafter. Usually, precision in measurement is not required. Measurement of ground resistance is necessary to verify the adequacy of a new grounding system with the calculated value, and to detect changes in an existing grounding system. It is important that specified or lower resistance be obtained, since all calculations for personnel and equipment safety are based on the specified grounding resistance. The margin of safety will be reduced if the resistance exceeds the specified value.

Three components constitute the resistance of a grounding system:

(1) The resistance of the grounding electrode conductor and grounding conductor connection to the electrode

(2) Contact resistance between the grounding electrode and the soil adjacent to it

(3) The resistance of the body of earth immediately surrounding the electrode

Grounding electrodes are usually of sufficient size or cross-section, and grounding connections are usually made by proven clamps or welding, so that their resistance is a negligible part of the total resistance. If the grounding electrode is free from paint or grease and the earth is packed firmly around the electrode, contact resistance is also negligible. Rust on an iron electrode has little or no effect.

When the current flows from a grounding electrode to earth, it radiates current in all directions. It can be considered that current flows through a series of concentric spherical like earth shells, all of equal thickness, surrounding the grounding electrode. The shell immediately surrounding the electrode has the smallest cross-sectional area and so offers greatest resistance. As the distance from the electrode increases, each shell becomes correspondingly larger in cross-section and offers less resistance. Finally, a distance from the electrode is reached where additional shells do not add significantly to the total ground resistance. Therefore, the resistance of the surrounding earth is the largest component of the resistance of a grounding system.

It is possible, however, to calculate the resistance of any system of grounding electrodes. But several factors can affect the calculated value due to considerable variation in soil resistivity at a given location. Soil resistivity depends on soil material, the moisture content, and the temperature. If all factors are considered, formulas for calculating the performance of grounding systems become very complicated and involve so many indeterminate factors that they are of little value. Many formulas have been developed, but they are only useful as general guides. The actual ground resistance of a grounding system can be determined only by measurement.

7.6.1 Methods of Measuring Ground Resistance. This section only covers commonly used methods of measuring ground resistance. The ohmic value measured is called resistance; however, there is a reactive component that should be taken into account when the ohmic value of the ground under test is less than 0.5 Ω, as in the case of large substation ground grids. This reactive component has little effect on grounds with an impedance higher than 1 Ω.

7.6.1.1 Equipment and Material. Equipment and material required for ground-resistance measurement are as follows:

(1) Ground resistance can be measured by commercially available, self-contained instruments, which give readings directly in ohms. These instruments are small in size and very easy to use because they require no external power source. They are equipped either with batteries or a generator. If necessary, however, approximate results can be obtained with a portable ac ammeter and voltmeter where power supply and transformer with nominal 120 V secondary (to isolate the grounding system under test from the grounding system of the power supply) is available at the location where measurements are to be made. However, it is not easy to obtain accurate results with an ammeter and voltmeter at energized stations.

(2) Two auxiliary test electrodes in addition to the ground electrode (or ground mat) under test.

(3) Flexible single-conductor cable AWG #14 or larger, at least 600 V rated, of sufficient length.

(4) Alligator clips for connecting test leads.

(5) Lineman gloves (optional).

(6) A field notebook.

It is recommended that manufacturers' instructions be followed when connecting the leads to the measuring instrument and taking measurements. Test circuits shown in the following paragraphs are for reference only.

7.6.1.2 Methods of Measurement. Four most commonly used methods of measuring and testing ground resistance are described as follows.

(1) *Fall of Potential Method.* This involves the passing of a current of known magnitude through the grounding electrode (or grounding network) under test and an auxiliary *current* electrode, and measuring the influence of this current in terms of voltage between the electrode under test and a second auxiliary *potential* electrode. (See Fig 125.)

For a large grounding network, both current and potential electrodes should be placed as far from the grounding network under test as practical (depending on the geography of the surroundings), so that they are outside the influence of the ground to be tested. A distance of 750–1000 ft or more from the grounding network is recommended for grounding mats with dimensions in the order of 300 ft × 300 ft. This is required to obtain measurements of adequate accuracy. The *potential* electrode, for large grounding networks (low-resistance grounds), should be driven in at a number of points. Resistance readings are then plotted for each point as a function of distance from the grounding network, and a curve is drawn. The value in ohms at which the plotted curve appears to level off is taken as the resistance of the grounding network under test. When it is found that the curve is not leveling off, the *current* electrode should be moved farther from the grounding electrode under test. However, for a high-resistance ground, there is no preferred placement of electrodes, and the most practical placement of electrodes should be chosen.

The resistance between the ground network (electrode) under test and the auxiliary electrodes should be measured as shown in Fig 125. The resistance measured should be no more than 500 Ω for increased accuracy in the measurement of low-resistance ground network. In order to obtain the lowest possible auxiliary

Fig 125
Test Circuit for Measuring Test Electrode Resistances and Resistance of the Large Grounding Network

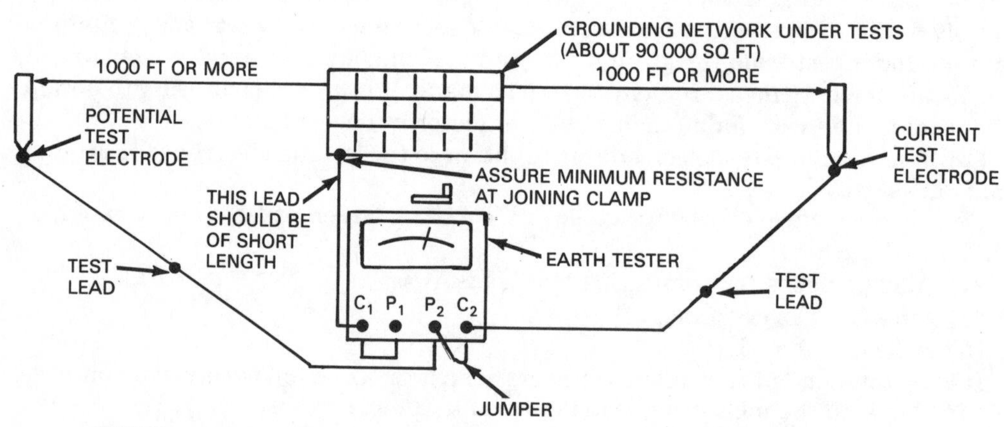

NOTES:
(1) For measuring test electrode resistance, remove connection of one test lead at each measurement but keep both jumpers connected (each test electrode should have less than 500 Ω resistance).
(2) Network:
(a) Remove jumper between P_2 and C_2.
(b) Take one reading with P_2 and C_2 connected as shown (except remove jumper).
(c) Reverse P_2 and C_2 connections at the instrument and take a second reading. Use average of the two readings.

electrode resistance, locate the electrodes in moist locations, such as drainage ditches or ponds, or drive two or more rods spaced 3 or 4 ft apart, or where practicable, use long rods driven to considerable depths.

After checking the auxiliary electrodes' resistance, connect test probes to the instrument as shown in Fig 125 for measuring ground resistance of the grounding network (electrode) under test. Reverse connections at the instrument and take another reading. The difference in both readings should be less than 15%, otherwise auxiliary electrodes should be moved farther away from the ground network (electrode) under test.

This method should be used for large substations, industrial plants, and generating stations where grounding network resistance is usually less than 1 Ω.

For a small ground mat or single-rod-driven electrode, the influence of the ground to be tested is assumed to be negligible at about 100–125 ft, so the *current* electrode can be placed at about 100–125 ft from the ground rod under test. To measure earth resistance of a single-rod-driven electrode or small ground mat, the *potential* electrode can be placed midway between *current* electrode and the ground electrode under test as shown in Fig 126. Readings with the circuit as connected are taken.

This method should be used for a single-rod electrode or small ground mat and where the earth electrode under test can be separated from the water-pipe system, which usually has negligible ground resistance.

(2) *Two-Point Method*. The two-point method is usually used to determine the resistance of a single grounding rod driven near a residence where it is necessary to know only that a given grounding electrode's resistance to earth is below a stipu-

Fig 126
Test Circuit for Measuring Earth Resistance of Ground Rod or Small Grid—Fall of Potential Method

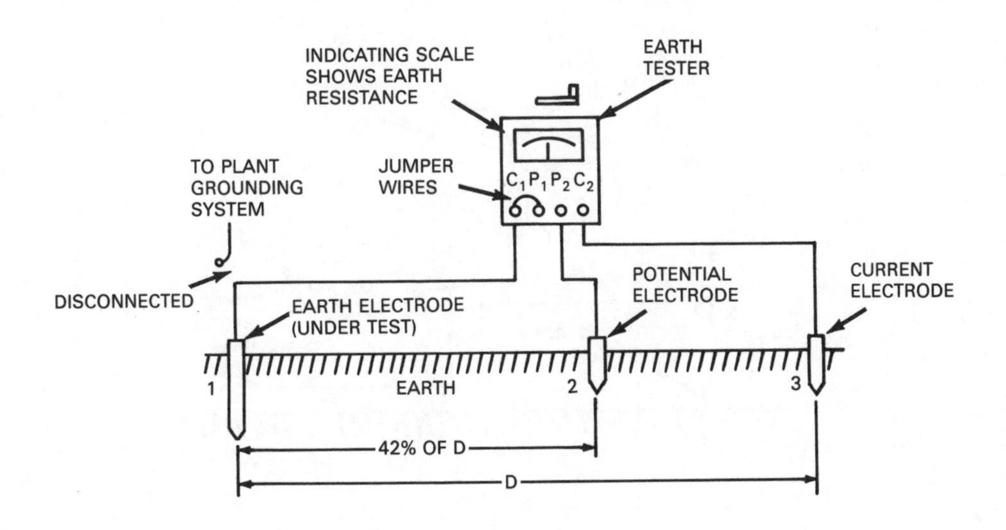

lated value, say 25 Ω or less. In this method the total resistance of the unknown and an auxiliary grounding rod, usually existing metallic water-pipe system (with no insulating joints), is measured. Since the water-pipe system's resistance is considered to be negligible, the resistance measured by the meter will be that of the grounding electrode under test. (See Fig 127.)

This method is subject to large errors for low-resistance grounding networks but is very useful and adequate where a *go, no-go* type of test is required.

(3) *Three-Point Method.* This method involves the use of two auxiliary electrodes as in the case of fall-of-potential method (see Fig 126). The resistance between each pair of grounding electrodes in series is measured and designated as $R_{1\text{-}2}$, $R_{1\text{-}3}$, and $R_{2\text{-}3}$, where $R_{1\text{-}2}$ is the resistance of the grounding electrode under test and one auxiliary electrode. The resistance of electrode under test can be obtained by solving for R_1:

$$R_1 = \frac{R_{1\text{-}2} + R_{1\text{-}3} - R_{2\text{-}3}}{2}$$

If two auxiliary electrodes are of higher resistance than the grounding electrode under test, small errors in the individual measurements may result in a large error. For this method the electrodes must be at least 20 ft or more apart, otherwise absurdities may arise in the calculations, such as zero or even negative resistance. Either alternating current of commercial frequency or direct current may be used. When direct current is used, the effect of stray alternating current is eliminated though stray direct current may give a false reading. If alternating current is used, stray alternating current of the same frequency as the test current may introduce an error; however stray direct currents have no effect. These effects may be mini-

Fig 127
Test Circuit for Measuring Earth Resistance of a Ground Rod—Two-Terminal Method

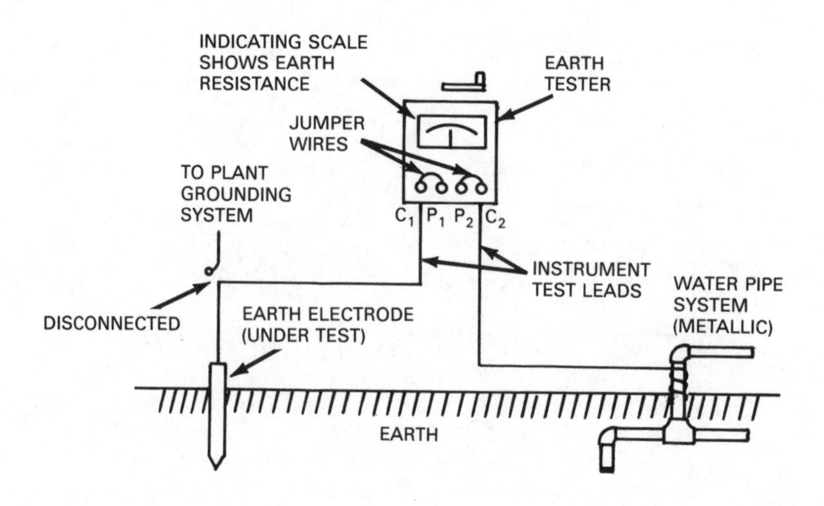

mized by taking a reading with the current flowing in one direction, then reversing the polarity and taking a reading with current flowing in the other direction. An average of these two readings will be an accurate value.

This method is not suitable for large substation grounds, and the fall-of-potential method is recommended.

(4) *Ratio Method*. This method uses two auxiliary electrodes as in the fall-of-potential method. The resistance of the electrode under test is compared with the known resistance of auxiliary electrodes. This method is not commonly used since it has limitations in measuring low-resistance grounding networks of large area. It is necessary to use the fall-of-potential method for accurate measurements.

It is preferable to measure grounding network resistance before a station is energized. When this is not possible, instruments designed for use at energized stations should be used and necessary precautions should be taken when connecting or disconnecting test leads. Where practicable, avoid locations that will cause the test leads to be parallel to transmission lines.

7.7 References. This standard shall be used in conjunction with the following publications:

[1] ANSI C2-1987, American National Standard National Electrical Safety Code.

[2] ANSI/IEEE Std 80-1986, IEEE Guide for Safety in Substation Grounding.

[3] ANSI/IEEE Std 81-1983, IEEE Guide for Measuring Earth Resistivity, Ground Impedance, and Earth Surface Potentials of a Ground System.

[4] ANSI/IEEE Std 142-1982, IEEE Recommended Practice for Grounding of Industrial and Commercial Power Systems.

[5] ANSI/NFPA 70-1987, National Electrical Code.

[6] ANSI/NFPA 78-1983, Lightning Protection Code.

[7] AIEE Committee Report. Voltage Gradients Through the Ground Under Fault Conditions. *AIEE Transactions* (*Power Apparatus and Systems*), pt III, vol 77, Oct 1958, pp 669–692.

[8] BEEMAN, D. L., Ed. *Industrial Power Systems Handbook*. New York: McGraw-Hill, 1955, chap 5–7.

[9] BISSON, A. J. and ROCHAU, E. A. Iron Conduit Impedance Effects in Ground Circuit Systems. *AIEE Transactions* (*Applications and Industry*), pt II, vol 73, July 1954, pp 104–107.

[10] COLMAN, W. E. and FROSTICK, H. G. Electrical Grounding and Cathodic Protection at the Fairiess Works. *AIEE Transactions* (*Applications and Industry*), pt II, vol 74, Mar 1955, pp 19–24.

[11] FAGAN, E. J. and LEE, R. H. The Use of Concrete-Encased Reinforcing Rods as Grounding Electrodes. *IEEE Transactions on Industry and General Applications*, vol IGA-6, July/Aug 1970, pp 337–348.

[12] HERTZBERG, L. B. The Water Utilities Look at Electrical Grounding. *IEEE Transactions on Industry and General Applications*, vol IGA-6, May/June 1970, pp 278-281.

[13] KAUFMANN, R. H. Some Fundamentals of Equipment-Grounding Circuit Design. *AIEE Transactions (Applications and Industry)*, pt II, vol 73, Nov 1954, pp 227-232.

[14] WIENER, P. A. Comparison of Concrete-Encased Grounding Electrodes to Driven Ground Rods. *IEEE Transactions on Industry and General Applications*, vol IGA-6, May/June 1970, pp 282-287.

[15] ZASTROW, O. W. Underground Corrosion and Electrical Grounding. *IEEE Transactions on Industry and General Applications*, vol IGA-3, May/June 1967, pp 237-243.

7.8 Bibliography

[B1] BRERETON, D. S. and HICKOCK, H. N. System Neutral Grounding for Chemical Plant Power Systems. *AIEE Transactions (Applications and Industry)*, pt II, vol 74, Nov 1955, pp 315-320.

[B2] BULLARD, W. R. Grounding Principles and Practice. Part IV—System Grounding. *Electrical Engineering*, vol 64, Apr 1945, pp 145-151.

[B3] CRAWFORD, L. E. and GRIFFITH, M. S. A Closer Look at "the Facts of Life" in Ground Mat Design. *IEEE Transactions on Industry Applications*, vol IA-15, May/June 1979, pp 241-250.

[B4] DALZIEL, C. F. Effects of Electric Shock on Man. *IRE Transactions on Medical Electronics*, vol PGME-5, July 1956, pp 44-62.

[B5] DUNKI-JACOBS, Jr. The Reality of High-Resistance Grounding. *IEEE Transactions on Industry Applications*, vol IA-13, no 5, Sept/Oct 1977, pp 469-475.

[B6] GIENGER, J. A., DAVIDSON, O. C., and BRENDEL, R. W. Determination of Ground Fault Current on Common AC Grounded-Neutral Systems in Standard Steel or Aluminum Conduit. *AIEE Transactions (Applications and Industry)*, pt II, vol 79, May 1960, pp 84-90.

[B7] HARVIE, R. A. Avoiding Hazards from Earth Currents in Industrial Plants, *IEEE Transactions on Industry Applications*, vol IA-13, May/June 1977, pp 207-214.

[B8] HARVIE, R. A. Hazards During Ground Faults on 480 V Grounded Systems. *IEEE Transactions on Industry Applications*, vol IA-10, Mar/Apr 1974, pp 190-196.

[B9] JENSEN, C. Grounding Principles and Practice. Part II—Establishing Grounds. *Electrical Engineering*, vol 64, Feb 1945, pp 68-74.

[B10] JOHNSON, A. A. Grounding Principles and Practice. Part III — Generator-Neutral Grounding Devices. *Electrical Engineering*, vol 64, Mar 1945, pp 92-99.

[B11] KALBACH, J. F. *Electrical Environment for Computers*. Presented at IEEE Industrial and Commercial Power Systems Conference, St Louis, MO, May 6, 1981, Conference Paper 81CH1674-1.

[B12] KAUFMANN, R. H. Let's Be More Specific About Equipment Grounding. *Proceedings of the American Power Conference*, 1962, vol 24, pp 913-922.

[B13] THACKER, H. B. Grounded Versus Ungrounded Low-Voltage Alternating Current Systems. *Iron and Steel Engineer*, Apr 1954.

[B14] WALSH, G. W. A Review of Lightning Protection and Grounding Practices. *IEEE Transactions on Industry Applications*, vol IA-9, no 2, Mar/Apr 1973.

[B15] WEST, R. B. Equipment Grounding for Reliable Ground-Fault Protection in Electrical Systems Below 600 V. *IEEE Transactions on Industry Applications*, vol IA-10, Mar/Apr 1974, pp 175-189.

8. Power Factor and Related Considerations

8.1 General. This chapter provides some basic information about power-factor improvement, including common and important applications. Necessary fundamentals are covered and selected references are included for appropriate detailed information on switching, harmonics, resonance, automatic control, measurements, metering arrangements, and capacitor protection. High-frequency systems and series capacitors are excluded because their applications are limited. The fundamentals presented in this chapter are based on the assumption that the power supply voltage wave is sinusoidal and, therefore, the mathematical analysis of harmonics is not covered in detail. This is a valid perspective for the nature of most loads in industrial service. However, with the increasing use of static power converters, the waveform often is not sinusoidal and some of the associated problems are covered in 8.13.

8.1.1 Emphasis on Capacitors. Adding capacitors is generally the most economical way to improve the plant power factor, especially in existing plants. Therefore, the application emphasis will be on capacitors. However, there are some cases in which synchronous motors are most economical, and they will also be covered.

A third method for improving power factor by means of a static power factor controller will be briefly discussed.

In addition to their relatively low cost, capacitors have other properties, such as ease of installation, practically no maintenance, and very low losses. Capacitors are manufactured in relatively small ratings (kvar) for economical manufacturing and engineering reasons. The individual units can, however, be combined into suitable banks to obtain a large range of ratings. Thus, capacitors can be added in small or large units to meet current operating requirements and avoid a larger than necessary installation in anticipation of increased future requirements.

8.1.2 Benefits of Power-Factor Improvement. Most benefits provided by a power-factor improvement stem from the reduction of reactive power in the system. This may result in:

(1) lower purchased-power costs if the utility enforces a power-factor clause
(2) release of system electrical capacity
(3) voltage improvement
(4) lower system losses

Maximum benefits are obtained when capacitors or synchronous motors are located at low-power-factor loads.

Although reducing the power bill is still the primary reason for improving the power factor and it is becoming more important because of conservation of energy, the function of releasing system capacity or improving system voltage, or both, are sometimes the decisive factors.

8.2 Typical Plant Power Factor. The unimproved power factor in plants depends on equipment design and operating conditions. Thus, in many cases it is not possible to accurately predict what should be expected in a new plant. The typical figures listed in this section are drawn from operating experience.

8.2.1 General Industry Applications. The power-factor values in Table 46 are by industry category and are representative of the normal part of industrial or commercial equipment assemblage. The values can be expected to change in some cases where converters are used.

8.2.2 Plant Applications. The values in Table 47 are those that may be expected per shift. The overall plant power factor will be different if there are partly loaded motors or idle shifts.

8.2.3 Utilization Equipment Applications

(1) *Motors*. The power factor of a partly loaded induction motor is poor, as indicated in Fig 135 in 8.9.1. This figure also shows the attractive improvement over the entire load range by using a capacitor of proper rating.

The T-frame motor, available since about 1964, generally has a lower power factor than the U-frame design. The comparison is shown in Fig 137 in 8.9.3.

More recently introduced high-efficiency motors may have much higher power factor characteristics in certain ratings, as described in 8.9.4.

Hermetic and wound-rotor type motors have a lower operating power factor than other induction motors of the same power and speed ratings.

(2) *Rectifiers*

(a) *Diode Type with No Phase Control*. Small, single-phase units have about 50% power factor at full load, while larger, multiphase units may have 95–98% power factor.

(b) *Static Converter Drives*. The power factor is roughly proportional to the ratio of dc output voltage to rated voltage. At partial loads the power factor is poor. A step-by-step procedure for determining the capacitor rating required for power-factor improvement is given in [9][25].

---

[25] The numbers in brackets correspond to those of the references listed at the end of this chapter; when preceded by B, they correspond to the bibliography at the end of this chapter.

Table 46
Typical Unimproved Power-Factor Values, by Industries

Industry	Power Factor
Auto parts	75-80
Brewery	75-80
Cement	80-85
Chemical	65-75
Coal mine	65-80
Clothing	35-60
Electroplating	65-70
Foundry	75-80
Forge	70-80
Hospital	75-80
Machine manufacturing	60-65
Metalworking	65-70
Office building	80-90
Oil-field pumping	40-60
Paint manufacturing	55-65
Plastic	75-80
Stamping	60-70
Steel works	65-80
Textile	65-75
Tool, die, jig	60-65

Table 47
Typical Operating Power-Factor Values, by Plant Operations

Operation	Power Factor
Air compressor	
external motors	75-80
hermetic motors	50-80
Metalworking	
arc welding	35-60
with commercially furnished capacitors	70-80
Machining	40-65
Melting	
arc furnace	75-90
induction furnace, 60 Hz	100
Stamping	
standard	60-70
high speed	45-60
Spraying	60-65
Welding	
arc	35-60
resistance	40-60
Weaving	
individual	60
multiple	70
brind	70-75

(3) *Electric Furnaces*. Arc furnaces have a poor power factor, typically 75-90%. Power-factor improvement may be a system problem. Induction furnaces have a power factor of typically 30-70%, switched capacitors are used to maintain near unity power factor.

(4) *Lamps*. Incandescent lamps operate at unity power factor. Fluorescent and other discharge lamps have a poor operating power factor of approximately 70%. With a corrected ballast the power factor will range from about 90% lagging to slightly leading.

(5) *Transformers*. These are not ordinarily considered to be loads, but they do contribute to lowering the system power factor. The transformer exciting current is usually 1-2% of the transformer rating in kilovolt-amperes and is independent of load. Reactive power is also required by the transformer leakage reactance. Such reactive power varies as the square of the load current (I^2_{XL}). At rated current the leakage reactance requires reactive power equal in magnitude to the transformer rating in kilovolt-amperes times the nameplate impedance in per unit.

8.3 Instruments and Measurements for Power-Factor Studies. When power-factor studies are made, sufficient data should be obtained to select the proper rating and location of capacitors or synchronous motors. If the study is for power billing rate purposes, then the power bills usually furnish sufficient information to determine the capacitor rating required.

Most utility rates include a demand charge, usually obtained from a demand register attachment on the watthour meter or by recording or printing-type instruments.

The power factor may be measured directly by indicating instruments or obtained from other indications, such as kilowatt, kilovolt-ampere, or kilovar readings. Average values may be obtained from kilowatt-hour, kilovar-hour, or kilovolt-ampere-hour readings.

Measurements by recording or graphic instruments are most desirable and useful because they provide a permanent record.

Indicating instruments are satisfactory for spot checking individual feeder circuits or loads. They can also be used in place of recording instruments if readings are taken at frequent intervals.

The preferred measurements are power in kilowatts, reactive power in kilovar, and voltage in volts, from which the apparent power in kilovolt-amperes and the power factor can be calculated. Voltage readings are especially desirable if automatic capacitor control with a voltage-responsive master element is contemplated. Direct measurement of the power factor is not particularly desirable, since power-factor interpretation may be misleading; for example, even at 95% power factor, the reactive power component (in kilovars) of the load is 33% of the active power component (in kilowatts).

When a power-factor meter is used, it is generally a polyphase instrument. Even then, power-factor readings are accurate only on balanced loads, and become increasingly inaccurate as the load unbalance increases.

For metering arrangement and connection, diagrams refer to Chapter 10.

If switchboard meters are provided, their readings may be taken as applying to that portion of the load that each serves.

Readings from such instruments often disclose great variations in the power factor between different parts of a plant. Such knowledge can be valuable in locating equipment for power-factor improvement to best advantage.

8.4 Power-Factor Economics. High plant power factors can yield direct savings. Some, such as reduced power bills and release of system capacity, are quite visible; others, such as decreased I^2R losses, are less visible but nonetheless real as are many indirect savings as a result of more efficient performance.

The cost of improving the power factor in existing plants, and of maintaining proper levels as load is added, depends on the power-factor value selected and the equipment chosen to supply the compensating reactive power. In general, medium-voltage capacitors cost less per kilovar than do low-voltage capacitors. However, the cost of installation may change the comparison. Industrial capacitor additions can provide a large payback through power-bill savings in one to four years.

The combination of reduced power billing and released system capacity by improving the power factor can be very attractive economically.

Some utility rates specify a minimum operating power factor, which may be as high as 90 or 95%.

Optimizing the system power-factor level consists of combining an engineer's sense of system-operating efficiency with a businessman's sense of investment profitability.

8.5 Power-Factor Fundamentals. Most utilization devices require two components of current. Breaking down the total current into meaningful components (active and reactive) permits simple analysis and understanding.

(1) The power-producing current or working current is current that is converted by the equipment into work, usually in the form of heat, light, or mechanical power. The unit of measurement of active power is the watt.

(2) Magnetizing current, also known as wattless, reactive, or nonworking current, is the current required to produce the flux necessary to the operation of electromagnetic devices. Without magnetizing current, energy could not flow through the core of a transformer or across the air gap of an induction motor. The unit of measurement of reactive power is the var.

The normal phasor relationship of these two components of current to each other, to the total current, and to the system voltage is illustrated in Fig 128. It shows that the active current and the reactive current add vectorially to form the total current, which can be determined from the expression

$$\text{total current } I_t = \sqrt{(\text{active current})^2 + (\text{reactive current})^2} \qquad \text{(Eq 1a)}$$

$$I_t = \sqrt{(I \cos \phi)^2 + (I \sin \phi)^2} \qquad \text{(Eq 1b)}$$

At a given voltage V, the active, reactive, and apparent power are proportional to current and are related as follows:

$$\text{apparent power in volt amperes} = \sqrt{(\text{active power})^2 + (\text{reactive power})^2} \qquad \text{(Eq 2a)}$$

$$VI = \sqrt{(VI \cos \phi)^2 + (VI \sin \phi)^2} \qquad \text{(Eq 2b)}$$

As evidenced when Fig 128 is compared with Fig 129, the phasor diagram for power is similar in form to that for current. Equations are based on fundamental frequency and assume zero (0) harmonic current.

Fig 128
Angular Relationship of Current and Voltage in AC Circuits

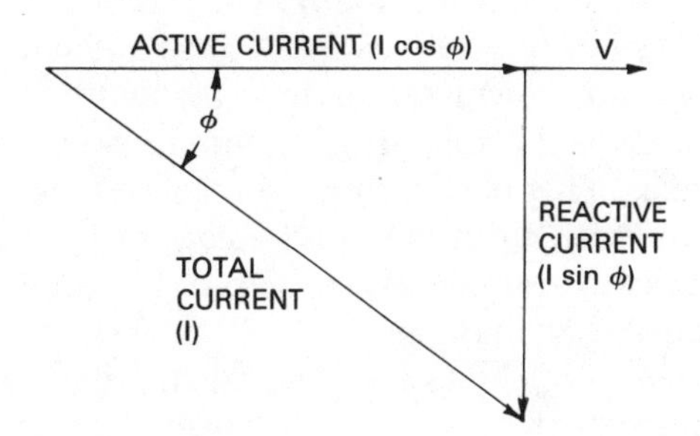

Fig 129
Relationship of Active, Reactive, and Total Power

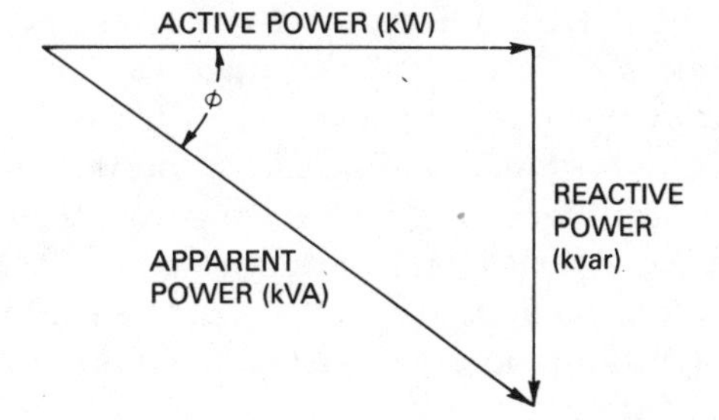

8.5.1 Definition of Power Factor. There are at least two commonly used definitions of power factor. One definition expresses power factor as the cosine of the phase displacement angle by which the circuit current leads or lags the circuit voltage (see Eq 3a). The power factor in this text is defined as the ratio of active power to apparent power in a circuit (see Eq 3b). It varies from one to zero, but is generally given in percent:

$$\begin{aligned} \text{power factor} &= \text{cosine of angle between active power and apparent power} \\ &= \cos\phi \end{aligned} \qquad \text{(Eq 3a)}$$

$$\text{power factor} = \frac{\text{active power}}{\text{apparent power}} = \frac{\text{kW}}{\text{kVA}} \qquad \text{(Eq 3b)}$$

$$\begin{aligned} \text{active power} &= \text{apparent power} \cdot \text{power factor} \\ \text{kW} &= (\text{kVA})\,(\text{pf}) \\ &= (\text{kVA})\,(\cos\phi) \end{aligned} \qquad \text{(Eq 4)}$$

In analytical procedures involving harmonics, the terms *displacement* factor and *distortion* factor are often used in connection with power factor considerations. The displacement factor results from lack of coincidence of the phase angle between current and voltage at any given frequency, mainly the fundamental frequency of the circuit. This is the power factor discussed in this section. The distortion factor represents the effect of wave-shape distortions (harmonics) and their influence on the *apparent* power factor of the circuit.

8.5.2 Leading and Lagging Power Factor. The power factor may be leading or lagging, depending on the direction of both the active and reactive power flows. If these flows are in the same direction, the power factor at that point of reference is lagging. If either power component flow is in an opposite direction, the power factor at that point of reference is leading. Since capacitors and overexcited synchronous motors are a source of reactive power their power factor is always leading.

An induction motor has a lagging power factor as it requires both active and reactive power to flow into the motor (same direction). An overexcited synchronous motor can supply reactive power to the system. The active power component flows into the motor and the reactive power flows into the power system (opposite directions), so the power factor is leading. In an actual power system, the overall system power factor may be lagging, even though some apparatus such as overexcited synchronous motors and capacitors may have a leading power factor.

8.5.3 How to Improve the Power Factor. The concept of a capacitor as a kilovar generator is helpful in understanding its use for power-factor improvement. A capacitor may be considered a kilovar generator because it supplies the magnetizing requirement (kilovars) to induction devices. This action may be explained in terms of the energy stored in capacitors and induction devices. As the voltage in ac circuits varies sinusoidally, it alternately passes through zero-voltage points and maximum-voltage points. As the voltage passes through zero voltage and starts toward maximum voltage, the capacitor stores energy in its electrostatic field, and the induction device gives up energy from its electromagnetic field. As the voltage passes through a maximum point and starts to decrease, the capacitor gives up energy and the induction device stores energy. Thus, when a capacitor and an induction device are installed on the same circuit, there will be an exchange of magnetizing current between them, that is, the leading current taken by the capacitor neutralizes the supply line of supplying magnetizing current to the induction device. The capacitor may be considered to be a kilovar generator, since it actually supplies magnetizing requirements in the induction device.

With recent developments in solid-state power technology, a static power-factor controller is available to improve the power factor of individual motors operating at less than full load. This controller might be cost-justifiable in some cases and may be considered on applications involving motors that, because of their duty-cycle, are required to operate at less than full load for prolonged periods. The controller adjusts its output voltage magnitude to maintain a constant load current phase angle as the shaft load of the motor changes. Effectively, the motor thus operates at full-load performance (including peak power factor and peak relative efficiency) as the controller output voltage decreases or increases in response to load variations. Typical performance curves comparing motor operation with and without the controller are shown in Fig 130. Although improved power factor for the motor results for operation at less than rated load, it should be noted that the power factor will never be better than the rated full-load power factor of the motor. No *bonus* power-factor improvement is available as with capacitors and synchronous motors. On the other hand, static power-factor controllers inherently have voltage-regulation capability and often a soft-start feature that can be of benefit in applications requiring reduced-voltage starting or involving frequent starting of shock-sensitive loads.

Fig 130
Typical Active and Reactive Power Savings With and Without Static Power-Factor Controller

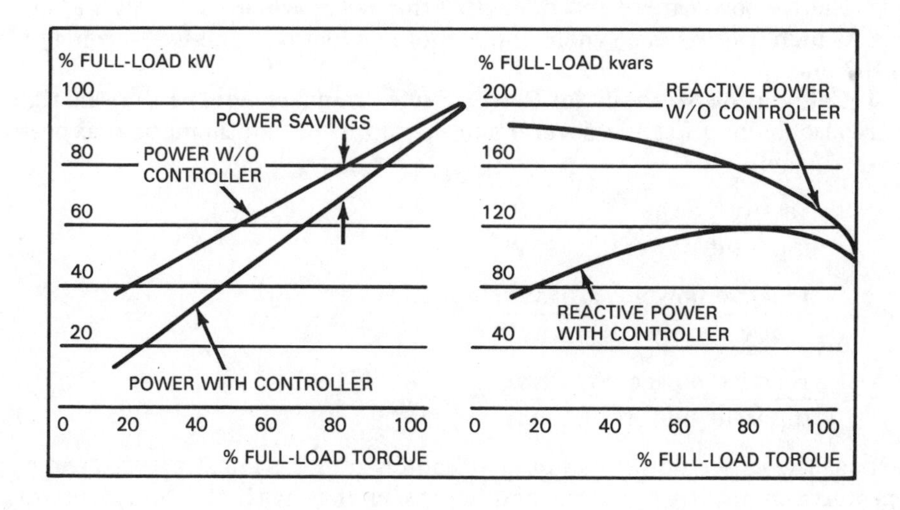

The economics of the static power-factor controller should be carefully evaluated for each prospective application. For example, since with the controller the power factor of each motor circuit will never be greater than the motor full-load power factor, it is likely that supplemental improvement will have to be provided in order to attain power factor levels of 90% and above for the overall plant. The supplemental improvement might be capacitor banks or capacitors switched with individual motors, assuming the close-coupling of the capacitors and the static controller does not result in excessive harmonic currents/voltages (see 8.13). The total supplemental amount of capacitors required will be less than without power-factor controllers, and in the case of bank correction, the smaller-sized banks may not have to be switched to prevent system overvoltage during periods of light plant loads.

When the reactive power component in a circuit is reduced, the total current is reduced. If the active power component does not change, as is usually the case, the power factor will improve as the reactive power component is reduced. When the reactive power component becomes zero, the power factor will be unity or 100%.

This is shown pictorially in Fig 131. The load requires an active current of 80 A, but because the motor requires a reactive current of 60 A, the supply circuit must carry the vector sum of these two components, which is 100 A. After a capacitor is installed to supply the motor reactive-current requirements, the supply circuit needs to deliver only 80 A to do exactly the same amount of work. The supply circuit is now carrying only active power, so no system kVA capacity is wasted in carrying reactive power.

Thus, for all practical purposes, the only way to improve the power factor is to reduce the reactive power component. This is usually done with capacitors. Synchronous motors are also used for power-factor improvement. The reactive power output that they are capable of supplying to the line is a function of excitation and motor load. The curves of Fig 132 show the reactive power that a typical synchronous motor is capable of delivering under various load conditions with normal excitation. For example, an 0.8 power factor (leading) synchronous motor can supply reactive power equivalent to 60% of its hp rating at 100% load, but will supply reactive power up to 75% of its hp if the motor is loaded to only 20% of its hp rating. At high overloads a synchronous motor may take reactive (lagging) power from the line.

8.5.4 Calculation Methods for Power-Factor Improvement. From the right triangle relationship of Fig 128 several simple and useful mathematical expressions may be written:

$$\cos\phi = \frac{\text{active power}}{\text{apparent power}} = \frac{\text{kW}}{\text{kVA}} \qquad \text{(Eq 5)}$$

$$\tan\phi = \frac{\text{reactive power}}{\text{active power}} = \frac{\text{kvar}}{\text{kW}} \qquad \text{(Eq 6)}$$

$$\sin\phi = \frac{\text{reactive power}}{\text{apparent power}} = \frac{\text{kvar}}{\text{kVA}} \qquad \text{(Eq 7)}$$

Because the active power component usually remains constant, and the apparent power and reactive power components change with the power factor, the

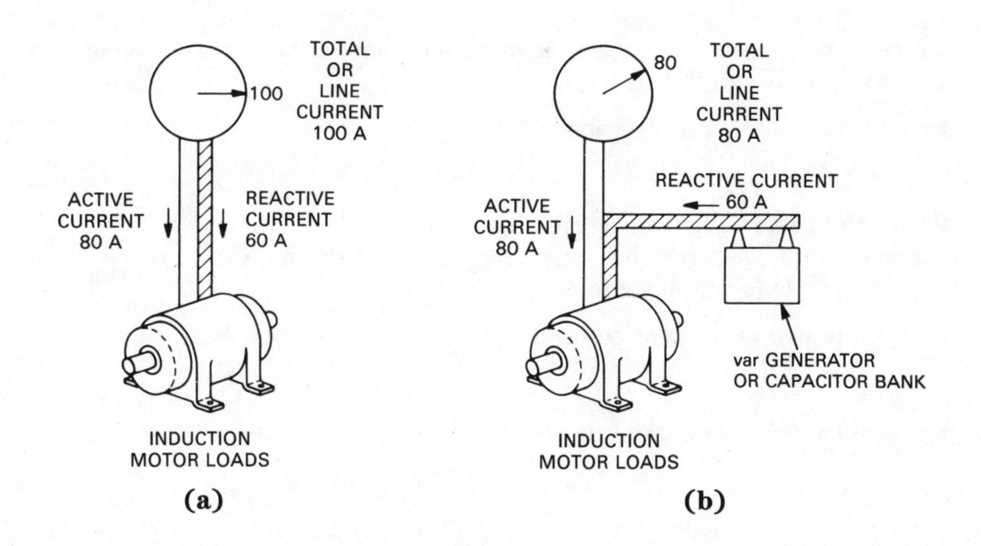

Fig 131
Schematic Arrangement Showing How Capacitors Reduce Total Line Current by Supplying Reactive Power Requirements Locally

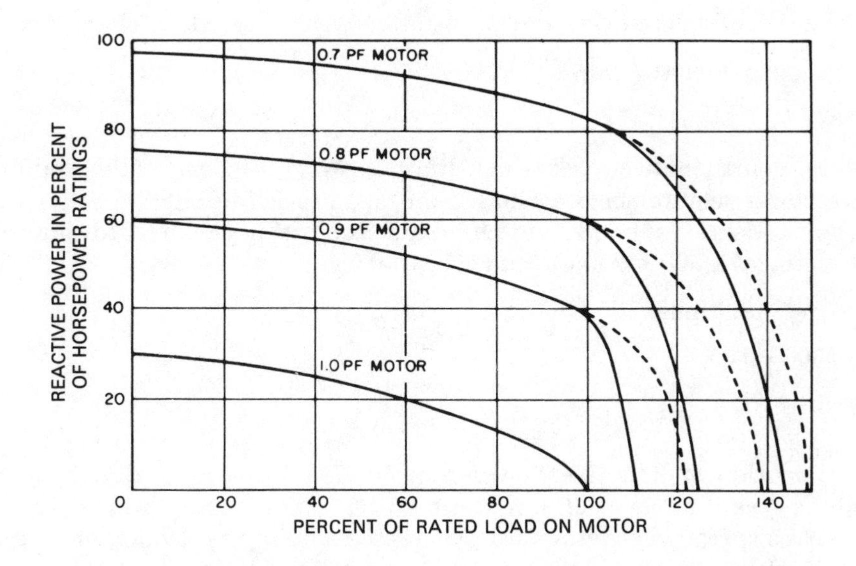

NOTE: Solid lines are based on reduction in excitation at overload to maintain rated full-load current; dashed lines represent rated excitation to maintain rated pullout torque.

Fig 132
Leading Reactive Power in Percent of Motor Horsepower Ratings for Synchronous Motors at Part Load and at Various Power-Factor Ratings

expression involving the active power component is the most convenient to use. This expression may be rewritten as

reactive power = active power · tan ϕ (Eq 8a)

kvar = (kW) (tan ϕ) (Eq 8b)

where the value of tan ϕ corresponds to the power factor angle (ϕ).

For example, assume that it is necessary to determine the capacitor rating to improve the load power factor:

reactive power at original power factor = active power · tan ϕ_1

= (kW) (tan ϕ_1) (Eq 9)

reactive power at improved power factor = active power · tan ϕ_2

= (kW) (tan ϕ_2) (Eq 10)

where ϕ_1 is the angle of the original power factor and ϕ_2 is the angle of the improved power factor. Therefore, the capacitor rating required to improve the power factor is

reactive power kvar = active power · (tan ϕ_1 - tan ϕ_2) (Eq 11a)

kvar = (kW) (tan ϕ_1 - tan ϕ_2) (Eq 11b)

For simplification (tan ϕ_1 - tan ϕ_2) is often written as Δtan. Therefore:

reactive power = active power · Δtan (Eq 12a)

kvar = (kW) (Δtan) (Eq 12b)

All tables, charts, and curves which have a kW multiplier for determining the reactive power requirements are based on Eq 12 (see Table 48).

Example. Using Table 48, find the capacitor rating required to improve the power factor of a 500 kW load from 0.76 to 0.93:

kvar = kW · multiplier

= 500 · 0.46

= 230

8.5.5 Location of Reactive Power Supply. The benefits derived by installing capacitors, synchronous machines, static power-factor (motor) controllers, or any other means for power-factor improvement result in the reduction of reactive power circulating in the system. Capacitors and synchronous machines should, therefore, be installed as close as possible to the load for which the power factor is being improved. A static power-factor controller, by its intended function, would be located electrically adjacent to the motor for which power-factor improvement is being provided. Figure 133 shows four common capacitor locations. Referring to Fig 133, the most desirable location for power-factor improvement is C_1, then, in the following order, C_2 (typically a plug-in busway), C_3, and C_4. The same principle

Table 48
Kilowatt Multipliers to Determine Reactive-Power Requirements for Power-Factor Improvement

Original Power Factor	Corrected Power Factor																				
	0.80	0.81	0.82	0.83	0.84	0.85	0.86	0.87	0.88	0.89	0.90	0.91	0.92	0.93	0.94	0.95	0.96	0.97	0.98	0.99	1.0
0.50	0.982	1.008	1.034	1.060	1.086	1.112	1.139	1.165	1.192	1.220	1.248	1.276	1.306	1.337	1.369	1.403	1.440	1.481	1.529	1.589	1.732
0.51	0.937	0.962	0.989	1.015	1.041	1.067	1.094	1.120	1.147	1.175	1.203	1.231	1.261	1.292	1.324	1.358	1.395	1.436	1.484	1.544	1.687
0.52	0.893	0.919	0.945	0.971	0.997	1.023	1.050	1.076	1.103	1.131	1.159	1.187	1.217	1.248	1.280	1.314	1.351	1.392	1.440	1.500	1.643
0.53	0.850	0.876	0.902	0.928	0.954	0.980	1.007	1.033	1.060	1.088	1.116	1.144	1.174	1.205	1.237	1.271	1.308	1.349	1.397	1.457	1.600
0.54	0.809	0.835	0.861	0.887	0.913	0.939	0.966	0.992	1.019	1.047	1.075	1.103	1.133	1.164	1.196	1.230	1.267	1.308	1.356	1.416	1.559
0.55	0.769	0.795	0.821	0.847	0.873	0.899	0.926	0.952	0.979	1.007	1.035	1.063	1.093	1.124	1.156	1.190	1.227	1.268	1.316	1.376	1.519
0.56	0.730	0.756	0.782	0.808	0.834	0.860	0.887	0.913	0.940	0.968	0.996	1.024	1.054	1.085	1.117	1.151	1.188	1.229	1.277	1.337	1.480
0.57	0.692	0.718	0.744	0.770	0.796	0.822	0.849	0.875	0.902	0.930	0.958	0.986	1.016	1.047	1.079	1.113	1.150	1.191	1.239	1.299	1.442
0.58	0.655	0.681	0.707	0.733	0.759	0.785	0.812	0.838	0.865	0.893	0.921	0.949	0.979	1.010	1.042	1.076	1.113	1.154	1.202	1.262	1.405
0.59	0.619	0.645	0.671	0.697	0.723	0.749	0.776	0.802	0.829	0.857	0.885	0.913	0.943	0.974	1.006	1.040	1.077	1.118	1.166	1.226	1.369
0.60	0.583	0.609	0.635	0.661	0.687	0.713	0.740	0.766	0.793	0.821	0.849	0.877	0.907	0.938	0.970	1.004	1.041	1.082	1.130	1.190	1.333
0.61	0.549	0.575	0.601	0.627	0.653	0.679	0.706	0.732	0.759	0.787	0.815	0.843	0.873	0.904	0.936	0.970	1.007	1.048	1.096	1.156	1.299
0.62	0.516	0.542	0.568	0.594	0.620	0.646	0.673	0.699	0.726	0.754	0.782	0.810	0.840	0.871	0.903	0.937	0.974	1.015	1.063	1.123	1.266
0.63	0.483	0.509	0.535	0.561	0.587	0.613	0.640	0.666	0.693	0.721	0.749	0.777	0.807	0.838	0.870	0.904	0.941	0.982	1.030	1.090	1.233
0.64	0.451	0.474	0.503	0.529	0.555	0.581	0.608	0.634	0.661	0.689	0.717	0.745	0.775	0.806	0.838	0.872	0.909	0.950	0.998	1.068	1.201
0.65	0.419	0.445	0.471	0.497	0.523	0.549	0.576	0.602	0.629	0.657	0.685	0.713	0.743	0.774	0.806	0.840	0.877	0.918	0.966	1.026	1.169
0.66	0.388	0.414	0.440	0.466	0.492	0.518	0.545	0.571	0.598	0.626	0.654	0.682	0.712	0.743	0.775	0.809	0.846	0.887	0.935	0.995	1.138
0.67	0.358	0.384	0.410	0.436	0.462	0.488	0.515	0.541	0.568	0.596	0.624	0.652	0.682	0.713	0.745	0.779	0.816	0.857	0.905	0.965	1.108
0.68	0.328	0.354	0.380	0.406	0.432	0.458	0.485	0.511	0.538	0.566	0.594	0.622	0.652	0.683	0.715	0.749	0.786	0.827	0.875	0.935	1.078
0.69	0.299	0.325	0.351	0.377	0.403	0.429	0.456	0.482	0.509	0.537	0.565	0.593	0.623	0.654	0.686	0.720	0.757	0.798	0.846	0.906	1.049
0.70	0.270	0.296	0.322	0.348	0.374	0.400	0.427	0.453	0.480	0.508	0.536	0.564	0.594	0.625	0.657	0.691	0.728	0.769	0.817	0.877	1.020
0.71	0.242	0.268	0.294	0.320	0.346	0.372	0.399	0.425	0.452	0.480	0.508	0.536	0.566	0.597	0.629	0.663	0.700	0.741	0.789	0.849	0.992
0.72	0.214	0.240	0.266	0.292	0.318	0.344	0.371	0.397	0.424	0.452	0.480	0.508	0.538	0.569	0.601	0.635	0.672	0.713	0.761	0.821	0.964
0.73	0.186	0.212	0.238	0.264	0.290	0.316	0.343	0.369	0.396	0.424	0.452	0.480	0.510	0.541	0.573	0.607	0.644	0.685	0.733	0.793	0.936
0.74	0.159	0.185	0.211	0.237	0.263	0.289	0.316	0.342	0.369	0.397	0.425	0.453	0.483	0.514	0.546	0.580	0.617	0.658	0.706	0.766	0.909
0.75	0.132	0.158	0.184	0.210	0.236	0.262	0.289	0.315	0.342	0.370	0.398	0.426	0.456	0.487	0.519	0.553	0.590	0.631	0.679	0.739	0.882
→ 0.76	0.105	0.131	0.157	0.183	0.209	0.235	0.262	0.288	0.315	0.343	0.371	0.399	0.429	0.460	0.492	0.526	0.563	0.604	0.652	0.712	0.855
0.77	0.079	0.105	0.131	0.157	0.183	0.209	0.236	0.262	0.289	0.317	0.345	0.373	0.403	0.434	0.466	0.500	0.537	0.578	0.626	0.685	0.829
0.78	0.052	0.078	0.104	0.130	0.156	0.182	0.209	0.235	0.262	0.290	0.318	0.346	0.376	0.407	0.439	0.473	0.510	0.551	0.599	0.659	0.802
0.79	0.026	0.052	0.078	0.104	0.130	0.156	0.183	0.209	0.236	0.264	0.292	0.320	0.350	0.381	0.413	0.447	0.484	0.525	0.573	0.633	0.776
0.80	0.000	0.026	0.052	0.078	0.104	0.130	0.157	0.183	0.210	0.238	0.266	0.294	0.324	0.355	0.387	0.421	0.458	0.499	0.547	0.609	0.750
0.81		0.000	0.026	0.052	0.078	0.104	0.131	0.157	0.184	0.212	0.240	0.268	0.298	0.329	0.361	0.395	0.432	0.473	0.521	0.581	0.724
0.82			0.000	0.026	0.052	0.078	0.105	0.131	0.158	0.186	0.214	0.242	0.272	0.303	0.335	0.369	0.406	0.447	0.495	0.555	0.698
0.83				0.000	0.026	0.052	0.079	0.105	0.132	0.160	0.188	0.216	0.246	0.277	0.309	0.343	0.380	0.421	0.469	0.529	0.672
0.84					0.000	0.026	0.053	0.079	0.106	0.134	0.162	0.190	0.220	0.251	0.283	0.317	0.354	0.395	0.443	0.503	0.646
0.85						0.000	0.027	0.053	0.080	0.108	0.136	0.164	0.194	0.225	0.257	0.291	0.328	0.369	0.417	0.477	0.620
0.86							0.000	0.026	0.053	0.081	0.109	0.137	0.167	0.198	0.230	0.264	0.301	0.342	0.390	0.450	0.593
0.87								0.000	0.027	0.055	0.083	0.111	0.141	0.172	0.204	0.238	0.275	0.316	0.364	0.424	0.567
0.88									0.000	0.028	0.056	0.084	0.114	0.145	0.177	0.211	0.248	0.289	0.337	0.397	0.540
0.89										0.000	0.028	0.056	0.086	0.117	0.149	0.183	0.220	0.261	0.309	0.369	0.512
0.90											0.000	0.028	0.058	0.089	0.121	0.155	0.192	0.233	0.281	0.341	0.484
0.91												0.000	0.030	0.061	0.093	0.127	0.164	0.205	0.253	0.313	0.456
0.92													0.000	0.031	0.063	0.097	0.134	0.175	0.223	0.283	0.426
0.93														0.000	0.032	0.066	0.103	0.144	0.192	0.252	0.395
0.94															0.000	0.034	0.071	0.112	0.160	0.220	0.363
0.95																0.000	0.037	0.079	0.126	0.186	0.329
0.96																	0.000	0.041	0.089	0.149	0.292
0.97																		0.000	0.048	0.108	0.251
0.98																			0.000	0.060	0.203
0.99																				0.000	0.143
																					0.000

applies to the location of synchronous motors as far as power-factor improvement is concerned.

Economics should be considered when determining the capacitor location. The cost of a switching device, where required, should be included in the cost comparison.

It is common practice to connect capacitors ahead of individual motors as shown in location C_1 in Fig 133. This provides power-factor improvement at the load and permits switching the capacitor and motor as a unit. More details are given in Section 8.9.

Power-factor improvement for small loads, or for those units that for some other reason may not lend themselves to having capacitors directly associated with the load, may be accomplished by connecting capacitors at a distribution point, such as a panel-board or plug-in busway at location C_2, or at a substation at location C_3. Automatic switching, controlled either by electrical sensors or by having interlocks with other equipment, has been successfully used at both locations C_2 and C_3.

A combination of capacitors at locations C_1 and C_2, and with bus-switched units as at location C_3, may prove to be the economical choice.

Fig 133
Possible Shunt Capacitor Locations

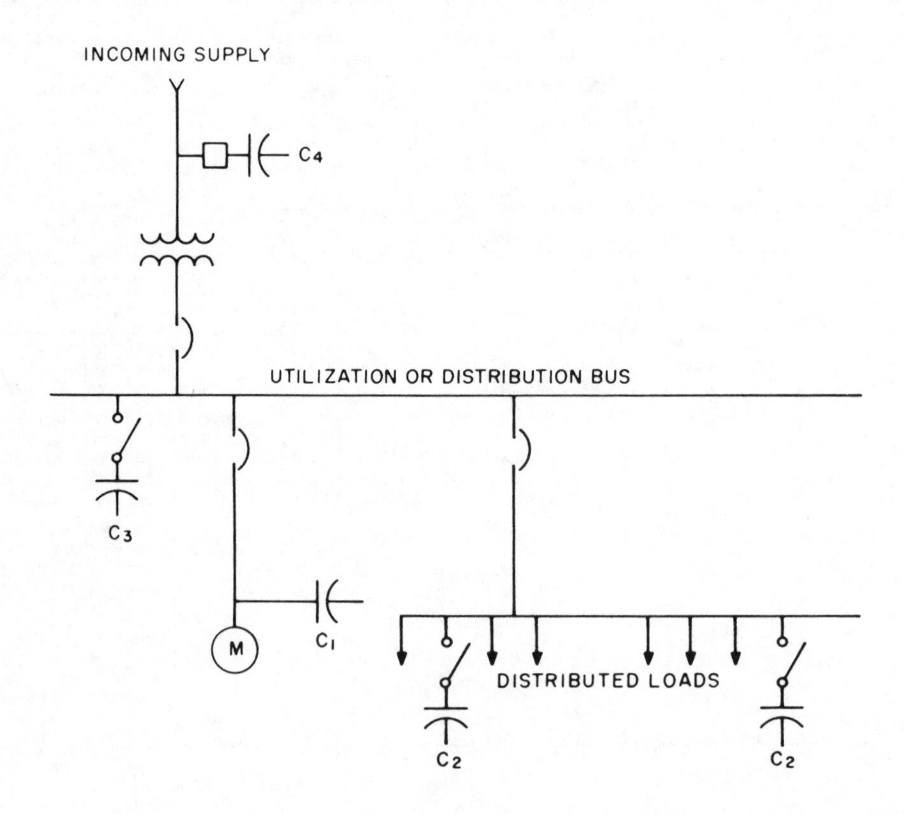

Large plants with extensive primary distribution systems often install capacitors at the primary voltage bus at location C_4 when utility billing encourages the user to improve power factor.

8.6 Release of System Capacity. The expression *release of capacity* means that as the power factor is improved, the current in the existing system will be reduced, permitting additional load to be served by the same system. Equipment such as transformers, cables, and generators may be thermally overloaded. Frequently, the active power rating of the prime mover corresponds to the apparent power rating of the generator. Thus, improvement of the power factor can release both active power and apparent power capacity.

Various expressions for determining the amount of capacity released by power-factor improvement, along with actual examples, curves, charts, and the economics, are covered in [7], [10], and [12]. Figure 134 shows curves to determine the capacity released.

Example. If a plant has a load of 1000 kVA at 70% power factor and 480 kvar of capacitors are added, the system electric capacity released is approximately 28.5%,

Fig 134
Percent Capacity Released and Approximate Combined Load Power Factor with Reactive Compensation

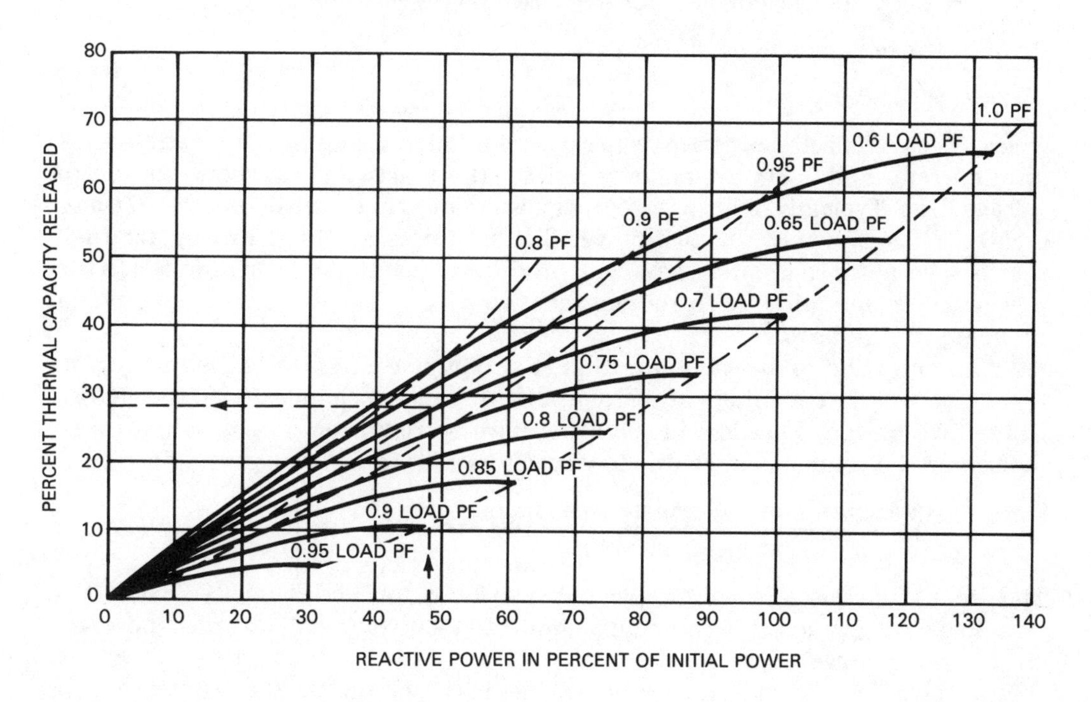

NOTE: ——— Original Load Power Factor ($\cos \phi_1$)
– – – – Final Power Factor ($\cos \phi_2$)

that is, the system can carry 28.5% more load (at 70% power factor) without exceeding the apparent power rating in kVA before the power factor was improved. The final power factor of the original load plus the additional load is approximately 90%.

8.7 Voltage Improvement. Although capacitors raise a circuit's voltage, it is rarely economical to apply them in industrial plants for that reason alone. The voltage improvement may, therefore, be regarded as an additional benefit.

The following approximate expression shows the importance of reducing the reactive power component of current in order to reduce the voltage drop:

$$\Delta V \cong RI \cos \phi \pm XI \sin \phi \qquad \text{(Eq 13)}$$

where ΔV is the voltage change, which may be a drop or rise in voltage. ΔV, R, X and I may be in absolute values of ΔV in volts, R and X in ohms, and I in amperes, or they may be in per-unit values with ΔV in per-unit volts. (Refer to Chapter 6 for an explanation of per-unit quantities.) ϕ is the power-factor angle. Plus is used when the circuit power factor is lagging and minus when it is leading.

ΔV is positive (voltage drop) for a circuit having a lagging power factor and usually negative (voltage rise) for the typical industrial circuit having a leading power factor. This may be rewritten as

$$\Delta V \cong R \cdot \text{active power current} \pm X \cdot \text{reactive power current} \qquad \text{(Eq 14)}$$

Perhaps the most useful form of Eq 13 is

$$\Delta V \cong I(R \cos \phi \pm X \sin \phi) \qquad \text{(Eq 15)}$$

where $R \cos \phi$ reflects the active power contribution to voltage drop per amperes of total current, and $X \sin \phi$ similarity reflects the reactive power contribution to voltage drop. Typically $X \sin \phi$ is many times greater than $R \cos \phi$, say 5–10 times greater. Thus, typically, reactive power flow produces a voltage drop magnitude that is several times greater than that produced by actual power flow. Since the power factor acts directly to reduce reactive power flow, it is most effective in reducing voltage drop.

An examination of Eq 14 shows that it is only necessary to know the system reactance and the capacitor rating to predict the voltage change due to the change in reactive power. Equation 14 may therefore be rewritten in a simple form to determine the voltage change due to capacitors at a transformer secondary bus:

$$\%\Delta V = \frac{\text{capacitor kvar} \cdot \text{\% transformer impedance}}{\text{transformer kVA}} \qquad \text{(Eq 16)}$$

The voltage increases when a capacitor is switched on and decreases when it is switched off. A capacitor permanently connected to the bus will provide a permanent boost in voltage.

Example. The percent change in voltage at the bus when the transformer is rated 1000 kVA with 6% impedance and with a capacitor bank rated 300 kvar, using Eq 14, is calculated as

$$\%\Delta V = \frac{300}{1000} \cdot 6$$

$$= 1.8\% \text{ voltage rise}$$

If excessive voltage becomes a problem, it is suggested that the transformer taps should be changed.

The voltage regulation of a system from no load to full load is practically unaffected by the amount of capacitors, unless the capacitors are switched; however, the addition of capacitors can raise the voltage level. The voltage rise due to capacitors in most industrial plants with modern power distribution systems and a single transformation is rarely more than a few percent.

8.8 Power System Losses. Although the financial return from conductor loss reduction alone is seldom sufficient to justify the installation of capacitors, it is an attractive additional benefit, especially in old plants with long feeders or in irrigation or other field pumping operations.

System conductor losses are proportional to current squared, and since current is reduced in direct proportion to power-factor improvement, the losses are inversely proportional to the square of the power factor:

$$\% \text{ power loss } \alpha\ 100 \left(\frac{\text{original pf}}{\text{improved pf}}\right)^2 \qquad \text{(Eq 17)}$$

$$\% \text{ loss reduction} = 100 \left[1 - \left(\frac{\text{original pf}}{\text{improved pf}}\right)^2\right] \qquad \text{(Eq 18)}$$

8.9 Selection of Capacitors with Induction Motors. Even where economics may not favor the individual motor-capacitor method because of a diversity among motors in operation and the higher unit cost of capacitors in small ratings, this method is gaining in popularity because of its operational advantages. It attains good power factors from the beginning of operations, without the need for power-factor surveys; it puts the right amount of capacitors at the correct location as production equipment is added, taken away, or moved about the plant; it unloads distribution facilities; and it assures that the capacitors are always on the line when (and only when) the motor is energized and therefore when the power-factor improvement is needed.

8.9.1 Effectiveness of Capacitors. The power factor of a squirrel-cage motor at full load is usually between 80 and 90%, depending upon the motor speed and type of motor. At light loads, however, the power factor drops rapidly as illustrated in Fig 135. Generally, induction motors do not operate at full load (often the drive is *overmotored*), and consequently they have low operating power factors. Even though the power factor of an induction motor varies materially from no load to full load, the motor reactive power does not change very much. This characteristic makes the squirrel-cage motor a particularly attractive application for capacitors. With a properly selected capacitor, the operating power factor is excellent over the entire load range of the motor, as shown in Fig 135. It is generally in excess of 95% of full load and higher at partial loads.

8.9.2 Limitations of Capacitor-Motor Switching. Capacitors have been applied to induction motors and switched with the motor as a unit with good results, except in a few applications. Experience has shown that when difficulties are encountered, it is usually because too large a capacitor rating has been used or the capacitors were misapplied on reversing applications. The three factors that limit the value of capacitors to be switched with a motor are

(1) excessive inrush current or reclosing
(2) transient torques
(3) overvoltage due to self-excitation

These limitations apply when the capacitor is connected to the load side of the motor starter as shown in Fig 136(a) and (b), and the capacitor and motor are switched as a unit.

One factor frequently not considered is that the capacitors can materially increase the motor time constant, which influences the safe time for reconnecting the motor-capacitor combination to the line. The effect of slow decay of voltage in

Fig 135
Motor Characteristics for Typical Medium-Sized and Medium-Speed Induction Motor

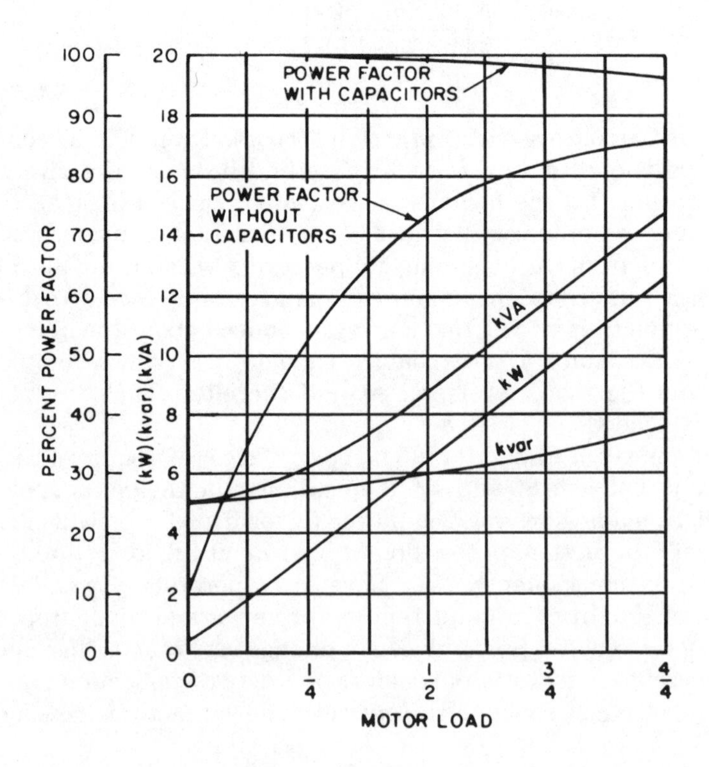

motors due to capacitors may have harmful effects where high-speed reclosing is used. In such cases reclosing should be delayed.

Example. A 700 hp, 4000 V, 900 r/min motor had an open-circuit time constant of 0.675 s. With the power capacitor rating of 155 kvar, the new time constant was 4.45 s. The time for the residual voltage to decay to 25%, a commonly accepted safe value before reconnection to the power source, was about 6 s. Thus, this fact becomes important in high-inertia drives and fast reclosing switching (see [8], [11]).

8.9.3 Selection of Capacitor Ratings. A good general rule to follow in selecting capacitor ratings is that the total kvar rating of capacitors that are connected on the load side of a motor controller should not exceed the value required to raise the no-load power factor of the motor to unity. [See Fig 136(a) and (b)..]

ANSI/NFPA 70-1987 [5] (National Electrical Code [NEC]) has now properly omitted tables of capacitor values for motors because of the difference in motor designs among manufacturers. A good rule to follow is to measure the motor no-load current in selecting the capacitor rating. This current can generally be measured by a clamp-on tang-type ammeter.

There may be great differences in the capacitor ratings that are used for a given motor rating, depending primarily on the motor speed; the slower the speed, the larger the capacitor rating that can be used. This is well demonstrated in the literature, [7], [8], [10], and [11], for a wide range of motor horsepower and speed

Fig 136
Electrical Location of Capacitors When Used with Induction Motors for Power-Factor Improvement

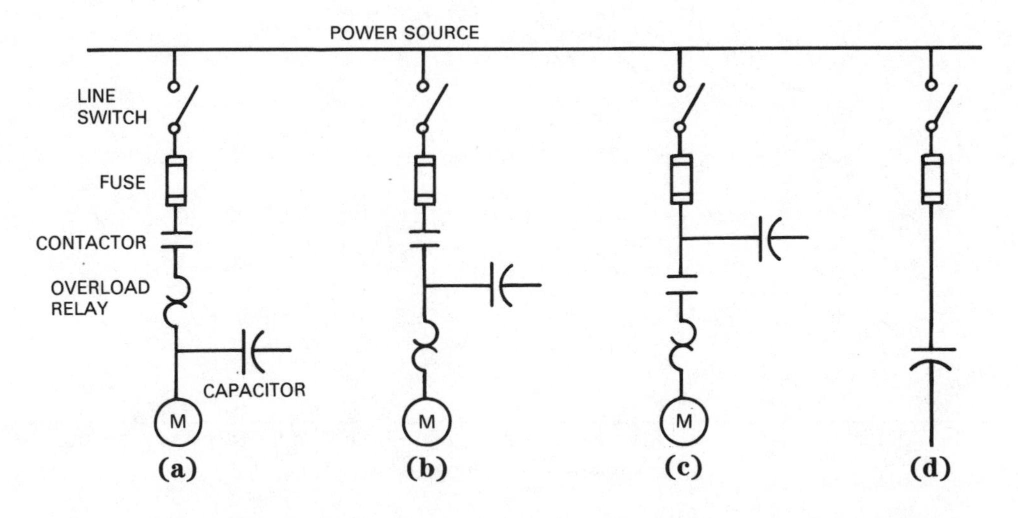

NOTE: (1) Overload relay in (a) should be selected on the basis of reduced current.
(2) Capacitor location (d) is normally an alternate to location (a), (b), or (c).

ratings. There will also be large differences in the recommended capacitor ratings for motors of different manufacturers and design generations, such as

(1) Pre-U-frame, generally before 1955
(2) U-frame, since 1955
(3) T-frame, since 1964
(4) U- and T-frame (high efficiency) since 1978

Figure 137 shows the differences in power factor for U- and T-frame designs.

When the motor capacitor rating is not known or when measurement of the motor no-load current is impractical, Tables 49–52 will serve as guides. In addition, capacitor manufacturers have publications available that provide tables recommending capacitor ratings for motors.

Hermetic motors are built with a minimum of copper and iron and have quite different characteristics from standard motors. The power factor is so poor that it is not unusual for no-load current to be about half the full-load value. Therefore, capacitor ratings are larger than for standard NEMA design B motors (see ANSI/NEMA MG1-1978 [4]).

Fig 137
Power Factor Versus Motor Horsepower Rating for U-Frame and T-Frame Designs

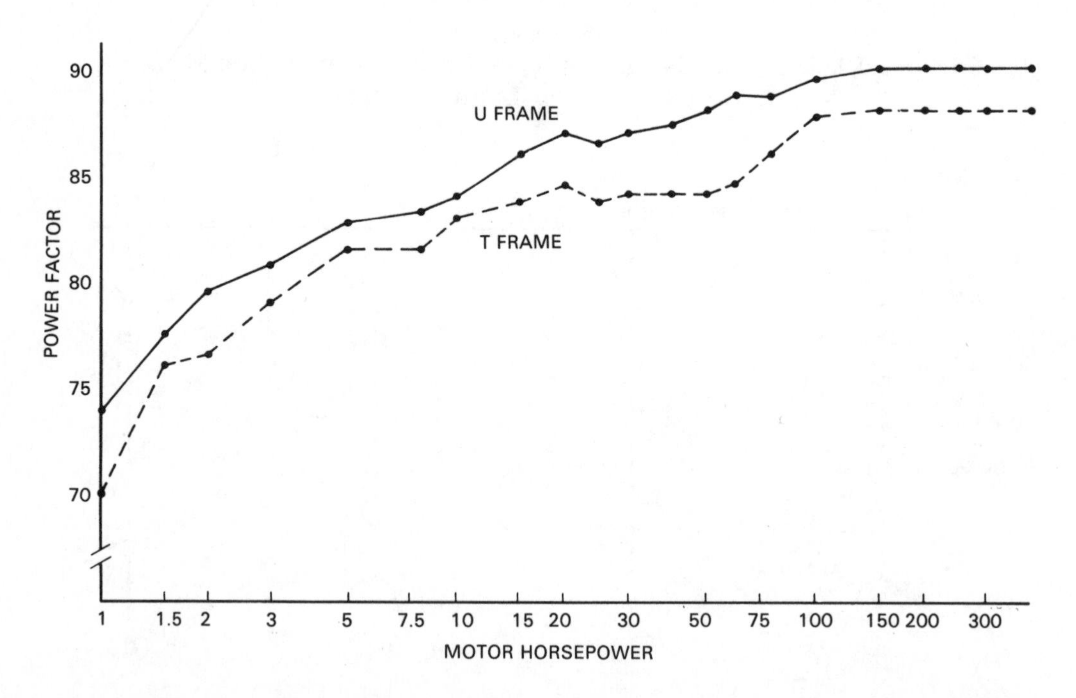

NOTE: Based on compilation of data from six major manufacturers for three-phase NEMA design B 1800 r/min totally enclosed 460 V motors at full load (1964 data) (per ANSI/NEMA MG1-1978 [4]).

Table 49
Suggested Maximum Capacitor Ratings for Pre-U-Frame NEMA Design B 230 V, 460 V, 575 V Squirrel-Cage Motors

	Nominal Motor Speed											
	3600 r/min		1800 r/min		1200 r/min		900 r/min		720 r/min		600 r/min	
Induction Motor Rating (hp)	Capacitor Rating (kvar)	Line Current Reduction (%)	Capacitor Rating (kvar)	Line Current Reduction (%)	Capacitor Rating (kvar)	Line Current Reduction (%)	Capacitor Rating (kvar)	Line Current Reduction (%)	Capacitor Rating (kvar)	Line Current Reduction (%)	Capacitor Rating (kvar)	Line Current Reduction (%)
3	1.5	14	1.5	15	1.5	20	2	27	2.5	35	3.5	41
5	2	12	2	13	2	17	3	25	4	32	4.5	37
7½	2.5	11	2.5	12	3	15	4	22	5.5	30	6	34
10	3	10	3	11	3.5	14	5	21	6.5	27	7.5	31
15	4	9	4	10	5	13	6.5	18	8	23	9.5	27
20	5	9	5	10	6.5	12	7.5	16	9	21	12	25
25	6	9	6	10	7.5	11	9	15	11	20	14	23
30	7	8	7	9	9	11	10	14	12	18	16	22
40	9	8	9	9	11	10	12	13	15	16	20	20
50	12	8	11	9	13	10	15	12	19	15	24	19
60	14	8	14	8	15	10	18	11	22	15	27	19
75	17	8	16	8	18	10	21	10	26	14	32.5	18
100	22	8	21	8	25	9	27	10	32.5	13	40	17
125	27	8	26	8	30	9	32.5	10	40	13	47.5	16
150	32.5	8	30	8	35	9	37.5	10	47.5	12	52.5	15
200	40	8	37.5	8	42.5	9	47.5	10	60	12	65	14
250	50	8	45	7	52.5	8	57.5	9	70	11	77.5	13
300	57.5	8	52.5	7	60	8	65	9	80	11	87.5	12
350	65	8	60	7	67.5	8	75	9	87.5	10	95	11
400	70	8	65	6	75	8	85	9	95	10	105	11
450	75	8	67.5	6	80	8	92.5	9	100	9	110	11
500	77.5	8	72.5	6	82.5	8	97.5	9	107.5	9	115	10

NOTE: Applies to three-phase, 60 Hz motors when switched with capacitors as a single unit (from ANSI/IEEE C37.012-1979 [2]).

Table 50
Suggested Maximum Capacitor Ratings for U-Frame NEMA Design B 230 V, 460 V, 575 V Squirrel-Cage Motors

	Nominal Motor Speed											
	3600 r/min		1800 r/min		1200 r/min		900 r/min		720 r/min		600 r/min	
Induction Motor Rating (hp)	Capacitor Rating (kvar)	Line Current Reduction (%)	Capacitor Rating (kvar)	Line Current Reduction (%)	Capacitor Rating (kvar)	Line Current Reduction (%)	Capacitor Rating (kvar)	Line Current Reduction (%)	Capacitor Rating (kvar)	Line Current Reduction (%)	Capacitor Rating (kvar)	Line Current Reduction (%)
2	1	17	1	20	1	23	1	24	—	—	—	—
3	1	11	1	16	1	19	2	24	—	—	—	—
5	1	9	2	15	2	19	2	20	—	—	—	—
7½	1	6	2	13	4	19	4	20	—	—	—	—
10	2	6	2	11	4	16	5	15	5	17	5	21
15	4	6	4	11	4	13	5	15	5	17	5	21
20	4	6	5	11	5	13	5	15	10	17	10	21
25	5	6	5	8	5	9	5	15	10	17	10	18
30	5	6	5	8	5	9	10	15	10	15	10	18
40	5	6	10	8	10	9	10	15	10	15	15	17
50	5	6	10	8	10	9	15	12	15	12	20	17
60	10	6	10	8	10	9	15	12	20	12	25	17
75	15	6	15	8	15	9	20	11	25	12	30	17
100	15	6	20	8	25	9	25	11	40	12	45	17
125	20	6	25	7	30	9	30	11	45	12	45	15
150	25	6	30	7	30	9	40	11	45	12	50	15
200	35	6	40	7	60	9	55	11	55	11	60	13
250	40	5	40	6	60	9	80	11	60	11	100	13
300	45	5	45	6	80	8	80	10	80	10	120	13
350	60	5	70	6	80	8	80	9	—	—	—	—
400	60	5	80	6	80	6	160	—	—	—	—	—
450	70	5	100	6	—	—	—	—	—	—	—	—
500	70	5	—	—	—	—	—	—	—	—	—	—

NOTE: Applies to three-phase, 60 Hz motors when switched with capacitors at a single unit.

Table 51
Suggested Maximum Capacitor Ratings for T-Frame NEMA Design B 230 V, 460 V, 575 V Squirrel-Cage Motors

	Nominal Motor Speed											
	3600 r/min		1800 r/min		1200 r/min		900 r/min		720 r/min		600 r/min	
Induction Motor Rating (hp)	Capacitor Rating (kvar)	Line Current Reduction (%)	Capacitor Rating (kvar)	Line Current Reduction (%)	Capacitor Rating (kvar)	Line Current Reduction (%)	Capacitor Rating (kvar)	Line Current Reduction (%)	Capacitor Rating (kvar)	Line Current Reduction (%)	Capacitor Rating (kvar)	Line Current Reduction (%)
3	1.5	14	1.5	23	2.5	28	3	38	3	40	4	40
5	2	14	2.5	22	3	26	4	31	4	40	5	40
7½	2.5	14	3	20	4	21	5	28	5	38	6	45
10	4	14	4	18	5	21	6	27	7.5	36	8	38
15	5	12	5	18	6	20	7.5	24	8	32	10	34
20	6	12	6	17	7.5	19	9	23	10	29	12	30
25	7.5	12	7.5	17	8	19	10	23	12	25	18	30
30	8	11	8	16	10	19	14	22	15	24	22.5	30
40	12	12	13	15	16	19	18	21	22.5	24	25	30
50	15	12	18	15	20	19	22.5	21	24	24	30	30
60	18	12	21	14	22.5	17	26	20	30	22	35	28
75	20	12	23	14	25	15	28	17	33	14	40	19
100	22.5	11	30	14	30	12	35	16	40	15	45	17
125	25	10	36	12	35	12	42	14	45	15	50	17
150	30	10	42	12	40	12	52.5	14	52.5	14	60	17
200	35	10	50	11	50	10	65	13	68	13	90	17
250	40	11	60	10	62.5	10	82	13	87.5	13	100	17
300	45	11	68	10	75	12	100	14	100	13	120	17
350	50	12	75	8	90	12	120	13	120	13	135	15
400	75	10	80	8	100	12	130	13	140	13	150	15
450	80	8	90	8	120	10	140	12	160	14	160	15
500	100	8	120	9	150	12	160	12	180	13	180	15

NOTE: Applies to three-phase, 60 Hz motors when switched with capacitors as a single unit.

Table 52
Suggested Capacitor Ratings, in Kilovars, for NEMA Design C, D, and Wound-Rotor Motors

Induction Motor Rating (hp)	Design C Motor 1800 and 1200 r/min	Design C Motor 900 r/min	Design D Motor 1200 r/min	Wound-Rotor Motor
15	5	5	5	5.5
20	5	6	6	7
25	6	6	6	7
30	7.5	9	10	11
40	10	12	12	13
50	12	15	15	17.5
60	17.5	18	18	20
75	19	22.5	22.5	25
100	27	27	30	33
125	35	37.5	37.5	40
150	37.5	45	45	50
200	45	60	60	65
250	54	70	70	75
300	65	90	75	85

NOTE: Applies to three-phase, 60 Hz motors when switched with capacitors as a single unit.

8.9.4 Self-Excitation Considerations. As already indicated, the magnetizing requirement of an induction motor can vary significantly with the design. Recently, for example, a new generation of higher efficiency induction motors has been introduced. Generally, these new motor designs operate less saturated than previous U- or T-frame designs, so they require less capacitance to improve the power factor. Using the same value of capacitors on certain ratings of these high-efficiency motors as recommended for a U- or T-frame design can overvoltage the motor significantly. Therefore, traditional capacitor sizing tables do not apply for these new motors.

The two saturation curves shown in Fig 138(a) and (b) describe the vast differences that can exist between motor designs. Note that for the extreme case and rating plotted, the magnetizing power for the higher efficiency motor design is only 6 kvar versus 14.4 kvar for the standard design. Thoughtful examination of the curves reveals what would happen to the motor (and capacitor) terminal voltage if switched together and, after steady-state operation with no load on the motor is established, the switching device opens. The motor is running at nearly synchronous speed and, therefore, prior to slowing down as a result of friction and windage losses, operates as a generator producing power at (nearly) the frequency of the system, that is, 60 Hz. The intersection of the (60 Hz) motor magnetizing curve and the straight line representing the capacitor (60 Hz) current/voltage characteristic then determines the approximate terminal voltage after switch opening.

The motor/capacitor network, with stored electrical and mechanical energy, will circulate a current between the motor and the capacitor that corresponds to their terminal voltage. In this manner such a network is said to *self-excite*.

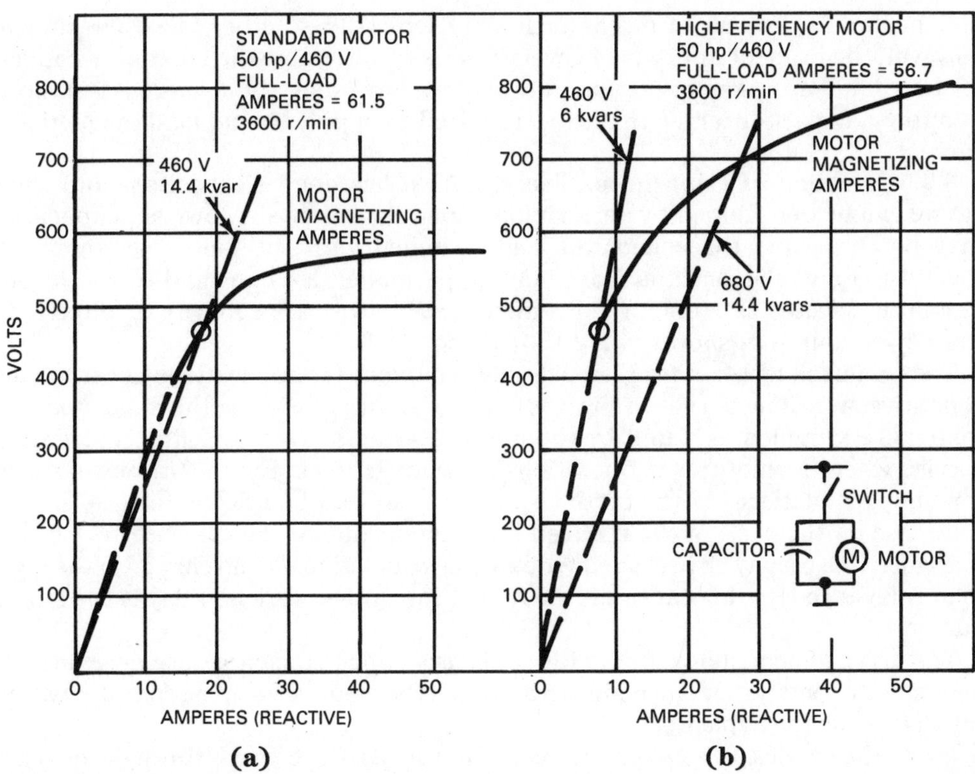

NOTE: kvar values indicated are for a specific rating from a single manufacturer; this represents an extreme case.

Fig 138
Typical Motor Saturation Characteristic for Standard and High-Efficiency Motors

With a properly sized capacitor providing just the necessary magnetizing power for either motor [Fig 138(a) or (b)] the *self-excitation* terminal voltage for such a switching condition is 460 V, as we would expect. If, on the other hand, we had applied the same size capacitor for the motor of Fig 138(b) as was required for the motor of Fig 138(a) (14.4 kvar), the resulting terminal voltage after switching would have been 680 V, clearly excessive.

Although less severe due to the *flatter* magnetization characteristic, a similar overvoltage situation would have occurred for the (standard) motor in Fig 138(a), had a more lenient criterion been used for sizing the capacitor for this machine, as is frequently done. Also, a comparison of the magnetization characteristics for older pre-U- and U-frame motors with the saturation curve for T-frame motors would, in general, yield discovery of a similar, although less severe, performance relationship.

In actual practice the self-excitation overvoltage problem is not as serious as suggested, due to losses in the electric system and the sudden slowing of the motor that occurs as a result of mechanical shaft load. Unless other facts are known, however, the prudent approach in every case is to determine the maximum capacitor kvar limits from the motor manufacturer in any application where it is desired to apply a capacitor larger than that required to supply the no-load magnetizing current.

8.9.5 Location of Capacitors. The optimum location of capacitors and their rating can be determined by performing a load flow analysis; however, capacitors may be connected to each motor and switched with the motor, as shown in Fig 136(a) and (b), or capacitors may be permanently connected to the feeder circuit at selected starters for convenience, as shown in Fig 136(c), or directly to the power source as shown in Fig 136(d).

The preferred location from an overall standpoint for applications not involving repetitive switching is that of Fig 136(a) or (b). In either case the capacitor and motor are switched as a unit by the motor starter, so the capacitor is always in service when the motor is in operation. The connection in Fig 136(a) may be used for new installations, as the motor overload relay can be selected at the time of purchase on the basis of the reduced line current due to the capacitors.

Figure 136(b) may be preferred for existing installations, as no change in the overload relay is required because the current through the overload relay is the motor current.

The arrangement shown in Fig 136(c) is used when capacitors are permanently connected to the system. Its main advantage is the elimination of a separate switching device for the capacitors.

When the capacitor is connected as in Fig 136(a), the current through the overload relay is less than the motor current alone. The percent line current reduction may range from 10 to 25%.

With less margin in present motor designs and with motors usually applied closer to the actual load requirements, there is less margin for overloading. Therefore the motor overload relay should be selected or changed for the lower motor current with capacitors.

The percent line current reduction may be approximated from the following expression:

$$\% \Delta I = 100\left(1 - \frac{\cos \phi_1}{\cos \phi_2}\right) \quad \text{(Eq 19)}$$

where

$\% \Delta I$ = percent line current reduction

$\cos \phi_1$ = power factor before installation of capacitor

$\cos \phi_2$ = power factor after installation of capacitor

8.9.6 Selection of Capacitors for Motors. The following points should be considered when selecting motors and accompanying capacitors:

(1) Select motors that have long hours of use, so that each capacitor has a high-duty factor and is likely to be on the line when needed.

(2) Select large motors first.

(3) Limit capacitor ratings to the values recommended in Section 8.9.3 and by the motor manufacturer.

(4) Power-factor improvement for a multiplicity of small loads or general load compensation may be accomplished by locating capacitors at a distribution point, such as a panelboard, plug-in busway, substation, or on a feeder, in a manner similar to that shown in Fig 136(d). These capacitors are usually permanently connected through a proper disconnecting means.

The method of allocating capacitors to various sizes of motors is

(1) Determine the total capacitance needed

(2) List the motors in the portion of the plant involved

(3) Assign capacitors to motors in descending order of horsepower rating until the total required amount of capacitance has been accumulated

Since it is not generally practical or economical to use capacitors with all motors, additional plant capacitor requirements will have to be supplied by capacitors at other locations.

A few such studies for each type of plant operation and desired power factor will yield a figure of the minimum size motor to which capacitors should be economically assigned.

8.9.7 Motor-Capacitor Applications to Avoid

(1) Motors that are subject to reversing or plugging [6]

(2) Motors that are restarted while still running and generating substantial back voltage

(3) Capacitors that are used with crane or elevator motors where the load may drive the motor; or on multispeed motors

(4) Open-transition reduced-voltage starters that are used with wye-delta connections and capacitors. Capacitors should be connected on the line side of contactors involved in any open-circuit transition for voltage change or speed.

It is possible to apply capacitors in these cases; however, technical investigations are required. It is suggested that the capacitors be switched separately or assigned with other loads under these applications.

8.9.8 Induction Versus Synchronous Motors. Induction motors with capacitors are often more economical than synchronous motors alone. Sometimes the type of drive will dictate the use of one type of motor, but where a free choice is available, an economic comparison should be made. The capacitor-induction-motor combination generally has the advantage of lower maintenance.

Generally, synchronous motors used for system power-factor improvement are of the 0.8 leading power-factor type, because the incremental cost of the reactive power produced is low. Synchronous motors are especially attractive for slow-speed drives.

The following expression may be used as a guide to obtain the capacitor rating required to deliver the same net reactive power as an 0.8 power-factor synchronous motor of the same hp rating at full load:

$$\text{capacitor rating kvar} \cong 1.12 \cdot \text{induction-motor hp rating} \qquad \text{(Eq 20)}$$

For induction-motor applications of up to 500 hp, it will be found necessary to add approximately 1.1–1.2 kvar of capacitors per hp to make the combination comparable to an 0.8 power-factor synchronous-motor application.

From the previous discussions relating to the selection of the proper capacitor rating, it is evident that not all the capacitors can be switched with the motor. Therefore, in making an economic comparison, a separate switching device should be included in the cost comparisons.

The equipment comparisons for initial costs should include:

synchronous motor
+ starter
+ exciter
+ efficiency (operating cost analysis)

versus

induction motor
+ starter
+ capacitors
+ separate capacitor switching devices, when needed
+ efficiency (operating cost analysis)

These economic comparisons may be summarized as follows:

(1) *Low-Voltage Systems. (1000 V or Less)*. The induction-motor method is more economical up to about 200 hp with full-voltage starters and to about 350 hp with reduced-voltage starters.

(2) *Medium-Voltage Systems*. The synchronous motor usually is more economical than the induction-motor equipment over the entire hp and speed ranges if a power circuit breaker or a contactor of adequate interrupting capacity is used to switch the excess capacitors.

8.10 Automatic Control Equipment. Automatic switching of capacitors is becoming more commonly used in industrial plants, but when used, it is usually required for one or more of these reasons:

(1) To control circuit loading

(2) To reduce voltage during light-load conditions and to improve voltage regulation under all load conditions

(3) To meet the requirements of a rate clause that sets maximum or minimum reactive power demand; or to comply with other utility requirements

The most common control types are

Current control	Single step (ON or OFF)
Voltage control	Generally single step (one or more capacitor blocks)
Reactive power control	Generally multistep (usually a series of capacitor blocks)
Time control	Time clock

These various control systems, the basis of selection, the characteristics of associated master elements, and the construction and use of load diagrams to select the bandwidth and amount of capacitors to be switched per step are described in References [7] and [10].

More automatic switching may be required in plants having a large portion of load consisting of thyristor motor drives because of the drive characteristics of reactive power versus active power.

8.11 Capacitor Standards and Operating Characteristics

8.11.1 Capacitor Ratings. Early industry standards[26] list ratings for shunt capacitor units from 240-26 600 V. Although only a limited number of ratings for low-voltage service (240-600 V) are listed, several manufacturers have additional ratings of 1-15 kvar to cover applications with motors. Larger units are available in ratings up to 600 kvar, 13 200 V, three-phase units.

8.11.2 Maximum Voltage. Where capacitors are to be operated above 100% of rated voltage, including harmonics, refer the application to the capacitor manufacturer.

8.11.3 Temperature. Capacitors are more sensitive to temperature limits than are many other devices. Therefore, manufacturers' precautions on installation should be followed. Capacitors should not be placed in hot locations near furnaces or resistors, exposed to sunshine in hot climates, or placed where air cannot circulate, unless special provision is made for cooling or for operating the capacitors below nameplate voltage. Neglect of these points will shorten capacitor life. (See ANSI/IEEE Std 18-1980 [3].)

8.11.4 Time to Discharge. The NEC requires capacitors to be discharged to a residual voltage of 50 V or less in 1 min for capacitors rated 600 V or less, and requires discharge to 50 V or less in 5 min for those rated above 600 V. This is usually accomplished with built-in discharge resistors. However, they are not required when capacitors are connected without disconnecting means directly to other discharge paths such as motors or transformers.

8.11.5 Effect of Harmonics on Capacitors. Capacitors have a substantial margin for harmonic currents and voltages. ANSI/IEEE Std 18-1980 [3] requires capacitors to carry 135% of rating in kvar, including that of the fundamental and harmonics.

If the voltage is approximately normal, it is unlikely that a capacitor would be overloaded by harmonics, although it can happen, as illustrated in Fig 139.

For specific effects of harmonics on capacitors, consult the capacitor manufacturer.

8.11.6 Operating Characteristics. The following relationships apply when capacitors are operated at other than their design-rated operating conditions:

(1) The reactive power varies approximately as the square of the applied voltage

(2) The reactive power varies approximately as the frequency

8.11.7 Overcurrent Protection. The NEC [5], Article 460, requires overcurrent protection for capacitors under 600 V, and Section 460-25 requires overcurrent protection for capacitors over 600 V nominal. An exception is provided for capacitors under 600 V that are protected by the motor overcurrent device. Fuses are

[26] NEMA CP1-1976, Standard for Shunt Capacitors, was rescinded in November 1982. Copies, *for information only*, can be obtained from NEMA, 2101 L Street, NW, Washington, DC 20037.

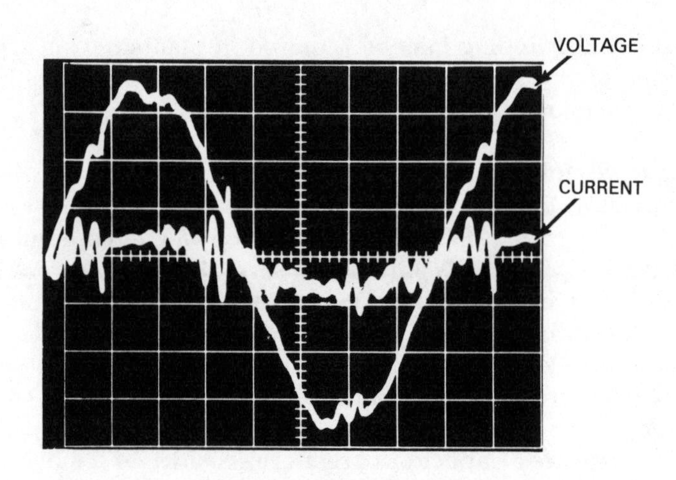

Fig 139
Oscillogram of 480 V Line Voltage and Capacitor Current Near a Thyristor-Controlled Furnace, 150 ft Away from a 1000 kVA Transformer, Capacitor Overheated

generally used and may be applied to each individual capacitor unit or to small groups of units in parallel. The blowing of a fuse of properly selected rating may provide an indication of overcurrent operating conditions as well as an actual fault in the capacitor.

A fuse for a capacitor is not for overload protection in the same sense as it is used for other electric apparatus, such as a motor. The current ratings of capacitor fuses range from 165 to 250% of the capacitor current rating to allow for inrush current; therefore, their overload protection is quite limited.

The type and rating of the fuse should be carefully selected for the proper time-current characteristics and voltage; only the fuse rating recommended by the capacitor manufacturer should be used. The fuse time-current characteristic is the most important factor, as it is a measure of fuse performance, whereas the current rating may be only a nominal value established for the particular type of fuse.

8.11.8 Low-Voltage Switching Devices. There is rarely any problem encountered in the interruption, closing, or repetitive operation of low-voltage air circuit breakers, molded-case circuit breakers, contactors, or switches associated with capacitor equipment for industrial service.

The manufacturer of the capacitor switching device should be consulted concerning the device capacitance current switching capability.

The NEC [5], Article 460, requires switching devices to be selected for at least 135% of the continuous-current rating of the capacitor and to have the proper interrupting rating for the system short-circuit capacity.

Table 53 is a convenient reference in selecting the various switching devices for low-voltage systems.

8.11.9 Medium-Voltage Switching Devices. The phenomenon of reconduction of the arc in capacitor switching devices sometimes creates a problem on high-

Table 53
Capacitor Rating Multipliers to Obtain Switching-Device* Rating

Type of Switching Device	Multiplier to Obtain Equivalent Capacitor Rating	Equivalent Current per kvar 240 V	480 V	600 V
Magnetic-type power circuit breaker	1.35	3.25	1.62	1.30
Molded-case circuit breakers				
magnetic type	1.35	3.25	1.62	1.30
others	≅ 1.5	≅ 3.61	≅ 1.8	≅ 1.44
Contactors, enclosed†	1.5	3.61	1.8	1.44
Safety switch	1.35	3.25	1.62	1.30
Safety switch (fusible)	1.65	3.98	1.98	1.58

*Switching device must have a continuous-current rating that is equal to or exceeds the current associated with the capacitor kvar rating times the indicated multiplier. Enclosed switch ratings at 40 °C (104 °F) ambient temperature.

†If manufacturers give specific ratings for capacitors, these should be followed.

voltage transmission lines or where large blocks of capacitors are switched, such as ratings of 10 000-15 000 kvar on 13.8 kV and higher voltage systems. Few industrial plants have capacitors of such large blocks, and few plants have many capacitors located on the primary distribution systems. The capacitors are usually located on the lower voltage utilization circuits.

The repetitive duty of switching devices can be a problem where capacitor banks must be switched so often that contact or mechanism maintenance is frequently required. For these applications, vacuum switch devices are recommended.

Vacuum circuit breakers are derated considerably when used in capacitor switching applications (see ANSI/IEEE C37.012-1979 [2]).

8.11.10 Selection of Cable Sizes. In selecting cables, allowance should be made for the same 1.35 multiplier and any additional allowance or derating for temperature.

8.11.11 Inspection of Capacitors. Capacitors should be inspected at intervals determined by experience. Ventilation, ambient temperature, buses, and applied voltage should be checked as well as capacitor appearance. Refer to manufacturers' manuals for detailed information and guidance.

8.12 Transients. The industry has not had much difficulty with transients, considering the extensive number of capacitor installations in service. One reason for this is that the capacitor bank acts as a sink for absorbing any voltage transients that might come along the power system. The capacitor bank rating for power-factor improvement will be so much larger than any of the capacitance in the circuit, cable, transformer, etc, that in the analysis of the circuit the capacitor looks like an infinite bus by comparison. This fact greatly simplifies circuit analysis.

8.12.1 Medium-Voltage Switching. With the increased use of vacuum switchgear and vacuum load-break switches, the phenomena of circuit reignition becomes important when capacitor banks are applied. The capacitor bank changes the natural frequency of the circuit, and switching near the bank will produce very high-frequency reignition transient voltages across the contacts of a switch that is opening. Consider the circuit of Fig 140. The source is separated from the switch that is opening by the inductance of the source line, L_1 and L_2. The distributed capacitance C_1, and $L_1 + L_2 + L_3$, determine the natural frequency of the circuit. When capacitors are applied to the circuit, the natural frequency of the circuit, as seen by the switch, is determined by C_2 and L_3. C_2 is large compared to C_1 and L_3 is very small compared to $L_1 + L_2$. The combination of these two values (C_2, L_3) produces a very high frequency when compared to the small capacitance of the cable C_1 and the large inductance of the system, $L_1 + L_2$. On the load side of the switch, the small capacitance of the cable C_3 and the load is connected to the switch by the small inductance of the cable L_4. As the switch starts to open, the load side oscillates at the natural frequency determined by L_4 and C_3. The line side oscillates at its natural frequency determined by C_1 and $L_1 + L_2 + L_3$. When the power factor improvement capacitors C_2 are applied to the system, a new natural frequency is present, determined by C_2 and L_3. Because L_3 is very small compared to $L_1 + L_2$, the natural frequency is very high. Once the current is interrupted by the opening of the switch, the two circuits oscillate at their natural frequency, and if the voltage difference between them is high enough to reestablish the flowing of the current, it will flow at the frequency governed by C_2 and L_3. At this frequency the current zero will occur very rapidly, and again a voltage build-up across the contacts will be very rapid with the possibility of reestablishing the current.

This particular problem is associated mainly with vacuum switches and vacuum circuit breakers, where a low voltage of current can be interrupted before a natural current zero. This can happen before the contacts have parted an appreciable distance. Therefore, as the voltage gradient across the contacts builds up (at this high frequency), it is possible to reestablish the flow of current. Vacuum switches and circuit breakers are designed to take these reignitions and most have surge suppression built into the equipment to control overvoltages.

Fig 140
Circuit for Switching with Shunt Capacitor Bank

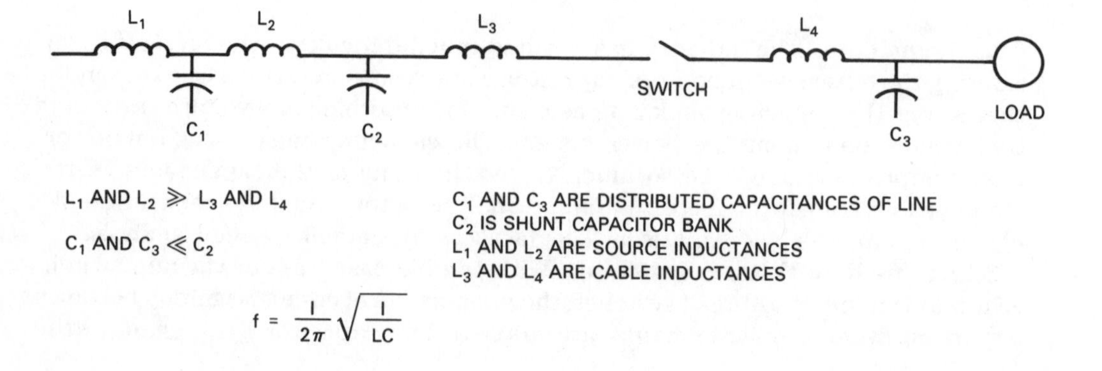

8.12.2 Static Power Converters. On systems where there are static power converters, capacitors for power-factor improvement tend to act as an infinite source and maintain the voltage during the commutation of current between the different phases by the static power converter. These commutation switching transients are damped or completely eliminated as shown by the oscilloscope pictures of Figure 141. Without capacitors, the transient voltage dip of approximately 50% of the crest value of the voltage wave is present.

Fig 141
Illustration of the Reduction of Rectifier Commutation Switching Transients by the Application of Power Capacitors Before the Addition of Capacitors
(a) Line-Voltage Distortions Caused by Chopped-Wave-Loads (Notches Varied with Loading, but *dV/dt* was Hundreds of Volts per Microsecond)
(b) Improvement with Capacitors Installed at Loads

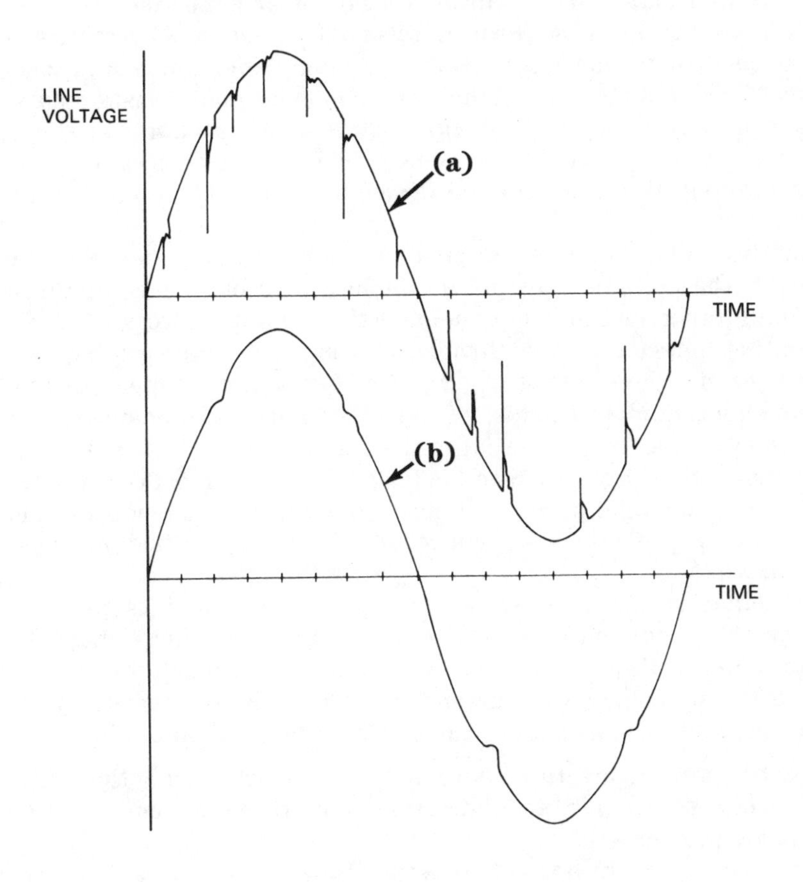

8.13 Resonances and Harmonics. Resonance is a special circuit condition in which the inductive reactance is equal to the capacitive reactance. Any circuit has a resonant condition at a frequency. The frequency at which the circuit is in resonanced is called the natural frequency of the circuit. In this case, we are talking about reactive elements between the point in the circuit that is used as reference and the source at the infinite bus. When there is no intentional capacitance added to the circuit, the natural frequency of most power circuits is in the kilohertz range. Since there is normally no source of currents in this range, the natural frequency of a circuit and the resonance associated with it is normally not a problem.

The addition of capacitors to the power system can either reduce or increase the harmonic and transient voltages. Figure 142 illustrates an application where capacitors accentuated harmonics. This shows the difference in harmonic reinforcement effected by the intervening busway.

8.13.1 Generation of Harmonic Voltages and Currents. A sinusoidal voltage across a nonlinear impedance will result in a nonsinusoidal current through the impedance. A sinusoidal current through a nonlinear impedance will result in a nonsinusoidal voltage across it. The nonsinusoidal voltages and currents associated with transformers operating in saturation are a familiar example of the generation of harmonics by a nonlinear impedance. The nonlinear magnetization curve, especially noticeable in saturation, along with hysteresis in this characteristic, create a nonlinear impedance. This effect is present in all iron-core devices, and the magnitude depends upon the design of the device and the voltage under which it is operated.

A rectifier is another common example of a nonlinear impedance. A sinusoidal voltage across the rectifier does not result in a sinusoidal current through the rectifier. Many other circuit components have varying degrees of nonlinearity inherent in their impedance. Arc furnaces, for example, impose a high harmonic duty on the system. When capacitors for power-factor improvement are applied on a circuit which has nonlinear loads, it is possible to have a resonant condition at a frequency where currents are present in the system. Loads that produce nonsinusoidal currents and voltages are listed in 3.10.3. The loads most commonly used in industrial and commercial power systems today that produce nonsinusoidal currents and voltages are static power converters. An example using this type of load will be discussed.

Although capacitors in themselves do not generate harmonics, the effects of a capacitor on the circuit impedance may cause the harmonic voltages to either decrease or increase. However, since the reactance of a capacitor is inversely proportional to the frequency, the current through a shunt capacitor per volt of impressed voltage is proportional to the order of the harmonic.

8.13.2 Static Power Converter Theory. A brief explanation of the static power converter theory is necessary to understand how these nonsinusoidal currents cause harmonic problems.

If we were to consider the dc load on a static power converter as being relatively constant over a short period of time (seconds or minutes), we could consider the dc current as being constant. A static power converter, Fig 143, commutates or

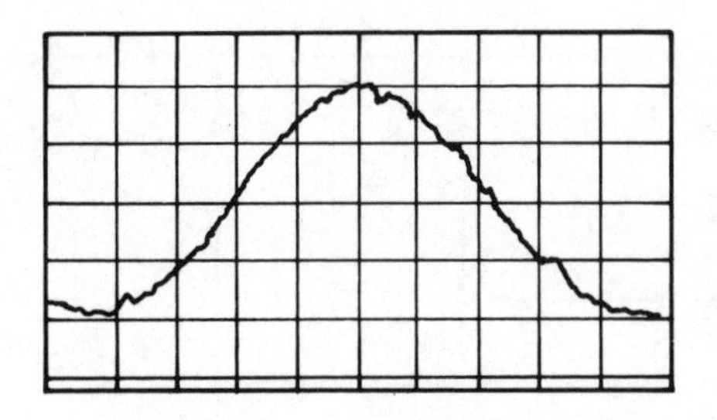
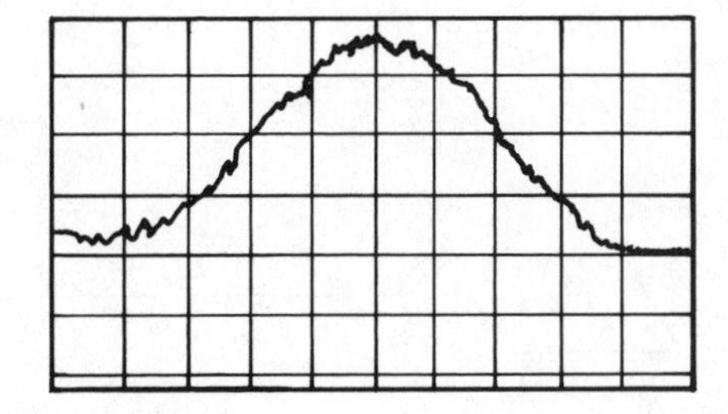

(a) Without Capacitors

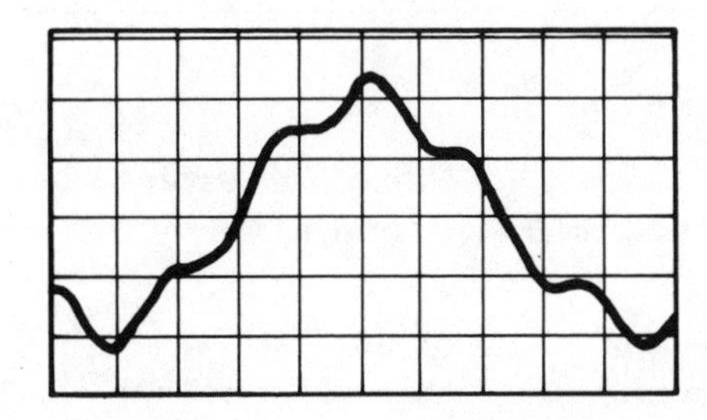
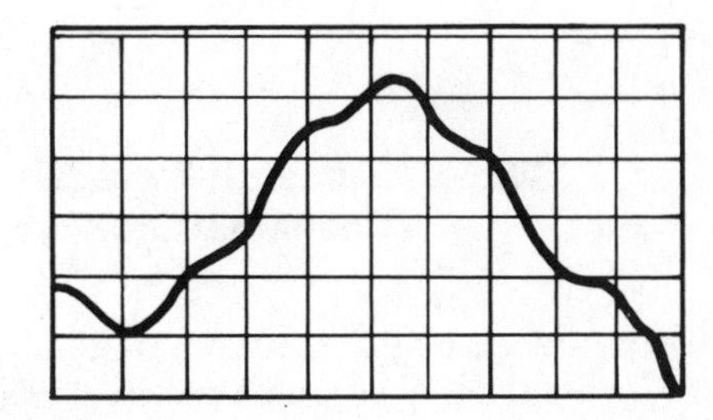

(b) With 350 kvar of Capacitors

NOTE: High-frequency effects have been practically eliminated but 7th harmonic accentuated. Left: oscillograms—at transformer 200 ft from load; right: oscillograms—at load.

Fig 142
Oscillograms Showing Transient Voltage and Harmonics of Line Voltage on 480 V Side of 2000 kVA Transformer Loaded Mostly with Thyristor Drives Having a Wide Range of Control Settings

switches this constant dc current sequentially among the three phases of the ac power system. Ideally, this constant current results in a square wave of current flowing on a three-phase ac power system, which would normally have a sinusoidal waveform. Through Fourier analysis, it can be shown that the ac power system now has to furnish harmonic currents of the order and magnitude of

$$h = pn \pm 1$$
$$I_h = I_1/h \quad \text{(Eq 21)}$$

where

h = order of harmonic
P = an integer, 1, 2 . . . (from the Fourier analysis)
n = number of pulses in rectifier current
I_1 = amplitude of fundamental current
I_h = amplitude of harmonic current

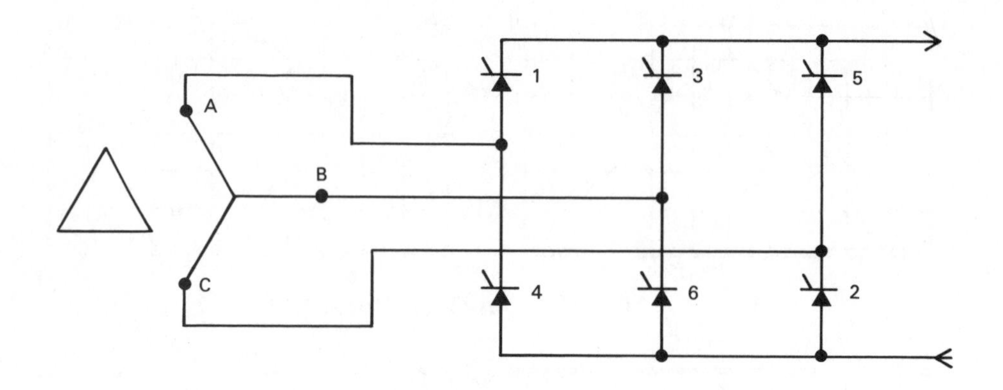

Fig 143
Three-Phase Full-Wave Bridge Circuit Six-Pulse Converter (Commutation Sequence: 1, 2, 3, 4, 5, and 6)

Thus a six-pulse rectifier will generate harmonics of the 5th, 7th, 11th, 13th, 17th, . . . orders. A 12-pulse rectifier will generate harmonics of the 11th, 13th, 23rd, 25th, . . . orders.

Other components that generate harmonics include inverters; overexcited transformers, generators, or motors; salient-pole motors; and solid-state speed controls.

In a symmetrical three-phase system, even-multiple harmonics of the fundamental are absent. The 3rd, 9th, 15th, 21th and 27th are zero-sequence harmonics. They can flow in a three-phase grounded-wye capacitor installation having a ground neutral—one reason why industrial capacitor banks should not be grounded.

The $(pn + 1)$th order of harmonics, for example, the 7th, 13th, 19th, and 25th harmonics, are positive-sequence harmonics. These are present in the balanced-system three-phase line-to-line loads.

The $(pn - 1)$th order of harmonics, for example, the 5th, 11th, 17th, and 23rd harmonics, are negative-sequence harmonics. These will cause additional heating in motors. In rectifier circuits the theoritical maximum magnitude of the harmonic current generated is the reciprocal of the harmonic number. For example, the 5th harmonic current is 20% of the fundamental.

8.13.3 Harmonic Resonance. When shunt capacitors are used on power systems having rectifiers, there is the possibility of resonance between the capacitor and the power system reactance. (It is convenient to consider the rectifier as a harmonic current generator; thus the capacitor and power system are in parallel as seen by the recitifier.) If reactances of the capacitor and the power system are nearly equal at one of the harmonic frequencies generated by the rectifier, the parallel combination approaches resonance and results in a high-impedance to the flow of that harmonic current. As a result, a relatively high harmonic voltage will exist by virtue of the harmonic current flowing through the apparent high impedance.

As resonance is approached, the magnitude of harmonic current in the system and capacitor becomes much larger than the harmonic current generated by the rectifier. The current may be high enough to blow capacitor fuses, one indication of the possibility of resonance.

The following expression may be used to determine the potential of harmonic resonance with a capacitor bank for radial systems:

$$h \cong \sqrt{\frac{\mathrm{kVA}_{SC}}{\mathrm{kvar}_C}} \qquad \text{(Eq 22)}$$

where

h = order of the harmonic
kVA_{SC} = system short-circuit duty
kvar_C = capacitor rating

This is graphically shown in Fig 144.

Eq 22 may also be approximated to

$$h \cong \sqrt{\frac{\mathrm{kVA}_T \cdot 100}{\mathrm{kvar}_C \cdot \%X}} \qquad \text{(Eq 23)}$$

where

kVA_T = transformer rating
$\%X = X_T + X_{\mathrm{syst}}$

with X_T being the transformer reactance in percent and X_{syst} the equivalent system reactance in percent of the transformer rating.

Figure 145 plots Eq 23 and is useful for indicating the critical capacitor ratings to avoid harmonic resonance. The 5th and 7th harmonics are common with six-phase rectifiers, and resonance at these harmonics is to be especially avoided.

In the case of the unit substation transformer used in industry (500–2000 kVA), the utility system reactance $\%X_{\mathrm{syst}}$ in Eq 23 may be assumed as ½ of 1% if the actual system short-circuit current capacity is not known. Thus values of X = 6–7% can be used as an approximation for $\%X$ in Eq 23. In applications involving large power transformers, the actual system short-circuit current capacity should be obtained. An alternate approach is included in [13].

Figure 145 is limited to applications on radial systems. For more complex arrangements, which utilize capacitors in a number of scattered locations, the analytical determination of harmonic distribution is much more difficult and recourse to a computer study of the harmonic current flow may be necessary. Added capacitor locations multiply the combinations by which resonances may occur. It may be difficult to accurately define in the planning stage the necessary harmonic impedanaces upon whch to base a meaningful harmonic analysis. Sometimes a wait-and-see attitude is adopted in such cases. If trouble becomes apparent, such as blown capacitor fuses or cell failures, then corrective steps should be taken.

8.13.4 Application Guidelines. If excessive harmonic currents or voltages are suspected, remedies fall into several general categories.

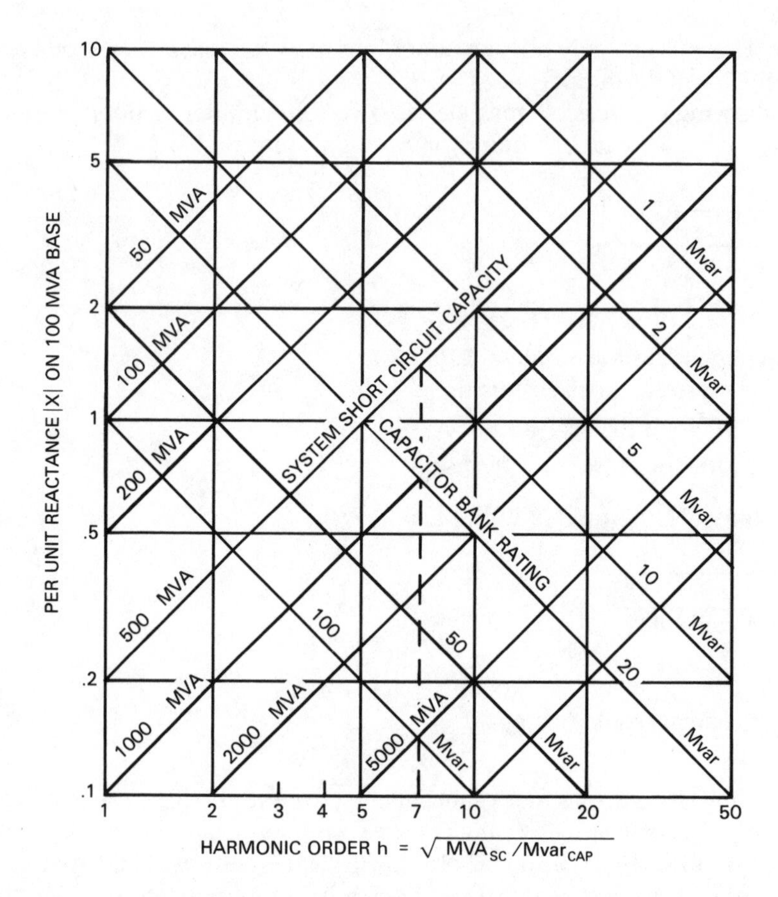

Reactance versus frequency plot of system inductance and power factor capacitors

Example: 500 MVA and 10 Mvar ≅ 7th harmonic order

Fig 144
Order of Resonant Harmonic As a Function of $\sqrt{kVA_{SC}/kvar_C}$, Based on Eq 22

(1) Detuning consists of changing the capacitance or inductance of the circuit so that the circuit natural frequency will not fall near an expected or integral multiple of the fundamental frequency. This usually takes the form of removing or adding capacitor units. If that does not solve the problem, then the addition of tuning reactors in series with the capacitor bank can be used to shift the parallel harmonic resonant point out of the range of predominant orders of harmonic current.

(2) Wye-connected capacitor banks should be ungrounded, eliminating a path for the zero-sequence harmonics that should flow through a grounded neutral. Another important reason for operating capacitors ungrounded is that grounding them could interfere with the performance of the plant ground-relaying system.

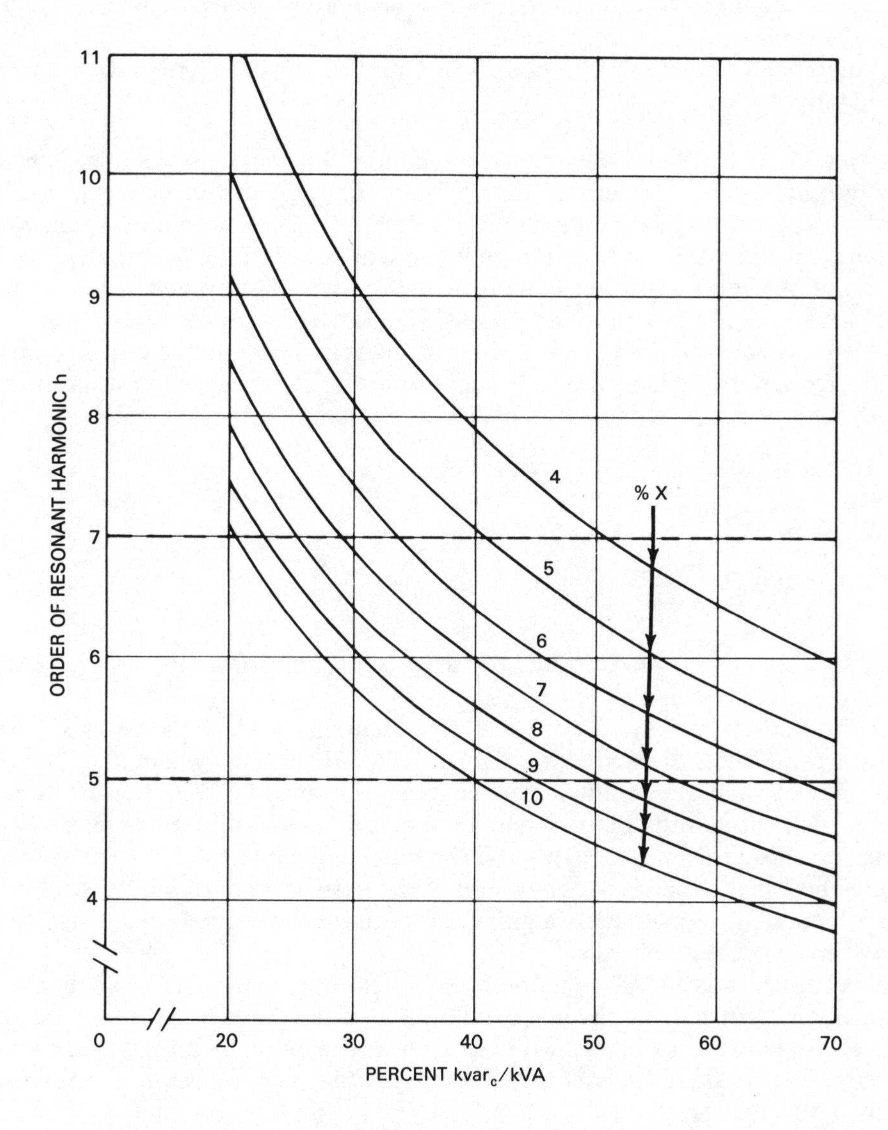

NOTES: (1) Both short-circuit impedance and capacitor size are in percent on transformer kVA base.
(2) $\%X = (\%X_T + \%X_{syst})$ on transformer base.

Fig 145
Order of Resonant Harmonic Versus Capacitor Size for Selected Short-Circuit Impedance

(3) Harmonic input to the system can be reduced by operating at a lower level on the saturation curve of transformers and motors.

(4) Increasing the number of phases of a rectifier or converter will reduce the harmonic input.

8.14 Capacitor Switching. Switching capacitors on an energized system will generally result in a transient voltage. The magnitude of the transient voltage depends upon the available short-circuit current, the amount of capacitance switched, and the point on the voltage wave where switching occurs. Except for switching of very large capacitor banks or the presence of extremely sensitive loads (semiconductor controls, computers, etc), these transients can generally be ignored.

The approximate maximum peak inrush current on energizing a discharged capacitor from normal line voltage, in the absence of other capacitors that may be charged, is, in per unit of capacitor peak rated current:

$$I_{\text{inrush/unit}} = \frac{I_{\text{inrush (peak)}}}{I_{\text{rated (peak)}}}$$

$$\cong 1 + \sqrt{\frac{\text{kVA}_{SC}}{\text{kvar}_C}} \quad \text{(Eq 24)}$$

where kVA_{SC} is the short-circuit duty of the system in kilovolt-amperes and kvar_C is the rating of the capacitor.

Generally, this peak current value is less than the system momentary short-circuit current. Thus the switching device selected for the system short-circuit-current capacity should be adequate. An exception may be a contactor having a limited inrush withstand ability. When a capacitor is energized in parallel with an existing capacitor, the inrush current will be much higher, as the inherent inductive reactance between banks is usually small in the case of bank-to-bank switching. If the inrush current is excessive, a small inductance or resistance can be inserted in series with the capacitor banks.

When a capacitor is energized alone, the capacitor acts initially as a short circuit, causing an initial inrush of current to build up a back voltage in the capacitor. This creates an oscillation at the natural frequency of the circuit until it is damped out by the circuit resistance. The amount of oscillation depends on where in the voltage wave the switch is closed.

The current and voltage transients occurring during capacitor energization vary from operation to operation, depending upon the exact voltage conditions across the contacts for each operation. Difficulties occur only for the most severe voltage conditions, usually for applications over 600 V.

8.15 References. This standard shall be used in conjunction with the following publications:

[1] ANSI C37.06-1979, American National Standard Preferred Ratings and Related Required Capabilities for AC High-Voltage Circuit Breakers Rated on a Symmetrical Current Basis.

[2] ANSI/IEEE C37.012-1979, IEEE Application Guide for Capacitance Current Switching for AC High-Voltage Circuit Breakers Rated on a Symmetrical Current Basis.

[3] ANSI/IEEE Std 18-1980, IEEE Standard for Shunt Power Capacitors.

[4] ANSI/NEMA MG1-1978, Motors and Generators.

[5] ANSI/NFPA 70-1987, National Electrical Code.

[6] BECK, C. D. and RHUDY, R. G. Plugging an Induction Motor. *IEEE Transactions on Industry and General Applications*, vol IGA-6, Jan/Feb 1970, pp 10-18.

[7] BLOOMQUIST, W. C., CRAIG, C. R., PARTINGTON, R. M., and WILSON, R. C. *Capacitors for Industry*. New York: Wiley, 1950. (Out of print. Available in paperback from University Microfilm, 300 North Zeeb Road, Ann Arbor, MI 38106.)

[8] DEMELLO, F. P. and WALSH, G. W. Reclosing Transients in Induction Motors with Terminal Capacitors. *AIEE Transactions (Power Apparatus and Systems)*, pt III, vol 79, Feb 1961, pp 1206-1213.

[9] JACOBS, A. P. and WALSH, G. W. Application Considerations for SCR DC Drives and Associated Power Systems. *IEEE Transactions on Industry and General Applications*, vol IGA-4, Jul/Aug 1968, pp 396-404.

[10] MARBURY, R. E. *Power Capacitors*. New York: McGraw-Hill, 1949.

[11] MOORE, R. C. and SCHWARTZBURG, W. E. Applying Capacitors at Motor Terminals. *Allis-Chalmers Electrical Review*, 4th Quarter 1958, pp 20-22.

[12] STANGLAND, G. The Economic Limit of Capacitor Application for Load Relief. *Power Engineering*, Nov 1950, pp 78-80.

[13] STRATFORD, R. P. Capacitors on AC System Having Large Rectifier Loads. *Industrial Power Systems Magazine*, vol 4, Mar 1961, pp 3-6.

8.16 Bibliography

[B1] ANSI/IEEE Std 519-1981, IEEE Guide for Harmonic Control and Reactive Compensation of Static Power Converters.

[B2] *Electric Utility Engineering Reference Book*, vol 3: *Distribution Systems*. Trafford, PA: Westinghouse Electric Corporation, 1965.

[B3] ERLICKI, M. S., SCHIEBER, D., and BEN URI, J. Power-Measurement Errors in Controlled Rectifier Circuits. *IEEE Transactions on Industry and General Applications*, vol IGA-2, Jul/Aug 1966, pp 309-311.

[B4] GREENWOOD, A. *Electrical Transients in Power Systems*. New York: Wiley, 1971.

[B5] IEEE Committee Report. Bibliography on Switching of Capacative Circuits Exclusive of Series Capacitors. *IEEE Transactions on Power Apparatus and Systems*, vol PAS-89, Jul/Aug 1970, pp 1203-1207.

[B6] LINDHORST, P. K. High-Efficiency Motors—A New Way to Save Energy. *Specifying Engineer*, Oct 1975, pp 94-96.

[B7] MILLER, O. F. Application Guide for Shunt Capacitors on Industrial Distribution Systems of Medium Voltage Levels. *IEEE Transactions on Industry Applications*, vol IA-12, no 5, Sept/Oct 1976.

[B8] NAILON, R. L. Use Capacitors with Large Motors. *Power*, Jul 1971, pp 81-83.

[B9] OSCARSON, G. L. Telephone Influence Factor and What to Do About It. *Power Generation*, Dec 1949, pp 68-71.

[B10] PALKO, E. Cutting Costs with Power Capacitors. *Plant Engineering*, Feb 14, 1972, pp 55-59.

[B11] PALKO, E. Installing Power-Factor Improvement. *Plant Engineering*, Jan 11, 1979, pp 78-83.

[B12] STACY, E. M., SELCHAU-HASEN, P. V., and BOICE, W. K. SCR Drives—General Electric Company, GET-6468A.

[B13] STEEPER, D. E. and STRATFORD, R. P. Reactive Compensation and Harmonic Suppression for Industrial Power Systems Using Thyristor Convertors. *IEEE Transactions on Industry Applications*, vol IA-12, May/Jun 1976, pp 232-254.

[B14] ZIMMERMAN, J. A. *Motor Starting with Capacitors*. IEEE Cement Industry Technical Conference, Toronto, May 1969.

9. Power Switching, Transformation, and Motor-Control Apparatus

9.1 Introduction. This chapter provides information on the requirements for and application of major apparatus utilized in an industrial electric distribution system. More detailed information on this apparatus is available from the American National Standards Institute (ANSI), National Electrical Manufacturers Association (NEMA), testing companies, such as Underwriters' Laboratory (UL) or Factory Mutual (FM), as well as in manufacturers' publications.

The engineer should make basic decisions in his choice of equipment for a particular electric system. He should include all facets of the actual project, such as continuity of service, reliability, safety, protection, coordination, initial installed costs, flexibility and costs for operating and maintenance, security, and procurement time to meet schedules. Energy cost and conservation should be considered in initial plant design and equipment. For information related to energy cost and conservation, refer to Chapter 12 of this standard.

9.1.1 Equipment Installation. Electric equipment should be installed to be safe and yet readily accessible to qualified personnel. Sufficient access and working space should be provided and maintained about all electric apparatus to permit ready and safe operation and maintenance of such equipment. Table 54 lists minimum working clearances in front of electric equipment.

Installations in industrial plants require that adequate aisles, hatchways, wall openings, etc, be provided for easy removal and replacement of all electric equipment. For metal-enclosed switchgear, extreme care should be exercised in setting and aligning leveling channels in order to prevent stressing insulators and bus structures and to provide for easy insertion and removal of circuit breakers. Floor area in front of switchgear should be level or have only a moderate slope away from the equipment for adequate drainage.

Special attention should be given to the floor under and in front of any equipment that is required to roll in and out of the enclosure. More detailed installation procedures are given in NEMA PB2.1-1979 [31][27], especially sections II and IV. If

[27] The numbers in brackets correspond to those of the references listed at the end of this chapter; when preceded by B, they correspond to the bibliography at the end of this chapter.

equipment is to be stored or installed in locations where internal condensation may occur, suitable internal heating should be provided (NEMA PB 2.1-1979 [31], Section III).

ANSI/NFPA 70-1987 [27] (National Electrical Code [NEC]), Article 450, outlines the installation requirements for transformers of all types. In some industrial plants, the unit substation transformer is located outside a pressure-ventilated switchgear room with a secondry throat connection for buswork through the wall of the room, and with a primary connection to an air-filled terminal box with a stress relieving termination for 15 kV and lower voltage cable. This pressure-ventilated switchgear room may also house motor-control centers, panelboards, and other electric equipment in addition to the switchgear. The following installation method is used:

(1) To protect electric equipment from accumulation of dirt, dust, and other foreign material

(2) To enable the use of less expensive and more readily maintainable general-purpose electric equipment enclosures in lieu of costly explosion-proof enclosures in areas classified as hazardous in accordance with Article 500 of the NEC [27]

(3) To prevent access by unauthorized people

9.1.2 Maintenance, Testing, and Safety. The use of sound engineering principles in the design of electric systems dictates consideration of maintenance, testing, and safety as factors always present during the life of the plant.

Table 54
Minimum Clear Working Space in Front of Electric Equipment

Voltage to Ground (volts)	Working Space (feet) for Conditions* (1)	(2)	(3)
0-150	3	3	3
151-600	3	3½	4
601-2500	3	4	5
2501-9000	4	5	6
9001-25 000	5	6	9
25 001-75 kV	6	8	10
Above 75 kV	8	10	12

*Conditions:

(1) Exposed live parts on one side and no live or grounded parts on the other side of the working space, or exposed live parts on both sides effectively guarded by suitable wood or other insulating materials. Insulated wire or insulated bus bars operating at not over 300 V shall not be considered live parts.

(2) Exposed live parts on one side and grounded parts on the other side. Concrete, brick, or tile walls will be considered as grounded surfaces.

(3) Exposed live parts on both sides of the work space [not guarded as provided in Condition (1)], with the operator between.

(*Exception*: Working space is not required in back of equipment such as deadfront switchboards or control assemblies where there are no renewable or adjustable parts, such as fuses or switches, and when all connections are accessible from other locations than the back.)

Based on ANSI/NFPA 70-1987 [27], Tables 110-16(a) and 110-34(a).

An effective preventive maintenance program should sustain production at required levels, protect valuable investment, and reduce down time and maintenance costs. Planned maintenance should include a series of tasks to be performed on each unit of equipment on a regularly scheduled basis in order to detect areas of potential failure and to provide a warning to plan for necessary repairs. Record keeping is an important part of the planned maintenance program. Inherent with any maintenance program is the testing of electric equipment and appropriate safety precautions for personnel and equipment. Chapter 5 discusses methods of testing protective equipment, such as circuit breakers, relays, etc. Testing procedures for other equipment are found in appropriate standards of the Institute of Electrical and Electronics Engineers (IEEE), ANSI, NFPA, and NEMA, as well as in manufacturers' literature.

9.1.3 Heat Losses. The heat generated by power losses in electric equipment, particularly transformers, motors, rectifiers, and motor-control centers should be considered in the initial design of a project. An estimate of the electric losses that show up as heat should be developed and added to the cooling requirements of the building. Table 55 lists the range of losses in electric equipment. Heat losses for specific equipment should be obtained from the manufacturer.

Table 55
Range of Losses in Power System Equipment

Component	Percent Energy Loss* (Full Load)
Outdoor Circuit Breakers (15-230 kV)	0.002 - 0.015
Generators	0.09 - 3.50
Medium-Voltage Switchgear (5 and 15 kV)	0.005 - 0.02
Current-Limiting Reactors (600 V - 15 kV)	0.09 - 0.30
Transformers	0.40 - 1.90
Load Break Switches	0.003 - 0.025
Medium-Voltage Starters	0.02 - 0.15
Busway (480 V and below)	0.05 - 0.50
Low-Voltage Switchgear	0.13 - 0.34
Motor-Control Centers	0.01 - 0.40
Cable	1.00 - 4.00
Motors	
1 - 10 hp	14.00 - 35.00
10 - 200 hp	6.00 - 12.00
200 - 1500 hp	4.00 - 7.00
1500 hp and up	2.30 - 4.50
Rectifiers (large)	3.00 - 9.00
Static Variable Speed Drivers	6.00 - 15.00
Capacitors (watts loss/var)	0.50 - 2.00
Lighting (lumens/watts)	8.00 - 9.00

*Percent energy loss is simply a ratio of power consumed internally in equipment to the total energy passed through it.

From H. N. Hickok, "Electrical Energy Losses in Power Systems," *IEEE Transactions on Industry Applications*, vol IA-14, no 5, Sept/Oct 1978.

Exhausting air from the equipment location and allowing cooler (even if filtered) air to enter may cause internal condensation and possibly create a hazardous atmosphere in the building. Alternate methods include air conditioning, ventilation with dehumidification, higher efficiency equipment, or outdoor location of the major heat-producing items. The amount of cooling air should be kept at the required minimum by use of the thermally-controlled power-actuated ventilation dampers. References are included in 9.7 to aid in evaluating the effects of solar radiation and the losses in power systems.

9.2 Switching Apparatus for Power Circuits. Switching apparatus can be defined as a device for opening and closing, or for changing the connections of a circuit. The general classification of switching apparatus as used in this chapter includes switches, fuses, circuit breakers, and contactors.

9.2.1 Switches. The types of switches normally applied for power circuits include the following:

(1) Disconnecting

(2) Load interrupter

(3) Safety switches for 600 V and lower power applications, including bolted-pressure switches

(4) Transfer switches for emergency power

9.2.1.1 A disconnecting switch is used for isolating a circuit or equipment from the source of power. It has no interrupting rating and is intended to be operated only after the circuit has been opened by other means. Interlocking is generally provided to prevent operation when the switch is carrying current. Latches may be required to prevent the switch from being opened by magnetic forces under heavy fault currents.

9.2.1.2 For services above 600 V an interrupter or load-break switch, generally associated with unit substations supplied from the primary distribution system, is a switch combining the functions of a disconnecting switch and a load interrupter for interrupting, at rated voltage, currents not exceeding the continuous-current rating of the switch. Load-break switches are of the air or fluid-immersed type. The interrupter switch is usually manually operated and has a quick-make quick-break mechanism independent of the speed of handle operation. Such switches usually have a close and latch rating to provide maximum safety in the event of closing in on a faulted circuit.

With a fused load-break switch combination, fast fault clearing and circuit isolation can be provided. This application, if properly coordinated to protect the transformer and to interrupt transformer magnetizing currents and load currents within the switch rating, may be more economical than a circuit breaker. Rollout fuse-switch assemblies, supplied by several manufacturers, have advantages such as the replacement of a fuse without the necessity of de-energizing the primary circuit, ability to fuse on the line side of the switch, and complete access to mechanical parts for checkout and maintenance.

It may be desirable, from a safety standpoint, to interlock the operation of an interrupter switch with the secondary circuit breaker in the case of a transformer application to minimize the chance of operating the interrupter switch at a time when the current exceeds the interrupting rating of the switch. Many interrupter switches, however, have interrupting ratings in excess of the transformer full-load current so that interlocking is not required.

9.2.1.3 For services of 600 V and below, safety switches are commonly used. These are enclosed and may be fused or unfused. This type of switch is operable by means of a handle from outside the enclosure and is so interlocked that the enclosure cannot be opened unless the switch is open or the interlock defeater is operated. Many safety switches have quick-make and quick-break features.

A safety switch for motors is rated in horsepower and voltage and is capable of interrupting the maximum operating current of a motor, that is, the stalled-rotor current of the same horsepower as the switch rating at the rated voltage. The NEC [27] recognizes six times full-load motor current as the stalled-rotor current.

The application of fused switches is limited by the NEC [27] to a constant current rating of at least 115% of the full-load current rating of the motor.

Safety switches with current-limiting fuses can be applied to circuits with up to 200 000 A rms symmetrical fault current if the switch-fuse combination has been suitably tested. Switches are labeled either by UL or the manufacturer to indicate the particular switch-fuse combination that must be used to obtain the specified rating. It should never be assumed that the application of a high-interrupting capacity fuse with a lower rated switch will result in an elevation of the switch rating.

A bolted-pressure switch consists of movable blades and stationary contacts with arcing contacts and a simple toggle mechanism for applying bolted pressure to both the hinge and jaw contacts in a manner similar to a bolted bus joint. The bolted-pressure switch, unlike a conventional fused switch, can be applied to 100% of its rating. The operation mechanism consists of a spring that is compressed by the operating handle and released at the end of the operating stroke to provide quick-make and quick-break switching action.

The electrical-trip bolted-pressure switch is basically the same as the manually operated bolted-pressure switch, except that a stored-energy latch mechanism and a solenoid trip release are added to provide automatic electrical opening and all other features normally required on low-voltage main and feeder circuits rated 600 A and above. These switches also are designed for use with ground-fault protection equipment that may be required by the NEC [27] and have a contact-interrupting rating of 12 times the continuous rating. These switches are available in ratings of 800, 1200, 1600, 2000, 2500, 3000, 4000, 5000 and 6000 A, 480 V ac, and are suitable for use on circuits having available fault currents of 200 000 A rms symmetrical, when applied in combination with current-limiting fuses.

9.2.1.4 Automatic transfer switches of double-throw construction are primarily used for emergency and standby power generation systems rated 600 V and less. These transfer switches do not normally incorporate overcurrent protection and are designed and applied in accordance with the NEC [27], particularly Articles 230, 517, and 700. They are available in ratings from 30-4000 A. For reliability, most automatic transfer switches rated above 100 A are mechanically held and are electrically operated from the power source to which the load is to be transferred.

These switches are applied to provide protection against failure of the normal service. In addition to utility failures, continuity of power to critical loads can also be disrupted by:

(1) An open circuit within the building area on the load side of the incoming service

(2) Overload or fault conditions

(3) Electrical or mechanical failure of the electric power distribution system within the building

Therefore many engineers advocate the use of lower current-rated transfer switches located near the load rather than one large transfer switch at the point of incoming service. For additional information, see ANSI/IEEE Std 446-1980 [24].

9.2.2 Fuses

9.2.2.1 Types and Rating Basis. A fuse is an overcurrent protective device with a circuit-opening fusible part that is heated and severed by the passage of overcurrent through it.

Fuses are available in a wide range of voltage, current, and interrupting ratings, current-limiting types, and for indoor and outdoor applications.

Fuses rated greater than 600 V have an interrupting capability based on asymmetrical current, although their published ratings are expressed in symmetrical amperes. Current-limiting fuses interrupt a short circuit within the first half-cycle, and their equivalent asymmetrical rating includes a 1.6 multiplier to provide for the maximum expected current asymmetry. Fuse ratings for 600 V and below are also published as symmetrical current values.

Current-limiting fuses 600 V and below are extremely fast in operation at very high values of fault current, and act to limit the current in less than one-quarter cycle to a value well below the available peak short-circuit current. Noncurrent-limiting fuses are widely applied above 600 V. They are generally available in higher current ratings, but lower interrupting ratings than current-limiting fuses. Several types of current-limiting fuses for 600 V and below are now available for ac service with interrupting ratings as high as 200 000 A rms symmetrical. For more information on fuses see Chapter 5.

9.2.2.2 Application Considerations. There is no general rule for deciding whether to use fuses or a circuit breaker. The designer should evaluate the total performance in terms of what the particular application demands. The following application considerations may be of assistance to the designer.

(1) *Interrupting Ratings*. Current-limiting fuses with 200 000 A rms symmetrical ratings are available for 600 V and below applications.

(2) *Component Protection*

(a) Current-limiting fuses permit minimal short-circuit current let-throughs, thus minimizing damage to lower interrupting capacity-rated and withstand-rated components.

(b) Current-limiting fuses can cause transient voltages in clearing faults that may be detrimental to the system components, such as motors, surge arresters, etc.

(c) Fuses in conjunction with shunt-trip switches can provide sensitive ground-fault protection.

(3) *Selective Coordination*. Fuse time-current clearing characteristics are both accurate and stable, and coordination is easily accomplished by referring to published manufacturers' fuse ratio charts.

(4) *Space Requirements*. Fusible switching devices take up more space; however, fuses alone are generally smaller than mechanical protective devices if switching is not required.

(5) *Economics*

(a) First cost, life cycle, and maintenance cost are lower for fusible equipment.

(b) The simpler mechanical operation of fusible equipment results in lower maintenance costs as there is no periodic maintenance required by either NEMA or manufacturers' recommendations.

(6) *Automatic Switching*. Fuses alone are not capable of automatic switching, but can be installed in suitable shunt-trip equipped switches to provide this service. Care must be exercised in applying a shunt-tripped switch. If the shunt-trip is actuated by an overcurrent relay, the switch must be capable of interrupting the fault duty to which it may be subjected.

9.2.3 Circuit Breakers. A circuit breaker is a device designed to open and close a circuit by nonautomatic means, and to open the circuit automatically on a predetermined overload of current without injury to itself when properly applied within its rating. Ordinarily, circuit breakers are required to operate only infrequently, although some classes of circuit breakers are suitable for more frequent operation. The interrupting and momentary ratings of a circuit breaker must be equal to or greater than the available system short-circuit currents.

Power circuit breakers are used extensively on utility and industrial (largely over 600 V) and industrial (predominantly 600 V and below) power distribution systems to provide essential switching flexibility and circuit protection.

Circuit breakers are available for the entire voltage range and may be furnished single-, double-, or triple-pole, and arranged for indoor or outdoor use. Circuit breakers for above 34.5 kV service are generally not enclosed or available for indoor location.

9.2.3.1 Circuit Breakers Over 600 V. The rated close and latch and interrupting current capabilities are very important factors for use in the application of circuit breakers over 600 V. The close and latch capability is a measure of the equipment's ability to withstand the mechanical stresses produced by the asymmetrical short-circuit current during the first cycle without mechanical damage, and is normally expressed as total rms current. An asymmetrical current consists of a dc component superimposed on an ac component. The dc component decays with time, depending upon the resistance and reactance, or the X/R of the circuit. The initial value of the dc component of the short-circuit current depends on the point of the normal voltage wave at which the fault occurs. The procedure to be used for short-circuit selection of power circuit breakers in the over 600 V class is covered in Chapter 5. Application data can be found in ANSI/IEEE C37.010-1979 [10].

For the rating of power circuit breakers in the over 600 V class refer to ANSI C37.06-1979 [2]. Circuit breakers currently being manufactured are rated on the

Table 56
Preferred Ratings for Indoor Oilless Circuit Breakers

	Rated Values									Related Required Capabilities			
											Current Values		
	Voltage		Insulation Level		Current		Transient Recovery Voltage				Max Symmetrical Interrupting Capability (8)	3-Second Short-Time Current Carrying Capability (9)	
			Rated Withstand Test Voltage								*K* Times Rated Short-Circuit Current		
Line No	Rated Max Voltage (1)* kV, rms	Rated Voltage Range Factor, *K* (2)	Low Frequency kV, rms	Impulse (3) kV, Crest	Rated Continuous Current at 60 Hz (4) Amperes, rms	Rated Short-Circuit Current (at Rated Max kV) (5) (6) kA, rms	Rated Time to Point *P* T_2 μs	Rated Interrupting Time (7) Cycles	Rated Permissible Tripping Delay *Y* Seconds	Rated Max Voltage Divided by *K* kV, rms	kA, rms	kA, rms	Closing and Latching Capability 1.6 *K* Times Rated Short-Circuit Current (9) (10) kA, rms
	Col 1	Col 2	Col 3	Col 4	Col 5	Col 6	Col 7	Col 8	Col 9	Col 10	Col 11	Col 12	Col 13
1	4.76	1.36	19	60	1200	8.8	See	5	2	3.5	12	12	19
2	4.76	1.24	19	60	1200	29	Note	5	2	3.85	36	36	58
3	4.76	1.24	19	60	2000	29	(11)	5	2	3.85	36	36	58
4	4.76	1.19	19	60	1200	41		5	2	4.0	49	49	78
5	4.76	1.19	19	60	2000	41		5	2	4.0	49	49	78
6	4.76	1.19	19	60	3000	41		5	2	4.0	49	49	78
7	8.25	1.25	36	95	1200	33		5	2	6.6	41	41	66
8	8.25	1.25	36	95	2000	33		5	2	6.6	41	41	66
9	15.0	1.30	36	95	1200	18		5	2	11.5	23	23	37
10	15.0	1.30	36	95	2000	18		5	2	11.5	23	23	37
11	15.0	1.30	36	95	1200	28		5	2	11.5	36	36	58
12	15.0	1.30	36	95	2000	28		5	2	11.5	36	36	58
13	15.0	1.30	36	95	1200	37		5	2	11.5	48	48	77
14	15.0	1.30	36	95	2000	37		5	2	11.5	48	48	77
15	15.0	1.30	36	95	3000	37		5	2	11.5	48	48	77
16	38.0	1.65	80	150	1200	21		5	2	23.0	35	35	56
17	38.0	1.65	80	150	2000	21		5	2	23.0	35	35	56
18	38.0	1.65	80	150	3000	21		5	2	23.0	35	35	56
19	38.0	1.0	80	150	1200	40		5	2	38.0	40	40	64
20	38.0	1.0	80	150	3000	40		5	2	38.0	40	40	64

*Numbers in parentheses refer to the notes below.

This material is reproduced with permission from American National Standard Preferred Ratings and Related Required Capabilities for AC High-Voltage Circuit Breakers Rated on a Symmetrical Current Basis, ANSI C37.06-1979, copyright 1979 by the American National Standards Institute. Copies of this standard may be purchased from ANSI, 1430 Broadway, New York, NY 10018.

Table 56 *(Continued)*

NOTES:

These ratings were prepared by EEI-AEIC-NEMA Joint Committee on Power Circuit Breakers.

For service conditions, definitions, and interpretation of ratings, tests, and qualifying terms, see ANSI/IEEE C37.04-1979 [8], ANSI/IEEE C37.09-1979 [9], and ANSI C37.100-1981 [14].

The interrupting ratings are for 60-Hz systems. Applications on 25-Hz systems should receive special consideration.

Current values have been rounded off to the nearest kiloampere (kA) except below 10 kA where two significant figures are used.

(1) The voltage rating is based on American National Standard Voltage Ratings for Electric Power Systems and Equipment (60 Hz), ANSI C84.1-1982 [7], where applicable, and is the maximum voltage for which the breaker is designed and the upper limit for operation.

(2) The rated voltage range factor, K, is the ratio of rated maximum voltage to the lower limit of the range of operating voltage in which the required symmetrical and asymmetrical current interrupting capabilities vary in inverse proportion to the operating voltage.

(3) 1.2×50 microsecond positive and negative wave. All impulse values are phase-to-phase and phase-to-ground and across the open contacts.

(4) The 25-Hz continuous current ratings in amperes are given herewith following the respective 60-Hz rating: 600–700; 1200–1400; 2000–2250; 3000–3500.

(5) To obtain the required symmetrical current interrupting capability of a circuit breaker at an operating voltage between $1/K$ times rated maximum voltage and rated maximum voltage the following formula shall be used:

Required Symmetrical Current Interrupting Capability

$$= \text{Rated Short-Circuit Current}\left(\frac{\text{Rated Maximum Voltage}}{\text{Operating Voltage}}\right)$$

For operating voltages below $1/K$ times rated maximum voltage, the required symmetrical current interrupting capability of the circuit breaker shall be equal to K times rated short-circuit current.

(6) With the limitation stated in 5.10 of ANSI/IEEE C37.04-1979 [8], all values apply for polyphase and line-to-line faults. For single phase-to-ground faults, the specific conditions stated in 5.10.2.3 of ANSI/IEEE C37.04-1979 [8] apply.

(7) The ratings in this column are on a 60-Hz basis and are the maximum time interval to be expected during a breaker opening operation between the instant of energizing the trip circuit and interruption of the main circuit on the primary arcing contacts under certain specified conditions. The values may be exceeded under certain conditions as specified in 5.7 of ANSI/IEEE C37.04-1979 [8].

(8) Current values in this column are not to be exceeded even for operating voltages below $1/K$ times rated maximum voltage. For voltages between rated maximum voltage and $1/K$ times rated maximum voltage, follow (5) above.

(9) Current values in this column are independent of operating voltage up to and including rated maximum voltage.

(10) If currents are to be expressed in peak amperes, multiply values in this column by a factor of 1.69 which is a ratio of 2.7/1.6.

(11) The rated values for T_2 are not standardized for indoor oil-less circuit breakers; however, E_2 = 1.88 times rated maximum voltage.

symmetrical basis. In specifying these circuit breakers, consideration should be given to the related values and required capabilities listed as headings in Table 56. This table lists preferred ratings for indoor oil-less circuit breakers. These ratings are applicable for service at altitudes up to 3300 ft. For service beyond 3300 ft, derating factors must be applied in accordance with ANSI/IEEE C37.04-1979 [8].

Power circuit breakers used for applications up through 15 kV are predominantly of the air-magnetic circuit breaker type, with increasing usage of vacuum-type interrupters, and some SF6 types. For voltages above 15 kV, the available types of circuit breakers include oil, compressed air or gas, and vacuum interrupters. In general, vacuum power circuit breakers are applied in accordance with the specific continuous- and short-circuit current requirements in the same manner as air-magnetic circuit breakers. It must be recognized, however, that under certain conditions, vacuum interrupters have characteristics that are different from air-magnetic power circuit breakers. Vacuum interrupters will sometimes, in special applications, force a premature current zero by opening the circuit in an unusually short time. When this occurs, a higher than normal transient recovery voltage occurs that can be of a magnitude that will impose excessive dielectric stress on the connected equipment. In some equipment this magnitude may be greater than the basic impulse insulation level of any connected device and failure may result.

When applying vacuum power circuit breakers, the following precautions should be taken.

(1) *Switching Unloaded Transformers*. When switching power transformers that are unloaded, that is, interrupting just the small magnetizing current on an infrequent basis (less than 50 operations per year), and where the basic impulse level is 95 kV or higher, no special attention is required. However, should a dry-type transformer be involved with a less than 95 kV basic impulse level rating, or else if all switching is highly repetitive, then the application should be checked with the transformer manufacturer.

(2) *Switching Loaded Transformers*. When a power transformer has a permanently connected load in kilovolt-amperes of 5% or more, no special consideration is needed.

(3) *Switching Motors*. When vacuum power circuit breakers are utilitzed to switch motors, the standard rotating-machine protection package of capacitors and surge arresters should always be used.

9.2.3.2 Circuit Breakers 600 V and Below. Circuit breakers rated 600 V and below have traditionally been divided into two classifications: power circuit breakers, sometimes known as metal frame breakers, and molded-case circuit breakers. A new type of circuit breaker that could be classified as an insulated case power circuit breaker is also available. This device offers many of the features of both the power and molded-case circuit breakers.

Power circuit breakers 600 V and below are open-construction assemblies on metal frames with all parts designed for accessible maintenance, repair, and ease of replacement. They are intended for service in switchgear compartments or other enclosures of dead front construction. Tripping units are field adjustable over a wide range and are interchangable within their frame sizes. The tripping units used have been the electromagnetic overcurrent direct-acting type; however, solid-state type tripping units are now available from most manufacturers and are widely used.

Power circuit breakers 600 V and below can be used with integral current-limiting fuses in drawout construction to meet interrupting current requirements up to 200 000 A rms symmetrical. When part of the circuit breaker, the fuses are combined with an integral-mounted open-fuse trip device to prevent single-phase damage to utilization equipment if one fuse should blow.

In current designs, 600 V and below air circuit breaker contacts often begin to part during the first cycle of short-circuit current, but have a multicycle total clearing time. Consequently, these breakers should be designed to interrupt the maximum available quarter-cycle asymmetrical current. However, since air circuit breakers 600 V and below are rated on a symmetrical current basis, the need for applying dc offset multipliers to determine their interrupting ratings is eliminated. (Note that caution should be applied when these air circuit breakers are supplied with short-time delay trips because increases in short-circuit stress on the breaker could result in both a lower breaker interrupting capacity rating and extensive equipment damage because withstand ratings could be exceeded. Manufacturers' literature should be consulted.) Molded-case circuit breakers fall into the same classification; thus, their momentary withstand and interrupting ratings are the same.

Power circuit breakers 600 V and below can be applied on a symmetrical basis if the system X/R ratio does not exceed 6.6. If the system X/R is higher, then the asymmetrical capability in the pole of the circuit breaker having maximum offset current should be checked against the maximum phase asymmetrical current available at the circuit breaker location. Molded-case circuit breakers are to be applied in accordance with system power factor ranges as specified in NEMA AB1-1975 [30].

A molded-case circuit breaker (NEMA AB1-1975 [30]) is a switching device and an automatic protective device assembled in an integral housing of insulating material. These breakers are generally capable of clearing a fault more rapidly than power circuit breakers and are available in the following general types.

(1) *Thermal Magnetic.* This type employs thermal tripping for overloads and instantaneous magnetic tripping for short circuits. These are the most widely applicable molded-case circuit breakers.

(2) *Magnetic.* This type employs only instantaneous magnetic tripping where only short-circuit interruption is required. The NEC [27] recognizes adjustable magnetic types for motor circuit protection.

(3) *Integrally Fused.* This type combines regular thermal magnetic protection against overloads and lower value short-circuit faults with current-limiting fuses responding to higher short-circuit currents. Interlocks are provided to ensure safe and proper operation.

(4) *High Interrupting Capacity.* This fuseless type provides interrupting capabilities for higher short-circuit currents than do standard constructed thermal magnetic circuit breakers. This line incorporates sturdier construction of contacts and mechanism, plus a special high-impact molded casing.

(5) *Current Limiting.* This type provides high interrupting capacity protection, plus it limits let-through current and energy to a value significantly lower than the

corresponding value for a conventional molded-case circuit breaker. Clearing time is also limited, and restoration of service is possible by resetting without replacement of any fusible elements or other parts.

With the advent of solid-state trip devices being incorporated into molded-cased circuit breakers, coordination with power circuit breakers is possible, but requires close scrutiny for proper application. Molded-case circuit breakers have an instantaneous override even when they are set on short-time delay. When a downstream circuit breaker sees a large fault, not only will it trip, but if the fault exceeds the instantaneous setting of the upstream circuit breaker, it will also trip. For system integrity, care must be exercised to ensure that coordination is maintained with upstream devices and that downstream equipment withstand ratings are not exceeded.

Power circuit breakers are designed for periodic planned maintenance. This design permits higher endurance ratings and repetitive duty capabilities and forms one basis for a broader range of applications. It is important to recognize that differences in test criteria between power circuit breakers and molded-case circuit breakers can be significant in the application of circuit breakers at or near their interrupting rating. A combination of circumstances can reduce the performance of molded-case circuit breakers to less than adequate if they are applied at or near their interrupting rating on a par with power circuit breakers.

Molded-case circuit breakers generally are not designed to be maintained in the field as are power circuit breakers. Many molded-case breakers are sealed to prevent tampering, thereby precluding inspection of the contacts. In addition, replacement parts are not generally available. Manufacturers recommend total replacement of the molded-case circuit breaker if a defect appears, of if the unit begins to overheat. Molded-case circuit breakers, particularly the larger sizes, are not suitable for repetitive switching.

The circuit breaker must be capable of closing, carrying, and interrupting the highest fault current within its rating at that location. It is essential to select a circuit breaker whose interrupting rating at the circuit voltage is equal to or greater than the available short-circuit current at the point of installation. The procedure to be used for short-circuit selection of 600 V and below power circuit breakers is covered in Chapter 6.

Manufacturers' publications give specific information on mechanical and electrical features of circuit breakers 600 V and below. Refer to Tables 57 and 58 for lists of standard ratings for 600 V and below power circuit breakers and molded-case circuit breakers. For service at altitudes above 6600 ft above sea level, derating factors must be applied in accordance with ANSI/IEEE C37.13-1981 [11].

NOTE (applicable to Tables 57 and 58): Solid-state trip devices for both overcurrent and ground-fault protection are readily available from most manufacturers, although the trip ranges given here are based on electromechanical direct or indirect acting types. Standards for solid-state trip devices should be forthcoming in later editions of this standard. In the meantime, it is recommended that the manufacturer be consulted relative to solid-state trip characteristics. Solid-state trip devices provide many advantages over conventional types and their use should be seriously considered. A major one is the inherent provision for sensitive ground-fault protection. For further discussion and trip characteristic curves, see Chapter 5. Particular attention must be given to coordination of load-side fuse devices with instantaneous trip devices of circuit breakers.

Table 57
Preferred Ratings for Low-Voltage AC Power Circuit Breakers with Instantaneous Direct-Acting Phase Trip Elements (See ANSI/IEEE C37.13-1981 [11])

Line No	System Nominal Voltage (volts)	Rated Maximum Voltage (volts)	Insulation (Dielectric) Withstand (volts)	Three-Phase Short-Circuit Current Rating, (symmetrical amperes) *	Frame Size (amperes)	Range of Trip-Device Current Ratings (amperes) †
	Col 1	Col 2	Col 3	Col 4	Col 5	Col 6
1	600	635	2200	14 000	225	40–225
2	600	635	2200	22 000	600	40–600
3	600	635	2200	22 000	800	100–800
4	600	635	2200	42 000	1600	200–1600
5	600	635	2200	42 000	2000	200–2000
6	600	635	2200	65 000	3000	2000–3000
7	600	635	2200	85 000	4000	4000
8	480	508	2200	22 000	225	40–225
9	480	508	2200	30 000	600	100–600
10	480	508	2200	30 000	800	100–800
11	480	508	2200	50 000	1600	400–1600
12	480	508	2200	50 000	2000	400–2000
13	480	508	2200	65 000	3000	2000–3000
14	480	508	2200	85 000	4000	4000
15	240	254	2200	25 000	225	40–225
16	240	254	2200	42 000	600	150–600
17	240	254	2200	42 000	800	150–800
18	240	254	2200	65 000	1600	600–1600
19	240	254	2200	65 000	2000	600–2000
20	240	254	2200	85 000	3000	2000–3000
21	240	254	2200	130 000	4000	4000

*Ratings in this column are rms symmetrical values for single-phase (2-pole) circuit breakers and three-phase average rms symmetrical values of three-phase (3-pole) circuit breakers. When applied on systems where rated maximum voltage may appear across a single pole, the short-circuit current ratings are 87% of these values. See ANSI/IEEE C37.13-1981 [11], 4.6.

†For preferred trip-device current ratings, see Table 22. Note that the continuous-current-carrrying capability of some circuit-breaker-trip-device combinations may be higher than the trip-device current rating. See ANSI/IEEE C37.13-1981 [11], 9.1.3.

Table 58
Preferred Ratings for Low-Voltage AC Power Circuit Breakers without Instantaneous Direct-Acting Phase Trip Elements (Short Time-Delay Element or Remote Relay) (See ANSI/IEEE C37.13-1981 [11])

						Range of Trip-Device Current Ratings (amperes)‡		
						Setting of Short-Time-Delay Trip Element		
Line No	System Nominal Voltage (volts)	Rated Maximum Voltage (volts)	Insulation (Dielectric) Withstand (volts)	Three-Phase Short-Circuit Current Rating or Short-Time Current Rating (symmetrical amperes)*†	Frame Size (amperes)	Minimum Time Band	Intermediate Time Band	Maximum Time Band
	Col 1	Col 2	Col 3	Col 4	Col 5	Col 6	Col 7	Col 8
1	600	635	2200	14 000	225	100–225	125–225	150–225
2	600	635	2200	22 000	600	175–600	200–600	250–600
3	600	635	2200	22 000	800	175–800	200–800	250–800
4	600	635	2200	42 000	1600	350–1600	400–1600	500–1600
5	600	635	2200	42 000	2000	350–2000	400–2000	500–2000
6	600	635	2200	65 000	3000	2000–3000	2000–3000	2000–3000
7	600	635	2200	85 000	4000	4000	4000	4000
8	480	508	2200	14 000	225	100–225	125–225	150–225
9	480	508	2200	22 000	600	175–600	200–600	250–600
10	480	508	2200	22 000	800	175–800	200–800	250–800
11	480	508	2200	42 000	1600	350–1600	400–1600	500–1600
12	480	508	2200	50 000	2000	350–2000	400–2000	500–2000
13	480	508	2200	65 000	3000	2000–3000	2000–3000	2000–3000
14	480	508	2200	85 000	4000	4000	4000	4000
15	240	254	2200	14 000	225	100–225	125–225	150–225
16	240	254	2200	22 000	600	175–600	200–600	250–600
17	240	254	2200	22 000	800	175–800	200–800	250–800
18	240	254	2200	42 000	1600	350–1600	400–1600	500–1600
19	240	254	2200	50 000	2000	350–2000	400–2000	500–2000
20	240	254	2200	65 000	3000	2000–3000	2000–3000	2000–3000
21	240	254	2200	85 000	4000	4000	4000	4000

*Short-circuit current ratings for breakers without direct-acting trip devices, opened by a remote relay, are the same as those listed here.

†Ratings in this column are rms symmetrical values for single-phase (2-pole) and three-phase average rms symmetrical values of three phase (3-pole) circuit breakers. When applied on systems where rated maximum voltage may appear across a single pole, the short-circuit current ratings are 87% of these values. See ANSI/IEEE C37.13-1981 [11], 4.6.

‡Note that the continuous-current-carrying capability of some circuit-breaker-trip-device combinations may be higher than the trip-device current rating. See ANSI/IEEE C37.13-1981 [11], 9.1.3

This material is reproduced with permission from American National Standard Preferred Ratings, Related Requirements, and Application Recommendations for Low-Voltage Power Circuit Breakers and AC Power Circuit Protectors, ANSI C37.16-1980, copyright 1980 by the American National Standards Institute. Copies of this standard may be purchased from ANSI, 1430 Broadway, New York, NY 10018.

Unusual service conditions as defined in ANSI/IEEE C37.04-1979 [8] and ANSI/IEEE C37.13-1981 [11] should be considered when applying power circuit breakers. Such conditions should be brought to the attention of the circuit breaker manufacturer at the earliest possible time.

9.2.3.3 Service Protectors. A service protector consists of a current-limiting fuse and nonautomatic circuit-breaker-type switching device in a single enclosure. Stored energy operation provides for manual or electrical closing. The service protector, utilizing basic circuit breaker principles, permits frequent repetitive operation under normal and abnormal current conditions up to 12 times the device's continuous-current rating. In combination with current-limiting fuses it is capable of closing and latching against fault currents up to 200 000 A rms symmetrical. During fault interruption, the service protector will withstand the stresses created by the let-through current of the fuses.

Service protectors are available at continuous-current ratings of 800, 1200, 1600, 2000, 3000, 4000, 5000, and 6000 A for use on 240 and 480 V ac systems, in two-pole and three-pole construction. An open fuse trip device to prevent the occurrence of single phasing after a fuse opening is included in the design of the service protector.

9.3 Switchgear

9.3.1 General Discussion. Switchgear is a general term covering switching and interrupting devices alone, or their combination with other associated control, metering, protective, and regulating equipment.

A power switchgear assembly consists of a complete assembly of one or more of the above noted devices and main bus conductors, interconnecting wiring, accessories, supporting structures, and enclosure. Power switchgear is applied throughout the electric power system of an industrial plant, but is principally used for incoming line service and to control and protect load centers, motors, transformers, motor control centers, panelboards, and other secondary distribution equipment.

Outdoor switchgear assemblies can be of the nonwalk-in (without enclosed maintenance aisle) or walk-in (with an enclosed maintenance aisle) variety. Switchgear for industrial plants is generally located indoors; such location is preferred because of easier maintenance, avoidance of weather problems, and shorter runs of feeder cable or bus duct. In outdoor applications the effect of external influences, principally the sun, wind, moisture, and local ambient temperatures should be considered in determining the suitability and current-carrying capacity of the switchgear. Further information on this evaluation is contained in ANSI/IEEE C37.24-1971 [13].

In many locations, the use of lighter colored (nonmetallic) paints will minimize the effect of solar energy loading and increase the rating of the equipment in outdoor locations. See ANSI/IEEE C37.24-1971 [13].

9.3.2 Classifications. An *open switchgear assembly* is one that does not have an enclosure as part of the supporting structure.

An *enclosed switchgear assembly* consists of a metal-enclosed supporting structure with the switchgear enclosed on the top and all sides with sheet metal (except for ventilating openings and inspection windows). Access within the enclosure is provided by doors or removable panels.

Metal-enclosed switchgear is universally used throughout industry for utilization and primary distribution voltage service, for ac and dc applications, and for indoor and outdoor locations.

9.3.3 Types. Specific types of metal-enclosed power switchgear used in industrial plants are defined as (1) metal-clad switchgear, (2) low-voltage power circuit breaker switchgear, and (3) interrupter switchgear. The metal-enclosed bus will also be discussed because it is frequently used in conjunction with power switchgear in modern industrial power systems.

9.3.4 Definitions. *Metal-clad switchgear* is metal-enclosed power switchgear characterized by the following necessary features.

(1) The main circuit switching and interrupting device is of the removable type arranged with a mechanism for moving it physically between connected and disconnected positions and equipped with self-aligning and self-coupling primary and secondary disconnecting devices.

(2) Major parts of the primary circuit, such as the circuit switching or interrupting devices, buses, potential transformers, and control power transformers, are enclosed by grounded metal barriers. Specifically included is an inner barrier in front of or a part of the circuit interrupting device to ensure that no energized primary circuit components are exposed when the unit door is opened.

(3) All live parts are enclosed within grounded metal compartments. Automatic shutters prevent exposure of primary circuit elements when the removable element is in the test, disconnected, or fully withdrawn position.

(4) Primary bus conductors and connections are covered with insulating material throughout. For special configurations, insulated barriers between phases and between phase and ground may be specified.

(5) Mechanical interlocks are provided to ensure a proper and safe operating sequence.

(6) Instruments, meters, relays, secondary control devices, and their wiring are isolated by grounded metal barriers from all primary circuit elements with the exception of short lengths of wire, such as at instrument transformer terminals.

(7) The door through which the circuit interrupting device is inserted into the housing may serve as an instrument or relay panel and may also provide access to a secondary or control compartment within the housing.

Auxiliary frames may be required for mounting associated auxiliary equipment, such as potential transformers, control power transformers, etc.

The term *metal-clad switchgear* can be properly used only if metal-enclosed switchgear conforms to the foregoing definition. All metal-clad switchgear is metal enclosed, but not all metal-enclosed switchgear can be correctly designated as metal-clad. The most prevalent type of switching and interrupting device used in metal-clad switchgear is the air-magnetic power circuit breaker over 1000 V.

Metal-enclosed 1000 V and below power circuit breaker switchgear is metal-enclosed power switchgear, including the following equipment as required:

(1) 1000 V and below power circuit breakers (fused or unfused)

(2) Bare bus and connections

(3) Instrument and control power transformers

(4) Instrument, meters, and relays
(5) Control wiring and accessory devices
(6) Cable and busway termination facilities

The 1000 V and below power circuit breakers are contained in individual grounded metal compartments and controlled either remotely or from the front of the panels. The circuit breakers are usually of the drawout type, but may be non-drawout. When drawout-type circuit breakers are used, mechanical interlocks must be provided to ensure a proper and safe operating sequence.

Metal-enclosed interrupter switchgear is metal-enclosed power switchgear including the following equipment as required:

(1) Interrupter switches
(2) Power fuses
(3) Bare bus and connections
(4) Instrument and control power transformers
(5) Control wiring and accessory devices

The interrupter switches and power fuses may be of the stationary or removable type. For the removable type, mechanical interlocks are provided to ensure a proper and safe operating sequence.

Metal-enclosed bus is an assembly of rigid electrical buses with associated connections, joints, and insulating supports, all housed within a grounded metal enclosure. Three basic types of metal-enclosed bus construction are recognized: nonsegregated phase, segregated phase, and isolated phase. The most prevalent type used in industrial power systems is the nonsegregated phase, which is defined as one in which all phase conductors are in a common metal enclosure without barriers between the phases. When metal-enclosed buses over 1000 V are used with metal-clad switchgear, the bus conductors and connections are covered with insulating material throughout. When metal-enclosed buses are associated with metal-enclosed 1000 V and below power circuit breaker switchgear or metal-enclosed interrupter switchgear, the primary bus conductors and connections are usually bare.

9.3.5 Ratings. The ratings of switchgear assemblies and metal-enclosed buses are designations of the operational limits of the particular equipment under specific conditions of ambient temperature, altitude, frequency, duty cycle, etc. Table 59 lists the rated voltages and insulation levels for ac switchgear assemblies discussed in this section. Table 60 lists similar ratings for metal-enclosed buses. Rated voltages and insulation levels for dc switchgear assemblies can be found by referring to ANSI/IEEE C37.20-1969 [12]. The definition of the ratings listed in Tables 59 and 60, and others subsequently discussed, can be found in ANSI C37.100-1981 [14].

Standard self-cooled continuous-current ratings of the main bus in metal-enclosed power switchgear are listed in Table 61. Metal-enclosed bus standard self-cooled continuous-current ratings are shown in Table 62.

The momentary and short-time short-circuit current ratings of power switchgear assemblies shall correspond to the equivalent ratings of the switching or interrupting devices used.

The limiting temperature for a power switchgear assembly or metal-enclosed bus (where applicable) is the maximum temperature permitted for:

Table 59
Rated Voltages and Insulation Levels for AC Switchgear Assemblies

Rated Voltage (rms)		Insulation Levels (kV)		
Rated Nominal Voltage	Rated Maximum Voltage	Power Frequency Withstand (rms)	DC Withstand*	Impulse Withstand
Metal-Enclosed Low-Voltage Power Circuit Breaker Switchgear				
Volts	Volts			
240	254	2.2	3.1	—
480	508	2.2	3.1	—
600	635	2.2	3.1	—
Metal-Clad Switchgear				
kV	kV			
4.16	4.76	19	27	60
7.2	8.25	36	50	95
13.8	15.0	36	50	95
34.5	38.0	80	†	150
Metal-Enclosed Interrupter Switchgear				
kV	kV			
4.16	4.76	19	27	60
7.2	8.25	26	37	75
13.8	15.0	36	50	95
14.4	15.5	50	70	110
23.0	25.8	60	†	125
34.5	38.0	80	†	150
Station-Type Cubicle Switchgear				
kV	kV			
14.4	15.5	50	†	110
34.5	38.0	80	†	150
69.0	72.5	160	†	350

*The column headed "DC Withstand" is given as a reference only for those using direct-current tests and represents values believed to be appropriate and approximately equivalent to the corresponding power frequency withstand test values specified for each voltage class of switchgear. The presence of this column in no way implies any requirement for a direct-current withstand test on alternating-current equipment. When making direct-current tests, the voltage should be raised to the test value in discrete steps and held for a period of 1 min.

†Because of the variable voltage distribution encountered when making direct-current withstand tests, the manufacturer should be contacted for recommendations before applying direct-current withstand tests to the switchgear. Potential transformers above 34.5 kV should be disconnected when testing with direct current. Refer to 6.8 of ANSI/IEEE C57.13-1978 [17] and in particular to 6.8.2, which reads "Periodic kenotron tests should not be applied to transformers of higher than 34.5 kV voltage ratings."

From ANSI/IEEE C37.20-1969 [12].

Table 60
Voltage Ratings for Metal-Enclosed Bus

Rated AC Voltage (kV rms)		Insulation Level (kV)			
		Power Frequency Withstand (rms)			
Nominal	Rated Maximum	(Dry 1 Minute)	(Dew 10 Seconds) *	DC Withstand (Dry) †	Impulse Withstand
0.6	0.63	2.2	—	3.1	—
4.16	4.76	19.0	15	27.0	60
13.8	15.00	36.0	24 (36)	50.0	95
14.4	15.50	50.0	30 (50)	70.0	110
23.0	25.80	60.0	40 (60)	‡	125
34.5	38.00	80.0	70 (80)	‡	150
69.0	72.50	160.0	140 (160)	‡	350

For applications of isolated phase bus to generators, the following voltage ratings apply:§

	Power Frequency Withstand (rms)			
Rated kV of Generator (rms)	(Dry 1 Minute)	(Dew 10 Seconds)	DC Withstand (Dry)	Impulse Withstand
14.4 to 24	50	50	70	110

*Applied to porcelain insulation only. Values in parentheses apply to "high creepage" designs.

†The column headed "DC Withstand" is given as a reference only for those using direct-current tests and represents equivalent to the corresponding power frequency withstand test values specified for each voltage class of bus. The presence of this column in no way implies any requirement for a direct-current withstand test on alternating-current equipment. When making direct-current tests the voltage should be raised to the test value in discrete steps and held for a period of 1 min.

‡Because of the variable voltage and distribution encountered when making direct-current withstand tests, the manufacturer should be contacted for recommendations before applying direct-current withstand tests to these voltage ratings. Potential transformers above 34.5 kV should be disconnected when testing with direct current. Refer to 6.8 of ANSI/IEEE C57.13-1978 [17], and in particular to 6.8.2, which reads "Periodic kenotron tests should not be applied to transformers of higher than 34.5 kV voltage rating."

§These ratings are applicable to generators rated 14.4 to 24 kV, which are directly connected to transformers without intermediate circuit breakers and where adequate surge protection is provided. These bus withstand ratings are compatible with or in excess of required withstand values of the generators.

From ANSI/IEEE C37.20-1969 [12].

Table 61
Continuous-Current Ratings of Buses for Metal-Enclosed Power Switchgear

Type of Assembly	Rated Continuous Current of Buses (amperes)
Metal-clad switchgear	1200, 2000, 3000
Metal-enclosed interrupter switchgear	600, 1200, 2000
Metal-enclosed bus	See Table 63
Station-type cubicle switchgear	2000, 3000, 4000, 5000

NOTE: The numerous combinations of circuit breaker sizes and ratings in similar housings make it impractical to provide current ratings of buses for metal-enclosed low-voltage power circuit breaker switchgear.
From ANSI/IEEE C37.20-1969 [12].

Table 62
Current Ratings for Metal-Enclosed Bus in Amperes

Voltage Ratings (kV)						
0.6 AC and All DC	2.4, 4.16	13.8	14.4	23.0	34.5	69.0
600	—	—	—	—	—	—
1200	1200	1200	1 200	1200	1200	1200
1600	—	—	—	—	—	—
2000	2000	2000	2 000	2000	2000	2000
—	—	—	2 500	2500	2500	2500
3000	3000	3000	3 000	3000	3000	3000
—	—	—	3 500	3500	—	—
4000	—	—	4 000	4000	—	—
—	—	—	4 500	4500	—	—
5000	—	—	5 000	5000	—	—
—	—	—	5 500	5500	—	—
6000	—	—	6 000	6000	—	—
—	—	—	6 500	—	—	—
—	—	—	7 000	—	—	—
—	—	—	7 500	—	—	—
—	—	—	8 000	—	—	—
—	—	—	9 000	—	—	—
—	—	—	10 000	—	—	—
—	—	—	11 000	—	—	—
—	—	—	12 000	—	—	—

(1) Any component, such as insulation, buses, instrument transformers, and switching and interrupting devices

(2) Air in cable termination compartments

(3) Any noncurrent-carrying structural parts

The hot-spot temperature rise and hot-spot total temperature to which insulating materials are subjected shall not exceed established values for the various classes of insulating materials. The hot-spot temperature rise and hot-spot total temperature of buses and connections for metal-enclosed power switchgear assemblies and buses shall not exceed the limits as established in Table 63. These temperature limitations apply for all bus continuous-current ratings shown in Tables 61 and 62. Additional information regarding limits of temperature rise for air surrounding enclosed power switchgear devices, power cables, parts handled by operating personnel, exposed surfaces, etc, can be obtained by referring to ANSI/IEEE C37.20-1969 [12].

9.3.6 Application Guides. After determining system requirements for continuity of service, reliability, security, and safety, the engineer should establish initial system capacity and provisions for future load growth.

From this data, he can establish maximum fault duty and select the type of power switching apparatus for the primary and secondary distribution systems. For the primary system, the choice is between circuit breaker and switch-fuse combinations. For the secondary system, the choice is between fused and unfused power circuit breakers.

The following steps are normally taken in applying switchgear equipment:

(1) Develop a single-line diagram

(2) Determine rating of power switching apparatus

(3) Select main bus rating

(4) Select current transformers

Table 63
Temperature Limits for Buses and Connections in Switchgear Assemblies

Type of Bus or Connection	Limit of Hottest Spot Temperature Rise (°C)	Limit of Hottest Spot Total Temperature (°C)
Buses and connections with copper-to-copper connecting joints	30	70
Buses and connections with silver-surfaced (or equivalent) connecting joints	65	105
Terminations to insulated cables (copper to copper)	30	70
Terminations to insulated cables (silver surfaced or equivalent)	45	85

NOTE: The temperature of the air surrounding all devices within an enclosed assembly, considered in conjunction with their rating and loading as used, shall not cause these devices to operate outside their normal temperature range when the enclosure is surrounded by air within the range of -30 °C to +40 °C.

From ANSI/IEEE C37.20-1969 [12].

(5) Select potential transformers
(6) Select metering, relaying, and control power
(7) Determine closing, tripping, and other control power requirements
(8) Consider special applications

Since open switchgear assemblies are rarely used in industrial installations, consideration will be given to metal-enclosed assemblies only.

Metal-enclosed switchgear is available for application at voltages up through 34.5 kV. Metal-clad switchgear is available for application at voltages from 2.4 kV through 34.5 kV; however, it is seldom used above 15 kV for economic reasons.

Metal-enclosed switchgear is adaptable to many applications because it is easily expanded and can be specified and designed with load location and load characteristics in mind. If metal-enclosed switchgear with drawout interrupting devices is applied, *maintenance is facilitated* because of the accessibility of most components. On the average, metal-enclosed switchgear represents a small percentage of total plant cost. Metal-enclosed switchgear is generally shipped factory assembled and reduces the amount of expensive field assembly.

Essentially all recognized basic bus arrangements, radial, double bus, circuit breaker and half, main, and transfer bus, sectionalized bus, synchronizing bus, and ring bus, are available in metal-enclosed switchgear to ensure the desired system reliability and flexibility. A choice is made based on an evaluation of initial cost, installation cost, required operating procedures, and total system requirements.

The continuous-current rating of the switchgear main bus must be no less than that of the highest rated circuit breaker. Table 61 lists standard main bus continuous-current ratings for metal-enclosed power switchgear. The rated continuous current of a switchgear assembly is the maximum current in rms amperes, at rated frequency, that can be carried continuously by the primary circuit components without causing temperatures in excess of limits specified in Table 63. The switchgear main bus will be designed and rated for the full current capacity specified and will not be tapered. Because power system facilities are increased to serve larger loads, it is advisable to consider future expansion when selecting the bus continuous-current rating, and the momentary and interrupting ratings of the power switching apparatus.

The switchgear assembly should have momentary and short-time ratings equal, respectively, to the close and latch capability and short-time rating of the circuit breaker.

Current transformers are used to develop scale replica secondary currents, separated from the primary current and voltage, to provide a readily usable current for application to instruments, meters, and relays. For switchgear applications they are manufactured in single and double secondary types, also in the tapped multiratio type. The double secondary is suitable where two transformers of the same ratio would otherwise be required at the same location, with a resulting saving in space. The primary current rating should be no less than 125% of the ultimate full-load current of the circuit.

The metering and relaying accuracy must be adequate for the imposed burdens. The current transformer accuracy and excitation characteristics must be checked for proper relay application. Tables 64 and 65 list standard ratios and relaying and metering accuracies for current transformers.

Voltage transformers are used to transform primary voltage to a nominal safe value, usually 120 V. The primary rating normally is that of the system voltage, though slightly higher ratings may be used, that is, a 14 400 V rating on a 13 800 V nominal system. These transformers are used to isolate the primary voltages from the instrumentation, metering, and relaying systems, yet provide replica scale values of the primary voltage. All ratings, such as impulse, dielectric, etc, should be adequate for the purpose. Table 66 lists standard voltage transformer ratios.

9.3.7 Control Power. Successful operation of switchgear embodying electrically operated devices is dependent on a reliable source of control power that at all times will maintain voltage at the terminals of such devices within their rated operating voltage range. See Table 67.

There are two primary uses for control power in switchgear: tripping power and closing power. Since an essential function of switchgear is to provide instant and unfailing protection in emergencies, the source of tripping power must always be available. The requirements for the security of the source of closing power are less rigid, and other options are available. For 1000 V and below circuit breakers, manual closing for circuit breakers up through 1600 A frame is not uncommon.

Four practical sources of tripping power are

(1) Direct current from a storage battery

(2) Direct current from a charged capacitor

(3) Alternating current from the secondaries of current transformers in the protected power circuit

Table 64
Standard Accuracy Class Ratings* of Current Transformers in Metal-Enclosed Low-Voltage Power Circuit Breaker Switchgear

Ratio	B 0.1	B 0.2
100/5	1.2	2.4†
150/5	1.2	2.4†
200/5	1.2	1.2
300/5	0.6	0.6
400/5	0.6	0.6
600/5	0.6	0.6
800/5	0.3	0.3
1200/5	0.3	0.3
1500/5	0.3	0.3
2000/5	0.3	0.3
3000/5	0.3	0.3
4000/5	0.3	0.3

*See ANSI/IEEE C57.13-1978 [17].
†Not in ANSI/IEEE C57.13-1978 [17].
From ANSI/IEEE C37.20-1969 [12].

Table 65
Standard Accuracy Class Ratings* of Current Transformers in Metal-Clad Switchgear

Ratio	60 Hz Standard Burden B 0.1	B 0.5	B 2.0	Relaying Accuracy
50/5 †	1.2	2.4	—	C or T10
75/5 †	0.6	2.4	—	C or T20
100/5	0.6	2.4	—	C or T20
150/5	0.6	2.4	—	C or T20
200/5	0.6	2.4	—	C or T20
300/5	0.6	2.4	2.4	C or T20
400/5	0.3	1.2	2.4	C or T50
600/5	0.3	0.3	2.4	C or T50
800/5	0.3	0.3	1.2	C or T50
1200/5	0.3	0.3	0.3	C100
1500/5	0.3	0.3	0.3	C100
2000/5	0.3	0.3	0.3	C100
3000/5	0.3	0.3	0.3	C100
4000/5	0.3	0.3	0.3	C100

*See ANSI/IEEE C57.13-1978 [17].
†These ratios and transformer accuracies do not apply for metal-clad switchgear assemblies having rated momentary currents above 60 000 A. Where such assemblies have a rated momentary current above 60 000 A, the minimum current transformer ratio shall be 100/5.
From ANSI/IEEE C37.20-1969 [12].

(4) Direct or alternating current in the primary circuit passing through direct-acting trip devices

Where a storage battery has been chosen as a source of tripping power, it can also supply closing power. The battery ampere-hour and inrush requirements have been reduced considerably with the advent of the stored-energy spring mechanism closing on power circuit breakers up through 34.5 kV. General distribution systems, whether ac or dc, cannot be relied upon to supply tripping power because outages are always possible. These are most likely to occur in times of emergency when the switchgear is required to perform its protective functions.

Other factors influencing the choice of control power are

(1) Availability of adequate maintenance for a battery and its charger

(2) Availability of suitable housing for a battery and its charger

(3) Advantages of having removable circuit breaker units interchangable with those in other installations

(4) Necessity for closing circuit breakers with the power system de-energized

The importance of periodic maintenance and testing of the tripping power source cannot be overemphasized. The most elaborate protective relaying system is useless if tripping power is not available to open the circuit breaker under abnormal conditions. Alarm monitoring for abnormal conditions of the tripping source and for circuits is a general requirement.

Space heaters are supplied as a standard feature on outdoor metal-enclosed switchgear. Often ambient temperatures or other environmental conditions dictate the use of space heaters in indoor switchgear as well. When space heaters are furnished, they should be continuously energized from an ac power source. Since

Table 66
Standard Voltage Transformer Ratios

2400/4160Y-120
2400-120
4200-120
4800-120
7200-120
8400-120
12 000-120
14 000-120

Table 67
Preferred Control Voltages and Their Ranges for 600 V and Below Power Circuit Breakers

Rated Voltage (volts)	Control Voltage (volts)	Power Supply (volts) Solenoid or Motor Operator	Power Supply (volts) Stored Energy Operator†	Tripping Voltage Range (volts)
Direct Current*				28
24‡	—	—	—	14-30‡
48	38-56	—	—	28-56
125	100-140	90-130§	100-140	70-140
250	200-280	180-260§	200-280	140-280
Alternating Current				
120	104-127**	—	104-127	104-127
240	208-254**	190-250§	208-254	208-254
480	416-508**	380-500§	416-508	416-508

NOTE: It is recommended that trip, closing, relay coils, etc, normally connected continuously to one direct-current potential should be connected to the negative wire of the control circuit to minimize electrolytic deterioration.

*Control from exciter circuits is not recommended.
†For driving motor for air compressors and compressed spring mechanisms.
‡Unless the circuit breaker is located close to the battery and relay and adequate electric conductors are provided between the battery and trip coil, 24 V tripping is not recomended.
**Includes heater circuits.
§Some operating mechanisms will not meet all the closing requirements over the full control voltage range. In such cases it will be necessary to provide for two ranges of closing voltage. Where applicable, the preferred method of obtaining the double range of closing voltage is by the use of tapped coils. Otherwise, it will be necessary for the user to designate one of the two closing voltage ranges listed below as representing the condition existing at the circuit breaker location due to battery or lead voltage drop or control power transformer regulation.

Rated Voltage	Closing Voltage Ranges for Power Supply
125 (dc)	90-115 or 105-130
250 (dc)	180-230 or 210-260
230 (dc)	190-230 or 210-250
460 (ac)	380-460 or 420-500

From ANSI C37.16-1980 [3].

heaters are usually needed when the switchgear is out of service, a separate source of heater power is desirable.

Standard air-magnetic or vacuum power circuit breakers are rated at 60 Hz, but can be applied as low as 50 Hz without derating. For a 25 Hz application, however, there is a derating factor that should be applied to the circuit breaker interrupting rating. Equipment manufacturers should be consulted to determine the proper derating factor for low-frequency power switchgear applications.

The application of metal-enclosed switchgear in contaminated atmospheres may create many problems if adequate precautions are not taken. Typical precautions include, but are not limited to:

(1) Location of equipment away from localized sources of contamination and potential sources of moisture, such as steam pipes and traps, water pipes, etc

(2) Isolation of equipment through the use of air-conditioning or pressurization equipment

(3) Development of an appropriate supplemental maintenance program

(4) Maintenance of adequate spare part replacements

9.4 Transformers

9.4.1 Classifications. Transformers have many classifications that are useful in the industry to distinguish or define certain characteristics of design and application. Some of these classifications are the following.

9.4.1.1 Distribution and Power. A classification according to the rating in kilovolt-amperes. The distribution type covers the range of 3–500 kVA, and the power type all ratings above 500 kVA.

9.4.1.2 Insulation. This grouping includes liquid and dry types. Liquid-insulated can further be defined by the types of liquid: mineral oil, nonflammable, or low-flammable liquids. The dry-type grouping includes the ventilated, cast coil, totally enclosed nonventilated, and sealed gas-filled types. A third grouping would include a combination liquid-, vapor-, and gas-filled unit.

9.4.1.3 Substation or Unit Substation. The title *substation* transformer usually denotes a power transformer with direct cable or overhead line termination facilities to distinguish it from a unit substation transformer designed for integral connection to primary or secondary switchgear, or both, through enclosed bus connections. The substation classification is further defined by the terms *primary* and *secondary*. The primary substation transformer has a secondary or load-side voltage rating of 1000 V or higher, whereas the secondary substation transformer has a load-side voltage rating of less than 1000 V.

Most transformer ratings and design features have been standardized by ANSI and NEMA, and these are listed as such in manufacturers' publications. The selection of other than standard ratings will usually result in higher costs.

9.4.2 Specifications. In specifying a transformer for a particular application, the following items comprising the rating structure should be included:

(1) Rating in kilovolt-amperes or megavolt-amperes

(2) Single-phase or three-phase

(3) Frequency

(4) Voltage ratings
(5) Voltage taps
(6) Winding connections, delta or wye
(7) Impedance (base rating)
(8) Basic impulse level (BIL)
(9) Temperature rise
and the desired construction details to be specified should include:
(10) Insulation medium, dry or liquid type (see 9.4.1.2)
(11) Indoor or outdoor service
(12) Accessories
(13) Type and location of termination facilities
(14) Sound level limitations if the installation site requires this consideration
(15) Manual or automatic load tap changing
(16) Grounding requirements
(17) Provisions for future cooling of the specified type

Consideration should be given to energy conservation features in the transformer specification. Although this is optional, it may in some instances be mandated by law or by individual company policy. Where efficiency is of concern, several cost analysis techniques are used to formalize procurement decisions with the goal of maximizing efficiency, or minimizing overall life-cycle cost. In either case, the following information about the transformer should be supplied to the prospective vendors:
(1) The cost in $/kW at which load and no-load losses are valued
(2) The percentage of the transformer rating at which losses will be evaluated during the bid comparison process

Given this information, a prospective vendor can then establish the optimum proportion of copper and iron to be used in the construction of the transformer. In this manner, the cost of inefficiency can be factored into the initial capital expenditure.

9.4.3 Power and Voltage Ratings. Ratings in kilovolt-amperes or megavolt-amperes will include the self-cooled rating at a specified temperature rise, as well as the forced-cooled rating if the transformer is to be so equipped. The standard self-cooled ratings, and self-cooled/forced-cooled relationships, are listed in Tables 68 and 69. As a minimum, the self-cooled rating should be at least equal to the expected peak demand, with an allowance for projected load growth.

The standard average winding temperature rise (by resistance test) for the modern liquid-filled power transformer is 65 °C, based on an average ambient of 30 °C (40 °C maximum) for any 24-hour period. Liquid-filled transformers may be specified with a 55 °C/65 °C rise to permit 100% loading with a 55 °C rise, and 112% loading at the 65 °C rise. In addition, 115 °C rise high-fire-point liquid-insulated transformers are available from some manufacturers.

In NEMA TR1-1980 [36], ANSI/IEEE C57.12.00-1980 [15], and ANSI/IEEE C57.12.01-1979 [16], average winding temperature rise (by resistance) for the modern dry-type transformer is 150 °C rise, based on an average ambient of 30 °C (40 °C maximum) for any 24-hour period. Low-loss high-efficiency dry-type transformers can be specified with 115 °C rise or 80 °C rise. These lower temperature rise

Table 68
Transformer Standard Base kVA Ratings

Single-Phase				Three-Phase			
3	75	1250	10 000	15	300	3 750	25 000
5	100	1667	12 500	30	500	5 000	30 000
10	167	2500	16 667	45	750	7 500	37 500
15	250	3333	20 000	75	1000	10 000	50 000
25	333	5000	25 000	112½	1500	12 000	60 000
37½	500	6667	33 333	150	2000	15 000	75 000
50	833	8333		225	2500	20 000	100 000

From ANSI/IEEE C57.12.00-1980 [15].

Table 69
Classes of Transformer Cooling Systems

Type Letters	Method of Cooling
OA	Oil-immersed, self-cooled
OW	Oil-immersed, water-cooled
OW/A	Oil-immersed, water-cooled/self-cooled
OA/FA	Oil-immersed, self-cooled/forced-air-cooled
OA/FA/FA	Oil-immersed, self-cooled/forced-air-cooled/forced-air-cooled
OA/FA/FOA	Oil-immersed, self-cooled/forced-air-cooled/forced air—forced-oil-cooled
OA/FOA/FOA	Oil-immersed, self-cooled/forced-air—forced-oil-cooled/forced air—forced-oil-cooled
FOA	Oil-immersed, forced-oil-cooled with forced-air cooler
FOW	Oil-immersed, forced-oil-cooled with forced-water cooler
AA	Dry-type, self-cooled
AFA	Dry-type, forced-air-cooled
AA/FA	Dry-type, self-cooled/forced-air-cooled

From ANSI/IEEE C57.12.00-1980 [15].

units have longer life expectancy and greater overload capabilities than the regular 150 °C rise design. A 115 °C rise dry-type transformer has approximately ten times the life expectancy of a 150 °C rise unit and has a 15% overload capability. An 80 °C rise unit has a 30% overload capability. Most modern dry-type transformers 30 kVA and larger are designed with a UL component recognized 220 °C insulation system.

Both liquid-insulated and dry-type transformers are available with low core and coil watt loss designs at higher initial prices, but with significantly lower overall operating costs due to the higher energy efficiency.

Transformers have certain overload capabilities, varying with ambient temperature, preloading, and overload duration. These capabilities are defined in ANSI/IEEE C57.92-1981 [18] and ANSI C57.96-1959 [6] for both the liquid-insulated and dry types.

The modern substation transformer for industrial plant service is an integral three-phase unit, as contrasted to three single-phase units. The advantages of the three-phase unit, such as lower cost, higher efficiency, less space, and elimination of exposed interconnections, have contributed to its widespread acceptance.

The transformer voltage ratings will include the primary and secondary continuous-duty levels at the specified frequency, as well as the basic impulse level for each winding. The continuous rating specified for the primary winding will be the nominal line voltage of the system to which the transformer is to be applied, and preferably within ±5% of the normally sustained voltage. The secondary or transformed voltage rating will be the value under no-load conditions. The change in secondary voltage experienced under load conditions is termed regulation, and is a function of the impedance of the system and the transformer and the power factor of the load. Table 70 illustrates the proper designation of voltage ratings for three-phase transformers.

The basic impulse level rating for a transformer winding signifies the design and tested capability of its insulation to withstand transient overvoltages from lightning and other surges. Standard values of basic impulse level established for each nominal voltage class are listed in Table 71. A description of test requirements for these values is given in ANSI/IEEE C57.12.00-1980 [15]. Transformer bushings may be specified with extra creepage distance and higher than standard basic impulse level ratings, if required by local conditions.

A cost advantage can be realized in a high-voltage transformer if it is protected by surge arresters of reduced rating that have been applied in accordance with the requirements for *effectively grounded* systems. This cost advantage is obtained by the use of a lower level of insulation dielectric. Refer to Chapter 4 for a discussion of arrester application for the protection of transformers.

9.4.4 Voltage Taps. Voltage taps are usually necessary to compensate for small changes in the primary supply to the transformer, or to vary the secondary voltage level with changes in load requirements. The most commonly selected tap arrangement is the manually adjustable no-load type, consisting of four 2½% steps or variations from the nominal primary voltage rating. These tap positions are usually numbered 1 through 5, with the number 1 position providing the greatest number of effective turns. Based on a specific incoming voltage, selection of a higher voltage tap (lower tap number) will result in a lowering of the output voltage. The changing of tap positions is performed manually only with the transformer de-energized. In addition to the no-load taps, and considered desirable when load swings are larger and more frequent or voltage levels more critical, the *automatic tap changing under load* feature is available. This provides an additional ±10% voltage adjustment automatically in incremental steps, with continuous monitoring of the secondary terminal voltage or of a voltage level remote from the transformer.

9.4.5 Connections. Connections for the standard two-winding power transformers are preferably delta-primary and wye-secondary. The wye-secondary, specified with external neutral bushing, provides a convenient neutral point for establishing a system ground, or can be run as a phase conductor for phase-to-neutral load. The delta-connected primary isolates the two systems with respect to the flow of zero-sequence currents resulting from third-harmonic exciting current or a secondary ground fault, and may be used without regard to whether the system to which the primary is connected is three-wire or four-wire.

In some installations a grounded primary wye–wye transformer connection is used to minimize the problem of ferroresonance. However, this connection introdu-

Table 70
Designation of Voltage Ratings of Three-Phase Windings (Schematic Representation)

Identi-cation	Nomenclature	Nameplate Marking	Typical Winding Diagram	Condensed Usage Guide
(2) (a)	E	2400	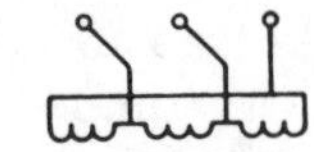	E shall indicate a winding which is permanently Δ connected for operation on an E volt system.
(2) (b)	E_1Y	4160Y	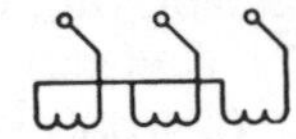	E_1Y shall indicate a winding which is permanently Y connected without a neutral brought out (isolated) for operation on an E_1 volt system.
(2) (c)	E_1Y/E	4160Y/2400	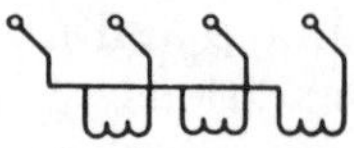	E_1Y/E shall indicate a winding which is permanently Y connected with a fully insulated neutral brought out for operation on an E_1 volt system, with E volts available from line to neutral.
(2) (d)	E/E_1Y	2400/4160Y		E/E_1Y shall indicate a winding which may be Δ connected for operation on an E volt system, or may be Y connected without a neutral brought out (isolated) for operation on an E_1 volt system.
(2) (e)	$E/E_1Y/E$	2400/4160Y/2400	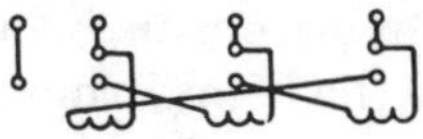	$E/E_1Y/E$ shall indicate a winding which may be Δ connected for operation on an E volt system or may be Y connected with a fully insulated neutral brought out for operation on an E_1 volt system with E volts available from line to neutral.
(2) (f)	E_1GrdY/E	67 000GrdY/38 700	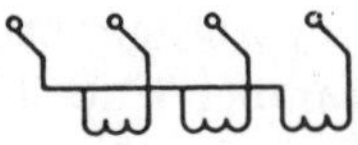	E_1GrdY/E shall indicate a winding with reduced insulation and permanently Y connected, with a neutral brought out and effectively grounded for operation on an E_1 volt system with E volts available from line to neutral.
(2) (g)	$E/E_1GrdY/E$	38 700/67 000GrdY/38 700	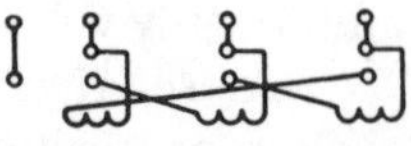	$E/E_1GrdY/E$ shall indicate a winding, having reduced insulation, which may be Δ connected for operation on an E volt system or may be connected Y with a neutral brought out and effectively grounded for operation on an E_1 volt system with E volts available from line to neutral.
(2) (h)	$V \times V_1$	7200 × 14 400 4160Y/2400 × 12 470Y/7200	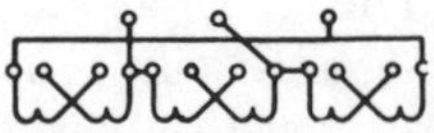 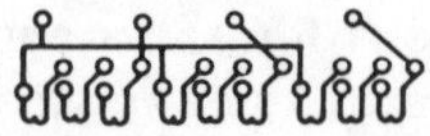	$V \times V_1$ shall indicate a winding, the sections of which may be connected in parallel to obtain one of the voltage ratings (as defined in a, b, c, d, e, f, and g) of V, or may be connected in series to obtain one of the voltage ratings (as defined in a, b, c, d, e, f, and g) of V_1. Windings are permanently Δ or Y connected.

Key: $E_1 = \sqrt{3}\,E$.

From ANSI/IEEE C57.12.00-1980 [15].

Table 71
Transformer Standard Basic Impulse Insulation Levels Usually Associated with Nominal System Voltages

Nominal System Line-to-Line Voltage (volts)	Insulation Class (kilovolts)	Basic Impulse Insulation Level (kilovolts)			
		Liquid Insulated		Dry Type	
		Power	Distribution	Ventilated	TENV Cast Coil Gas-Filled Sealed
120–1 200	1.2	45	30†	10	10
2 400	2.5	60	45†	20	20
4 160	5.0	75	60†	30	30
4 800	5.0	75	60†	30	30
6 900	8.7	95	75†	45	45
7 200	8.7	95	75†	45	45
12 470	15.0	110	95†	60	60
13 200	15.0	110	95†	60	60
13 800	15.0	110	95†	60	60
14 400	15.0	110	95†	60	60
22 900	25.0	150	150	110	110
23 000	25.0	150	150	110	110
26 400	34.5	200	200	125	125
34 500	34.5	200	200	150	150
43 800	46.0	250	250	—	—
46 000	46.0	250	250	—	—
67 000	69.0	350	350	—	—
69 000	69.0	350	350	—	—
92 000	92.0	450	—	—	—
115 000	115.0	550	—	—	—
		350‡	—	—	—
		450‡	—	—	—
138 000	138.0	650	—	—	—
		550‡	—	—	—
		450‡	—	—	—
161 000	161.0	750	—	—	—
		650‡	—	—	—
		550‡	—	—	—

†Ratings are also applicable to primary and secondary unit substation transformers.

‡Optional reduced levels applicable if equivalent reduced rating arresters are properly applied on the system.

Ventilated dry-type transformers, totally-enclosed nonventilated dry-type transformers, cast coil dry-type transformers, gas-filled sealed dry-type transformers, and liquid-insulated-type transformers are available with basic impulse levels higher than indicated at increased prices. For dry-type transformers 15 kV units can be specified to have 95 kV BIL, 25 kV units can be specified to have 150 kV BIL, and 34.5 kV units can be specified to have 200 kV BIL.

Based on ANSI/IEEE C57.12.00-1980 [15], NEMA 201-1982 [37], NEMA 210-1982 [38], and ANSI/IEEE C57.12.01-1979 [16].

ces the new problem of having to cope with zero-sequence quantities during conditions of circuit unbalance. There are two methods for balancing zero-sequence ampere turns:

(1) Shell-form construction can be used to provide a low reluctance return path for the single-phase zero-sequence flux. This construction would include the five- or four-legged core. The five-legged core is a three-phase core with five legs. Coils are mounted on three of the legs with the remaining two serving as a return path for magnetic flux. Thus, the five-legged core has a path for undesirable zero-sequence flux during unbalanced conditions.

(2) A delta-connected tertiary winding could circulate the required balancing ampere turns. In using this tertiary method, the high voltage to tertiary zero-sequence impedance should have a high enough value to keep the flow of current from blowing the transformer fuse. In addition, it must be low enough to keep the transformer parts from overheating.

9.4.6 Impedance. The impedance voltage of a transformer is the voltage required to circulate rated current through one of two specified windings of a transformer when the other winding is short-circuited, and with the windings connected as they would be for rated voltage operation.

Impedance voltage is normally expressed as a percent value of the rated voltage of the winding in which the voltage is measured on the transformer self-cooled rating in kilovolt-amperes. The percent impedance voltage levels considered as standard for two-winding transformers rated up through 10 000 kVA are listed in Table 72, and a value specified above or below those listed will usually result in higher costs. For transformer ratings above 10 000 kVA or 67 kV, a percent impedance voltage may be selected and considered standard if it lies within a published minimum and maximum range. The percent impedance voltage of a two-winding transformer shall have a tolerance of 7.5% of the specified value. For three-winding or autotransformers, the manufacturing tolerance is ±10%. When considering a low-impedance voltage level, it should be remembered that the standard transformer is designed with a limited ability to withstand the stresses imposed by external faults. Refer to ANSI C57.12.00-1980 [15] for short-circuit requirements and ANSI C57.12.90-1980 [5] for short-circuit test levels. A combined primary system and transformer impedance voltage permitting rms symmetrical fault magnitudes in excess of these standards should be avoided.

With respect to impedance, transformers are generally considered suitable for parallel operation if their impedances match within 10%. The importance of minimizing the mismatch becomes greater as the total load approaches the combined capacity of the paralleled transformers, since load division is inversely proportional to the internal impedance. The impedance mismatch should be checked throughout the entire range of taps (both no-load and load).

9.4.7 Insulation Medium. The selection of the insulation medium is dictated mainly by the installation site and cost. For outdoor installations, the mineral oil-insulated transformer has widespread use due to its lowest cost and inherent weatherproof construction. When located close to combustible buildings, safeguards are required as specified by the NEC [27], Article 450-27. For indoor installations refer to the NEC [27], Article 450-26.

Table 72
Standard Impedance Values for Three-Phase Transformers

High-Voltage Rating (volts)	Kilovolt Rating	Percent Impedance Voltage	
Secondary Unit Substation Transformers*			
2400-13 800	112.5-224	Not less than 2.0	
2400-13 800	300-500	Not less than 4.5	
2400-13 800	750-2500	6.75 ‡	
22 900	All	6.75 ‡	
34 400	All	7.25	
Liquid-Immersed Transformers, 501-30 000 kVA†			
		Low Voltage, 480 V	Low Voltage, 2400 V and Above
2400-22 900		6.75 ‡	6.5 **
26 400, 34 400		7.25	7.0
43 800		7.75	7.5
67 000			8.0
115 000			8.5
138 000			9.0

NOTES: (1) Ratings separated by hyphens indicate that all intervening standard ratings are included. Ratings separated by a comma indicate that only those listed are included.
(2) Percent impedance voltages are at self-cooled rating and as measured on rated voltage connection.
*From NEMA 210-1982 [38].
†From ANSI C57.12.10-1977 [4].
‡Three-phase transformers 5000 kVA and smaller with high-voltage windings rated 25 kV and below are commonly used in industrial applications and are normally built with impedance voltages of 5.75%.
**Three-phase transformers 5000 kVA and smaller with high-voltage windings rated 25 kV and below are commonly used in industrial applications and are normally built with impedance voltages of 5.5%.

For indoor installations, the discontinuance of the use of askarel (PCB) liquid-filled transformers has promoted the use of high-fire-point liquids being employed, such as polyalpha olefins, silicones, and high molecular weight hydrocarbons. They are being used in the applications previously applying to askarel transformers with the tacit approval of insurance and safety authorities as specified in the NEC [27], Article 450-23. In general, these high-fire-point liquids increase the cost of the transformer compared to mineral oil. These liquids should receive essentially the same care and maintenance that applies to conventional mineral oil-insulated transformers. Transformers insulated with a dielectric fluid identified as nonflammable shall be permitted to be installed indoors per Article 450-24 of the NEC [27].

Due to possible environmental pollution, the users of askarel-insulated transformers should consult the manufacturer of the transformer or the manufacturer of the liquid for selection as well as proper safeguard in the disposal of used liquid (see ANSI/IEEE C57.102-1974 [19]).

The ventilated dry-type transformer has application in industrial plants for indoor installation where floor space, weight, and regard for liquid maintenance and safeguards would be important factors. Since the ANSI/IEEE C57.12.01-1979 [16] basic impulse level for the ventilated dry-type transformer winding is usually less than that of the liquid- or gas-filled dry-type, surge arresters should be included for the primary winding in order to obtain additional protection.

The totally enclosed nonventilated dry-type transformer, the cast coil (where both the high- and low-voltage coils are cast), and the sealed or gas-filled dry-type transformer, although more expensive than ventilated dry or mineral oil-insulated units, are especially suitable for adverse environments. They require little maintenance, need no fire-proof vaults, and generally have lower losses than comparable ventilated or mineral oil-filled units. The same applies to the high-fire-point liquid insulated transformers; however, when they are installed in combustible buildings or areas, automatic fire extinguishing systems or vaults are required.

On oil-insulated transformers a sealed-tank construction and welded cover is standard practice with manufacturers. Optional oil-preservation systems may be specified as follows:

(1) A gas-oil seal that consists of an auxiliary tank mounted on the transformer. This seal provides for the safe expansion and contraction of the transformer gas and oil without exposing the transformer oil to the atmosphere.

(2) An automatic gas seal that maintains a constant positive nitrogen pressure within the tank. A combination regulating valve and pressure relief operates with a cylinder of high-pressure nitrogen to control the proper functioning of this seal.

9.4.8 Accessories. Accessories furnished with the transformer include those identified as standard and optional in manufacturers' publications. The standard items will vary with different types of transformers. Some of the optional items that offer protective features include the following:

(1) Winding temperature equipment in addition to the standard top-oil temperature indicator. This device is calibrated for use with specific transformers and automatically takes into account the hottest spot temperature of the windings, ambient temperature, and load cycling. For this reason it provides a more accurate, continuous, and automatic measure of the transformer loading and overloading capacity. It may have contacts that can be set to alarm and even subsequently trip a circuit breaker.

(2) The pressure relay for sensitive high-speed indication of liquid-filled transformer internal faults. Since the device is designed to operate on the rate of change of internal pressure, it is sensitive only to that resulting from internal faults and not to pressure changes due to temperature and loading.

(3) Alarm contacts can be included on the standard devices such as temperature indicators, liquid-level and pressure vacuum gauges, and pressure relief devices for more effective utilization of such devices.

(4) Surge arresters mounted directly on the transformer tank provide maximum surge protection for the transformer. The type of arrester specified and its voltage rating must be coordinated with the voltage parameters of the system on which it is applied. Refer to Chapter 4 for a detailed discussion of surge arrester application.

9.4.9 Termination Facilities. Termination facilities are available to accommodate most types of installation. For the unit substation arrangement, indoors or outdoors, the incoming and outgoing bushings are usually side-wall mounted and enclosed in a throat or transition section for connection to adjacent switchgear assemblies. Tank-wall mounted enclosures, oil or air insulated, with or without potheads or cable clamps, are available for direct cable termination. The size and number of conductors must be specified, along with minimum space for stress cone termination, if required. For the nonunit substation type of outdoor installation, cover-mounted bushings provide the simplest facility for overhead lines.

9.4.10 Sound Levels. The transformer sound level is of importance in certain installations, and the maximum standard levels are listed in NEMA TR1-1980 [36]. These can be reduced to some extent by special design, so the transformer manufacturer should be consulted regarding the possible reduction for a particular type and rating.

9.5 Unit Substations

9.5.1 General Discussion. A unit substation consists of the following sections:

(1) A primary section that provides for the connection of one or more incoming high-voltage circuits, each of which may or may not be provided with a switching device or a switching and interrupting device.

(2) A transformer section that includes one or more transformers with or without automatic load-tap-changing equipment. The use of automatic load-tap-changing equipment is not common in unit substations.

(3) A secondary section that provides for the connection of one or more secondary feeders, each of which is provided with a switching and interrupting device.

9.5.2 Types. Unit substation sections are normally subassemblies for connection in the field and are usually one of the following types (the application of these to industrial power systems is described in Chapter 2):

(1) *Radial.* One primary feeder to a single stepdown transformer with a secondary section for the connection of one or more outgoing radial feeders (see Figs 1 and 2 in Chapter 2).

(2) *Primary Selective and Primary Loop.* Each stepdown transformer connected to two separate primary sources through switching equipment to provide a normal and alternate source. Upon failure of the normal source, the transformer is switched to the alternate source (see Figs 3 and 4 in Chapter 2).

(3) *Secondary Selective.* Two stepdown transformers, each connected to a separate primary source. The secondary of each transformer is connected to a separate bus through a suitable switching and protective device. The two sections of bus are connected by a normally open switching and protective device. Each bus has provisions for one or more secondary radial feeders (see Fig 5 in Chapter 2).

(4) *Secondary Spot Network.* Two stepdown transformers, each connected to a separate primary source. The secondary side of each transformer is connected to a common bus through a special type of circuit breaker called a network protector, which is equipped with relays to trip the circuit breaker on reverse power flow to the transformer and reclose the circuit breaker upon restoration of the correct voltage, phase angle, and phase sequence at the transformer secondary. The bus has provisions for one or more secondary radial feeders (see Fig 6 in Chapter 2).

(5) *Distributed Network.* A single stepdown transformer having its secondary side connected to a bus through a special type of circuit breaker called a network protector, which is equipped with relays to trip the circuit breaker upon restoration of the correct voltage, phase angle, and phase sequence at the transformer secondary. The bus has provisions for one or more secondary radial feeders and one or more tie connections to a similar unit substation (see Fig 146).

(6) *Duplex (Breaker and a Half Scheme).* Two stepdown transformers, each connected to a separate primary source. The secondary side of each transformer is connected to a radial feeder. These feeders are joined on the feeder side of the power circuit breakers by a normally open tie circuit breaker. This type is used mostly on electric utility primary distribution systems (see Fig 147).

9.5.3 Selection and Location. Considerations in the selection and location of unit substations are apparent power and voltage ratings, allowance for future growth, appearance, atmospheric conditions, and outdoor versus indoor location. NEMA 201-1981 [37] and NEMA 210-1982 [38] contain useful information, which is also applicable for general substation design when the components are purchased separately.

9.5.4 Advantages of Unit Substations. The engineering of the components is coordinated by the manufacturer, the costs of field labor and installation time are reduced, substation appearance is improved, and the substation is safer to operate. The operating costs are reduced due to the reduced power losses from shorter secondary feeders, and a power system using unit substations is flexible and easy to expand.

Unit substations are available for either indoor or outdoor location. In some applications, the heat-producing transformer is located outdoors and connected by a metal-enclosed bus duct to indoor switchgear.

Fig 146
Distributed Network

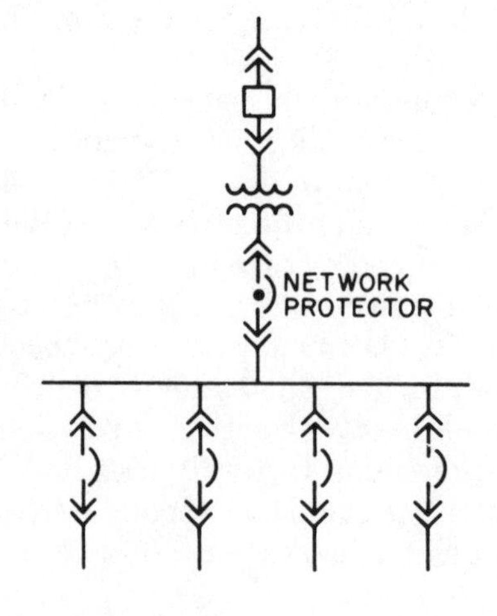

Fig 147
Duplex
(Circuit Breaker and a Half Scheme)

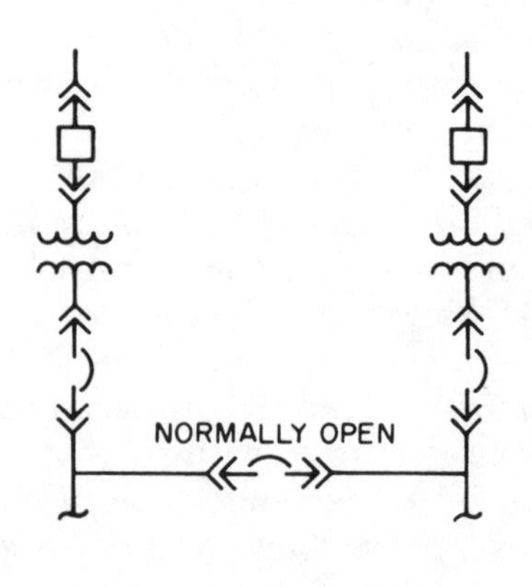

Primary unit substations may be located outdoors, particularly when the primary supply is above 34.5 kV. The high-voltage equipment, including the design, all components, the supporting structures, and installation drawings, may be obtained as a package. There is a trend toward metal-enclosed equipment above 34.5 kV in a unit substation arrangement because of the increasing requirements for safety, compactness, appearance, and reduction of installation labor and time.

Most secondary unit substations are located indoors to reduce costs and improve voltage reduction by placing the transformer as close as possible to the center of the load in the area being supplied.

9.5.5 Application Guides

(1) The transformer main secondary circuit breakers or fused switches and connections should have a continuous-current rating that is approximately 25% greater than the continuous-current rating of the transformer. This is necessary since transformers are often required to carry short-time overloads above their nameplate ratings for a short time, such as during plant startup. When selecting the continuous-current rating of the transformer main secondary circuit breaker, or fused switches and connections, consideration should also be given to whether or not the transformer has, or will have in the future, a continuous forced-air-cooled rating, dual temperature rating, or other extension of the basic continuous rating.

(2) The effects of solar radiation and atmospheric conditions should be considered in the selection of outdoor equipment. ANSI/IEEE C37.24-1971 [13] gives guidance for evaluating the effect of solar radiation. The lighter the color of the exterior paint, the lower the effect of solar radiation on the equipment.

(3) When connected to circuits that are subject to lightning or switching surges, substations should be equipped with surge-protective equipment, which is selected to limit voltage surges to values below the impulse withstand ratings of the transformers and the switching equipment.

NEMA 201-1982 [37] and NEMA 210-1982 [38] contain sections that give additional information on unit substation application.

9.6 Motor-Control Equipment

9.6.1 General Discussion. The majority of motors utilized by industrial firms are integral horsepower of induction squirrel-cage design powered from three-phase ac 600 V and below distribution systems. The choice of an integral horsepower controller depends on a number of factors.

(1) *Power System.* Use dc or ac; single-phase or three-phase? What is the voltage and frequency? Will the system permit the large inrush currents during the full-voltage starting without excessive voltage drop?

(2) *Motor.* Is the controller to be used with dc, squirrel-cage induction, wound-rotor induction, or synchronous motor? What is the horsepower? Will the motor be *plugged* or reversed frequently? What is the acceleration time from start to full speed?

(3) *Load.* Is the load geared, belt-driven, or direct coupled? Loaded or unloaded start?

(4) *Operation*. Is operation to be manual or automatic?

(5) *Protection*. Are fuses or circuit breakers to be used for short-circuit protection? To size the thermal overload elements, the full-load current of the motor and the ambient temperature at the motor and controller should be known.

(6) *Environment*. Will the motor and controller be subjected to excessive vibration, dirt, dust, oil, or water? Will either be located in a hazardous or corrosive area?

(7) *Cable Connections and Space*. Will there be the required space for cable entrance, bending radius, and terminations, and for reliable connections to line and load buses?

To answer these questions for proper application of motor controllers, the specifying engineer should seek the assistance of the application engineers from the utility and manufacturers. In addition, process engineers and operating personnel associated with the installation should be consulted.

9.6.2 Starters Over 600 V. Starters for motors from 2300–13 200 V are designed as integrated complete units based on maximum horsepower ratings for use with squirrel-cage, wound-rotor, synchronous, and multispeed motors for full- or reduced-voltage starting. Alternating-current magnetic-fused-type starters, NEMA class E2 (see ANSI/NEMA ICS2-1983 [26]), employ current-limiting power fuses and magnetic air-break contactors. Each starter will be completely self-contained, prewired, and with all components in place. Air-break contactors will be current rated based on motor horsepower requirements. Combination starters will provide an interrupting fault capacity of 260 MVA symmetrical on a 2300 V system, and 520 MVA symmetrical on a 4160 or 4800 V system. This starter will conform to ANSI/NEMA ICS2-1983 [26], class E-2 controllers, and applicable IEEE and ANSI standards. There is also a UL certification standard on this equipment. Combinations of motor controllers and switchgear are available as assemblies.

9.6.3 Starters 600 V and Below. ANSI/NEMA ICS2-1983 [26] summarizes the NEMA standard for magnetic controller ratings of 115 through 575 V. In ac motor starters, contactors are generally used for controlling the circuit to the motor. Starters should be carefully applied on circuits and in combination with associated short-circuit protective devices (circuit breakers, fusible disconnects) that will limit the available fault current and the let-through energy to a level the starter can safely withstand. These withstand ratings should be in accordance with ANSI/UL 508-1983 [29], and ANSI/NEMA ICS1-1983 [25] and ANSI/NEMA ICS2-1983 [26], which cover industrial controls, systems, and devices. Some of the common motor starting devices used in industry are as follows.

9.6.3.1 Across-the-Line Starter

(1) *Manual*. Provides overload protection, but not undervoltage protection. One- or two-pole single-phase for motor ratings to 3 hp. Single- or polyphase motor control for motor ratings up to 5 hp at 230 V single-phase, 7½ hp at 230 V three-phase, and 10 hp at 460 V three-phase. Operating control available in toggle, rocker, or push-button design.

(2) *Magnetic, Nonreversing*. For full-voltage frequent starting of ac motors, suitable for remote control with push-button station, control switch, or with automatic pilot devices. Undervoltage protection is obtained by using momentary contact-starting push-button in parallel with interlock contact in starter and

series-connected stop push button. Available in single-phase construction up to 15 hp at 230 V, and three-phase ratings up to 1600 hp at 460 V.

(3) *Magnetic, Reversing.* For full-voltage starting of single-phase and polyphase motors where application requires frequent starting and reversing or plugging operation. It consists of two contactors wired to provide phase reversal, mechanically and electrically interlocked to prevent both contactors from being closed at the same time.

9.6.3.2 Combination Across-the-Line Starter

(1) *Magnetic, Nonreversing.* For full-voltage starting of polyphase motors. It provides motor overcurrent protection with thermal overload relays. Available with a nonfuse disconnect; provides branch circuit protection when specified with a fusible disconnect or circuit breaker. Check available fault current before deciding which fuses or circuit breakers will be used. It also provides undervoltage protection and is suitable for remote control. Readily available up to NEMA size 5 from most manufacturers and size 9 from several.

Reduced voltage starting is accomplished by inserting a resistance in series with the armature winding. As counter EMF builds up in the armature, the external starting resistance can be gradually reduced and then removed as the motor comes up to speed. Speed control of dc motors can be accomplished by varying resistance in the shunt or series fields, or in the armature circuit.

(2) *Magnetic, Reversing.* Same as reversing starter, except equipped with nonfusible disconnect, fusible disconnect, or circuit breaker.

9.6.3.3 Reduced Voltage Starter

(1) *Autotransformer, Manual.* For limiting starting current and torque on polyphase induction motors to comply with power supply regulations or to avoid excessive shock to the driven machine. Overload and undervoltage protection are provided. Equipped with mechanical interlock to assure proper starting sequence. Taps are provided on the autotransformer for adjusting starting torque and current. Since the autotransformer controller reduces the voltage by transformation, the starting torque of the motor will vary almost directly as does the line current, even though the motor current is reduced directly with the voltage impressed on the motor.

(2) *Autotransformers, Magnetic.* Same as manual, but suitable for remote control. It has timing relay for adjustment of time at which full voltage is applied.

To overcome the objection of the open-circuit transition associated with an autotransformer starter, a circuit known as the Korndorfer connection is in common use. This type of starter requires a two-pole and a three-pole start contactor. The two-pole contactor opens first on the transition from start to run, opening the connections to the neutral of the autotransformer. The windings of the transformer are then momentarily used as series reactors during the transfer, allowing a closed-circuit transition. Although it is somewhat more complicated, this type of starter is frequently used on high-inertia centrifugal compressors to obtain the advantages of low line current surges and closed circuit transition.

(3) *Primary Resistor or Reactor Type.* Automatic reduced voltage starter designed for geared or belted drive where sudden application of full-voltage torque must be avoided. Inrush current is limited by the value of the resistor or reactor;

starting torque is a function of the square of the applied voltage. Therefore if the initial voltage is reduced to 50%, the starting torque of the motor will be 25% of its full-voltage starting torque. A compromise must be made between the required starting torque and the inrush current allowed on the system. It provides both overload and undervoltage protection and is suitable for remote control. The resistor or reactor is shorted out as speed approaches rated rpm.

(4) *Part Winding Type*. Used on light or low-inertia loads where the power system requires limitations on the increments of current inrush. It consists of two magnetic starters, each selected for one of the two motor windings, and a time-delay relay controlling the time at which the second winding is energized. It provides overload and undervoltage protection and is suitable for remote control.

(5) *Wye–Delta Type*. This type of starter is most applicable to starting motors that drive high-inertia loads with resulting long acceleration times. When the motor has accelerated on the wye connection, it is automatically reconnected by contactors for normal delta operation.

In selecting the type of reduced-voltage starter, consideration should be given to the motor-control transition from starting to running. In a closed-circuit transition, power to the motor is not interrupted during the starting sequence, whereas on open-circuit transition it is interrupted. Closed-circuit transition is recommended for all applications to minimize inrush voltage disturbances.

A comparison of starting currents and torques produced by various kinds of reduced-voltage starters is shown in Table 73.

9.6.3.4 Slip-Ring Motor Controller. The wound-rotor or slip-ring motor functions in the same manner as the squirrel-cage motor, except that the rotor windings are connected through slip rings and brushes to external circuits with resistance to vary motor speed. Increasing the resistance in the rotor circuit reduces motor speed and decreasing the resistance increases motor speed. Some variation of this type controller employs silicon-controlled rectifiers (SCRs) in place of contactors and resistors. Some even rectify the secondary current and invert it to line frequency to supply back into the input, raising the efficiency appreciably.

9.6.3.5 Controller for DC Motors. These motors have favorable speed-torque characteristics, and their speed is easily controlled.

9.6.3.6 Multispeed Controller. These controllers are designed for the automatic control of two-, three-, or four-speed squirrel-cage motors of either the consequent-pole or separate-winding types. They are available for constant-horsepower, constant-torque, or variable-torque three-phase motors used on fans, blowers, refrigeration compressors, and similar machinery.

9.6.3.7 Solid-State Reduced Voltage Motor Starter

(1) *Introduction*. Solid-state motor starters can control the starting cycle and provide reduced voltage starting for standard ac squirrel-cage motors. Solid-state control electronics combined with power SCRs ensure long life, low maintenance, and eliminate burnout of parts such as power contacts and ac coils associated with electromechanical starters. They provide an adjustable controlled acceleration and eliminate high power demands during starting. In some applications other types of power semiconductors can be utilized. These starters are available in standard models for motors rated from 10–600 hp.

Table 73
Comparison of Different Reduced-Voltage Starters

	Autotransformer*			Primary Resistor or Reactor		Part Winding‡		Wye Delta
	50% Tap	65% Tap	80% Tap	65% Tap	80% Tap	2-Step	3-Step	
Starting current drawn from line as percentage of that which would be drawn upon full-voltage starting†	28%	45%	67%	65%	80%	60%‡	25%	33⅓%
Starting torque developed as percentage of that which would be developed on full-voltage starting	25%	42% Increases slightly with speed	64%	42% Increases greatly with speed	64%	50%	12½%	33⅓%
Smoothness of acceleration	Second in order of smoothness			Smoothness of reduced-voltage types. As motor gains speed, current decreases. Voltage drop across resistor decreases and motor terminal voltage increases		Fourth in order of smoothness		Third in order of smoothness
Starting current and torque adjustment	Adjustable within limits of various taps			Adjustable within limits of various taps				Fixed

*Closed transition.
†Full-voltage start usually draws between 500 and 600 percent of full-load current.
‡Approximate values only. Exact values can be obtained from motor manufacturer.

(2) *Principle of a Solid-State Motor Starter*. One type of motor reduced voltage starter uses six SCRs in a full-wave configuration to vary the input voltage from zero to full on, so that the motor accelerates smoothly from zero to full running speed. The SCRs are activated by an electronic control section that has an initial step voltage adjustment. This adjustment, combined with a ramped voltage and current limit override, provides constant current (torque) to the motor until it reaches full speed. See Fig 148 for schematic diagram.

Variations in the design of starting circuit are as follows:

(1) Three power diodes replace the three return conducting SCRs. The control circuit is simple and each SCR is protected against reverse voltage by its associated diode. This half-wave configuration could produce harmonics that generate added heat in the motor windings. Thermal protective devices should be properly sized to prevent this additional heat from damaging the motor.

(2) SCRs are used only during the starting phase. At full voltage, a run contactor closes, and the circuit operates as a conventional electromechanical starter.

(3) A starter with linear timed acceleration uses a closed-loop feedback system to maintain the motor acceleration at a constant rate. The required feedback signal is provided by a dc tachometer coupled to the motor.

(3) *Starter Performance Comparison*. The solid-state reduced voltage motor starter maintains a constant level of kilovolt-amperes and reduces sudden torque

Fig 148
Typical Schematic Diagram of a Solid-State Motor Starter

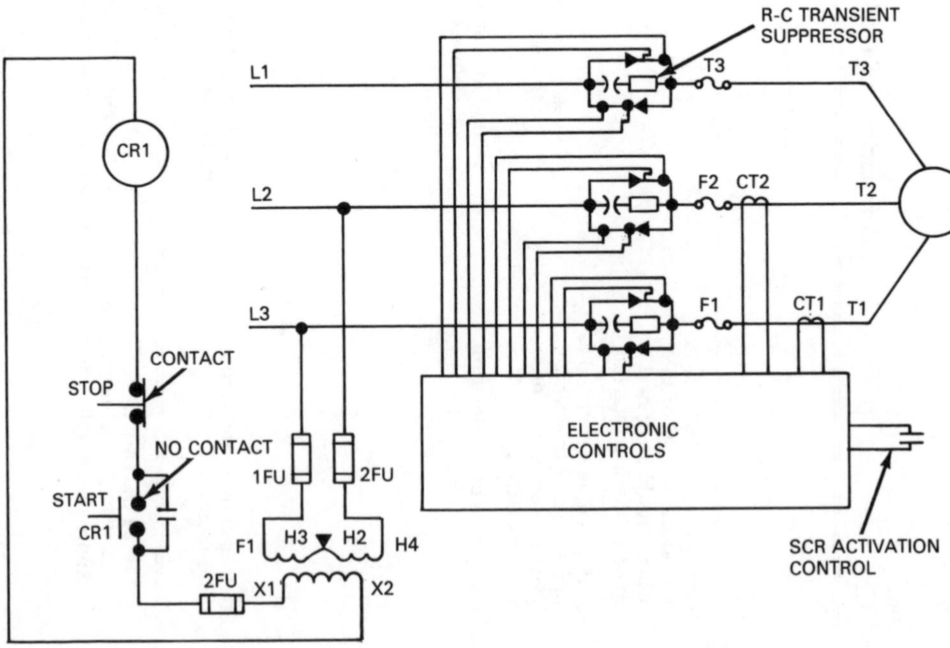

surges to the motor. The current limiter in conjunction with the acceleration ramp holds the current constant at a preset level during the start-up period. When the start cycle is complete, the motor is running at almost full voltage with, essentially, a pure sine wave in each phase.

9.6.4 Motor-Control Center. Most centers are tailor-made assemblies of conveniently grouped control equipment primarily used for power distribution and associated control of motors. They contain all necessary buses, incoming line facilities, and safety features to provide the utmost in convenience for space and labor saving and adaptability to everchanging conditions with a minimum of effort and a maximum of safety. ANSI/NEMA ICS2-1983 [26] governs the type of enclosure and wiring; NEMA type 1, 2, 3 and 12 enclosures are generally available. Wiring of motor-control centers conforms to two NEMA classes and three types. Class I provides for no wiring by the manufacturer between compartments of the center. Class II requires prewiring by the manufacturer with interlocking and other control wiring completed between compartments of the center. With type A, no terminal blocks are provided; with type B, all connections within individual compartments are made to terminal blocks; and with type C, all connections are made to a master terminal block located in the horizontal wiring trough at the top or bottom of the center. The ideal wiring specification for minimum field installation time and labor is NEMA class II, type C wiring. The wiring specifications most frequently used by industrial contractors are NEMA class I, type B wiring. Refer to ANSI/NEMA ICS2-1983 [26] for definitions of wiring class and type.

ANSI/NEMA ICS2-1983 [26] specifies that a control center shall carry a short-circuit rating defined as the maximum available rms symmetrical current in amperes permissible at line terminals. The available short-circuit current at the line terminals of the motor-control center is computed as the sum of maximum available current of the system at the point of connection and the short-circuit current contribution of the motors connected to the control center. It is common practice by many manufacturers to show only the short-circuit rating of the bus-work on the nameplate. As a result, it is very important to establish the actual rating of the entire unit and, in particular, the plug-in units (that is, circuit breakers, disconnects, starters, etc), especially for applications where available fault currents exceed 10 000 A.

9.6.5 Control Circuits. Conventional starters for 600 V and below are factory wired with coils of the same voltage rating as the phase voltage to the motor. However, there are many cases in which it is desirable or necessary to use control circuits and devices of lower voltage rating than the motor. In such cases, control transformers are used to step the voltage down to permit the use of lower voltage coil circuits. Control transformers can be supplied by manufacturers as separate units with provisions for mounting external to the controller, or they can be incorporated in the controller enclosure, and wired in with an operating coil of proper voltage rating. Such transformers can be obtained with fused or otherwise protected secondaries to meet code requirements on control-circuit overcurrent protection. Selection of the proper control transformer for a controller is a simple matter of matching the characteristics of the control circuit to the specifications of the transformer. The line voltage of the supply to the motor determines the required primary rating of the transformer. The secondary must be rated to provide the desired control-circuit voltage to match the voltage of the contactor operating coil. The continuous secondary current rating of the transformer should be sufficient for the magnetizing current of the operating coil and should also be able to handle the inrush current. In addition the control transformer should be of sufficient capacity to supply power requirements of control devices associated with the particular control circuit, that is, indicating lamps, solenoids, etc.

Two forms of control, undervoltage release and undervoltage protection, can be provided in the motor starters. In the first, if the voltage drops below a set minimum, or if the control-circuit voltage fails, the contactor will drop out but will reclose as soon as the voltage is restored. With undervoltage protection, low voltage or failure of the control-circuit voltage will cause the contactor to drop out, but the contactor will not reclose upon restoration of voltage. On some occasions it may be desirable to measure the duration of the voltage dip and if the undervoltage lasts less than some predetermined time, the motor is not disconnected. This feature is called time delay undervoltage protection.

9.6.6 Overload Protection. Motor starters are equipped with overload relays. These relays may be of the magnetic or the thermal type. If of the magnetic type, a dash pot provides the necessary delay time for the flow of starting current. If of the thermal type, this time delay is derived from the behavior of certain subcomponents of the overload in response to internally generated heat corresponding to line current. The magnetic type requires a particular coil to adapt the relay for a given size motor. In the thermal type, adaptation is accomplished by a heater of a particular size, corresponding to the motor rating. Motor-load current flowing through a compensated heater element simulates the motor overtemperature characteristics.

ANSI/NEMA ICS2-1983 [26] divides overload relays into classes: class 30, 20, and 10. The class is defined by the maximum time in seconds in which the relay must function on 6 times its ultimate trip current (ultimate trip current is 1.25 times the full-load motor current for motors having a service factor of 1.15, and 1.15 times the full-load motor current for motors having a service factor of 1.0).

The types of overload relays to be used on a particular application depend on required reliability, type of load, ambient conditions, motor type and size, safety factors, acceleration time, and probability of an overload.

In the NEC [27], Table 430-37 lists the minimum number of motor overcurrent units required to protect single- and polyphase motors connected to various supply systems. The minimum number of overload relays required for any three-phase motor application is three.

9.6.7 Solid-State Control. Solid-state controls are frequently applied with variable-speed systems, such as pump motor drives. Most solid-state systems consist of these basic sections: the sensor, the programmer, and the adjustable-speed unit.

(1) The sensor generates an electric signal proportional to the changing system conditions (pressure, flow, etc). This dc signal is applied to the programmer.

(2) The programmer determines the automatic starting and stopping sequence of each motor and sets the speed range of each adjustable-speed drive. One type of solid-state programmer consists of plug-in printed-circuit cards and a mother

board. The input signal is applied to the plug-in cards through printed circuits on the mother board. The card functions are power supply, speed programming control, sequencing control, and metering.

(3) The adjustable-speed unit controls the electric energy to an ac motor to change speed by means of a frequency-control system, a voltage-control system, or an impedance-control system. This unit contains the thyristors (silicon-controlled rectifiers) that provide the power control function to change motor speed.

9.7 References. The following publications shall be used in conjunction with this standard.

[1] ANSI C2-1987, National Electrical Safety Code.

[2] ANSI C37.06-1979, American National Standard Preferred Ratings and Related Required Capabilities for AC High-Voltage Circuit Breakers Rated on a Symmetrical Current Basis.

[3] ANSI C37.16-1980, American National Standard Preferred Ratings, Related Requirements, and Application Recommendations for Low-Voltage Power Circuit Breakers and AC Power Circuit Protectors.

[4] ANSI C57.12.10-1977, American National Standard Requirements for Transformers 230 000 Volts and Below 833/958 through 8333/10 417 kVA, Single-Phase, and 750/862 through 60 000/80 000/100 000 kVA, Three-Phase.

[5] ANSI C57.12.90-1980, American National Standard Distribution and Power Transformer Short-Circuit Test Code.

[6] ANSI C57.96-1959, American National Standard Guide for Loading Dry-Type Distribution and Power Transformers.

[7] ANSI C84.1-1982, American National Standard Voltage Ratings for Electric Power Systems and Equipment (60 Hz).

[8] ANSI/IEEE C37.04-1979 (R 1982), IEEE Standard Rating Structure for AC High-Voltge Circuit Breakers Rated on a Symmetrical Current Basis.

[9] ANSI/IEEE C37.09-1979, IEEE Standard Test Procedure for AC High-Voltage Circuit Breakers Rated on a Symmetrical Current Basis.

[10] ANSI/IEEE C37.010-1979, IEEE Application Guide for AC High-Voltage Circuit Breakers Rated on a Symmetrical Current Basis.

[11] ANSI/IEEE C37.13-1981, IEEE Standard for Low-Voltage AC Power Circuit Breakers Used in Enclosures.

[12] ANSI/IEEE C37.20-1969 (R 1981), IEEE Standard for Switchgear Assemblies Including Metal-Enclosed Bus.

[13] ANSI/IEEE C37.24-1971 (R 1984), IEEE Guide for Evaluating the Effect of Solar Radiation on Outdoor Metal-Clad Switchgear.

[14] ANSI/IEEE C37.100-1981, IEEE Standard Definitions for Power Switchgear.

[15] ANSI/IEEE C57.12.00-1980, IEEE Standard General Requirements for Liquid-Immersed Distribution, Power, and Regulating Transformers.

[16] ANSI/IEEE C57.12.01-1979, IEEE Standard General Requirements for Dry-Type Distribution and Power Transformers.

[17] ANSI/IEEE C57.13-1978, IEEE Standard Requirements for Instrument Transformers.

[18] ANSI/IEEE C57.92-1981, IEEE Guide for Loading Mineral Oil-Immersed Power Transformers Up to and Including 100 MVA with 55 °C or 65 °C Winding Rise.

[19] ANSI/IEEE C57.102-1974, IEEE Guide for Acceptance and Maintenance of Transformer Askarel in Equipment.

[20] ANSI/IEEE Std 100-1984, IEEE Standard Dictionary of Electrical and Electronics Terms.

[21] ANSI/IEEE Std 142-1982, IEEE Recommended Practice for Grounding of Industrial and Commercial Power Systems.

[22] ANSI/IEEE Std 241-1983, IEEE Recommended Practice for Electric Power Systems in Commercial Buildings.

[23] ANSI/IEEE Std 242-1986, IEEE Recommended Practice for Protection and Coordination of Industrial and Commercial Power Systems.

[24] ANSI/IEEE Std 446-1980, IEEE Recommended Practice for Power Systems for Industrial and Commercial Applications.

[25] ANSI/NEMA ICS1-1983, General Standards for Industrial Control and Systems.

[26] ANSI/NEMA ICS2-1983, Industrial Control Devices, Controllers, and Assemblies.

[27] ANSI/NFPA 70-1987, National Electrical Code.

[28] ANSI/NFPA 70B-1983, American National Standard Recommended Practice for Electrical Equipment Maintenance.

[29] ANSI/UL 508-1983, Safety Standard for Industrial Control Equipment.

[30] NEMA AB1-1975 (R 1981), Molded-Case Circuit Breakers.

[31] NEMA PB2.1-1979, Instructions for Safe Handling, Installation, Operation, and Maintenance of Deadfront Distribution Switchboards Rated 600 Volts or Less.

[32] NEMA SG2-1981, High-Voltage Fuses.

[33] NEMA SG3-1981, Low-Voltage Power Circuit Breakers.

[34] NEMA SG4-1975 (R 1980), AC High-Voltge Circuit Breakers.

[35] NEMA SG5-1981, Power Switchgear Assemblies.

[36] NEMA TR1-1980, Transformers, Regulators, and Reactors.

[37] NEMA 201-1982, Primary Unit Substations.

[38] NEMA 210-1982, Secondary Unit Substations.

10. Instruments and Meters

10.1 Introduction. This chapter covers instruments and meters utilized in industrial power distribution systems. (See Figs 149, 150, and 151 for examples.) Metering and instrumentation are essential to satisfactory plant operation. The amount required depends on the size and complexity of a plant, as well as on economic factors.

An instrument is defined as a device for measuring the value of a quantity under observation. Instruments may be either indicating or recording types.

A meter is defined as a device that measures and registers the integral of a quantity with respect to time. The term meter is also commonly used with other words, as in varmeter, voltmeter, frequency meter, even though these devices, technically speaking, should be classed as instruments.

Because ac or dc instruments and meters are usually constructed differently, they should not be used interchangeably. In general, dc instruments and meters cannot be used on ac circuits. However, ac instruments and meters, depending on their construction, may be adequate for dc purposes. The manufacturer's instructions or the nameplate should dictate usage.

10.2 Basic Objectives. Instrumentation and meters are used in plants for such purposes as operating, monitoring, billing, accounting, planning, ensuring safe operations, conserving energy, and maintaining equipment. They provide information relative to the magnitude of an electrical load, energy consumption, load characteristics, load factor, power factor, voltage, etc. The electrical equipment of a plant requires certain performance checks prior to placing instruments and meters into service—determining whether the voltages are correct, the insulation is in proper condition, connections have been properly made, etc. After the equipment is in service, additional periodic checks are necessary to ensure that the equipment remains in proper operating condition or to locate problems. Care should be taken to ensure the compatibility of these devices to their application so that the user is not injured or the instruments damaged. All instrumentation should be checked and recalibrated periodically.

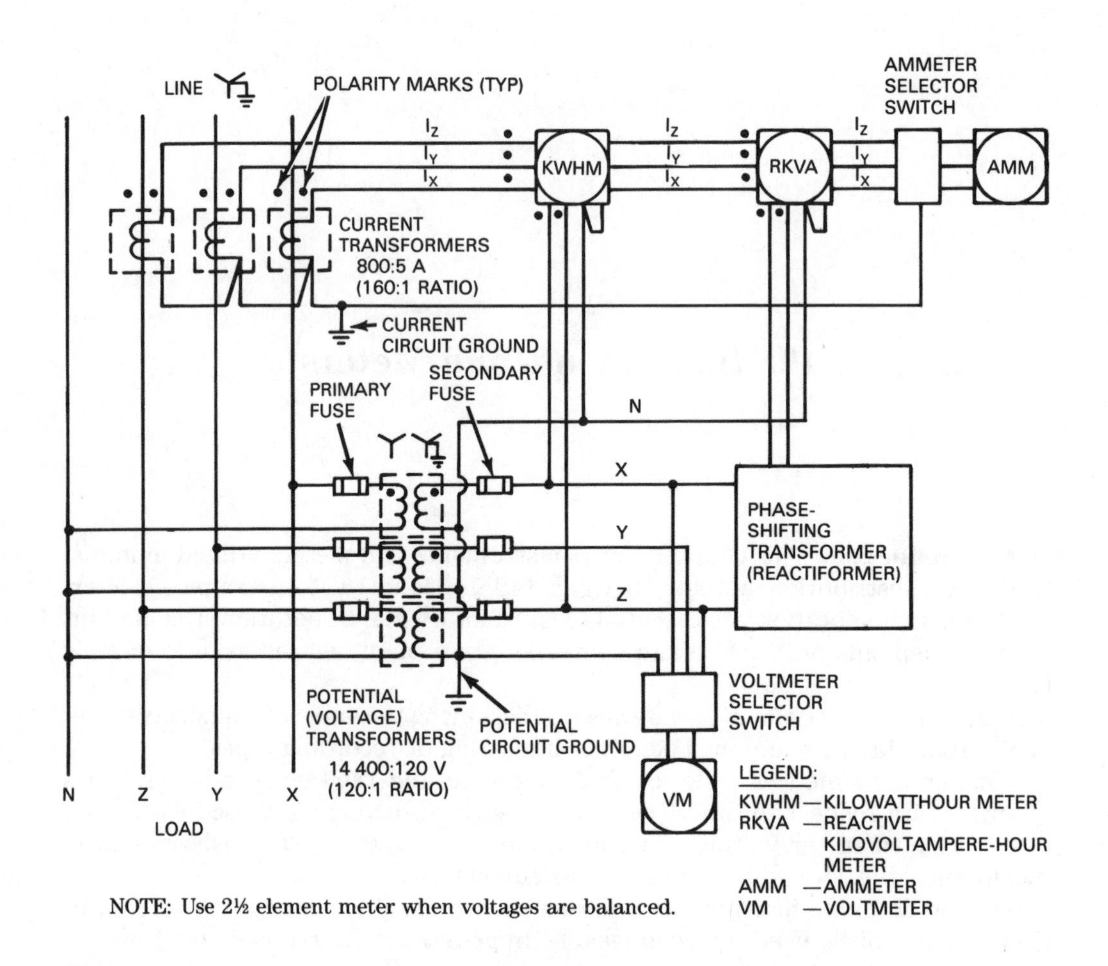

NOTE: Use 2½ element meter when voltages are balanced.

Fig 149
Sample Metering Scheme
(3-Phase, 4-Wire, High Current and Voltage)

10.3 Switchboard and Panel Instruments. Switchboard and panel instruments are permanently mounted, and most are single-range devices used in the continuing operation of a plant. The current coils of most instruments are rated 5 A; their potential coils are rated 120 V. Whenever the current and voltage of a circuit exceed the rating of the instruments, current and potential (voltage) transformers are required.

In general, *switchboard instruments* are physically larger, have longer scale lengths, are more tolerant of transients and vibrations, and are more accurate than an equivalent *panel instrument*. For example, an ammeter for switchboard use might be 4-5 in square and have a scale length of 6 in and an accuracy of 1%. An

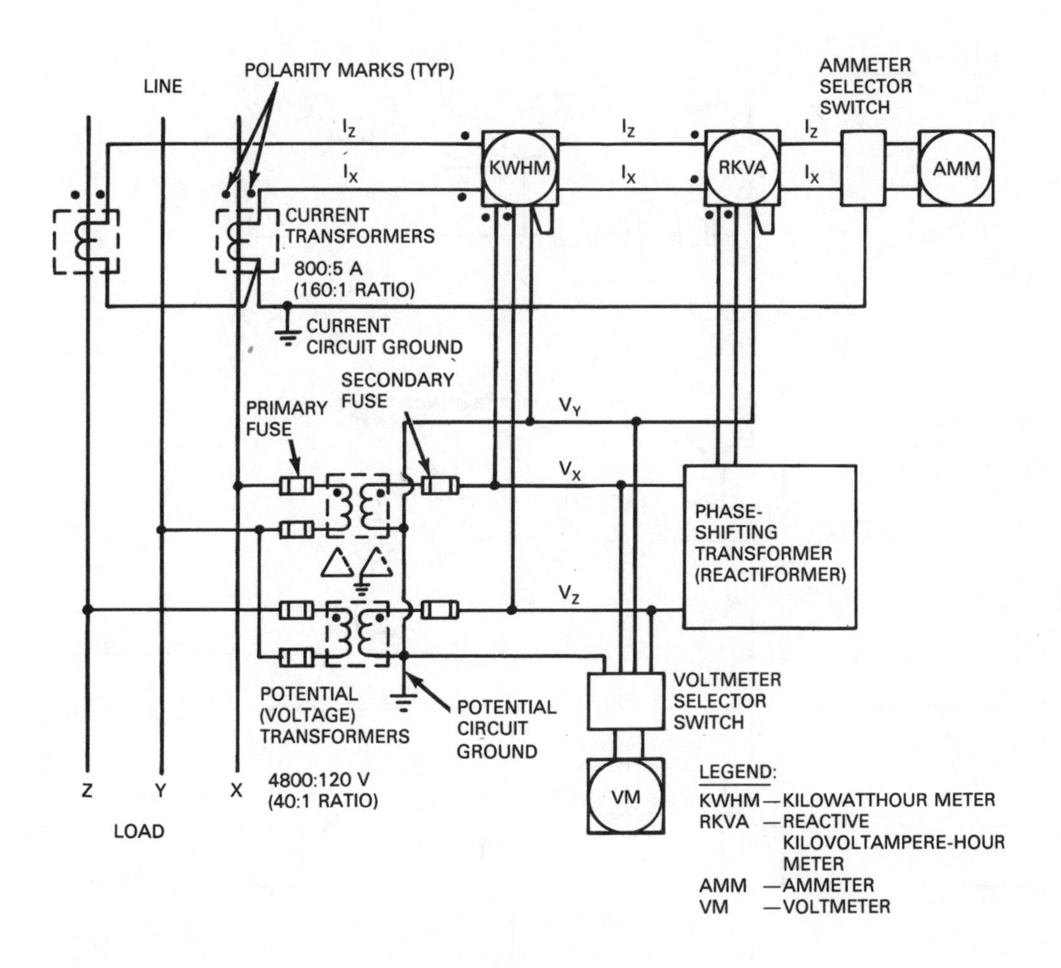

Fig 150
Sample Metering Scheme
(3-Phase, 3-Wire, High Current and Voltage)

equivalent panel ammeter might have a diameter of 2–3 in, a scale length of 1½ in, and an acccuracy of 2%. With both types of instrument, it is always best to specify the size, scale length, and accuracy needed. Some of the common instruments are discussed below.

10.3.1 Ammeters. Ammeters are used to measure the current that flows in a circuit. If the current is less than 5 A, an ammeter is directly connected in the circuit to be measured. If the current is high, the ammeter is connected to a current transformer or to a shunt. Selector switches are often installed so that an ammeter may be reconnected from one phase to another.

10.3.2 Voltmeters. Voltmeters are used to measure the potential difference between conductors or terminals. A voltmeter is connected directly across the

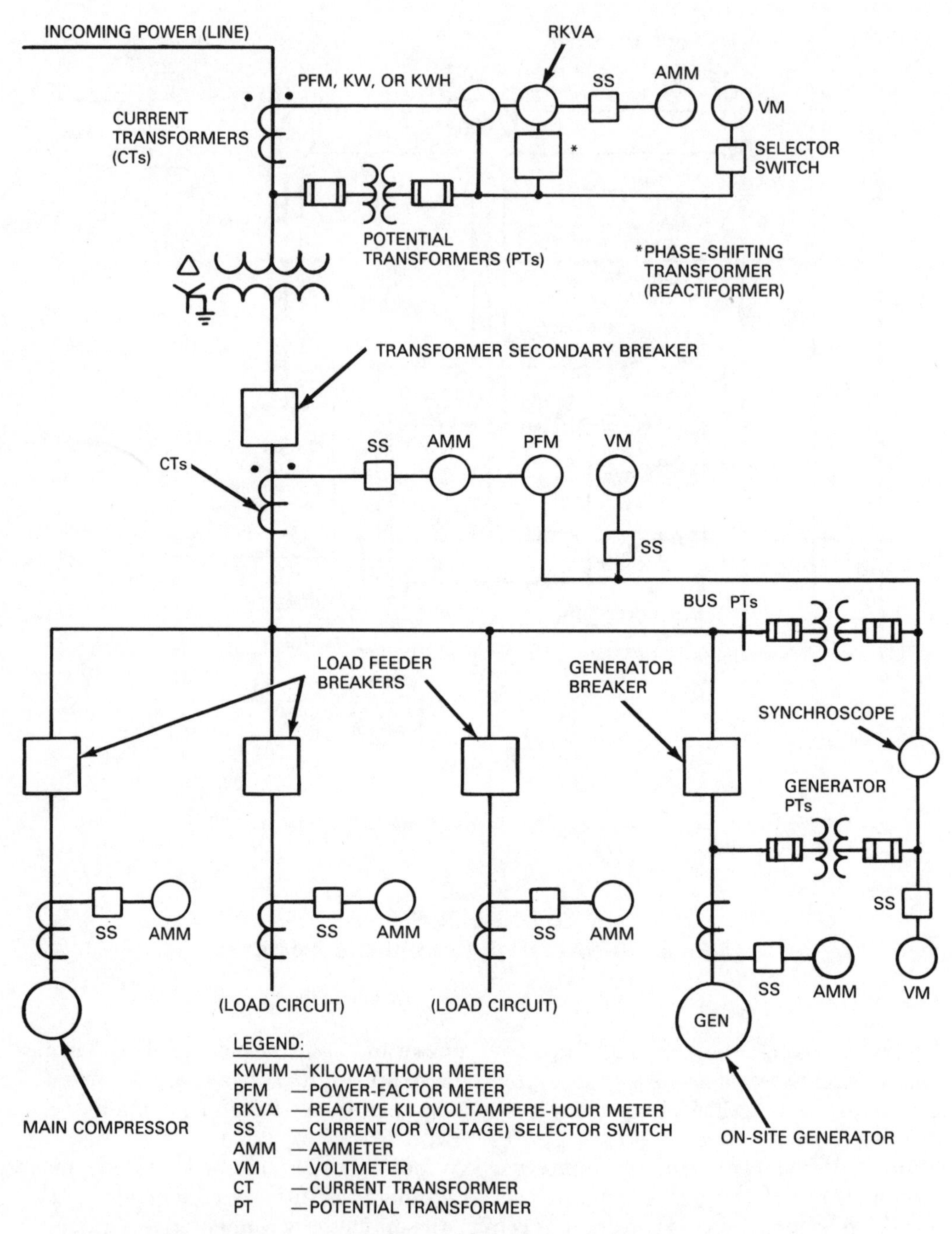

NOTE: No relaying is shown in this figure. If metering PTs and CTs are also used for relay purposes, check connected load (burden).

Fig 151
Primary Voltage Substation Sample Metering Layout

points between which the potential difference is to be measured. In general, when more than 120 V is monitored, voltage (potential) transformers would be required.

10.3.3 Wattmeters. A wattmeter measures the magnitude of electric power being delivered to a load. Proper application of this instrument requires that the polarity of voltages, phasing of voltages, and currents applied to the wattmeter be taken into account.

10.3.4 Varmeters. A varmeter measures reactive power. Varmeters usually have the zero point at the center of the scale, since reactive power may be leading or lagging. The varmeter has an advantage over a power-factor meter in that the scale is linear; thus small variations in reactive power can be read.

10.3.5 Power-Factor Meters. A power-factor meter indicates the power factor of a load. The meter indicates unity power factor at center scale, leading power factor to the left of center, and lagging power factor to the right of center. Power-factor meters are reasonably accurate only when adequately loaded. When accuracy is desired throughout the load range, a wattmeter and a varmeter should be used in combination. It should be noted that a power-factor meter can monitor the power factor of only one phase at a time. This often leads to erroneous conclusions if the phase loads are not similar and if only one reading is taken. The proper selection of a power-factor meter or other instrument intended to monitor multiphase systems depends on the system to be monitored—for example, 3-phase 3-wire; 3-phase, 4-wire Wye; 3-phase, 4-wire delta, etc.

10.3.6 Frequency Meters. The frequency of an ac power supply can be measured directly by a frequency meter. Two commonly used types are the pointer-indicating and the vibrating-reed. These instruments are connected in the same way voltmeters are.

10.3.7 Synchroscopes. A synchroscope shows the phase-angle difference between two systems and is utilized wherever two generators or systems are to be connected in parallel. A synchroscope has the appearance of a switchboard instrument except that the pointer is free to revolve 360 degrees. When the frequency of the system being synchronized is too low, the pointer rotates in one direction; when it is too high, the pointer rotates in the opposite direction. When the frequency is the same, the pointer stands still. When the voltages are equal and the pointer indicates a zero angular difference, the circuits are said to be *in phase*, and the systems may be safely paralleled.

10.3.8 Elapsed-Time Meters. Elapsed-time meters have a small, synchronous motor that drives cyclometer dials. The dials register the cumulative amount of time a circuit or electrically driven piece of machinery is in operation.

10.4 Portable Instruments. A portable instrument has the same functions as a switchboard instrument but is typically installed in a case that can be easily moved. Ordinarily, portable instruments have many ranges and functions. They are useful for special tests or for augmenting other measuring instruments mounted on a switchboard. Portable current and voltage (potential) transformers are also available for situations in which the self-contained range of the portable instrument is not sufficient for the values to be measured. They thus provide flexible instrumentation for various conditions.

10.4.1 Volt-Ohm Meter (VOM), Multitester, or Multimeter. This instrument can indicate a wide range of voltages, resistances (in ohms), and currents (in milliamperes). It is particularly useful for investigating circuit problems.

10.4.2 Clamp-on Ammeters. A clamp-on ammeter uses a split-core current transformer to encircle a conductor and determine the amount of ac flowing. It usually has several current ranges.

10.5 Recording Instruments. Most direct-reading, indicating instruments are available as recording or curve-drawing instruments for portable or switchboard use. Charts may be strip or circular. The record may be continuous, or readings can be taken at regular intervals. The chart is moved at a constant speed by a spring or electrical clock. Recording instruments have special design problems that indicating instruments do not have, one of which is the need to provide sufficient torque to overcome pen friction without impairing the accuracy of the recording.

10.6 Miscellaneous Instruments

10.6.1 Temperature Indicators. Temperature-indicating and temperature-control devices include liquid, gas, or saturated-vapor thermometers, resistance thermometers, bimetal thermometers, and radiation pyrometers. Some may be obtained with electric contacts for use on an alarm device or relay circuit. Their application determines the type of sensor required.

10.6.2 Megohmmeters. A megohmmeter tests the insulation resistance of electric cables, insulators, buses, motors, and other electric equipment. It consists of either a hand-cranked or motor-driven dc generator and resistance indicator. It is calibrated in megohms and is available in different voltage ratings, usually 500, 1000, or 2500 V dc.

A resistance test of the electrical insulation, prior to placing equipment in service or during routine maintenance, will indicate the condition of the insulation. Wet or defective insulation can be detected very readily. A high reading, however, does not necessarily mean that the equipment's insulation can withstand rated potential since the megohmmeter's voltage normally is not equal to the equipment's rated potential. A high-potential test is commonly performed after the equipment has passed the megohmmeter test. Periodic testing and plotting the resistance readings will show trends that indicate possible problems.

10.6.3 Ground Ohmmeters. A ground ohmmeter measures the resistance to earth of ground electrodes. It is calibrated in ohms, usually 0–300. Some types also provide for the measurement of soil resistance.

10.6.4 Oscillographs. An oscillograph is an instrument for observing and recording rapidly changing values of short duration, such as the waveform of alternating voltage, current, or power transients. They are available for frequencies up to 10 kHz.

10.6.5 Oscilloscopes. Oscilloscopes are electronic instruments used to study very high frequencies or phenomena of short duration. They can be used to study transients that occur in power circuits. These instruments use electronic controls and an electron beam, thereby eliminating the inertia of mechanical instruments. Oscilloscopes can be used for frequencies up to millions of hertz. A storage scope will display this waveform for a short period, and a camera can be used with the oscilloscope to record the waveform permanently.

10.7 Meters. Meters are devices that distinguish and register the integral of a quantity with respect to time.

10.7.1 Kilowatthour Meters. A kilowatthour meter measures the amount of energy consumed by a load. AC kilowatthour meters use an induction-disk type of mechanism; the disk revolves at a speed proportional to the rate at which energy passes through the meter. The metered kilowatthours are indicated on a set of dials driven by the revolving disk through a gear train. Recently, solid-state kilowatthour meters have been developed.

The kilowatthour meter may be used to calculate the power being used by a load at the moment of testing. To calculate power, count the seconds for a given number of revolutions of the disk, and then use this formula:

$$\text{Power (in kilowatts)} = \frac{3.6 \cdot r \cdot \text{Kh} \cdot \text{multiplier}}{\text{seconds}}$$

Kh is the meter disk constant, and r the number of revolutions. The Kh will be noted on the kilowatthour meter. The multiplier is 1 unless a meter is installed with instrument transformers. If current transformers are installed, the multiplier is equal to the ratio of the current transformer. For example, 400:5 current transformers have a ratio of 80 to 1, and so the meter multiplier would be 80. If voltage (potential) transformers are also installed, the meter multiplier is the product of the current transformer ratio and the voltage transformer ratio. A meter connected to 400:5 (80 ratio) current transformers and 14 400:120 (120 ratio) voltage transformers would have an overall multiplier of 80 · 120, or 9600.

Kilowatthour meters come in several classes. Below is a listing of the common classes along with the maximum current each can safely monitor.

Class 10	10 A
Class 20	20 A
Class 100	100 A
Class 200	200 A
Class 320	320 A

High-current services would require a Class 10 or Class 20 meter employed with current transformers. For example, a 1000 A service would use 1000:5 (200:1 ratio) current transformers and a Class 10 (or Class 20) meter.

The following kilowatthour meter application data can be used as a general guideline only. The number of phases being metered, the number of wires, the degree of phase-to-phase current, and power-factor balance all have an effect on the number of stators (or coils) the kilowatthour meter should have. An *unbalanced* condition would exist if the phase-to-phase differences in either load current or load power factor were great.

The following data define the number of stators and, if required by the service voltage or load size, the number of current and voltage (potential) transformers required to properly meter the more commonly encountered services:

Service Voltage	Stators	CTs	PTs	
1-phase, 2-wire	1	1	1	
1-phase, 3-wire	1	2	1	
1-phase, 3-wire	2	2	2	
1-phase, 3-wire (wye)	2	2	2	
3-phase, 3-wire (delta)	2	2	2	
3-phase, 4-wire (wye)	2½	3	2	*balanced conditions*
3-phase, 4-wire (wye)	3	3	3	*unbalanced conditions*

Other factors used in selecting kilowatthour meters include:

Type of mountings: socket, bottom-connected
Voltage: 120, 240, 480, 240/120, etc
Register: clock, cyclometer (like an odometer)
Type of load current bypass: automatic, manual

The probability of error in selecting or connecting a kilowatthour meter is high, especially if instrument transformers are required. If there is any doubt, consult a metering specialist. The high probability for error also applies to kilovarhour and demand meters.

10.7.2 Kilovarhour Meters. A kilovarhour (kvarh) meter measures the amount of reactive energy—the integral of reactive power—drawn by a load. The internal mechanisms of the kilovarhour meter are similar to those of a kilowatthour (kWh) meter. However, the potential applied to this meter is shifted 90 electrical degrees. A standard kilowatthour meter and a phase-shifting transformer (reactiformer) can be connected to function as a kilovarhour meter.

To calculate kilovar demand, apply the timing formula defined in 10.7.1. Data derived from the kilovarhour meter and a kilowatthour meter may be used to calculate power factor by applying the following trigometric formula:

$$\cos \arctan \left(\frac{\text{kvarh}}{\text{kWh}} \right) \cdot 100 = \text{power factor in percent}$$

Most kilovarhour meters have a ratchet-type assembly to prevent them from running backwards. For this reason, depending upon the connection, they can record only lagging or leading flow.

10.7.3 *Q*-Meters. Identified by a coined name assigned by certain utilites but at variance with the definition in ANSI/IEEE Std 100-1984 [3]* this *Q*-meter is a kilowatthour meter with voltages displaced 60 electrical degrees (lagging) from the standard connection. Separate voltage phase-shifting transformers (reactiformers)

* The numbers in brackets correspond to those of the references listed at the end of this chapter; when preceded by B, they correspond to the bibliography at the end of this chapter.

are not needed. A *Q*-meter reading does not by itself equal a standard defined quantity. But, provided that the actual power factor is between limits of 0.50 lagging and 0.866 leading, coincident readings of both a *Q*-meter and a standard kilowatthour meter are data for a calculation that determines reactive energy (kilovarhours) and power factor.

10.7.4 Demand Meters. Demand meters register the average usage of power during a specified interval. They record the demand for each interval or, by means of a maximum indicating pointer, indicate the maximum demand that has occurred since the meter was last reset.

A lagged demand meter indicates demand by a thermally driven pointer on a scale. The internal thermal characteristics of the meter determine the time interval. Basically, a red demand pointer indicates the load through the course of the high-load period. This pointer moves a black maximum pointer upscale with it, and the black pointer will stay at the maximum value until the meter is read and reset. The value of a constant load is reasonably approximated by this type of meter after two time intervals.

A printing demand meter records the average power during a specific interval. A kilowatthour meter equipped with a contactor device provides information on energy usage, with each impulse (contact closure) representing the usage of a specified amount of energy. The printing demand meter records the total number of impulses received during a given time interval. The record may be on printed paper tape, a chart, punched tape, magnetic tape, or a computer memory chip. A computer is required to process the information from the magnetic tape or memory chip.

10.7.5 Voltage-Squared Meters. Identified by a coined name (which is misleading because the meter reading directly determines effective voltage and not the square of voltage), a V^2 (voltage-squared) meter is an integrating meter similar in construction to a kilowatthour meter. The effective (rms) voltage for a time interval is the dial reading difference divided by the interval.

10.7.6 Ampere-Squared Meters. Identified by a coined name (which is misleading because the meter reading directly determines effective current and not the square of current), an A^2 (ampere-squared) meter is an integrating meter similar in construction to a kilowatthour meter. The effective (rms) current for a time interval is the dial reading difference divided by the interval.

10.8 Auxiliary Devices

10.8.1 Current Transformers. Current transformers insulate the instrument circuit from the primary voltage. Care should be taken to ensure that the current transformer is insulated for the full system voltage. For example, current transformers in the 600 V class would be used for 480 V systems, and current transformers in the 15 kV class would be used for 13.8 kV systems.

Current transformers also reduce current through the connected instruments to values within the rating of the instrument elements — usually 5 A. The best ratio choice would be the one that would supply 5 A to the instruments when the circuit being monitored is carrying a current equal to the highest anticipated load under normal conditions.

Usually, a current transformer will generate a dangerously high potential when the secondary current circuit is opened. Therefore, a shorting bar, test switch, or current jack should be provided to short-circuit the transformer secondary when the connected instrument is being tested. If a test switch or current jack is installed, it would be possible to connect portable meters whenever needed.

Current transformers must have a secondary circuit ground. The ground will establish a firm ground reference point and will restrict the buildup of static voltages caused by the high-voltage conductor(s).

The accuracy of a current transformer is affected by the burden (load) connected to the secondary coil. For this reason, it is best to keep the burden as low as possible. Choose a current transformer that can operate accurately with the burden of the connected meters and relays.

Protective-relaying current-transformer burdens are commonly designated as follows:

Type	Maximum VA Burden	Maximum External Impedance
B = 0.1	2.5	0.1 Ω
B = 0.2	5.0	0.2 Ω
B = 0.5	12.5	0.5 Ω
B = 1.0	25.0	1.0 Ω
B = 2.0	50.0	2.0 Ω

The accuracy of a current transformer or voltage transformer is usually stated as a percent at rated burden; that is, 0.3 at B = 0.1 means 0.3% accuracy at a maximum burden of 2.5 VA at 5 A.

10.8.2 Voltage (Potential) Transformers. Voltage (potential) transformers provide a secondary voltage compatible to the rating of the instrument's potential coil. Switches should be provided in the secondary circuit of the voltage transformer to disconnect the instrument for testing. The connected load should not exceed the VA rating of the transformer if accuracy is to be maintained.

For safety, the secondary winding of a voltage transformer should be grounded. In most industrial applications, the primary and secondary circuits are fused.

10.8.3 Shunts. In dc measurements of current or energy, shunts are used to carry the main current to be measured. Ordinarily, the leads should be calibrated with the shunt with which they are to be used. The dc ammeter actually measures the millivolt drop across its shunt and is calibrated in terms of the current rating of its associated shunt.

10.8.4 Transducers. A transducer is a device that is used to transform one or more analog inputs into another analog value more suitable for usage in instrumentation. The output is related to the input(s) in a prescribed relationship.

10.8.5 Computers. A computer is a device that performs mathematical operations upon information fed into it. The internal programs of a computer determine the flow of information from the computer. This information may be used to control or monitor an operation. Computers record and display information of an operation in useful ways.

10.9 Typical Installations. The following combinations of instruments and meters are commonly used in industrial switchgear.

10.9.1 High-Voltage Equipment (Above 600 V)

(1) *Utility supply and main feeder positions*

Voltmeter
Ammeter
Wattmeter
Varmeter or power-factor meter
Demand meter (optional)

(2) *Plant feeders*

Ammeter
Watthour meter (demand attachment optional)
Test block for portable instruments (optional)

(3) *Generators*

Voltmeter
Ammeter
Watthour meter
Varmeter or power-factor meter (optional)
Synchroscope
Frequency meter (optional)
Recording ammeter and voltmeter (optional)

(4) *Synchronous motors*

ac and dc voltmeters (optional or dc only)
ac and dc ammeters
Wattmeter (optional)
Varmeter or power-factor meter (optional)
Watthour meter (optional)
Elapsed-time meter (optional)

(5) *Large induction motors*

Ammeter
Voltmeter (optional)
Watthour meter (optional)
Elapsed-time meter (optional)

10.9.2 Low-Voltage Equipment (Below 600 V)

(1) *Utility supply and main feeder positions*

Voltmeter
Ammeter
Wattmeter
Varmeter or power-factor meter
Demand meter (optional)

(2) *Plant feeders*

Ammeter
Voltmeter (optional)
Watthour meter (optional)

(3) *Generators*

Voltmeter

Ammeter
Watthour meter
Varmeter or power-factor meter (optional)
Synchroscope
Frequency meter
Recording ammeter and voltmeter (optional)
(4) *Synchronous motors*
ac and dc voltmeters (optional or dc only)
ac and dc meters
Wattmeter (optional)
Varmeter or power-factor meter (optional)
Watthour meter (optional)
Elapsed-time meter (optional)
(5) *Large induction motors*
Ammeter
Voltmeter (optional)
Watthour meter (optional)
Elapsed-time meter (optional)

Because of added cost, these instruments are used in low-voltage feeders only when justified by a potential saving in operations or maintenance. If permanent instrumentation is not used, checks are made periodically with portable instruments.

10.10 References. This standard shall be used in conjunction with the following publications:

[1] ANSI C12.1-1982, Code for Electricity Metering.

[2] ANSI/IEEE C57.13-1978 (R 1986), IEEE Standard Requirements for Instrument Transformers.

[3] ANSI/IEEE Std 100-1984, IEEE Standard Dictionary of Electrical and Electronics Terms.

[4] ANSI/IEEE Std 242-1986, Recommended Practice for Protection and Coordination of Industrial and Commercial Power Systems, Chapter 4.

[5] *Electric Utility Engineering Reference Book*, vol 3: *Distribution Systems*, Chapter 11. Trafford, PA: Westinghouse Electric Corporation, 1965.

11. Cable Systems

11.1 Introduction. The primary function of cable is to carry energy reliably between source and utilization equipment. In carrying this energy there are heat losses generated in the cable that must be dissipated. The ability to dissipate these losses depends on how the cables are installed, and this affects their ratings.

Cables may be installed in raceway, cable trays, underground in duct or direct buried, preassembled on a messenger, in cable bus, or as open runs of cable.

Selection of conductor size requires consideration of load current to be carried and loading cycle, emergency overloading requirements and duration, fault clearing time and interrupting capacity of the cable overcurrent protection or source capacity, and voltage drop for the particular installation conditions.

Insulations can be classified in broad categories as solid insulations, taped insulations, and special-purpose insulations. Cables incorporating these insulations cover a range of maximum and normal operating temperatures and exhibit varying degrees of flexibility, fire resistance, and mechanical and environmental protection.

Installation of cables requires care to avoid excessive pulling tensions that could stretch the conductor or insulation shield or rupture the cable finish when pulled around bends. The minimum bending radius of the cable or conductors should not be exceeded during pulling around bends, at splices, and particularly at terminations to avoid damage to the conductors.

Provisions should be made for the proper terminating, splicing, and grounding of cables. Minimum clearances must be maintained between phases and between phase and ground for the various voltage levels. The terminating compartments should be designed and constructed to prevent condensation from forming. Condensation or contamination on high-voltage terminations could result in tracking over the terminal surface with possible flashover.

Many users test cables after installation and periodically test important circuits. Test voltages are usually direct current of a level recommended by the cable manufacturer for his particular cable. Usually this test level is well below the dc strength of the cable, but it is possible for accidential flashovers to weaken or rupture the cable insulation due to the higher transient overvoltages that can occur from reflections of the voltage wave. ANSI/IEEE Std 400-1980 [8][28] provides a detailed discussion on cable testing.

The application and sizing of cables rated up to 35 kV is governed by ANSI/NFPA 70-1987 [12] (National Electrical Code [NEC]). Cable use may also be covered in state and local regulations recognized by the local electrical inspection authority having jurisdiction in a particular area.

The various tables in this chapter are intended to assist the electrical engineer in laying out and understanding, in general terms, his cable requirements.

11.2 Cable Construction

11.2.1 Conductors. The two conductor materials in common use are copper and aluminum. Copper has historically been used for conductors of insulated cables due primarily to its desirable electrical and mechanical properties. The use of aluminum is based mainly on its favorable conductivity-to-weight ratio, the highest of the electrical conductor materials, its ready availability, and the lower cost of the primary metal.

The need for mechanical flexibility usually determines whether a solid or a stranded conductor is used, and the degree of flexibility is a function of the total number of strands. The NEC [12] requires conductors of size 8 AWG and larger to be stranded. A cable may be either a single insulated conductor or an assembly of insulated conductors, with or without an overall covering.

Stranded conductors are available in various configurations, such as stranded concentric, compressed, compact, rope, and bunched, with the latter two generally specified for flexing service. Bunched stranded conductors, not illustrated, consist of a number of individual strand members of the same size that are twisted together to make the required area in circular mils for the intended service. Unlike the individual strands in the concentric-lay strands, illustrated in Fig 152, no attempt is made during manufacture of bunch strand to control the position of each strand with respect to another. This type of conductor is usually found in portable cords.

11.2.2 Comparison Between Copper and Aluminum. Aluminum requires larger conductor sizes to carry the same current as copper. The equivalent aluminum cable when compared to copper in terms of ampacity will be lighter in weight and larger in diameter. The properties of these metals are given in Table 74.

The 36% difference in thermal coefficients of expansion and the different electrical nature of the oxide films of copper and aluminum require consideration in connector designs. An aluminum oxide film forms immediately on exposure of fresh aluminum surface to air. Under normal conditions it slowly builds up to a thickness in the range of 3–6 nanometers and stabilizes at this thickness. The oxide film is essentially an insulating film or dielectric material and provides aluminum with its corrosion resistance. Copper produces its oxide rather slowly under nor-

[28] The numbers in brackets correspond to those of the references listed at the end of this chapter; when preceded by B, they correspond to the bibliography at the end of this chapter.

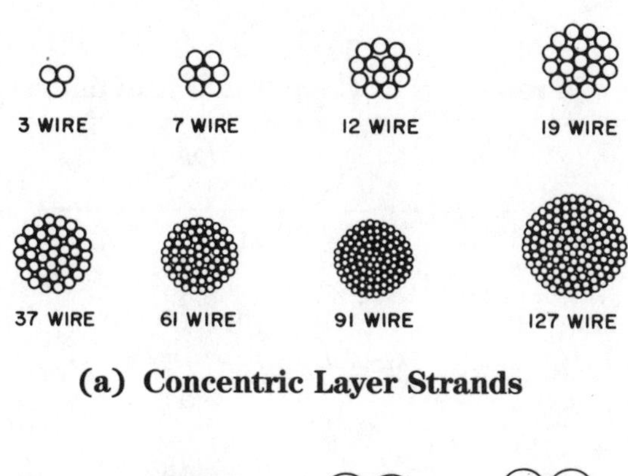

(a) Concentric Layer Strands

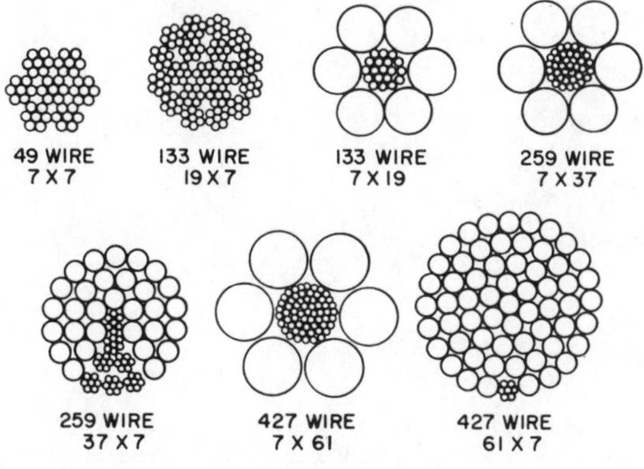

(b) Concentric Rope-Lay Strands

Fig 152
Conductor Stranding

mal conditions, and the film is relatively conducting, presenting no real problem at connections.

Approved connector designs for aluminum conductors essentially provide increased contact areas and lower unit stresses. These terminals possess adequate strength to ensure that the compression of the aluminum strands exceeds their yield strength and that a brushing action takes place that destroys the oxide film to form an intimate aluminum contact area yielding a low-resistance connection. Recently developed aluminum alloys provide improved terminating and handling compared to EC grades.

Water should be kept from entering the strand space in aluminum conductors at all times. Any moisture within a conductor, either copper or aluminum, is likely to cause corrosion of the conductor metal or impair insulation effectiveness.

11.2.3 Insulation. Basic insulating materials are either organic or inorganic, and there are a wide variety of insulations classed as organic. Mineral-insulated cable employs the one inorganic insulation (MgO) that is generally available.

Table 74
Properties of Copper and Aluminum

	Copper Electrolytic	Aluminum EC
Conductivity, % IACS† at 20 °C	100.0	61.0
Resistivity, Ω × cmil/ft at 20 °C	10.371	17.002
Specific gravity at 20 °C	8.89	2.703
Melting point, °C	1083	660
Thermal conductivity at 20 °C, (cal × cm)/(cm² × °C × s)*	0.941	0.58
Specific heat, cal/(g × °C)*		
for equal weights	0.092	0.23
for equal direct-current resistance	0.184	0.23
Thermal expansion, in; equal to constant ×10⁻⁶ length in inches × °F	9.4	12.8
steel = 6.1		
18-8 stainless = 10.2		
brass = 10.5		
bronze = 15		
Relative weight for equal direct-current resistance and length	1.0	0.50
Modulus of elasticity, (lb/in²) × 10⁶	16	10

* In this table, ca denotes the gram calorie.
† International annealed copper standard

Insulations in common use are
(1) Thermosetting compounds, solid dielectric
(2) Thermoplastic compounds, solid dielectric
(3) Paper-laminated tapes
(4) Varnished cloth, laminated tapes
(5) Mineral insulation, solid dielectric granular

Most of the basic materials listed in Table 75 are modified by compounding or mixing with other materials to produce desirable and necessary properties for manufacturing, handling, and end use. The thermosetting or rubber-like materials are mixed with curing agents, accelerators, fillers, and anti-oxidants in varying proportions, and crosslinked polyethylene (XLPE) is included in this class. Generally, smaller amounts of materials are added to the thermoplastics in the form of fillers, anti-oxidants, stabilizers, plasticizers, and pigments.

(1) *Insulation Comparison.* The aging factors of heat, moisture, and ozone are among the most destructive of organic based insulations, so the following comparisons are a gauge of the resistance and classifications of these insulations:

(a) *Relative Heat Resistance.* The comparison in Fig 153 illustrates the effect of a relatively short period of exposure at various temperatures on the hardness

Table 75
Commonly Used Insulating Materials

Common Name	Chemical Composition	Properties of Insulation Electrical	Physical
Thermosetting			
Crosslinked polyethylene	Polyethylene	Excellent	Excellent
EPR	Ethylene propylene rubber (copolymer and terpolymer)	Excellent	Excellent
Butyl	Isobutylene isoprene	Excellent	Good
SBR	Styrene butadiene rubber	Excellent	Good
Oil base	Complex rubber-like compound	Excellent	Good
Silicone	Methyl chlorosilane	Good	Good
TFE*	Tetrafluoroethylene	Excellent	Good
ETFE†	Ethylene Tetrafluoroethylene	Excellent	Excellent
Neoprene	Chloroprene	Fair	Good
Class CP rubber‡	Chlorosulfonated polyethylene	Good	Good
Thermoplastic			
Polyethylene	Polyethylene	Excellent	Good
Polyvinyl chloride	Polyvinyl chloride	Good	Good
Nylon	Polyamide	Fair	Excellent

*For example, Teflon or Halon.
†For example, Tefzel.
‡For example, Hypalon.

Fig 153
Typical Values for Hardness Versus Temperature

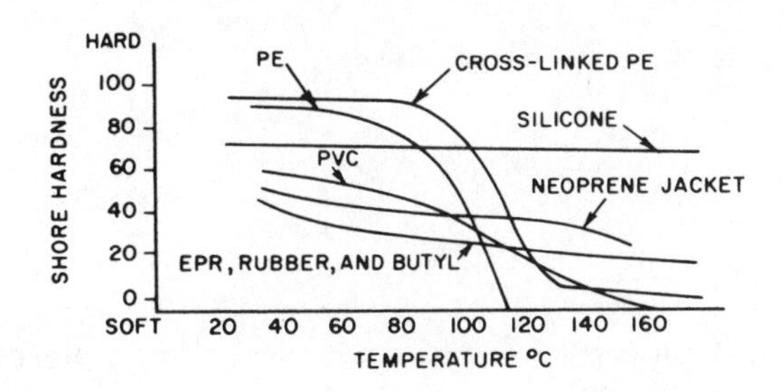

characteristic of the material at that temperature. Basic differences between thermoplastic and thermosetting type insulation, excluding aging effect, are evident.

(b) *Heat Aging*. The effect on elongation of an insulation (or jacket) when subjected to aging in a circulating air oven is an acceptable measure of heat resistance. The air oven test at 121 °C called for in some specifications is severe, but provides a relatively quick method of grading materials for possible use at high conductor temperatures or in hot-spot areas. The 150 °C oven aging is many times more severe and is used to compare materials with superior heat resistance. Temperature ratings of insulations in general use are shown in Table 76.

Table 76
Rated Conductor Temperatures

Insulation Type	Maximum Voltage Class (kV)	Maximum Operating Temperature (°C)	Maximum Overload* Temperature (°C)	Maximum Short-Circuit Temperature (°C)
Paper (solid-type) multiconductor and single conductor, shielded	9	95	115	200
	29	90	110	200
	49	80	100	200
	69	65	80	200
Varnished cambric	5	85	100	200
	15	77	85	200
	28	70	72	200
Polyethylene (natural)†	5	75	95	150
	35	75	90	150
SBR rubber	2	75	95	200
Butyl rubber	5	90	105	200
	35	85	100	200
Oil-base rubber	35	70	85	200
Polyethylene (crosslinked)†	35	90	130	250
EPR rubber†	35	90	130	250
Chlorosulfonated polyethylene‡	2	90	130	250
Polyvinyl chloride	2	60	85	150
	2	75	95	150
	2	90	105	150
Silicone rubber	5	125	150	250
Ethylene tetrafluoroethylene§	2	150	200	250

* Operation at these overload temperatures shall not exceed 100 h/yr. Such 100 h overload periods shall not exceed five.
† Cables are available in 69 kV and higher ratings.
‡ For example, Hypalon.
§ For example, Tefzel.

(c) *Ozone and Corona Resistance.* Exposure to accelerated conditions, such as higher concentrations of ozone (as standardized by the Insulated Cable Engineers Association (ICEA), EM-60 [23]-[26], for butyl, 0.03% ozone for 3 h at room temperature), or air oven tests followed by exposure to ozone, or exposure to ozone at higher temperatures, aids in measuring the ultimate ozone resistance of the material. Insulations exhibiting superior ozone resistance under accelerated conditions are silicone, polyethylene, XLPE, EPR, and polyvinyl chloride. In fact, these materials are, for all practical purposes, inert in the presence of ozone. However, this is not the case with corona discharge.

The phenomenon of corona discharge produces concentrated and destructive thermal effects along with formation of ozone and other ionized gases. Although corona resistance is a property associated with cables over 600 V, in a properly designed and manufactured cable, damaging corona is not expected to be present at operating voltage. Materials exhibiting less susceptibility than polyethylene and XLPE to such discharge activity are the ethylene propylene rubbers (EPRs).

(d) *Moisture Resistance.* Insulations such as XLPE, polyethylene, and EPR exhibit excellent resistance to moisture as measured by such standard industry tests as ICEA EM-60 [23]-[26]. The electrical stability of these insulations in water as measured by capacitance and power factor is impressive. A degradation phenomenon called *treeing* has been found to be aggravated by the presence of water. This phenomenon appears to occur in solid dielectric insulations and is more prevalent in polyethylene and XLPE. The capacitance and percent power factor of the ICEA EM-60 [23]-[26] test of natural polyethylene and some XLPEs are lower than those of EPR or other elastomeric power cable insulations.

(2) *Insulations in General Use.* Insulations in general use for 2 kV and above are shown in Table 76. Solid dielectrics of both plastic and thermosetting types are being used more and more commonly, while the laminated-type constructions, such as paper-lead cables, are declining in popularity in industrial service.

The generic names given for these insulations cover a broad spectrum of actual materials, and the history of performance on any one type may not properly be related to another in the same generic family.

11.2.4 Cable Design. The selection of power cable for particular circuits or feeders develops around the following considerations:

(1) *Electrical.* Dictates conductor size, type and thickness of insulation, correct materials for low- and medium-voltage designs, consideration of dielectric strength, insulation resistance, specific inductive capacitance (dielectric constant), and power factor

(2) *Thermal.* Compatible with ambient and overload conditions, expansion, and thermal resistance

(3) *Mechanical.* Involves toughness and flexibility, consideration of jacketing or armoring, and resistance to impact, crushing, abrasion, and moisture

(4) *Chemical.* Stability of materials on exposure to oils, flame, ozone, sunlight, acids, and alkalies

The installation of cable in conformance with the NEC [12] and state and local codes under the jurisdiction of a local electrical inspection authority requires evidence of approval for use in the intended service by a nationally recognized testing laboratory, such as Underwriters Laboratories Inc (UL) labeled cable. Some NEC [12] types are discussed below.

11.2.4.1 Low-Voltage Cables. Low-voltage power cables are generally rated at 600 V, regardless of the use voltage, whether 120, 208, 240, 277, 480, or 600 V.

The selection of 600 V power cable is oriented more to physical rather than to electrical service requirements. Resistance to forces, such as crush, impact, and abrasion, becomes a predominant factor, although good electrical properties for wet locations are also needed.

The 600 V compounds of XLPE are usually filled (carbon black or mineral fillers) to further enhance the relatively good toughness of conventional polyethylene. The combination of crosslinking the polyethylene molecules through vulcanization plus fillers produces superior mechanical properties. Vulcanization eliminates polyethylene's main drawback of a relatively low melting point of 105 °C. The 600 V cable consists simply of the conductor with a single extrusion of insulation in the specified thickness.

Rubberlike insulations such as EPR and SBR have been provided with outer jackets for mechanical protection, usually of polyvinyl chloride, neoprene, or CP rubber, such as Hypalon. However, the newer EPR insulations have improved physical properties that do not require an outer jacket for mechanical protection. A guide list of the more commonly used 600 V cables follows. Cables are classified by conductor operating temperatures and coverings with NEC [12] insulation thicknesses.

(1) EPR insulated with or without jacket, Type RHW for 75 °C maximum operating temperature in wet or dry locations, and Type RHH for 90 °C in dry locations only.

(2) XLPE insulated, without jacket, Type XHHW, for 75 °C maximum operating temperature in wet locations, and 90 °C in dry locations only.

(3) Polyvinyl chloride insulated, nylon jacketed, Type THWN, for 75 °C maximum operating temperature in wet or dry locations, and Type THHN for 90 °C in dry locations only.

(4) Polyvinyl chloride insulated, without jacket, Type THW, for 75 °C maximum operating temperature in wet or dry locations.

(5) Metal-clad or interlocked armor cable, Type MC; individual insulated conductors are usually type XHHW or RHH/RHW, and the cable has the rating of the conductors used; for use in any raceway, in cable tray, as open runs of cable, direct buried, or as aerial cable on a messenger.

(6) Tray cable, Type TC; multiconductor with an overall flame-retardant nonmetallic jacket; individual conductors may be any of the above, and cable takes rating of insulation selected; for use in cable trays, raceways, or where supported in outdoor locations by a messenger wire.

The preceding cables are suitable for installation in conduit, duct, or other raceway, and, when specifically approved for the purpose, may be installed in cable tray, or direct buried, provided NEC [12] requirements are satisfied.

Cables (2) and (4) are usually restricted to conduit or duct. Single conductors may be furnished paralleled or multiplexed, as multiconductor cables with overall nonmetallic jacket or as preassembled aerial cable on a messenger.

Note that the temperatures listed are the maximum rated operating temperatures as specified in the NEC [12].

11.2.4.2 Power-Limited Circuit Cables. When the power in the circuit is limited to the levels defined in the NEC [12], Article 725, for remote-control, signaling, and power-limited circuits, then the wiring method may utilize Power-Limited Circuit Cable or Type PLTC, Power-Limited Tray Cable. These cables, which are rated 300 V, include copper conductors for electrical circuits and thermocouple alloys for thermocouple extension wire.

Similarly Power-Limited Fire-Protective Signaling-Circuit Cable may be used on circuits that comply with the power limitations of the NEC [12], Article 760.

11.2.4.3 Medium-Voltage Cables. Type MV medium-voltage power cables have solid extruded dielectric insulation and are rated from 2001 to 35 000 V.

These single conductor and multiple conductor cables are available with nominal voltage ratings of 5, 8, 15, 25, and 35 kV.

EPR and XLPE are the usual insulating compounds for Type MV cables; however, polyethylene and butyl rubber are also authorized as insulations. The maximum operating temperatures are 90 °C for EPR and XLPE, 85 °C for butyl rubber, and 75 °C for polyethylene.

Type MV cables may be installed in raceways in wet or dry locations. The cable must be specifically approved for installation in cable tray, direct burial, exposure to sunlight, or for messenger supported wiring.

Multiconductor medium-voltage cables that also comply with the requirements for Type MC Metal-Clad cables will be labeled as Type MV or MC and may be installed as open runs of cable.

11.2.4.4 Shielding of Medium-Voltage Cable. For operating voltages below 2 kV, nonshielded constructions are normally used, while above 2 kV, cables are required to be shielded to comply with the NEC [12] and ICEA [23]-[26]. The NEC [12] does provide for the use of nonshielded cables up to 8 kV, provided the conductors are listed by a nationally recognized testing laboratory and are approved for the purpose. Where nonshielded conductors are used in wet locations, the insulated conductor(s) must have an overall nonmetallic jacket or a continuous metallic sheath, or both. Refer to the NEC [12] for specific insulation thicknesses for wet or dry locations.

Since shielded cable is usually more expensive than nonshielded cable, and the more complex terminations require a larger space in the terminal boxes, the nonshielded cable has been used extensively at 2400 and 4160 V and occasionally at 7200 V. However, any of the following conditions may dictate the use of shielded cable:

(1) Personnel safety
(2) Single conductors in wet locations
(3) Direct earth burial
(4) Where the cable surface may collect unusual amounts of conducting materials (salt, soot, conductive pulling compounds)

Shielding of an electric power cable is defined as the practice of confining the electric field of the cable to the insulation surrounding the conductor by means of conducting or semiconducting layers, or both, which are in intimate contact with or bonded to the inner and outer surfaces of the insulation. In other words, the outer insulation shield confines the electric field to the space between the conductor and the shield. The inner or strand stress relief layer is at or near the conductor potential. The outer or insulation shield is designed to carry the charging currents and in many cases fault currents. The conductivity of the shield is determined by its cross-sectional area and the resistivity of the metal tapes or wires employed in conjunction with the semiconducting layer.

The stress control layer at the inner and outer insulation surfaces, by its close bonding to the insulation surface, presents a smooth surface to reduce the stress concentrations and minimize void formation. Ionization of the air in such voids can progressively damage certain insulating materials to eventual failure.

Insulation shields have several purposes:

(1) Confine the electric field within the cable

(2) Equalize voltage stress within the insulation, minimizing surface discharges

(3) Protect cable from induced potentials

(4) Limit electromagnetic or electrostatic interference (radio, TV)

(5) Reduce shock hazard (when properly grounded)

Fig 154 illustrates the electrostatic field of a shielded cable.

The voltage distribution between a nonshielded cable and a grounded plane is illustrated in Fig 155. There it is assumed that the air is the same, electrically, as the insulation, so that the cable is in a uniform dielectric above the ground plane to permit a simpler illustration of the voltage distribution and field associated with the cable.

In a shielded cable (Fig 154) the equipotential surfaces are concentric cylinders between conductor and shield. The voltage distribution follows a simple logarithmic variation, and the electrostatic field is confined entirely within the insulation. The lines of force and stress are uniform and radial, and cross the equipotential surfaces at right angles, eliminating any tangential or longitudinal stresses within the insulation or on its surface.

The equipotential surfaces for the nonshielded system (Fig 156) are cylindrical but not concentric with the conductor, and cross the cable surface at many different potentials. The tangential creepage stress to ground at points along the cable may be several times the normal recommended for creepage distance at terminations in dry locations for unshielded cable operating on 4160 V systems.

Surface tracking, burning, and destructive discharges to ground could occur under these conditions. However, properly designed nonshielded cables as described in the NEC [12] limit the surface energies available which could compromise the cable from any of these effects.

Typical cables supplied for shielded and nonshielded applications are illustrated in Fig 157.

Fig 154
Electric Field of Shielded Cable

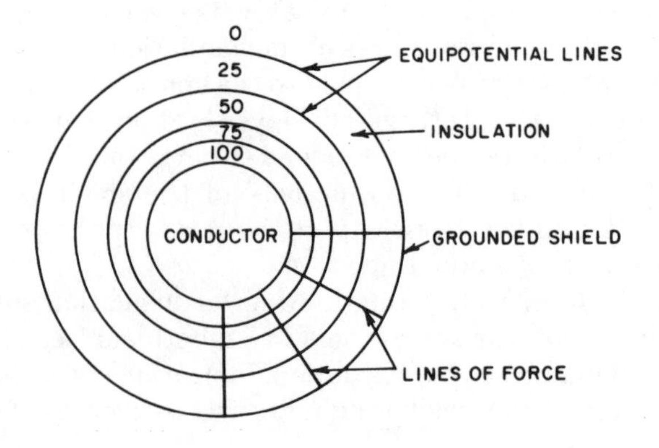

11.3 Cable Outer Finishes. Cable outer finishes or outer coverings are used to protect the underlying cable components from the environmental and installation conditions associated with intended service. The choice of cable outer finishes for a particular application is based on the same performance categories as for insulations, specifically electrical, thermal, mechanical, and chemical considerations. Except for highly protected and essentially room temperature operating environments, the usual conditions associated with cable operation cannot be met with a single outer finishing material. Therefore, combinations of metallic and nonmetallic finishes are required to provide the total protection needed for installation and

Fig 155
Electric Field of Conductor on Ground Plane in Uniform Dielectric

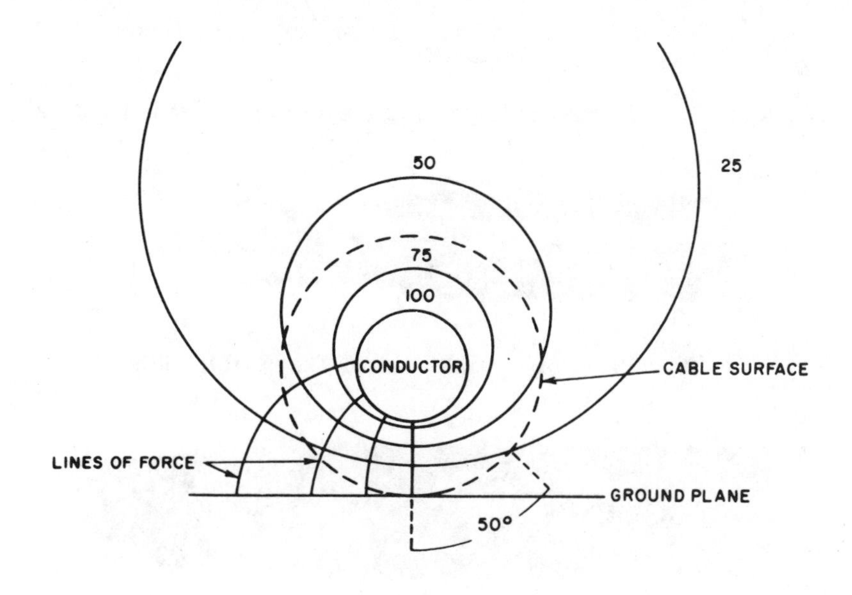

Fig 156
Electric Field of Nonshielded Cable on Ground Plane

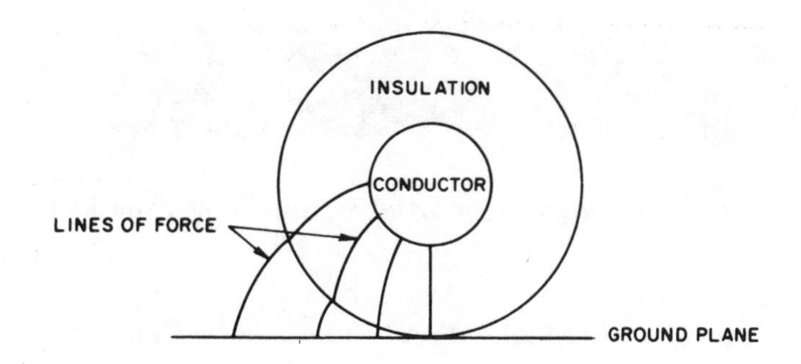

(a) Single-Conductor Cable (600 V or 5 kV Nonshielded)

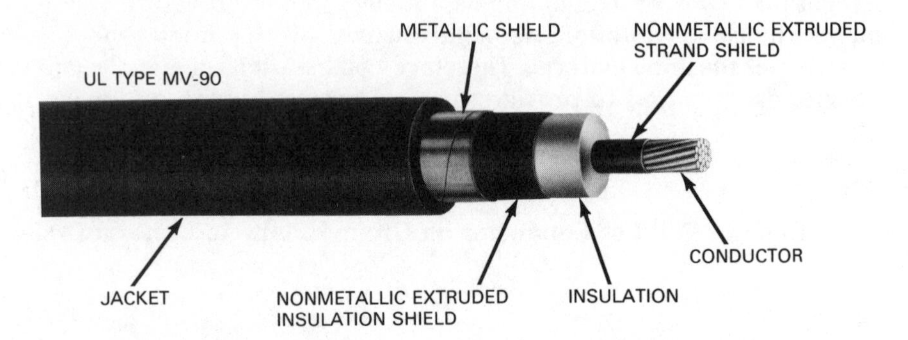

(b) Type MV Medium-Voltage Single Conductor Cable (5–35 kV)

(c) Type TC Power and Control Tray Cable (600 V)

(d) Type MC Metal-Clad Power and Control Cable (600 V–35 kV)

(e) Messenger-Supported Wiring (600 V–35 kV)

Fig 157
Commonly Used Shielded and Nonshielded Constructions

operation. Specific industry requirements for these materials are covered in standards of the ICEA [23]-[26], the American Society for Testing and Materials (ASTM), and UL.

11.3.1 Nonmetallic Finishes

(1) *Extruded Jackets*. There are outer coverings, either thermoplastic or vulcanized, which may be extruded directly over insulation or over electrical shielding systems of metal sheaths or tapes, copper braid, or semiconducting layers with copper drain wires or spiraled copper concentric wires, or over multiconductor constructions. Commonly used materials include polyvinyl chloride, NBR/PVC (nitrile butadiene/polyvinyl chloride), XLPE, polychloroprene (neoprene), chlorosulfonated polyethylene, and polyurethane. While the detailed characteristics may vary with individual manufacturers' compounding, these materials provide a high degree of moisture, chemical, and weathering protection, are reasonably flexible, provide some degree of electrical isolation, and are of sufficient mechanical strength to protect the insulating and shielding components from normal service and installation damage. Materials are available for service temperatures from -55 °C to +115 °C (Table 77).

(2) *Fiber Braids*. This category includes braided, wrapped, or served, and synthetic or natural fiber materials selected by the cable manufacturer to best meet the intended service. While asbestos fiber has been the material most commonly used in the past, fiberglass is now used extensively for employee health reasons. Some special industrial applications may require synthetic or cotton fibers applied in braid form. All fiber braids require saturants or coating and impregnating materials to provide some degree of moisture and solvent resistance as well as abrasive and weathering resistance.

Glass braid is used on cables to minimize flame propagation, smoking, and other hazardous or damaging products of combustion, which may be evolved by some extruded jacketing materials.

Table 77
Properties of Jackets and Braids

Material	Abrasion Resistance	Flexibility	Low Temperature	Heat Resistance	Fire Resistance
Neoprene	Good	Good	Good	Good	Good
Class CP rubber*	Good	Good	Fair	Excellent	Good
Crosslinked polyethylene	Good	Poor	Poor	Excellent	Poor
Polyvinyl chloride	Fair	Good	Fair	Good	Fair
Polyurethane	Excellent	Good	Good	Good	Poor
Glass braid	Fair	Good	Good	Excellent	Excellent
Nylon	Excellent	Fair	Good	Good	Fair
ETFE	Excellent	Poor	Excellent	Good	Fair

NOTE: Chemical resistance and barrier properties depend on the particular chemicals involved, and the question should be referred to the cable manufacturer.

*For example, Hypalon.

11.3.2 Metallic Finishes. This category of materials is widely used where a high degree of mechanical, chemical, or short-time thermal protection of the underlying cable components may be required in the intended service. Commonly used materials are interlocked galvanized steel, aluminum, or bronze armor; extruded lead or aluminum; strip formed, welded, and corrugated aluminum or copper; and spirally applied round or flat armor wires. The use of any of these materials, alone or in combination with others, will reduce flexibility of the overall cable. This characteristic is sacrificed to obtain the other benefits.

Installation and operating conditions may involve localized compressive loadings, occasional impact from external sources, vibration and possible abrasion, heat shock from external sources, extended exposure to corrosive chemicals, and condensation.

(1) *Interlocked Armor*. The unprotected interlocked armor provides a high degree of mechanical protection without significantly reducing flexibility. While not entirely impervious to moisture or corrosive agents, interlocked armor does provide protection from thermal shock by acting as a heat sink for short-time localized exposure.

Where corrosion and moisture resistance are required in addition to mechanical protection, an overall jacket of extruded material may be used.

The use of interlocked galvanized steel armor should be avoided on single-conductor ac power circuits due to its high hysteresis and eddy current losses. This effect, however, is minimized in three-conductor cables with armor overall and with aluminum armor on single-conductor cables.

Commonly used interlocked armor materials are galvanized steel, aluminum for less weight and general corrosion resistance, and marine bronze and other alloys for highly corrosive atmosphere.

(2) *Corrugated Metal Sheath*. Longitudinally welded and corrugated metal sheaths (corrugations or bellows formed perpendicular to the cable axis) have been used for many years in direct-burial communications cables, but only since 1960 has this method of cable core protection been applied to control and power cable. The sheath material may be of copper, aluminum, copper alloy, or a bimetallic composition with the choice of material selected to best meet the intended service.

The corrugated metal sheath offers mechanical protection equal to interlocked armor but at a lower weight. The aluminum or copper sheath may also be used as the equipment grounding conductor, either alone or in parallel with a grounding conductor within the cable.

The sheath is made from a metal strip that is longitudinally formed around the cable, welded into a continuous, impervious metal cylinder, and corrugated for pliability and increased radial strength. This sheath offers maximum protection from moisture and liquid or gaseous contaminants. An extruded nonmetallic jacket must be used over the metal sheath for direct burial, embedment in concrete, or in areas that are corrosive to the metal sheath.

(3) *Lead*. Pure or lead alloy is used for industrial power cable sheaths for maximum cable protection in underground manhole and tunnel or underground duct distribution systems subject to flooding. While not as resistant to crushing loads as

interlocked armor or a corrugated metal sheath, its very high degree of corrosion and moisture resistance makes lead attractive in the above applications. Protection from installation damage can be provided by an outer jacket of extruded material.

Pure lead is subject to work hardening and should not be used in applications where flexing may be involved. Copper- or antimony-bearing lead alloys are not as susceptible to work hardening as is pure lead, and may be used in applications involving limited flexing. Lead or its alloys should never be used for repeated flexing service.

One problem encountered today with the use of lead sheathed cable is in the area of splicing and terminating. Finding personnel experienced in the *art* of wiping lead sheath joints is not as easy as it was many years ago, posing an installation problem for many potential users. However, many insulation systems do not require lead sleeves at splices and treat the lead like any other metallic sheath.

(4) *Aluminum or Copper*. Extruded aluminum or copper or die-drawn aluminum or copper sheaths are used in certain applications for weight reduction and moisture penetration protection. While more crush-resistant than lead, aluminum sheaths are subject to electrolytic attack when installed underground. Under these conditions aluminum sheathed cable should be protected with an outer extruded jacket.

Mechanical splicing sleeves are available for use with aluminum sheathed cables, and sheath joints can be made by inert gas welding, provided that the underlying components can withstand the heat of welding without deterioration. Specifically designed hardware is available for terminating the sheath at junction boxes and enclosures.

(5) *Wire Armor*. A high degree of mechanical protection and longitudinal strength can be obtained with the use of spirally wrapped or braided round steel armor wire. This type of outer covering is frequently used in submarine cable and vertical riser cable for support. As noted for steel interlocked armor, this form of protection should be used only on three-conductor power cables to minimize sheath losses.

While not properly *armor*, spirally laid tinned copper wires, round or flat, are used in direct-buried cables as concentrics or neutrals. In potentially corrosive environments, these should be protected with an extruded jacket.

11.3.3 Single- and Multiconductor Constructions. The single-conductor cables are usually easier to handle and can be furnished in longer lengths as compared to multiconductor cables. The multiconductor constructions give smaller overall dimensions in comparison with the equivalent number of single-conductor cables, which can be an advantage where space is important.

Sometimes the outer finish can influence whether the cable should be supplied as a single- or multiconductor cable. For example, as mentioned previously, the use of steel interlock or steel wire armor on ac cables is practical on multiconductor constructions, but should be avoided over single-conductor cables. It is also economical to apply the more rugged finishes over multiconductor constructions rather than over each of the single-conductor cables.

11.3.4 Physical Properties of Materials for Outer Coverings. Depending on the environment and application, the selection of outer finishes can be complex as to the degree of protection needed. For a general appraisal, Table 77 lists the relative properties of various commonly used materials.

11.4 Cable Ratings

11.4.1 Voltage Rating. The selection of the cable insulation (voltage) rating is made on the basis of the phase-to-phase voltage of the system in which the cable is to be applied, and the general system category depending on whether the system is grounded or ungrounded and the time in which a ground fault on the system is cleared by protective equipment. It is possible to operate cables on ungrounded systems for long periods of time with one phase grounded due to a fault. This results in line-to-line voltage stress across the insulation of the two ungrounded conductors. Therefore such cable must have greater insulation thickness than a cable used on a grounded system where it is impossible to impose full line-to-line potential on the other two unfaulted phases for an extended period of time.

Consequently 100% voltage rated cables are applicable to grounded systems provided with protection that will clear ground faults within 1 minute. 133% rated cables are required on ungrounded systems where the clearing time of the 100% level category cannot be met, and yet there is adequate assurance that the faulted section will be cleared within 1 hour. 173% voltage level insulation is used on systems where the time required to de-energize a grounded section is indefinite.

11.4.2 Conductor Selection. The selection of conductor size is based on the following considerations:

(1) Load-current criteria as related to loadings, NEC [12] requirements, thermal effects of the load current, mutual heating, losses produced by magnetic induction, and dielectric losses

(2) Emergency overload criteria

(3) Voltage-drop limitations

(4) Fault-current criteria

11.4.3 Load-Current Criteria. The NEC [12] ampacity tables for low- and medium-voltage cables must be used where the NEC has been adopted. These are derived from IEEE S-135 [27].

All ampacity tables show the minimum size conductor required, but conservative engineering practice, future load growth considerations, voltage drop, and short-circuit heating may make the use of larger conductors necessary.

Large groups of cables should be carefully considered, as deratings due to mutual heating may be limiting. Conductor sizes over 500–750 kcmil require consideration of paralleling two or more smaller size cables because the current-carrying capacity per circular mil of conductor decreases for ac circuits due to skin effect and proximity effect. The reduced ratio of surface to cross-sectional area of the larger-sized conductors is a factor in the reduced ability of the larger cable to dissipate heat. When cables are used in multiple, consideration should be given to the phase placement of the cables to minimize the effects of maldistribution of current in the cables, which will reduce ampacity. Although the material cost of cable may be less for two smaller conductors, this saving may be offset by higher installation costs.

The use of load factor in underground runs takes into account the heat capacity of the duct bank and surrounding soil, which responds to the average heat losses.

The temperatures in the underground section will follow the average loss, thus permitting higher short-period loadings. The load factor is the ratio of average load to peak load and is usually measured on a daily basis for the average load. The peak load is usually the average of a ½ to 1 h period of the maximum loading that occurs in 24 h.

For direct-buried cables, the average surface temperature is limited to 60–70 °C, depending on soil condition, to prevent moisture migration and thermal runaway.

Cables should be derated when in proximity to other loaded cables or heat sources, or when the ambient temperature exceeds the ambient temperature on which the ampacity (current-carrying capacity) tables are based.

The normal ambient temperature of a cable installation is the temperature the cable would assume at the installed location with no load being carried on the cable. A thorough understanding of this temperature is required for a proper determination of the cable size required for a given load. For example, the ambient temperature for a cable exposed in the air isolated from other cables is the temperature of that cable before load is applied, assuming, of course, that this temperature is measured at the same time of day and with all other conditions exactly the same as they will be when the required load is being carried. It is assumed that for cables in air, the space around the cable is large enough so that the heat generated by the cable can be dissipated without raising the temperature of the room as a whole. Unless exact conditions are specified, the following ambients are commonly used for calculation of the current-carrying capacity.

(1) *Indoors*. The NEC [12] ampacity tables are based upon an ambient temperature of 30 °C for low-voltage cables. In most parts of the United States, 30 °C is too low for summer months, at least for some parts of the building. The NEC [12] type MV cable ampacity tables use 40 °C for air ambient temperature. In any specific case where the conditions are accurately known, the measured temperature should be used; otherwise, use 40 °C. Refer to the NEC [12], Article 318, for cables installed in cable tray.

Sources of heat adjacent to the cables under the most adverse condition should be taken into consideration in figuring the current-carrying capacity. This is usually done by correcting the ambient temperature for these localized hot spots. These may be caused by steam lines or heat sources adjacent to the cable, or they may be due to sections of the cable running through boiler rooms or other hot locations. Rerouting may be necessary to avoid this problem.

(2) *Outdoors*. An ambient temperature of 40 °C is commonly used as the maximum for cables installed in the shade and 50 °C for cables installed in the sun. In using these ambient temperatures, it is assumed that the maximum load occurs during the time when the ambient will be as specified. Some circuits probably do not carry their full load during the hottest part of the day or when the sun is at its brightest, so that an ambient temperature of 40 °C for outdoor cables is probably reasonably safe for such conditions. See the NEC [12], Article 310 ampacity tables and associated notes, for the procedure to be used for outdoor installations. Refer to the NEC [12], Article 318, for cables installed in cable tray.

(3) *Underground*. The ambient temperature used for underground cables varies in different sections of the country. For the northern sections, an ambient of

20 °C is commonly used. For the central part of the country, 25 °C is commonly used, while for the extreme south and southwest, an ambient of 30 °C may be necessary. The exact geological boundaries for these ambient temperatures cannot be set up, and the maximum ambient should be measured in the earth at a point away from any sources of heat at the depth at which the cable will be buried. Changes in the earth ambient will lag changes in the air ambient by several weeks.

The thermal characteristics of the medium surrounding the cable are of primary importance in determining the current-carrying capacity of the cable. The type of soil in which the cable or duct bank is buried has a major effect on the current-carrying capacity of cables. Porous soils, such as gravel and cinder fill, usually result in higher temperatures and lower ampacities than normal sandy or clay soil. The type of soil and its thermal resistivity should be known before the size of the conductor is calculated.

The moisture content of the soil has a major effect on the current-carrying capacity of cables. In dry sections of the country, cables may have to be derated or other precautions taken to compensate for the increase in thermal resistance due to the lack of moisture. On the other hand, in ground that is continuously wet or under tidewater conditons, cables may carry higher than normal currents. Shielding for even 2400 V circuits is necessary for continuously wet or alternately wet and dry conditions; since whenever there is a change from dry to the *naturally shielded* wet cables, there will be an abrupt voltage gradient stress, similar to what occurs at the end of shielded cables terminated without a stress cone, except where non-shielded cables are specifically designed for this service. Alternate wet and dry conditions have also been found to accelerate the progress of *treeing*.

Ampacities in the NEC [12] tables take into account the grouping of adjacent circuits. For ambient temperatures different from those shown in the tables, derating factors to be applied are shown in Article 310 and the notes to the ampacity tables.

11.4.4 Emergency Overload Criteria. Normal loading limits of insulated wire and cable are based on many years of practical experience and represent a rate of deterioration that results in the most economical and useful life of such cable systems. The rate of deterioration is expected to develop a useful life of about 20 to 30 years. The life of cable insulation is about halved and the average rate of thermally caused service failures about doubled for each 5–15 °C increase in normal daily load temperature. Additionally, sustained operation over and above maximum rated operating temperatures or ampacities is not a very effective or economical expedient, because the temperature rise is directly proportional to the conductor loss, which increases as the square of the current. The greater voltage drop might also increase the risks to equipment and service continuity.

As a practical guide, the ICEA has established maximum emergency overload temperatures for various types of insulation [23]–[26]. Operation at these emergency overload temperatures should not exceed 100 h per year, and such 100 h overload periods should not exceed five during the life of the cable. Table 78 gives uprating factors for short-time overloads for various types of insulated cables. The uprating factor, when multiplied by the nominal current rating for the cable in a

Table 78
Uprating for Short-Time Overloads*

Insulation Type	Voltage Class (kV)	Conductor Operating Temperature (°C)	Conductor Overload Temperature (°C)	Uprating Factors for Ambient Temperature 20 °C Cu	20 °C Al	30 °C Cu	30 °C Al	40 °C Cu	40 °C Al	50 °C Cu	50 °C Al
Paper (solid type)	9	95	115	1.09	1.09	1.11	1.11	1.13	1.13	1.17	1.17
	29	90	110	1.10	1.10	1.12	1.12	1.15	1.15	1.19	1.19
	49	80	100	1.12	1.12	1.15	1.15	1.19	1.19	1.25	1.25
	69	65	80	1.13	1.13	1.17	1.17	1.23	1.23	1.38	1.38
Varnished cambric	5	85	100	1.09	1.08	1.10	1.10	1.13	1.13	1.17	1.17
	15	77	85	1.05	1.05	1.07	1.07	1.09	1.09	1.13	1.13
	28	70	72								
Polyethylene (natural)	35	75	95	1.13	1.13	1.17	1.17	1.22	1.22	1.30	1.30
SBR rubber	0.6	75	95	1.13	1.13	1.17	1.17	1.22	1.22	1.30	1.30
	5	90	105	1.08	1.08	1.09	1.09	1.11	1.11	1.14	1.14
Butyl RHH	15	85	100	1.09	1.08	1.10	1.10	1.13	1.13	1.17	1.17
	35	80	95	1.09	1.09	1.11	1.11	1.14	1.14	1.20	1.20
Oil-base rubber	35	70	85	1.11	1.11	1.14	1.14	1.20	1.20	1.29	1.29
Polyethylene (cross-linked)	35	90	130	1.18	1.18	1.22	1.22	1.26	1.26	1.33	1.33
Silicone rubber	5	125	150	1.08	1.08	1.09	1.09	1.10	1.10	1.12	1.11
EPR rubber	35	90	130	1.18	1.18	1.22	1.22	1.26	1.26	1.33	1.33
Chlorosulfonated polyethylene †	0.6	75	95	1.13	1.13	1.17	1.17	1.22	1.22	1.30	1.30
Polyvinyl chloride	0.6	60	85	1.22	1.22	1.30	1.30	1.44	1.44	1.80	1.79
	0.6	75	95	1.13	1.13	1.17	1.17	1.22	1.22	1.30	1.30

* To be applied to normal rating determined for such installation conditions.
† For example, Hypalon.

particular installation, will give the emergency or overload current rating for the particular insulation type.

A more detailed discussion on emergency overload and cable protection is contained in ANSI/IEEE Std 242-1986 [6], Chapter 11.

11.4.5 Voltage-Drop Criteria. The supply conductor, if not of sufficient size, will cause excessive voltage drop in the circuit, and the drop will be in direct proportion to the circuit length. Proper starting and running of motors, lighting equipment, and other loads having heavy inrush currents should be considered. The NEC [12] recommends that the steady-state voltage drop in power, heating, or lighting feeders be no more than 3%, and the total drop including feeders and branch circuits be no more than 5% overall.

11.4.6 Fault-Current Criteria. Under short-circuit conditions, the temperature of the conductor rises rapidly. Then, due to thermal characteristics of the insulation, sheath, surrounding materials, etc, it cools off slowly after the short-circuit condition is removed. The ICEA has recommended a transient temperature limit for each type of insulation for short-circuit duration times not in excess of 10 s.

Failure to check the conductor size for short-circuit heating could result in permanent damage to the cable insulation due to disintegration of insulation material, which may be accompanied by smoke and generation of combustible vapors. These vapors will ignite, if sufficiently heated, and possibly start a serious fire. Less seriously, the insulation or sheath of the cable may be expanded to produce voids leading to subsequent failure. This becomes especially serious in 5 kV and higher voltage cables.

In addition to the thermal stresses, mechanical stresses are set up in the cable through expansion upon heating. As the heating is rapid, these stresses may result in undesirable cable movement. However, on modern cables, reinforcing binders and sheaths considerably reduce the effect of such stresses. Within the range of temperatures expected with coordinated selection and application, the mechanical aspects can normally be discounted except with very old or lead-sheathed cables.

During short-circuit or heavy pulsing currents, single-conductor cables will be subjected to forces tending to either attract or repel the individual conductors with respect to each other. Therefore such cables laid in trays, or racked, should be secured to prevent damage caused by such movements.

The minimum conductor size requirements for various rms short-circuit currents and clearing times are shown in Table 79. The ICEA initial and final conductor temperatures (see ICEA P-32-382 [21]) are shown for the various insulations. Table 76 gives conductor temperatures (maximum operating, maximum overload, and maximum short-circuit current) for various insulated cables.

The passage of excessive fault currents through the shield (drain) conductors can damage these conductors. Manufacturers (ICEA P-45-482 [5]) recommend that such ground-fault current-time not exceed 2000 A for ½ s. Some lighter duty shield constructions may have a lower current limit; check with the cable manufacturer. To limit ground-fault shield conductor exposure, the recommended practice is to employ low-resistance grounded supply systems for maximum ground-fault current of from 400 to 2000 A and suitably sensitive relaying. Without such limiting, it is likely that a ground fault could require replacement of substantial lengths of

Table 79
Minimum Conductor Sizes, in AWG or kcmil, for Indicated Fault Current and Clearing Times

Total RMS Current (amperes)	Polyethylene and Polyvinyl Chloride, 75–150 °C				Oil Base and SBR, 75–200 °C				Crosslinked Polyethylene and EPR, 90–250 °C			
	1/2 Cycle (0.0083 s)		10 Cycles (0.166 s)		1/2 Cycle (0.0083 s)		10 Cycles (0.166 s)		1/2 Cycle (0.0083 s)		10 Cycles (0.166 s)	
	Cu	Al	Cu	Al	Cu	Al	Cu	Al	Cu	Al	Cu	Al
5000	10	8	4	2	10	8	4	3	12	10	4	3
15 000	6	4	2/0	4/0	6	4	1/0	3/0	6	4	1	3/0
25 000	3	2	4/0	350	4	2	3/0	250	4	3	3/0	250
50 000	1/0	2/0	400	700	1	2/0	350	500	2	1/0	300	500
75 000	2/0	4/0	600	1000	1/0	3/0	500	750	1/0	3/0	500	700
100 000	4/0	300	800	1250	3/0	250	700	1000	2/0	4/0	600	1000

cable. Grounding of shield (drain) conductors at all splice and termination points will direct fault currents into multiple paths and reduce shield damage.

A more detailed discussion of fault current and cable protection is contained in ANSI/IEEE Std 242-1986 [6].

11.5 Installation. There are a variety of ways to install power distribution cables in industrial plants. The engineer's responsibility is to select the method most suitable for each particular application. Each mode has characteristics which make it more suitable for certain conditions than for others, that is, each mode will transmit power with a unique degree of reliability, safety, economy, and quality for any specific set of conditions. These conditions include the quantity and characteristics of the power being transmitted, the distance of transmission, and the degree of exposure to adverse mechanical and environmental conditions.

11.5.1 Layout. The first consideration, of course, in laying out wiring systems is to keep the distance between the source and the load as short as possible. This consideration should be tempered by many other important factors to arrive at the lowest cost system that will operate within the reliability, safety, economy, and performance required. Some other factors that should be considered for various routings are the cost of additional cable and raceway versus the cost of additional supports; inherent mechanical protection provided in one alternative versus additional protection required in another; clearance for and from other facilities; and the need for future revision.

11.5.2 Open Wire. This mode was used extensively in the past. Although it has now been replaced in most applications, it is still quite often used for primary power distribution over large areas where conditions are suitable.

Open-wire construction consists of uninsulated conductors on insulators that are mounted on poles or structures. The conductor may be bare or it may have a covering for protection from corrosion or abrasion.

The attractive features of this method are its low initial cost and the fact that damage can be detected and repaired quickly. On the other hand, the uninsulated conductors are a safety hazard and are also highly susceptible to mechanical damage and electrical outage from birds, animals, lightning, etc. There is increased hazard where crane or boom truck use may be involved. In some areas contamination on insulators and conductor corrosion can result in high maintenance costs.

Due to the large conductor spacing, open-wire circuits have a higher reactance, which results in a higher voltage drop. This problem is reduced with higher voltage and higher power-factor circuits.

Exposed open-wire circuits are more susceptible to outages from lightning than are other modes. The effects may be minimized, however, by the use of overhead ground wires and lightning arresters.

11.5.3 Aerial Cable. Aerial cable is finding increased use in industrial wiring. The greatest gain is in replacing open wiring, where it provides greater safety and reliability and requires less space. Properly protected cables are not a safety hazard and are not easily damaged by casual contact. They are, however, open to the same objections as open wire so far as vertical clearance is concerned. Aerial cables are frequently used in place of the more expensive conduit systems, where the latter's high degree of mechanical protection is not required. They are also generally more economical for long runs of one or two cables than are cable tray installations. It is cautioned that aerial cable having a portion of the run in conduit must be derated to the in-conduit ampacity for this condition.

Aerial cables may either be self-supporting or messenger-supported. They may be attached to pole lines or structures. Self-supporting aerial cables have high tensile strength for this application. Multiple single conductors: Types MV, RHH and RHW without outer braids, and THW; or multiconductor cables: Types MI, MC, SE, UF, TC, MV, or other factory-assembled multiconductor control, signal, or power cables that are identified for the use may be messenger-supported; see the NEC [12] Article 321, for requirements. Cables may be messenger-supported either by spirally wrapping a band around the cables and the messenger or by pulling the cable into rings suspended from the messenger. The spiral wrap method is used for factory-assembled cable, while both methods are used for field assembly. A variety of spinning heads is available for application of the spiral wire banding in the field. The messenger used on factory assembled messenger-supported wiring is required to be copper-covered steel or a combination of copper-covered steel and copper, and the assembly must be secured to the messenger by a flat copper binding strip. Single-insulated conductors should be cabled together. Factory preassembled aerial cables are especially subject to installation damage from high stress at support sheaves when *pulled in*.

Self-supporting cable is suitable for only relatively short spans. Messenger-supported cable can span large distances, depending on the weight of the cable and the tensile strength of the messenger. The supporting messenger provides high strength to withstand climatic rigors or mechanical shock. It may also serve as the grounding conductor of the power circuit.

A convenient feature available in one form of factory-assembled aerial cable makes it possible to form a slack loop to connect a circuit tap without cutting the cable conductors. This is done by reversing the direction of spiral of the conductor cabling every 10–20 ft.

11.5.4 Direct Attachment. This is a low-cost method where adequate support surfaces are available between the source and the load. It is most useful in combination with other methods such as branch runs from cable trays and when adding new circuits to existing installations. Its use in commercial buildings is usually limited to low-energy control and telephone circuits.

This method employs multiconductor cable attached to surfaces such as structural beams and columns. A cable with metallic covering should be used where exposed to adverse mechanical conditions. Otherwise, plastic or rubber jacketed cable is satisfactory, provided this is approved by local or regional codes. For architectural reasons it is usually limited to service areas, hung ceilings, and electric shafts.

11.5.5 Cable Trays. A cable tray is a unit or assembly of units or sections, and associated fittings, made of metal or other noncombustible material forming a continuous rigid structure used to support cables. These supports include ladders, troughs, and channels, and are becoming increasingly popular in industrial electric systems for the following reasons: low installed cost, system flexibility, improved reliability, accessibility for repair or addition of cables, and space saving when compared with conduit where a larger number of circuits having common routing are involved.

Cable trays are available in a number of types and materials. Mechanical load carrying ability and careful consideration should be given to the selection of the best system suited to the intended installation to provide the lowest installed cost. For other uses, see the NEC [12], Article 318.

Covers, either ventilated or nonventilated, may be used where additional mechanical protection is required or for additional electrical shielding where communication circuits are involved. Where cable trays are continuously covered for more than 6 ft with solid, unventilated covers, the cable ampacity rating must be derated as required by Section 318 of the NEC [12]. Barrier strips for separation of voltages and special coatings or materials for corrosion protection are available.

Seals or fire stops may be required when passing through walls, partitions, or elsewhere to minimize flame propagation.

Initial planning of a cable tray system should consider occupancy requirements as given in the NEC [12] and allow additional space for future system expansion.

In stacked tray installations it is good practice to separate voltages, locating the lowest voltage cables in the bottom tray and succeedingly higher voltage cables in ascending order of trays. In a multiphase system, all phase conductors should be installed closely grouped in the same tray.

11.5.6 Cable Bus. Cable bus is used for transmitting large amounts of power over relatively short distances. It is a more economical replacement of conduit or busway systems, but more expensive than cable tray. It also offers higher reliability and safety and lower maintenance than open-wire or bus systems.

Cable bus is a cross between cable tray and busway. It uses insulated conductors in an enclosure that is similar to the cable tray with covers. The conductors are supported at maintained spacings by some form of nonmetallic spacer blocks. Cable buses are furnished either as components for field assembly or as completely assembled sections. The completely assembled sections are best if the run is short enough so that splices may be avoided. Multiple sections requiring joining may preferably employ the continuous conductors.

The spacing of the conductors is such that their maximum rating in air may be attained. This spacing is also close enough to provide low reactance, resulting in minimum voltage drop.

11.5.7 Conduit. Among conduit systems, rigid steel affords the highest degree of mechanical protection available in above-ground conduit systems. Unfortunately, this is also a relatively high-cost system. For this reason their use is being superseded, where possible, by other types of conduit and wiring systems. Where applicable, rigid aluminum, intermediate grade steel conduit, thin-wall EMT, intermediate metal conduit, plastic, fiber, and cement ducts may be used.

Conduit systems offer some degree of flexibility in permitting replacement of existing conductors with new ones. However, in case of fire or faults it may be impossible to remove the conductors. In this case it is necessary to replace both conduit and wire at great cost and delay. Also, during fires conduits may transmit corrosive fumes into equipment where these gases can do much damage. To keep flammable gases out of such areas, seals should be installed.

With magnetic conduits, an equal number of conductors of each phase should be installed; otherwise, losses and heating will be excessive. For example, a single-conductor cable should not be used in steel conduit.

NOTE: Refer to the NEC [12] for code regulations on conduit use.

Underground ducts are used where it is necessary to provide a high degree of mechanical protection. Two cases where they are used are when overhead conduits are subject to extreme mechanical abuse or when the cost of going underground is less than providing overhead supports. In the latter case, direct burial may be satisfactory under certain circumstances.

Underground ducts use rigid steel, plastic, fiber, and asbestos-cement conduits encased in concrete, or precast with multihole concrete with close fitting joints. Clay tile is also used to some extent. Where the added mechanical protection of concrete is not required, heavy wall versions of fiber and asbestos-cement conduits are direct buried as are rigid steel and plastic.

Cables used in underground conduits must be suitable for use in wet areas. Some cost savings can be realized by using flexible plastic conduits with factory-installed conductors.

Where a relatively long distance between the point of service entrance into a building and the service entrance protective device is unavoidable, the conductors must be placed beneath the building under at least 2 in of concrete or they must be placed in conduit or duct and enclosed by concrete or brick not less than 2 in thick and are thus considered outside the building by the NEC [12].

11.5.8 Direct Burial. Cables may be buried directly in the ground where permitted by codes when the need for future maintenance along the cable run is not anticipated nor the protection of conduit required. The cables used must be suitable for this purpose, that is, resistant to moisture, crushing, soil contaminants, and insect and rodent damage. Direct buried cables rated over 2000 V nominal shall be shielded and provide an exterior ground path for personnel safety in the event of accidental dig-in. Refer to Tables 300-5 and 710-3(b) of the NEC [12] for minimum depth requirements. The cost savings of this method over duct banks can vary from very little to a considerable amount. While this system cannot readily be added to or maintained, the current-carrying capacity is usually greater than that of cables in ducts. Buried cable should have selected backfill for suitable heat dissipation. It should be used only where chances of its being disturbed are small or it should be suitably protected if used where these chances exist. Relatively recent advances in the design and operating characteristics of cable fault location equipment and subsequent repair methods and material have diminished the maintenance problem.

11.5.9 Hazardous Locations. Wire and cable installed in locations where fire or explosion hazards may exist must comply with Articles 500–517 of the NEC [12]. The authorized wiring methods are dependent upon the Class and Division of the specific area; see Table 80. The wiring method must be approved for the Class and

Table 80
Wiring Methods for Hazardous Locations

Wiring Method	Class I Division 1	Class I Division 2	Class II Division 1	Class II Division 2	Class III Division 1 or 2
Threaded rigid metal conduit	X	X	X	X	X
Threaded steel intermediate metal conduit	X	X	X	X	X
Rigid metal conduit				X	X
Intermediate metal conduit				X	X
Electrical metallic tubing				X	X
Rigid nonmetallic conduit					X
Type MI mineral insulated cable	X	X	X	X	X
Type MC metal-clad cable		X		X	X
Type SNM shielded nonmetallic cable		X		X	X
Type MV medium-voltage cable		X			
Type TC power and control tray cable		X			
Type PLTC power-limited tray cable		X			
Enclosed gasketed busways or wireways		X			
Dusttight wireways				X	X

Based on the NEC [12].

Division, but is not dependent upon the Group, which defines the hazardous substance.

Equipment and the associated wiring system approved as intrinsically safe is permitted in any hazardous location for which it has been approved, and the requirements of Articles 500-517 of the NEC [12] are not applicable. However, the installation must prevent the passage of gases or vapors from one area to another. Intrinsically safe equipment and wiring is not capable of releasing sufficient electrical or thermal energy under normal or abnormal conditions to cause ignition of a specific flammable or combustible atmospheric mixture in its most easily ignitable concentration.

Seals must be provided in the wiring system to prevent the passage of the hazardous atmosphere along the wiring system from one Division to the other or from a Division to a nonhazardous location. The sealing requirements are defined in Articles 501-503 of the NEC [12]. The use of multiconductor cables with a gas/vapor-tight continuous outer sheath, either metallic or nonmetallic, which will not transmit gases or vapors through the cable core in excess of the allowable limits, can significantly reduce the sealing requirements in a Class I, Division 2 hazardous location.

11.5.10 Installation Procedures. Care should be taken in the installation of raceways to make sure that no sharp edges exist to cut or abrade the cable as it is pulled in. Another important consideration is to not exceed the maximum allowable tensile strength or the manufacturer's recommendation for the side-wall pressure of a cable. These forces are directly related to the force exerted on the cable when it is pulled in. This can be decreased by shortening the length of each pull and reducing the number of bends. The force required for pulling a given length can be reduced by the application of a pulling compound on cables in conduit and the use of rollers in cable trays.

If the cable is to be pulled by the conductors, the maximum tension in pounds is limited to 0.008 times the area of the conductors in circular mils, within the constructions. This tension may be further reduced if pulled by grips over the outside covering. A reasonable figure for most jacketed constructions would be 1000 lb per grip, but the calculated conductor tension should not be exceeded.

Side-wall pressures on most single-conductor constructions limit pulling tensions to approximately 450 lb times cable diameter, in inches, times radius of bend, in feet. Triplexed and paralleled cables would use their single-conductor diameters and a figure of 225 lb and 675 lb, respectively, instead of the 450 lb factor for single-conductor construction.

For duct installations involving many bends, it is preferable to feed the cable into the end closest to the majority of the bends (in that the friction through the longer duct portion without the bends is not yet a factor) and pull from the other end. Each bend gives a multiplying factor to the tension it sees; therefore the shorter runs to the bends will keep this increase in pulling tensions to a minimum. However, it is best to calculate pulling tensions for installation from either end of the run and install from the end offering the least tension.

The minimum bending radius for metal-taped cables is normally based on twelve times the cable diameter, although cables with nonmetallic covering can be bent to at least half this radius without disrupting the cable components. The minimum bending radius is applicable to bends of even a fraction of an inch in length, not just the average of a long length being bent.

When installing cables in wet underground locations, the cable ends should be sealed to prevent entry of moisture into the conductor strands. These seals should be left intact or, if disrupted during pulling should be remade, until splicing, terminating, or testing is to be done. This practice is recommended to avoid unnecessary corrosion of the conductors and to safeguard against the generation of steam under overload, emergency loadings, or short-circuit conditions after the cable is placed in operation.

11.6 Connectors

11.6.1 Types Available. Connectors are classified as thermal or pressure, depending upon the method of attaching them to the conductor.

Thermal connectors use heat to make soldered, silver-soldered, brazed, welded, or cast-on terminals. Soldered connections have been used with copper conductors for many years, and their use is well understood. Aluminum connections may also be soldered satisfactorily with the proper materials and technique. However, soldered joints are not commonly used with aluminum. Shielded arc welding of aluminum terminals to aluminum cable makes a satisfactory termination for cable sizes larger than 4/0 AWG. Torch brazing and silver soldering of copper cable connections are in use, particularly for underground connections with bare conductors such as are found in grounding mats. Thermite welding kits utilizing carbon molds are also in use for making connections with bare copper cable for ground mats and for junctions that will be below grade. These are satisfactory as long as the conductors to be joined are dry and the charge and tool are proper. The thermite welding process has also proved satisfactory for attaching connectors to insulated power cables.

Mechanical and compression pressure connectors are used for making joints in electric conductors. Mechanical-type connectors obtain the pressure to attach the connector to the electric conductor from an integral screw, cone, or other mechanical parts. A mechanical connector thus applies force and distributes it suitably through the use of bolts or screws and properly designed sections. The bolt diameter and number of bolts are selected to produce the clamping and contact pressures required for the most satisfactory design. The sections are made heavy enough to carry rated current and withstand the mechanical operating conditions. These are frequently not satisfactory with aluminum, since only a portion of the strands are distorted by this connector.

Compression connectors are those in which the pressure to attach the connector to the electric conductor is applied externally, changing the size and shape of the connector and conductor.

The compression connector is basically a tube with the inside diameter slightly larger than the outer diameter of the conductor. The wall thickness of the tube is designed to carry the current, withstand the installation stresses, and withstand the mechanical stresses resulting from thermal expansion of the conductor. A joint is made by compressing the conductor and tube into another shape by means of a specially designed die and tool. The final shape may be indented, cup, hexagon, circular, or oval. All methods have in common the reduction in cross-sectional area by an amount sufficient to assure intimate and lasting contact between the connector and the conductor. Small connectors can be applied with a small hand tool. Larger connectors are applied with a hydraulic compression tool.

A properly crimped joint deforms the conductor strands sufficiently to have a good electrical conductivity and mechanical strength, but not so much that the crimping action overcompresses the strands, thus weakening the joint.

Mechanical and compression connectors are available as tap connectors. Many types have an independent insulating cover. After a connection is made, the cover is assembled over the joint to insulate, and in some cases to seal against the environment.

11.6.2 Connectors for Aluminum. Aluminum conductors are different from copper in several ways, and these differences should be considered in specifying and using connectors for aluminum conductors (see Table 74). The normal oxide coating on aluminum is of relatively high electrical resistance. Aluminum has a coefficient of thermal expansion higher than that of copper. The ultimate and the yield strength properties and the resistance to creep of aluminum are different from the corresponding properties of copper. Corrosion is possible under some conditions because aluminum is anodic to other commonly used metals, including copper, when electrolytes even from humid air are present.

(1) *Mechanical Properties and Resistance to Creep.* Creep has been defined as the continued deformation of the material under stress. The effect of excessive creep resulting from the use of an inadequate connector that applies excessive stress could be relaxation of contact pressure within the conductor, and a resulting deterioration and failure of the electric connection. In mechanical connectors for

aluminum, as for copper, proper design can limit residual unit-bearing loads to reasonable values, with a resulting minimum plastic deformation and creep subsequent to that initially experienced on installation. Connectors for aluminum wire can accommodate a range of conductor sizes, provided that the design takes into account the residual pressure on both minimum and maximum conductors.

(2) *Oxide Film.* The surface oxide film on aluminum, though very thin and quite brittle, has a high electrical resistance and therefore must be removed or penetrated to ensure a satisfactory electric joint. This film can be removed by abrading with a wire brush, steel wool, emery cloth, or similar abrasive tool or material. A plated surface, whether on the connector or bus, should never be abraded. It can be cleaned with a solvent or other means which will not remove the plating.

Some aluminum fittings are factory-filled with a connection aid compound, usually containing particles that aid in obtaining low contact resistance. These compounds act to seal connections against oxidation and corrosion by preventing air and moisture from reaching contact surfaces. Connection to the inner strands of a conductor requires deformation of these strands in the presence of the sealing compound to prevent the formation of an oxide film.

(3) *Thermal Expansion.* The linear coefficient of thermal expansion of aluminum is greater than that of copper and is important in the design of connectors on aluminum conductors. Unless provided for in the design of the connector, the use of metals with coefficients of expansion less than that of aluminum can result in high stresses in the aluminum during heat cycles, causing additional plastic deformation and significant creep. Stresses can be quite high, not only because of the differences of coefficients of expansion, but also because the connector may operate at an appreciably lower temperature than the conductor. This condition will be aggravated by the use of bolts that are of a dissimilar metal or have different thermal expansion characteristics from those of the terminal.

(4) *Corrosion.* Direct corrosion from chemical agents affects aluminum no more severely than it does copper, and in most cases, less. However, since aluminum is more anodic than other common conductor metals, the opportunity exists for galvanic corrosion in the presence of moisture and a more cathodic metal. For this to occur, a wetted path must exist between external surfaces of the two metals in contact to set up an electric cell through the electrolyte (moisture), resulting in erosion of the more anodic of the two, in this instance, the aluminum.

Galvanic corrosion can be minimized by the proper use of a joint compound to keep moisture away from the points of contact between dissimilar metals. The use of relatively large aluminum anodic areas and masses minimizes the effects of galvanic corrosion.

Plated aluminum connectors should be protected by taping or other sealing means.

(5) *Types of Connections Recommended for Aluminum.* UL has listed connectors approved for use on aluminum. Such connectors have successfully withstood UL performance tests (see ANSI/UL 486B-1982 [17]), which have recently been increased in severity. Both mechanical and compression-type connectors are available. The most satisfactory connectors are specifically designed for aluminum con-

ductors to prevent any possible troubles from creep, the presence of oxide film, and the differences of coefficients of expansion between aluminum and other metals. These connectors are usually satisfactory for use on copper conductors in noncorrosive locations. The connection of an aluminum connector to a copper or aluminum pad is similar to the connection of bus bars. When both the pad and the connector are plated and the connection is made indoors, few precautions are necessary. The contact surfaces should be clean; if not, a solvent should be used. Abrasive means are undesirable since the plating may be removed. In normal application, steel, aluminum, or copper alloy bolts, nuts, and flat washers may be used. A light film of a joint compound is acceptable, but not mandatory. When either of the contact surfaces is not plated, the bare surface should be cleaned by wire brushing and then coated with a joint compound. Belleville washers are suggested for heavy-duty applications where cold flow or creep may occur, or where bare contact surfaces are involved. Flat washers should be used wherever Belleville washers or other load-concentrating elements are employed. The flat washer must be located between the aluminum lug, pad, or bolt and the outside edge of the Belleville washer with the neck or crown of the Belleville against the bolting nut to obtain satisfactory operation. In outdoor or corrosive atmosphere, the above applies with the additional requirement that the joint be protected. An unplated aluminum-to-aluminum connection can be protected by the liberal use of compound.

In an aluminum-to-copper connection, a large aluminum volume compared to the copper is important as is the placement of the aluminum above the copper. Again, coating with a joint compound is the minimum protection; painting with a zinc chromate primer or thoroughly sealing with a mastic or tape is even more desirable. Plated aluminum should be completely sealed against the elements.

(6) *Welded Aluminum Terminals.* For aluminum cables in 250 kcmil and larger sizes carrying high currents, excellent terminations can be made by welding special terminals to the cable. This is best done by the inert-gas shielded metal arc method. The use of inert gas eliminates the need for any flux to be used in making the weld. The welding-type terminal is shorter than a compression terminal because the barrel for holding the cable can be very short. It has the advantage of requiring less room in junction or terminal boxes of equipment. Another advantage is the reduced resistance of the connection. Each strand of the cable is bonded to the terminal, resulting in a continuous metal path for the current from every strand of the cable to the terminal.

Welding of these terminals to the cables may also be done with the tungsten electrode type of welding equipment with ac power. The tungsten arc method is slower but for small work gives somewhat better control.

The tongues or pads of the welding-type terminals, like the large compression type, are available with bolt holes to conform to NEMA standards for terminals to be used on equipment.

(7) *Procedure for Connecting Aluminum Conductors (Fig 158)*

(a) When cutting cable, avoid nicking the strands. Nicking makes the cable subject to easy breakage [Fig 158(a)].

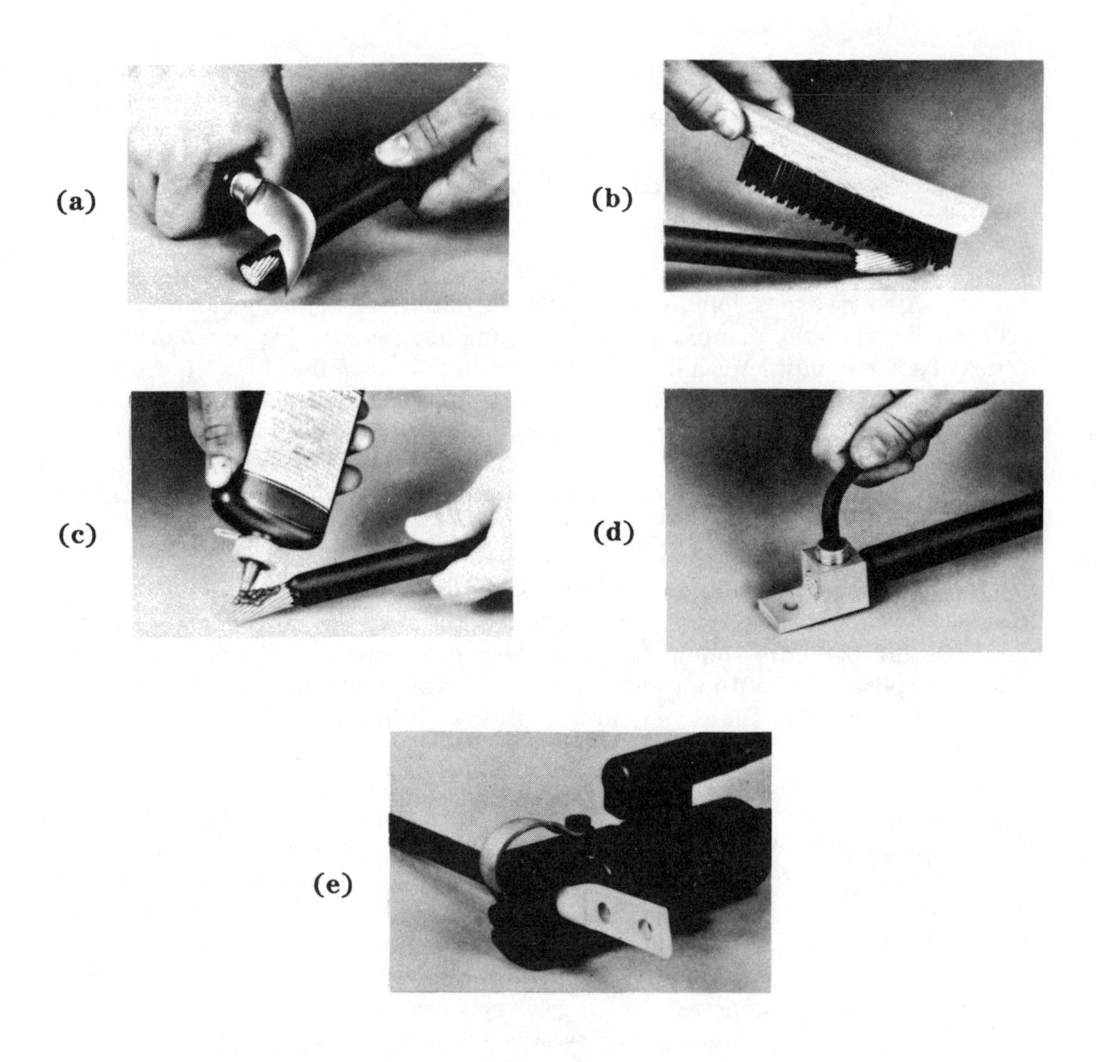

Fig 158
Procedures for Connecting
Aluminum Conductors

(b) Contact surfaces must be cleaned. The abrasion of contact surfaces is helpful even with new surfaces, and is essential with weathered surfaces. Do not abrade plated surfaces [Fig 158(b)].

(c) Apply joint compound to conductor if the connector does not already have it [Fig 158(c)].

(d) Use only connectors specifically tested and approved for use on aluminum conductors.

(e) For mechanical-type connectors tighten the connector with a screwdriver or wrench to the required torque. Remove excess compound [Fig 158 (d)]. The cube terminal connector has the disadvantage of the mounting screw being located under the conductor, requiring conductor removal to tighten or check the mounting screw.

NOTE: This type of aluminum connector should be used only where permitted by local codes.

(f) For compression-type connectors, crimp the connector using proper tool and die. Remove excess compound [Fig 158(e)].

(g) Always use a joint compound compatible with the insulation and as recommended by the manufacturer. The oxide-film penetrating or removing properties of some compounds aid in obtaining high initial conductivity. The corrosion-inhibiting and sealing properties of some compounds help ensure the maintenance of continued high conductivity and prevention of corrosion.

(h) In making an aluminum-to-copper connection that is exposed to moisture, place the aluminum conductor above the copper. This prevents soluble copper salts from reaching the aluminum conductor, which could result in corrosion. If there is no exposure to moisture, the relative position of the two metals is not important.

(i) When using insulated conductors outdoors, extend the conductor insulation or covering as close to the connector as possible to minimize weathering of the joint. Outdoors, joints should be completely protected by tape or other means whenever possible. If outdoor joints are to be covered or protected, the protection should completely exclude moisture, as the retention of moisture could lead to severe corrosion.

11.6.3 Connectors for Various Voltage Cables. Standard mechanical or compression-type connectors are recommended for all primary voltages provided the bus is uninsulated. Welded connectors may also be used for conductors sized in circular mils. Up to 600 V, standard connector designs present no problem for insulated or uninsulated conductors. The standard compression-type connectors are recommended for use on insulated conductors up to 5 kV. Above 5 kV, stress considerations make it desirable to use tapered-end compression connectors or semiconducting tape construction to give the same effect.

11.6.4 Performance Requirements. Electric connectors for industrial plants are designed to meet the requirements of the NEC [12]. They are evaluated on the basis of their ability to pass secureness, heating, heat-cycling, and pull-out tests as outlined in ANSI/UL 486A-1982 [16] and ANSI/UL 486B-1982 [17]. These standards recently have been revised to incorporate more stringent requirements for aluminum terminating devices. The reader is cautioned to use only those lugs meeting the current UL standards.

Connectors must be able to meet electrical and mechanical operating requirements. Electrically, the connectors must carry the current without exceeding the temperature rise of the conductors being joined. Joint resistance not appreciably higher than that of an equal length of conductor being joined is recommended to assure continuous and satisfactory operation of the joint. In addition, the connector should be able to withstand momentary overloads or short-circuit currents to the same degree as the conductor itself. Mechanically, a connector must be able to withstand the effects of the environment within which it is operating. If outdoors, it must stand up against temperature extremes, wind, vibration, rain, ice, sleet, chemical attack, etc. If used indoors, any vibration from rotating machinery, corrosion caused by plating or manufacturing processes, high temperatures from furnaces, etc, should not materially affect the performance of the joint.

11.7 Terminations

11.7.1 Purpose. A termination for an insulated power cable should provide certain basic electrical and mechanical functions. These essential requirements include the following:

(1) Electrically connect the insulated cable conductor to electric equipment, bus, or uninsulated conductor

(2) Physically protect and support the end of the cable conductor, insulation, shielding system, and overall jacket, sheath, or armor of the cable

(3) Effectively control electrical stresses to provide both internal and external dielectric strength to meet desired insulation levels for the cable system

The current-carrying requirements are the controlling factors in the selection of the proper type and size of connector or lug to be used. Variations in these components are related, in turn, to the base material used to make up the conductor within the cable, the type of termination used, and the requirements of the electric system.

The physical protection offered by the termination will vary considerably, depending on the requirements of the cable system, the environment, and the type of termination used. The termination should provide an insulating cover at the cable end to protect the cable components (conductor, insulation, and shielding system) from damage by any contaminants that may be present, including gases, moisture, and weathering.

Shielded high-voltage insulated cables are subject to unusual electrical stresses where the cable shield system is ended just short of the point of termination. The creepage distance that must be provided between the end of the cable shield, which is at ground potential, and the cable conductor, which is at line potential, will vary with the magnitude of the voltage, the type of terminating device used, and to some degree, the kind of cable used. The net result is the introduction of both radial and longitudinal voltage gradients, which impose dielectric stress of varying magnitude at the end of the cable. The termination provides a means of reducing and controlling these stresses within the working limits of the cable insulation and materials used to make up the terminating device itself.

11.7.2 Definitions. The definitions for cable terminations are contained in IEEE Std 48-1975 [28].

A Class 1 high-voltage cable termination, or more simply, a Class 1 Termination, provides:

(1) some form of electric stress control for the cable insulation shield terminus;

(2) complete external leakage insulation between the high-voltage conductor(s) and ground;

(3) a seal to prevent the entrance of the external environment into the cable and to maintain the pressure, if any, within the cable system. This classification encompasses what was formerly referred to as a pothead.

A Class 2 Termination is one that provides only (1) and (2): some form of electric stress control for the cable insulation shield terminus and complete external leakage insulation, but no seal against external elements. Terminations within this classification would be, for example, stress cones with rain shields or special outdoor insulation added to give complete leakage insulation, and the more recently introduced slip-on terminations for cables having extruded insulation when not providing a seal as in Class 1.

A Class 3 Termination is one that provides only (1): some form of electric stress control for the cable insulation shield terminus. This class of termination would be primarily for indoor use. Typically, this would include hand-wrapped stress cones (tapes or pennants) and the slip-on stress cones.

Refer to IEEE Std 48-1975 [28].

11.7.3 Cable Terminations. The requirements imposed by the installation location dictate the termination design class. Generally, the least imposing is an indoor or weather-protected installation, such as within a building or inside a protective housing. Here the termination is subjected to a minimum exposure to the elements, that is, sunlight, moisture, contamination. (IEEE Std 48-1975 [28] refers to what is now called a Class 3 Termination, as an *indoor termination*.)

Outdoor installations expose the termination to a broad range of elements and require that features be included in its makeup to withstand this exposure (the present Class 1 Termination defined by IEEE Std 48-1975 [28] was previously called *outdoor termination*). In some areas, the air can be expected to carry a high percentage of gaseous contaminants, liquid, or solid particles that may be conducting, either alone or in the presence of moisture. These environments impose an even greater demand on the termination for protection of the cable end (prevent entrance into the cable) from the damaging contaminants and for the termination itself to withstand exposure to these contaminants. The termination may be required to perform its intended function while partially or fully immersed in a liquid or gaseous dielectric. These exposures impose upon the termination the necessity of complete compatibility between the liquids and exposed parts of the termination, including any gasket-sealing materials. Generally speaking, cork gaskets have been used in the past, but the more recent materials such as tetrafluoroethylene (for example, Teflon) and silicone will provide superior gasketing characteristics when in direct contact with askarel. The gaseous dielectrics may be nitrogen or any of the electronegative gases, such as sulfur hexafluoride, that are used for the dielectric property of the gas to fill electric equipment.

11.7.3.1 Nonshielded Cable. These cables usually have a copper or aluminum conductor with a rubber or plastic-type insulation system with no shield. Terminations for these cables generally consist of a lug and over-taping. The lug is fastened to the cable by one of several methods described in 11.6, and tape is applied over the lower portion of the barrel of the lug and down onto the cable insulation. Tapes used for this purpose are selected on the basis of compatibility with the cable insulation and suitability for application in the environmental exposure anticipated.

11.7.3.2 Shielded Cable. Cables rated above 600 V may have a copper or aluminum conductor with either an extruded solid-type insulation (such as EP rubber or XLPE, etc) or a laminated insulation system (such as oil-impregnated paper tapes, varnished-cloth tapes, etc). A shielding system must be used on solid dielectric cables rated 2 kV and higher unless the cable is specifically listed or approved for nonshielded use; see 11.2.4.4

When terminating shielded cable, the shielding is terminated back from the conductor sufficient to provide the necessary creepage distance between the conductor and shielding. This abrupt ending of the shield introduces longitudinal stress over the surface of the exposed cable insulation. The resultant combination of radial and longitudinal electric stress at the termination of the cable shield results in maximum stress occurring at this point. However, these stresses can be controlled and reduced to values within safe working limits of the materials used to make up the termination. The most common method of reducing these stresses is to gradually increase the total thickness of insulation at the termination of the shield by adding insulation (tape or premolded rubber) to form a cone. The cable shielding is carried up to the cone surface and terminated at a point approximately ⅛ in behind the largest diameter of the cone. A typical geometric construction using tape is illustrated in Fig 159. This form is commonly referred to as a stress-relief cone or geometric stress cone. This function can also be performed by use of a high dielectric constant material, as compared to that of the cable insulation, either in tape form or premolded tube, applied to the insulation in this area. This method results in a low-stress profile. This is referred to as capacitive stress control. It is advisable to consult individual manufacturers of cable, terminating, and splicing materials for their recommendations in terminating and splicing shielded cables.

11.7.3.3 Termination Classes. A Class 1 Termination is designed to handle the electrical functions as defined and outlined in 11.7.2. A Class 1 Termination is used in areas that may have exposure to moisture or contaminants, or both. As pointed out in 11.7.3, the least severe requirements are those for a completely weather-protected area (within a building or sealed protective housing). In this case the material of the termination device to provide the external leakage insulation function would be in the form of a track-resistant insulation, such as silicone rubber tape or tube. The track-resistant surface would not necessarily need the skirts (fins or rain shields). The design of the termination to provide stress control and cable conductor seal can be the same for a weather-protected, low contamination area or high contamination areas. When a Class 1 Termination is installed in an area exposed to the elements (outdoors), the design of the termination will vary mainly with the external leakage insulation function. This will be in the form of silicone rubber, EPDM rubber, or porcelain insulation with rain shields. Of these forms, porcelain has the better resistance to long-term exposure in highly contaminated areas and to electrical stress with arc tracking. Because of this, for example, they are usually the choice along coastal areas where the atmosphere is laden with salt. The choice in other weather-exposed areas is usually based on such factors as ease of installation, time of installation, overall long-term corrosion resistance of components, device cost, and past history. Typical Class 1 Terminations are shown in Figs 160 and 161.

A Class 2 Termination is different from a Class 1 Termination only in that it does not seal the cable end to prevent entrance of the external environment into the

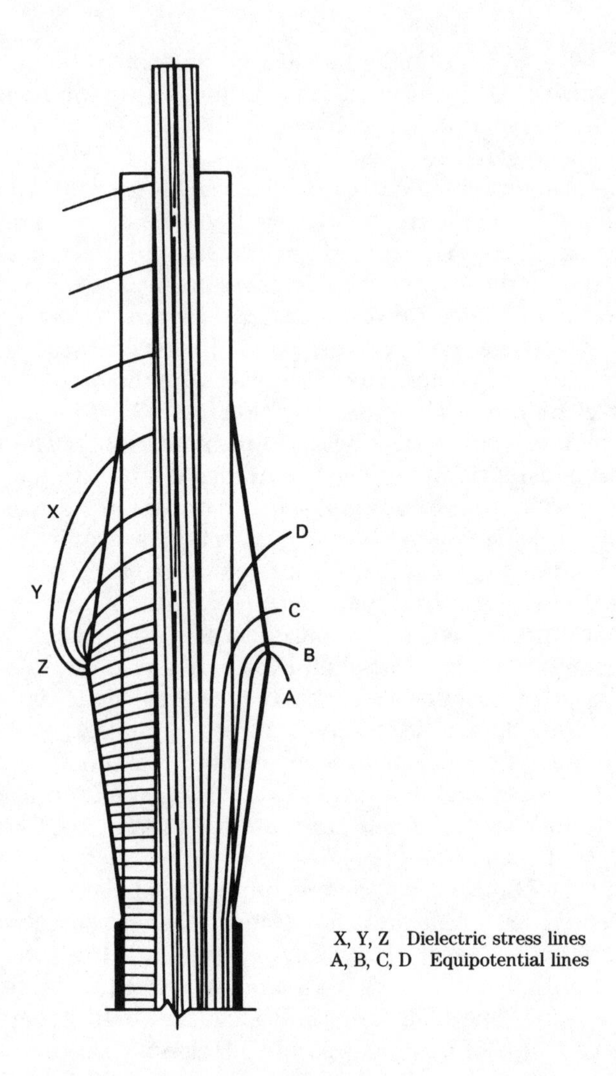

Fig 159
Stress-Relief Cone

cable or maintain the pressure, if any, within the cable. Therefore, a Class 2 Termination should not be used where moisture can enter into the cable. For a nonpressurized cable, typical of most industrial power cable systems using solid dielectric insulation, this seal is usually very easy to make. In the case of a poured porcelain-type terminator (commonly known as a *pothead*), the seal is normally built into the device. For a tape-type or slip-on-type terminator, the seal against external elements can be obtained by using tape (usually silicone rubber) to seal the conductor between the insulation and connector, assuming that the connector itself has a closed end.

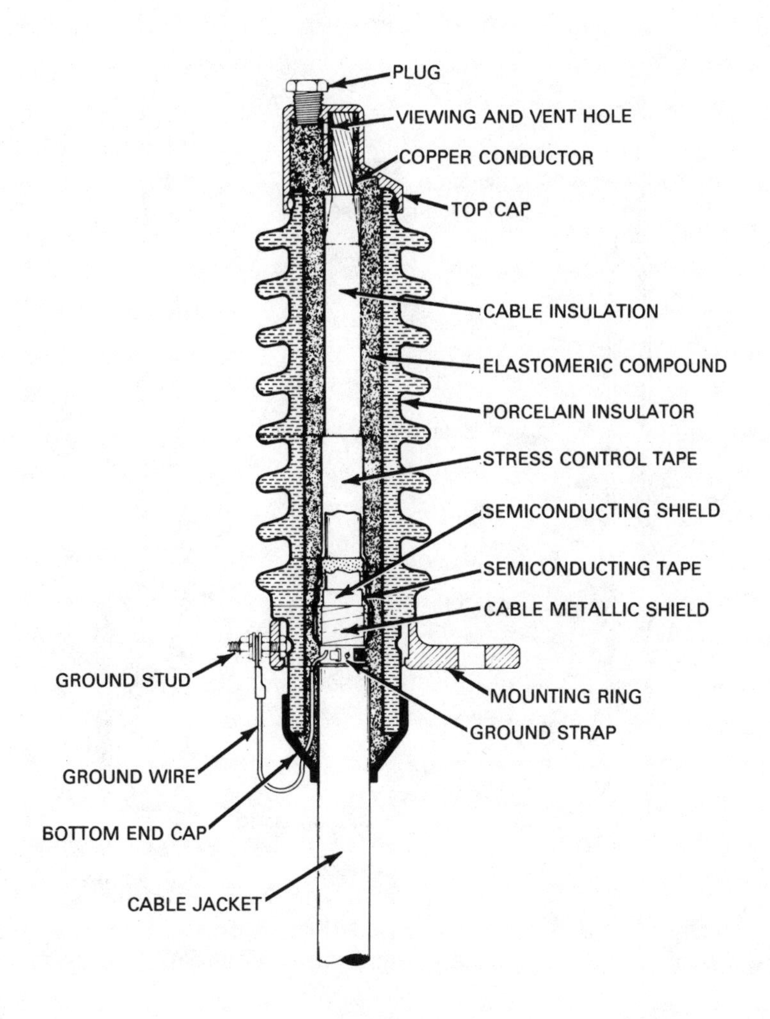

Fig 160
Typical Class 1 Porcelain Terminator (For Solid Dielectric Cables)

The Class 3 Termination only provides some form of stress control. Formerly it was known as an indoor termination and was used in weather-protected areas. Before selecting a Class 3 Termination, consideration should be given to the fact that while it is not directly exposed to the elements, there is no guarantee of the complete absence of some moisture or contamination. As a result, the lack of external leakage insulation between the high-voltage conductor(s) and ground (or track-resistant material), and a seal to prevent the moisture from entering the cable, can result in shortened life of the termination. A typical Class 3 Termination is shown in Fig 162.

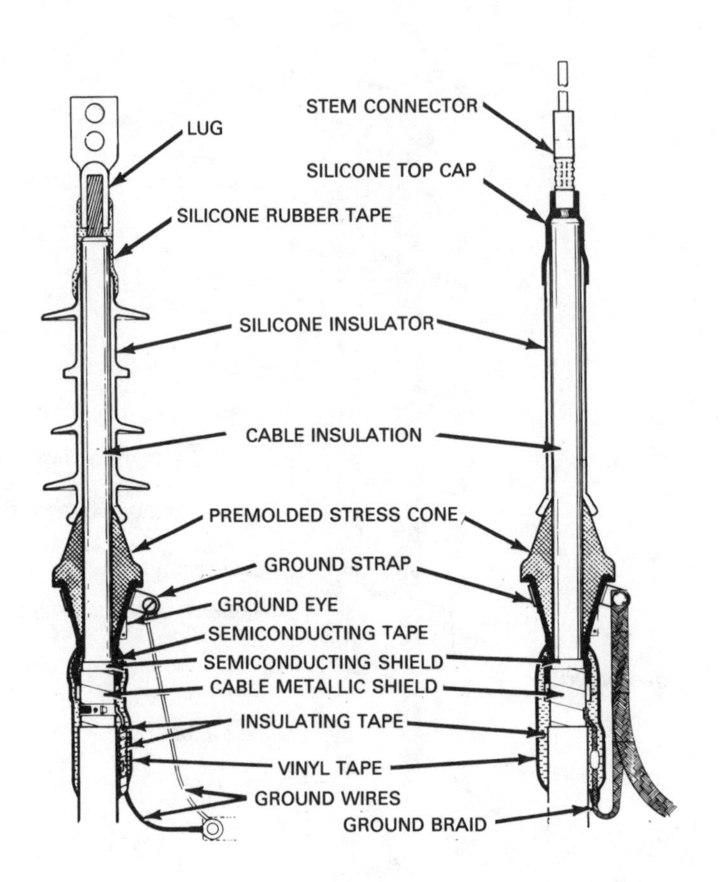

Fig 161
Typical Class 1 Molded Rubber Terminators (For Solid Dielectric Cables)

11.7.3.4 Other Termination Design Considerations. Termination methods and devices are available in ratings of 5 kV and above for either single-conductor or three-conductor installations and for indoor, outdoor, or liquid-immersed applications. Mounting variations include bracket, plate, flanged, and free-hanging types.

Both cable construction and type of application should be considered in the selection of a termination method or device. Voltage rating, desired basic impulse insulation level, conductor size, and current requirements are also considerations in the selection of the termination device or method. Cable construction is the controlling factor in the selection of the proper entrance sealing method and the stress relief materials or filling compound.

Application, in turn, is the prime consideration for selecting the type of termination device or method, method of mounting, and desired aerial connectors. Cable system requirements may be arranged in two general groups, nonpressurized and pressurized. Most industrial power cable systems are of the nonpressurized type using solid dielectric insulation.

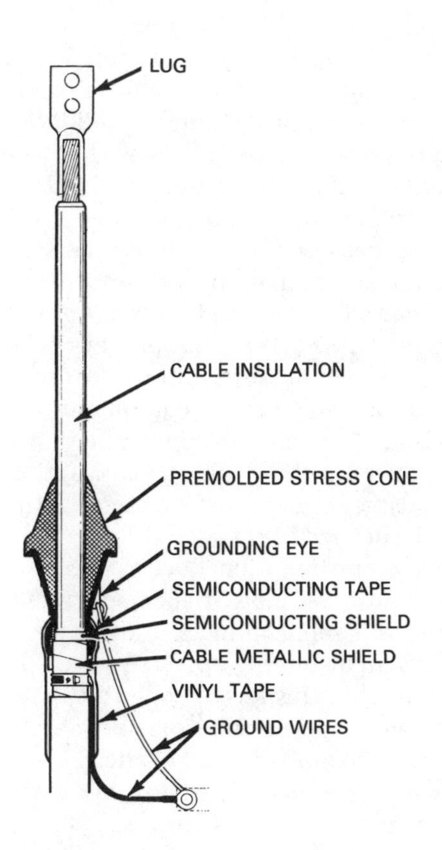

Fig 162
Typical Class 3 Molded Rubber Terminator (For Solid Dielectric Cable)

11.7.3.5 Termination Devices and Methods. The termination hardware used on a pressurized cable system, which can also be used on a nonpressurized system, includes a hermetically sealed feature used to enclose and protect the cable end. A typical design consists of a metallic body with one or more porcelain insulators with fins (also called skirts or rain shields). The body is designed to accept a variety of optional cable entrance fittings, while the porcelain bushings, in turn, are designed to accommodate a number of cable sizes and aerial connections. These parts are field-assembled to the prepared cable ends, with stress relief cones required for shielded cables, and the assembled unit is filled with an insulating compound. Considerable skill is required for proper installation of this Class 1 Termination, particularly in filling and cooling out, to avoid shrinkage and void formation in the fill material. Similar devices are available that incorporate high dielectric filling compounds, such as oil and thermosetting polyurethane resin, which do not require heating.

Advances in the art of terminating single conductor cables include several types of units designed to reduce the required cable end preparation, installation time, and eliminate the *hot fill with compound* step. Elastomeric materials are applied directly to the cable end. This type is offered both with and without a metal-porcelain housing and is applicable only to solid dielectric cables. The other type consists of a metal-porcelain housing filled with a gelatinlike substance designed to be partially displaced as the termination is installed on the cable. This latter unit may be used on any compatible nonpressurized type cable.

Advantages of the preassembled terminations include simplified installation procedures and reduced installation time. Accordingly, they offer a high degree of consistency to the overall quality and integrity of the installed system.

These preassembled Class 1 Termination devices are available in ratings of 5 kV and above for most types of applications. The porcelain-housed units include flanged mounting arrangements for equipment mounting and liquid-immersed applications. Selection of preassembled termination devices is essentially the same as for compound poured devices with the exception that those units using solid elastomeric materials generally must be sized, with close tolerance, to the cable diameters to provide proper fit.

Another category of termination devices incorporates preformed stress-relief cones (Fig 161). The most common preformed stress cone is a two-part elastomeric assembly consisting of a semiconducting lower section formed in the shape of a stress-relief cone and an insulating upper section. With the addition of high-voltage insulation protection from this stress cone to the termination lug (in the form of track-resistant silicone tape or tube or silicone insulators or fins for weather-exposed areas) and by sealing the end of the cable, the resultant termination is a Class 2 device, for use in areas exposed to moisture and contamination, but not required to hold pressure.

Taped terminations, although generally more time-consuming to apply, are very versatile. Generally, taped terminations are used at 15 kV and below; however, there are instances where they are used on some cable types at 69 kV. On non-shielded cables, the termination is made up with only a lug and a seal, such as tape. Termination of shielded cables requires the use of a stress-relief cone and cover tapes in addition to the lug. The size and location of the stress cone is controlled primarily by the operating voltage and location of the termination, that is, weather-exposed or weather-protected.

A creepage distance of 1 in/kV of nominal system voltage is commonly used for weather-protected areas, and 2–3 in for weather-exposed areas. Additional creepage distance may be gained by using a nonwetting-type of insulation, fins, skirts, or rain hoods between the stress cone and conductor lug. For weather-exposed areas, this insulation is usually a track-resistant material, such as silicone rubber or porcelain.

Insulating tapes for the stress-relief cone are selected to be compatible with the cable insulation, and tinned copper braid and semiconducting tape is used as a conducting material for the cone. A solid copper strap or solder-blocked braid should be used for the ground connection to prevent water wicking along the braid.

Some of the newer terminations do not require a stress cone. They utilize a stress-relief or stress-grading tape or tube. The stress-relief or stress-grading tape or tube is then covered with another tape or a heat-shrinkable tube for protection against the environment. The exterior tape or tube may also provide a track-resistant surface for greater protection in contaminated atmospheres.

11.7.4 Cable Connectors. Outer coverings for these cables may be nonmetallic, such as neoprene, polyethylene, or polyvinyl chloride, or metallic, such as lead, aluminum, or galvanized steel, or both, depending upon the installation environment. The latter two metallic coverings are generally furnished in a helical or corrugated form, with sections joined either by an interlocking arrangement or by continuous weld. The terminations available for use with these cables provide a means of securing the outer covering and may include conductor terminations. The techniques for applying them vary with the type of cable, its construction, its voltage rating, and the requirements for the installation.

The outer covering of multiconductor cables should be secured at the point of termination using cable connectors approved both for the cable and the installation conditions.

Type MC metal-clad cables with a continuously welded and corrugated sheath or an interlocking tape armor require, in addition to cable terminators, an arrangement to secure and ground the armor. Fittings available for this purpose are generally called armored cable connectors. These armored cable connectors provide mechanical termination and electrically ground the armor. This is particularly important on the continuous corrugated aluminum sheath since the sheath is the grounding conductor. In addition, the connector may provide a water-tight seal for the cable entrance to a box, compartment, pothead, or other piece of electric equipment.

These connectors are sized to fit the cable armor and are designed for use on the cable alone, with brackets, or with locking nuts or adaptors for application to other pieces of equipment.

11.7.5 Separable Insulated Connectors. These are two-part devices used in conjunction with high-voltage electric apparatus. A bushing assembly is attached to the high-voltage apparatus (transformer, switch, or fusing device, etc), and a molded plug-in connector is used to terminate the insulated cable and connect the cable system to the bushing. The dead-front feature is obtained by fully shielding the plug-in connector assembly.

Two types of separable insulated connectors for application at 15 kV and 25 kV are available, one load break and the other nonload break. Both are essentially of a molded construction design for use on solid dielectric insulated cables (rubber, crosslinked polyethylene, etc) and are suitable for submersible application. The connector section of the device assumes an elbow-type (90°) configuration to facilitate installation, improve separation, and save space. See ANSI-IEEE Std 386-1985 [7].

Electric apparatus may be furnished with a *universal bushing wall* only for future installation of bushings for either the load break or nonload break dead-front assemblies. Shielded elbow connectors may be furnished with a *voltage detection tap* to provide a means of determining whether or not the circuit is energized.

11.7.6 Performance Requirements. Design test criteria have been established for terminators and are listed in ANSI/IEEE Std 48-1975 [28], which outlines short-time ac 60 Hz and impulse withstand requirements. Also listed in this design standard are maximum dc field proof test voltages. Individual types may safely withstand higher test voltages, and the manufacturer should be contacted for such information. All devices employed to terminate insulated power cables should meet these basic requirements. Additional performance requirements may include thermal load cycle capabilities of the current-carrying components, environmental performance of completed units, and long-time overvoltage withstand capabilities of the device.

11.8 Splicing Devices and Techniques. Splicing devices are subjected to a somewhat different set of voltage gradients and dielectric stress from that of a cable termination. In a splice, as in the cable itself, the highest stresses are around the conductor and connector area and at the end of the shield.

Splicing design should recognize this fundamental consideration and provide the means to control these stresses to values within the working limits of the materials used to make up the splice. In addition, on shielded cables the splice is in the direct line of the cable system and must be capable of handling any ground currents or fault currents that may pass through the cable shielding.

The connectors used to join the cable conductors together must be electrically capable of carrying full rated load, emergency overload, and fault currents without overheating as well as being mechanically strong enough to prevent accidental conductor pullout or separation.

Finally, the splice housing or protective cover should provide adequate protection to the splice, giving full consideration to the nature of the application and its environmental exposure.

(1) *600 V and Below.* An insulating tape is applied over the conductor connection to electrically and physically seal the joint. The same taping technique is employed in the higher voltages, but with more refinement to cable end preparation and tape applications.

Insulated connectors are used where several relatively large cables must be joined together. These terminators, called *moles* or *crabs*, are, fundamentally, insulated buses with provision for making a number of tap connections that can be very easily taped or covered with an insulating sleeve. Connectors of this type enable a completely insulated multiple connection to be made without the skilled labor normally required for careful *crotch* taping or the expense of special junction boxes. One widely used type is a preinsulated multiple outlet joint in which the cable connections are made mechanically by compression cones and clamping nuts. Another type is a more compact preinsulated multiple joint in which the cable connections are made by standard compression tooling, which indents the conductor to the tubular cable sockets. Also available are range-taking tap connectors having an independent insulating cover. After the connection is made, the cover is snapped closed to insulate the joint.

Insulated connectors lend themselves particularly well to underground services and industrial wiring where a large number of multiple-connection joints must be made.

(2) *Over 600 V.* Splicing of unshielded cables consists of assembling a connector, usually soldered or pressed onto the cable conductors, and applying insulating

tapes to build up an insulation wall to a thickness of 1½ to 2 times that of the factory-applied insulation on the cable. Care should be exercised in applying the connector and insulating tapes to the cables, but it is not as critical with unshielded cables as with shielded cables.

Aluminum conductor cables require a waterproof joint to prevent moisture entry into the stranding of the aluminum conductors. Splices on solid dielectric cables are made with uncured tapes, which will fuse together after application and provide a waterproof assembly. It is necessary, however, to use a moistureproof adhesive between the cable insulation and the first layer of insulating tapes. Additional protection may be obtained through the use of a moistureproof cover over the insulated splice. This cover may consist of additional moistureproof tapes and paint or a sealed weatherproof housing of some form.

11.8.1 Taped Splices (Fig 163). Taped splices for shielded cables have been used quite successfully for many years. Basic considerations are essentially the same as for unshielded cables. Insulating tapes are selected not only on the basis of dielectric properties but also for compatibility with the cable insulation. The characteristics of the insulating tapes should also be suitable for the application of the splice. This latter consideration gives attention to such details as providing a moisture seal for splices subjected to water immersion or direct burial, thermal stability of tapes for splices subjected to high ambient and operating temperatures, and ease of handling for applications of tapes on wye- or tee-type splices.

Connector surfaces should be smooth and free from any sharp protrusions or edges. The connector ends are tapered, and indentations or distortion caused by pressing tools are filled and shaped to provide a round, smooth surface. Semiconducting tapes are recommended for covering the connector and exposed conductor stranding to provide a uniform surface over which insulating tapes can be applied. Cables with a solid-type insulation are tapered, and those with tape-type insulation are stepped to provide a gradual transition between conductor-connector diameter and cable insulation diameter prior to the application of insulating tapes. This is done to control the voltage gradients and resultant voltage stress to values within the working limits of the insulating materials. The splice should not be overinsulated to provide additional protection since this could restrict heat dissipation at the splice area and risk splice failure.

A tinned copper braid is used to provide the shielding function over the splice area. Grounding straps are applied to at least one end of the splice for grounding purposes, and a heavy braid jumper is applied across the splice to carry available ground-fault current. Refer to 11.9.1 for single-point grounding to reduce sheath losses.

Final cover tapes or weather barriers are applied over the built-up splice to seal it against moisture entry. A splice on a cable with lead sheath is generally housed in a lead sleeve, which is solder-wiped to the cable sheath at each end of the splice. These lead sleeves are filled with compound in much the same manner as potheads.

Hand-taped splices may be made between lengths of dissimilar cables if proper precautions are taken to ensure the integrity of each cable's insulating system and the tapes used are compatible with both cables. One example of this would be a splice between a rubber-insulated cable and an oil-impregnated paper insulated

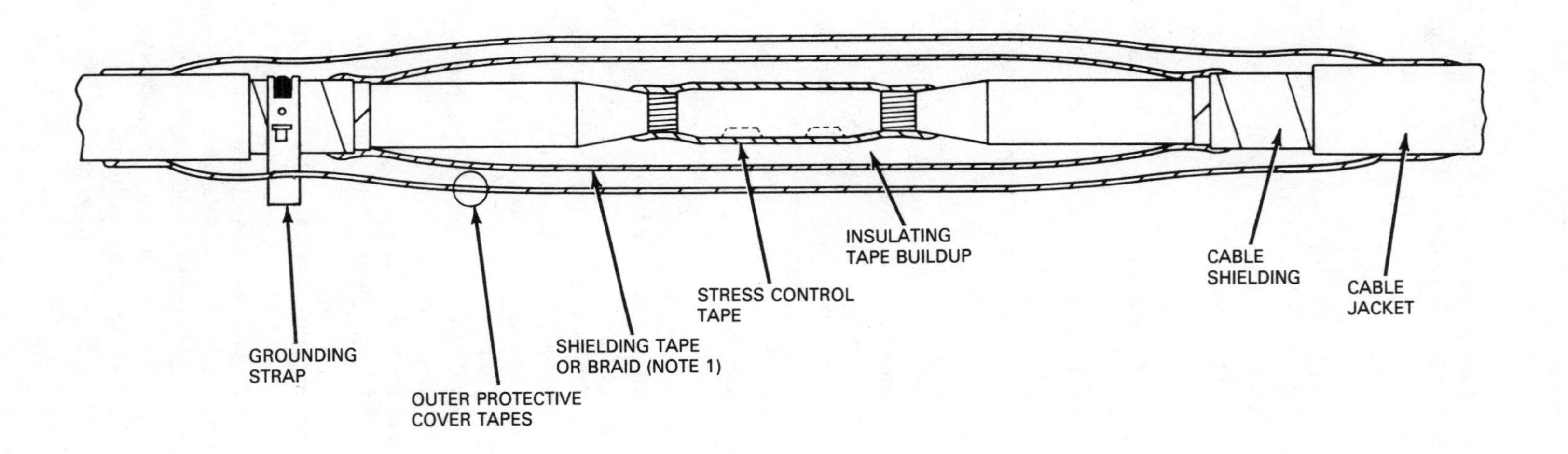

NOTES: (1) Heavy braid jumper or perforated strip should be used across splice to carry possible ground-fault current. Stress-control tape should cover strands completely, lapping slightly onto insulation taper.
(2) Consult individual cable supplier for recommended installation procedures and materials.

Fig 163
Typical Taped Splice in Shielded Cable

cable. Such a splice should have an oil barrier to prevent the oil impregnate in the paper cable from coming in contact with the insulation on the rubber cable. In addition, the assembled splice should be made completely moistureproof. This requirement is usually accomplished by housing the splice in a lead sleeve with wiped joints at both ends. A close-fitting lead nipple is placed on the rubber cable and is tape- or epoxy-sealed to the jacket of the rubber cable. The solder wipe is made to this lead tube.

Three-way wye and tee splices and the several other special hand-taped splices that can be made all require special design considerations. In addition, a high degree of skill on the part of the installer is a prime requirement for proper makeup and service reliability.

11.8.2 Preassembled Splices. Similar to the preassembled terminators are several types of factory-made splices. The most elementary is an elastomeric unit consisting of a molded housing sized to fit the cables involved, a connector for joining the conductors, and tape seals for sealing the ends of the molded housing to the cable jacket. Other versions of elastomeric units include an overall protective metallic housing that completely encloses the splice. These preassembled elastomeric-type splices are available in two-way, three-way tee-type, and multiple configurations for application up to 35 kV and can be used on most cables having an extruded solid-type insulation.

The preassembled splice provides a waterproof seal to the cable jacket and is suitable for submersible, direct-burial, and other applications where the splice housing must provide protection for the splice to the same degree that the cable jacket provides protection to the cable insulation and shielding system. An advantage of these preassembled splices is the reduction in time needed to complete the splice after cable end preparation. However, the solid elastomeric materials used for the splice must be sized, with close tolerance, to the cable diameters to provide proper fit.

11.9 Grounding of Cable Systems. For safety and reliable operation, the shields and metallic sheaths of power cables must be grounded. Without such grounding, shields would operate at a potential considerably above ground. Thus they would be hazardous to touch, and would incur rapid degradation of the jacket or other material intervening between shield and ground. This is caused by the capacitive charging current of the cable insulation, which is of the order of 1mA/ft of conductor length. This current normally flows, at power frequency, between the conductor and the earth electrode of the cable — normally the shield. In addition, the shield or metallic sheath provides the fault return path in the event of insulation failure, permitting rapid operation of the protection devices.

The grounding conductor and its attachment to the shield or metallic sheath, normally at a termination or splice, need to have an ampacity no lower than that of the shield. In the case of a lead sheath, the ampacity should be ample to carry the available fault current and duration without overheating. Attachment to shield or sheath is frequently by means of solder, which has a low melting point; thus an adequate area of attachment is required.

There is much disagreement as to whether the cable shield lengths should be grounded at both ends or at only one end. If grounded at only one end, any possible fault current must traverse the length from the fault to the grounded end, imposing high current on the usually very light shield conductor. Such a current could readily damage or destroy the shield and require replacement of the entire cable rather than only the faulted section. With both ends grounded, the fault current would divide and flow to both ends, reducing the duty on the shield, with consequently less chance of damage. There are modifications of both systems. In one, single-ended grounding may be attained by insulating the shields at each splice or sectionalizing point, and grounding only the source end of each section. This limits possible shield damage to only the faulted section. Multiple grounding, rather than just two-end grounding, is simply the grounding of the cable shield or sheath at all access points, such as at manholes or pull boxes. This also limits possible shield damage to only the faulted section.

11.9.1 Sheath Losses. Currents are induced in the mulitgrounded shields and sheaths of cables by the current flow in the power conductor. These currents increase with the separation of the power conductors, and increase with decreasing shield or sheath resistance. With three-conductor cable this sheath current is negligible, but with single-conductor cables separated in direct-burial or separate ducts it can be appreciable. For example, with three single-conductor 500 kcmil cables, flat, on 8 in centers, with twenty spiral 16 AWG copper shield wires, the ampacity is reduced by about 20% by this shield current. With single-conductor lead-sheathed cables in separate ducts, this current is important enough that single-end grounding is obligatory. As an alternate, the shields are insulated at each splice (at approximately 500 ft intervals) and crossbonded to provide sheath transposition. This neutralizes the sheath currents, but still provides double-ended grounding. Of course, these sheaths and bonding jumpers must be insulated; their voltage differential from ground may be in the 30-50 V range. For details on calculating sheath losses in cable systems, consult [35].

Difficulties may arise from current attempting to flow via the cable shield, unrelated to cable insulation failures. To prevent this, all points served by a multiple-grounded shield cable need to be interconnected with an ample grounding system. (Insulation between shield sections at splices of single-end grounded shield systems needs enough dielectric strength to withstand possible abnormal voltages as well.) This system requires interconnecting grounding conductors of suitably low impedance that fault or lightning currents will follow this path rather than the cable shield. Cable-shield ground connections should be made to this system, which should also connect to the grounded element of the source supplying the energy to the cable. Duct runs, or direct-burial routes, generally include a heavy grounding conductor to ensure such interconnection.

For further details, the reader is referred to Chapter 7, as well as to ANSI/IEEE Std 142-1982 [5] and ANSI C2-1987 [3] (National Electrical Safety Code [NESC]), Part 3.

11.10 Protection from Transient Overvoltage. Cables up through 35 kV used in industrial service have insulation strengths well above that of essentially all other types of electric equipment of similar voltage ratings. This is to compensate for installation handling and possibly a higher deterioration rate than insulation that is exposed to less severe ambient conditions. This high insulation strength may or may not exist in splices or terminations, depending on their design and construction. Except for deteriorated points in the cable itself, the splices or terminations are most affected by overvoltages of lightning and switching transients. The terminations of cable systems not provided with surge protection may flash over due to switching transients. In this event, the cable proper would be subjected to possible wave reflections of even higher levels, possibly damaging the cable insulation; however, this is unlikely in this medium-voltage class.

Like other electric equipment, the means employed for protection from these overvoltages is usually surge arresters. These may be used for protection of associated equipment as well as the cable. Distribution- or intermediate-type arresters are used, applied at the junctions of open-wire lines and cables and at terminals where switches may be open. Surge arresters are not required at intermediate positions along the cable run as they are frequently required on open-wire lines.

It is recommended that surge arresters be connected between the conductor and the cable shielding system with short leads to maximize the effectiveness of the arrester. Similarly recommended is the direct connection of the shields and arrester ground wires to a substantial grounding system to prevent surge current propagation through the shield.

Aerial messenger-supported, fully insulated cables and spacer-type cables are subject to direct lightning strokes, and a number of such cases are on record. The incidence rate is, however, rather low, and in most cases no protection is provided. Where, for reliability, such incidents must be guarded against, a grounded shield wire similar to that used for bare aerial circuits should be installed on the poles a few feet above the cable. Down-pole grounding conductors need to be carried past the cable messenger with a lateral offset of about 18 in to guard against side flashes consequent to direct strokes. Metal bayonets, where used to support the grounded shielding wire, should also be kept no less than 18 in clear of the cables or messengers.

11.11 Testing

11.11.1 Application and Utility. Testing, particularly of elastomeric and plastic (solid) insulations, is a useful method of checking the ability of a cable to withstand service conditions for a reasonable future period. Failure to pass test will either cause in-test breakdown of the cable or otherwise indicate the need for its immediate replacement.

Whether or not to routinely test cables is a decision each user has to make. The following factors should be taken into consideration:

(1) If there is no alternate source for the load supplied, testing should be done when the load equipment is not in operation.

(2) The costs of possible service outages due to cable failures should be weighed against the cost of testing. With solid-type insulation, in-service cable failures may be reduced by about 90% by dc maintenance testing.

(3) Personnel with adequate technical capability should be available to do the testing and evaluate observations and results.

The procedures outlined herein are intended as a guide, and many variations are possible. At the same time, variations made without sound technical basis can negate the usefulness of the test or even damage equipment.

With solid dielectric cable types (elastomeric and plastic), the principal failure mechanism results from *ac corona cutting* during service at locations of either manufacturing defects, installation damage, or accessory workmanship shortcomings. Initial tests reveal only gross damage, improper splicing or terminating, or cable imperfections. Subsequent use on alternating current usually causes progressive enlargement of such defects proportional to their severity.

Oil-paper (laminated) cable with lead sheath fails usually from water entrance at a perforation in the sheath, generally within three to six months after the perforation occurs. Periodic testing, unless very frequent, is therefore likely to miss many of these cases, making the method less effective with this type of cable.

Testing is not useful in detecting possible failure from moisture-induced tracking across termination surfaces, since this develops principally during periods of precipitation, condensation, or leakage failure of the enclosure or housing. However, terminals should be examined regularly for signs of tracking, and the condition corrected whenever detected.

11.11.2 Alternating Current Versus Direct Current. Cable insulation can, without damage, sustain application of dc potential equal to the system basic impulse insulation level for very long periods. In contrast, most cable insulations will sustain degradation from ac overpotential, proportional to a high power of overvoltage to time (and frequency) of the application. Hence it is desirable to utilize direct current for any testing that will be repetitive. While the manufacturers use alternating current for the original *factory* test, it is almost universal practice to employ direct current for any subsequent testing. All discussion of field testing hereafter applies to dc high-voltage testing.

11.11.3 Factory Tests. All cable is tested by the manufacturer before shipment, normally with alternating voltage for a 5 min period. Unshielded cable is immersed in water (ground) for this test; shielded cable is tested using the shield as the ground return. Test voltages are specified by the manufacturer, by the applicable specification of the ICEA [23] - [26], or by other specifications such as those of the Association of Edison Illuminating Companies (AEIC); see AEIC CS5-1982 [1] and AEIC CS6-1982 [2]. In addition, a test may be made using direct voltage of two to three times the rms value used in the ac test. On cable rated 3000 V and above, corona tests also may be made.

11.11.4 Field Tests. As well as having no deteriorating effect on good insulation, dc high voltage is most convenient for field testing since the test power sources or test sets are relatively light and portable.

Voltages for such testing should fulfill both of the following requirements:

(1) Not be high enough to damage sound cable or component insulation

(2) Be high enough to indicate incipient failure of unsound insulation that may fail in service before the next scheduled test

Test voltages and intervals require coordination to attain suitable performance. One large industrial company with cable testing experience of over 20 years has reached over 90% reduction of cable system service failures through use of ICEA specified voltages [23] - [26]. These are applied at installation, after about three years of service, and every five to six years thereafter. The majority of test failures occur at the first two tests; test (or service) failures after eight years of satisfactory service are less frequent. The importance of uninterrupted service should also influence the test frequency for specific cables. Tables 81 and 82 specify cable field test voltages.

The AEIC has specified test values for 1968 (see AEIC CS5-1982 [1] and AEIC CS6-1982 [2]) and later cables of approximately 20% higher values than the ICEA values [23] - [26].

ANSI/IEEE Std 400-1980 [8] specifies much higher voltages than either ICEA or AEIC. These much more severe test voltages as shown in Table 83 are intended to reduce cable failures during operation by overstressing the cables during shutdown testing and causing weak cables to fail at that time.

Cables to be tested should have their ends free of equipment and clear from ground. All conductors not under test should be grounded. Since equipment to which cable is customarily connected may not withstand the test voltages allowable for cable, either the cable should be disconnected from this equipment, or the test voltage should be limited to levels that the equipment can tolerate. The latter constitutes a relatively mild test on the cable condition, and the predominant leakage current measured is likely to be that of the attached equipment. In essence, this tests the equipment, not the cable.

In field testing, in contrast to the *go–no go* nature of factory testing, the leakage current of the cable system should be closely watched and recorded for signs of approaching failure. The test voltage may be raised continuously and slowly from zero to the maximum value, or it may be raised in steps, pausing for 1 min or more at each step. Potential differences between steps are of the order of the ac rms rated voltage of the cable. As the voltage is raised, current will flow at a relatively high rate to charge the capacitance, and to a much lesser extent to supply the

Table 81
ICEA Specified DC Cable Test Voltages (kV), Pre-1968 Cable

Insulation Type	Grounding	Maintenance Test Rated Cable Voltage 5 kV	15 kV	25 kV	35 kV
Elastomeric: butyl, oil base, EPR	Grounded	27	47	—	—
	Ungrounded	—	67	—	—
Polyethylene, including cross-linked polyethylene	Grounded	22	40	67	88
	Ungrounded	—	52	—	—

Table 82
ICEA Specified DC Cable Test Voltages (kV), 1968 and Later Cable*

Insulation Type	Insulation Level (%)	Rated Cable Voltage 5 kV 1	5 kV 2	15 kV 1	15 kV 2	25 kV 1	25 kV 2	35 kV 1	35 kV 2
Elastomeric:	100	25	19	55	41	80	60	—	—
butyl and oil base	133	25	19	65	49	—	—	—	—
Elastomeric:	100	25	19	55	41	80	60	100	75
EPR	133	25	19	65	49	100	75	—	—
Polyethylene,	100	25	19	55	41	80	60	100	75
including cross-linked polyethylene	133	25	19	65	49	100	75	—	—

NOTE: Columns 1 — Installation tests, made after installation, before service; columns 2 — maintenance tests, made after cable has been in service.

* These test values are lower than for pre-1968 cables because the insulation is thinner. Hence the ac test voltage is lower. The dc test voltage is specified as three times the ac test voltage, so it is also lower than for older cables.

dielectric absorption characteristics of the cable, as well as to supply the leakage current. The capacitance charging current subsides within a second or so, the absorption current subsides much more slowly and would continue to decrease for 10 min or more, ultimately leaving only the leakage current flowing.

At each step, and for the 5-15 min duration of the maximum voltage, the current meter (normally a microammeter) is closely watched. If, except when the voltage is being increased, the current starts to increase, slowly at first, then more rapidly, the last remnants of insulation at a weak point are failing, and total failure will occur shortly thereafter unless the voltage is reduced. This is characteristic of about 80% of all elastomeric-type test failures.

In contrast to this *avalanche* current increase to failure, sudden failure (flashover) can occur if the insulation is already completely (or nearly) punctured. In the latter case, voltage increases until it reaches the sparkover potential of the air gap length, then flashover occurs. Polyethylene cables exhibit the latter characteristic for all failure modes. Conducting leakage paths, such as at terminations or through the body of the insulation, exhibit a constant leakage resistance independent of time or voltage.

One advantage of step testing is that a 1 min *absorption-stabilized* current may be read at the end of each voltage step. The calculated resistance of these steps may be compared as the test progresses to higher voltage. At any step where the calculated leakage resistance decreases markedly (say to 50% of that of the next lower voltage level), the cable could be near failure and the test should be discontinued short of failure as it may be desirable to retain the cable in serviceable condition

Table 83
IEEE Std 400-1980 [8] Specified DC Cable Test Voltages (kV), Installation and Maintenance

L–L System Voltage (kV)	BIL (kV)	Test Voltage (kV)	
		100% Insulation Level	133% Insulation Level
2.5	60	40	50
5	75	50	65
8.7	95	65	85
15	110	75	100
23	150	105	140
28	170	120	
34.5	200	140	

NOTE: These test voltages should not be used without the cable manufacturers' concurrence as the cable warranty will be voided.

until a replacement can be made ready. On any test in which the cable will not withstand the prescribed test voltage for the full test period (usually 5 min) without current increase, the cable is considered to have failed the test and is subject to replacement as soon as possible.

The polarization index is the ratio of the current after 1 min to the current after 5 min of maximum voltage test, and on good cable it will be between 1.25 and 2. Anything less than 1.0 should be considered a failure, and between 1.0 and 1.25 only a marginal pass.

After completion of the maximum test voltage step, the supply voltage control dial should be returned to zero and the charge in the cable allowed to drain off through the leakage of the test set and voltmeter circuits. If this requires too long a time, a bleeder resistor of 1 MΩ/10 kV of test potential can be added to the drainage path, discharging the circuit in a few seconds. After the remaining potential drops below 10% of the original value, the cable conductor may be solidly grounded. All conductors should be left grounded when not on test during the testing of other conductors and for at least 30 min after the removal of a dc test potential. They may be touched only while the ground is connected to them; otherwise the release of absorption current by the dielectric may again raise their potential to a dangerous level.

11.11.5 Procedure. Load is removed from the cables either by diverting the load to an alternate supply or by shutdown of the load served. The cables are deenergized by switching; they are tested to ensure voltage removal, then grounded and disconnected from their attached switching equipment. (In case they are left connected, lower test potentials are required.) Surge arresters, potential transformers, and capacitors should be disconnected.

All conductors and shields should be grounded. The test set is checked for operation, and after its power is turned off, the test lead is attached to the conductor to be tested. At this time (and not before) the ground should be removed from that conductor, and the bag or jar (see 11.11.6) applied over all of its terminals, covering all uninsulated parts at both ends of the run. The test voltage is then slowly applied, either continuously or in steps as outlined in 11.11.4. At completion of the maximum test voltage duration, the charge is drained off, the conductor grounded, and the test lead removed for connection to the next conductor. This procedure is repeated for each conductor to be tested. Grounds should be left on each tested conductor for no less than 30 min.

11.11.6 Direct-Current Corona and Its Suppression. Starting at about 10–15 kV and increasing at a high power of the incremental voltage, the air surrounding all bare conductor portions of the cable circuit becomes ionized from the test potential on the conductor and draws current from the conductor. This ionizing current indication is not separable from that of the normal leakage current, and reduces the apparent leakage resistance value of the cable. Wind and other air currents tend to blow the ionized air away from the terminals, dissipating the space charge and allowing ionization of the new air, thus increasing the *direct corona current*, as this may be called.

Enclosing the bare portions of both end terminations in plastic or glass jars, or plastic bags, prevents the escape of this ionized air; thus it becomes a captive space charge. Once formed, it requires no further current, so the *direct corona current* disappears. With this treatment, testing up to about 100 kV is possible. Above 100 kV, larger bags or a small bag inside a large one are required. In order to be effective, the bags should be blown up so that no part of the bag touches the conductor.

An alternate method to minimize corona is to completely tape all bare conductor surfaces with standard electrical insulating tape. This method is superior to the bag method for corona suppression, but it requires more time to adequately tape all exposed ends.

11.11.7 Line-Voltage Fluctuations. The very large capacitance of the cable circuit makes the microammeter extremely sensitive to even minor variations in 120 V 60 Hz supply to the test set. Normally, it is possible only to read average current values or the near-steady current values. A low-harmonic-content constant-voltage transformer improves this condition moderately. Complete isolation and stability are attainable only by use of a storage battery and 120 V 60 Hz inverter to supply the test set.

11.11.8 Resistance Evaluation. High-voltage cable exhibits extremely high insulation resistance, frequently many thousands of megohms. While insulation resistance alone is not a primary indication of the condition of the cable insulation, the comparison of the insulation resistances of the three-phase conductors is useful. On circuits under 1000 ft long, a ratio in excess of 5:1 between any two conductors is indicative of some questionable condition. On longer circuits, a ratio of 3:1 should be regarded as maximum. Comparison of insulation resistance values with previous tests may be informative; but insulation resistance varies inversely with temperature, with winter insulation resistance measurements being much higher than those obtained under summer conditions. An abnormally low insulation resistance is frequently indicative of a faulty splice, termination, or a weak spot in the insulation. (Higher than standard test voltages have been found practical to locate these by causing a test failure where the standard voltage will not cause breakdown. Fault location methods may be used to locate the failure.)

11.11.9 Megohmmeter Test. Since the insulation resistance of a sound high-voltage cable circuit is generally in the order of thousands to hundreds of thousands of megohms, a megohmmeter test will reveal only grossly deteriorated insulation conditions of high-voltage cable. For low-voltage cable, however, the megohmmeter tester is quite useful, and is probably the only practicable test. Sound 600 V cable insulation will normally withstand 20 000 V or higher direct current. Thus a 1000 V or 2500 V meghommeter is preferable to the lower 500 V testers for such cable testing.

For this low-voltage class, temperature-corrected comparisons of insulation resistances with other phases of the same circuit, with previous readings on the same conductor, and with other similar circuits are useful criteria for adequacy. Continued reduction in a cable's insulation resistance over a period of several tests is indicative of degrading insulation; however, a megohmmeter will rarely initiate final breakdown of such insulation.

11.12 Locating Cable Faults. In an industrial plant a wide variety of cable faults can occur. The problem may be in a communication circuit or in a power circuit, either in the low- or high-voltage class. Circuit interruption may have resulted, or operation may continue with some objectionable characteristic. Regardless of the class of equipment involved or the type of fault, the one common problem is to determine the location of the fault so that repairs can be made.

The vast majority of cable faults encountered in an industrial power system occur between conductor and ground. Most fault-locating techniques are made with the circuit de-energized. In ungrounded or high-resistance-grounded low-voltage systems, however, the occurrence of a single line-to-ground fault will not result in automatic circuit interruption, and therefore the process of locating the fault may be carried out by special procedures with the circuit energized.

11.12.1 Influence of Ground-Fault Resistance. Once a line-to-ground fault has occurred, the resistance of the fault path can range from almost zero up to millions of ohms. The fault resistance has a bearing on the method used to locate the failure. In general a low-resistance fault can be located more readily than one of high resistance. In some cases the fault resistance can be reduced by the application of voltage sufficently high to cause the fault to break down with sufficient current to cause the insulation to carbonize. The equipment required to do this is quite large and expensive, and its success is dependent to a large degree on the type of insulation involved. Large users indicate that this method is useful with paper and elastomeric cables, but generally of little use with plastic types.

The fault resistance that exists after the occurrence of the original fault depends on the type of cable insulation and construction, the location of the fault, and the cause of the failure. A fault that is immersed in water will generally exhibit a variable fault resistance and will not consistently arc over at a constant voltage. Damp faults behave in a similar manner until the moisture has been vaporized. In contrast, a dry fault will normally be much more stable and consequently can more readily be located.

For failures that have occurred in service, the type of system grounding and available fault current, as well as the speed of relay protection, will be influencing factors. Because of the greater carbonization and conductor vaporization, a fault resulting from an in-service failure can generally be expected to be of a lower resistance than one resulting from overpotential testing.

11.12.2 Equipment and Methods. A wide variety of commercially available equipment and a number of different approaches can be used to locate cable faults. The safety considerations outlined in Section 11.11 should be observed.

The method used to locate a cable fault depends on

(a) Nature of fault
(b) Type and voltage rating of cable
(c) Value of rapid location of faults
(d) Frequency of faults
(e) Experience and capability of personnel

(1) *Physical Evidence of the Fault.* Observation of a flash, sound, or smoke accompanying the discharge of current through the faulted insulation will usually locate a fault. This is more probable with an overhead circuit than with underground construction. The discharge may be from the original fault or may be intentionally caused by the application of test voltages. The burned or disrupted appearance of the cable will also serve to indicate the faulted section.

(2) *Megohmmeter Instrument Test.* When the fault resistance is sufficiently low that it can be detected with a megohmmeter, the cable can be sectionalized and each section tested to determine which contains the fault. This procedure may require that the cable be opened in a number of locations before the fault is isolated to one replaceable section. This could, therefore, involve considerable time and expense, and might result in additional splices. Since splices are often the weakest part of a cable circuit, this method of fault locating may introduce additional failures at a subsequent time.

(3) *Conductor Resistance Measurement.* This method consists of measuring the resistance of the conductor from the test location to the point of fault by using either the Varley loop or the Murray loop test. Once the resistance of the conductor to the point of fault has been measured, it can be translated into distance by using handbook values of resistance per unit length of the size and type of conductor involved, correcting for temperature as required. Both of these methods give good results that are independent of fault resistance, provided the fault resistance is low enough that sufficient current for readable galvanometer deflection can be produced with the available test voltage. Normally a low-voltage bridge is used for this resistance measurement. For distribution systems using cables insulated with organic materials, relatively low resistance faults are normally encountered. The conductor resistance measurement method has its major application on such systems. Loop tests on large conductor sizes may not be sensitive enough to narrow down the location of the fault.

High-voltage bridges are available for higher resistance faults but have the disadvantage of increased cost and size as well as requiring a high-voltage dc power supply. High-voltage bridges are generally capable of locating faults with a resistance to ground of up to 1 or 2 MΩ, while a low-voltage bridge is limited to the application where this resistance is several kilohms or less.

(4) *Capacitor Discharge*. This method consists of applying a high-voltage high-current impulse to the faulted cable. A high-voltage capacitor is charged by a relatively low-current capacity source such as that used for high-potential testing. The capacitor is then discharged across an air gap or by a timed-closing contact into the cable. The repeated discharging of the capacitor provides a periodic pulsing of the faulted cable. The maximum impulse voltage should not exceed 50% of the allowable dc cable test voltage since voltage doubling can occur at open-circuit ends. Where the cable is accessible, or the fault is located at an accessible position, the fault may be located simply by sound. Where the cable is not accessible, such as in duct or directly buried, the discharge at the fault may not be audible. In such cases, detectors are available to trace the signal to the point of fault. The detector generally consists of a magnetic pickup coil, an amplifier, and a meter to display the relative magnitude and direction of the signal. The direction indication changes as the detector passes beyond the fault. Acoustic detectors are also employed, particularly in situations where no appreciable magnetic field external to the cable is generated by the tracing signal.

In applications where relatively high-resistance faults are anticipated, such as with solid dielectric cables or through compound in splices and terminations, the impulse method is the most practical method presently available and is the one most commonly used.

(5) *Tone Signal*. (May be used on energized circuits.) A fixed-frequency signal, generally in the audio frequency range, is imposed on the faulted cable. The cable route is then traced by means of a detector, which consists of a pickup coil, receiver, and head set or visual display, to the point where the signal leaves the conductor and enters the ground return path. This class of equipment has its primary application in the low-voltage field and is frequently used for fault location on energized ungrounded-type circuits. On systems over 600 V the use of a tone signal for fault location is generally unsatisfactory because of the relatively large capacitance of the cable circuit.

(6) *Radar System*. A short-duration low-energy pulse is imposed on the faulted cable and the time required for propagation to and return from the point of fault is monitored on an oscilloscope. The time is then translated into distance in order to locate the point of fault. Although this type of equipment has been available for a number of years, its major application in the power field has been on long-distance high-voltage lines. In older equipment the propagation time is such that it cannot be displayed with good resolution for relatively short cables encountered in industrial systems. However, recent equipment advances have largely overcome this problem. The major limitation to this method is the inability to adequately determine the difference between faults and splices on multitapped circuits. An important feature of this method is that it will locate an *open* in an otherwise unfaulted circuit.

11.12.3 Selection. The methods listed represent some of the means available to industrial plant power system operators to locate cable faults. They range from very simple to relatively complex. Some require no equipment, others require equipment that is inexpensive and can be used for other purposes, while still others require special equipment. As the complexity of the means used to locate a fault increases, so does the cost of the equipment, and also the training and experience required of those who are to use it.

In determining which approach is most practical for any particular plant, the size of the plant and the amount of circuit redundancy that it contains should be considered. The importance of minimizing the outage time of any particular circuit should be evaluated. The cable installation and maintenance practices and number and time of anticipated faults will determine the expenditure for test equipment that can be justified. Equipment that requires considerable experience and operator interpretation for accurate results may be satisfactory for a plant with frequent cable faults but ineffective where the number of faults is so small that adequate experience cannot be obtained. Because of these factors, many companies employ firms that offer the service of cable-fault locating. Such firms are established in large cities and cover a large area with mobile test equipment.

While the capacitor discharge is most widely used, no single method of cable-fault locating can be considered to be most suitable for all applications. The final decision on which method or methods to use should depend upon evaluation of the merits and disadvantages of each in relation to the particular circumstances of the plant in question. As a last resort, opening splices in manholes and testing the cable between manholes can be used to locate the faulted cable.

11.13 Cable Specification. Once the correct cable has been determined, it can be described in a cable specification. Cable specifications generally start with the conductor and progress radially through the insulation and coverings. The following is a check list that can be used in preparing a cable requirement:

(1) Number of conductors in cable, and phase identification required
(2) Conductor size (AWG, kcmil) and material
(3) Insulation type (rubber, polyvinyl chloride, polyethylene, EPR, etc)
(4) Voltage rating
(5) Shielding system
(6) Outer finishes
(7) Installation (cable tray, direct burial, wet location, exposure to sunlight or oil, etc)
(8) Applicable UL listing
(9) Test voltage and partial discharge voltage

An alternate method of specifying cable is to furnish the ampacity of the circuit (amperes), the voltage (phase-to-phase, phase-to-ground, grounded, or ungrounded), and the frequency, along with any other pertinent system data. Also required is the method of installation anticipated and the installation conditions (ambient temperature, load factor, etc). For either method, the total number of lineal feet of conductors required, the quantity desired shipped in one length, the pulling eyes, and whether it is desired to have several single-conductor cables paralleled on a reel should also be given.

11.14 References. This standard shall be used in conjunction with the following references:

[1] AEIC CS5-1982, Specifications for Thermoplastic and Crosslinked Polyethylene Insulated Shielded Power Cables Rated 5 Through 46 kV.[29]

[2] AEIC CS6-1982, Specifications for Ethylene Propylene Rubber Insulated Shielded Power Cables Rated 5 Through 69 kV.

[3] ANSI C2-1987, National Electrical Safety Code.

[4] ANSI/IEEE Std 100-1984, IEEE Standard Dictionary of Electrical and Electronics Terms.

[5] ANSI/IEEE Std 142-1982, IEEE Recommended Practice for Grounding of Industrial and Commercial Power Systems.

[6] ANSI/IEEE Std 242-1986, IEEE Recommended Practice for Protection and Coordination of Industrial and Commercial Power Systems.

[7] ANSI/IEEE Std 386-1985, IEEE Standard for Separable Insulated Connector Systems for Power Distribution Systems Above 600 V.

[8] ANSI/IEEE Std 400-1980, IEEE Guide for Making High-Direct-Voltage Tests on Power Cable Systems in the Field.

[9] ANSI/IEEE Std 404-1985, IEEE Standard for Cable Joints for Use with Extruded Dielectric Cable Rated 5000 V Through 46 000 V and Cable Joints for Use with Laminated Dielectric Cable Rated 2500 V Through 500 000 V.

[10] ANSI/IEEE Std 446-1980, IEEE Recommended Practice for Emergency and Standby Power Systems for Industrial and Commercial Applications.

[11] ANSI/IEEE Std 592-1977, IEEE Standard for Exposed Semiconducting Shields on Premolded High-Voltage Cable Joints and Separable Insulated Connectors.

[12] ANSI/NFPA 70-1987, National Electrical Code.

[13] ANSI/UL 44-1985, Safety Standard for Rubber-Insulated Wires and Cables.

[14] ANSI/UL 62-1985, Safety Standard for Flexible Cord and Fixture Wire.

[15] ANSI/UL 83-1985, Safety Standard for Thermoplastic-Insulated Wires and Cables.

[16] ANSI/UL 486A-1982, Safety Standard for Wire Connectors and Soldering Lugs for Use with Copper Conductors.

[17] ANSI/UL 486B-1982, Safety Standard for Wire Connectors for Use with Aluminum Conductors.

[29] AEIC publications are available from the Association of Edison Illuminating Companies, 51 East 42 Street, New York, NY 10017.

[18] ANSI/UL 493-1983, Safety Standard for Thermoplastic-Insulated Underground Feeder and Branch-Circuit Cables.

[19] ANSI/UL 1569-1985, Safety Standard for Metal-Clad Cables.

[20] ANSI/UL 1581-1985, Reference Standard for Electrical Wires, Cables, and Flexible Cords.

[21] ICEA P-32-382, Short-Circuit Characteristics of Insulated Cable. Revised March 1969.[30]

[22] ICEA P-45-482, Short-Circuit Performance of Metallic Shielding and Sheaths of Insulated Cable, 2nd Edition, Aug 1979.

[23] ICEA S-19-81/NEMA WC3-1980, Rubber-Insulated Wire and Cable for the Transmission and Distribution of Electrical Energy. Revision No 2, Dec 1984.

[24] ICEA S-61-402/NEMA WC5-1973, Thermoplastic-Insulated Wire and Cable for the Transmission and Distribution of Electrical Energy, Revision No 11, Dec 1984.

[25] ICEA S-66-524/NEMA WC7-1982, Cross-Linked-Thermosetting-Polyethylene-Insulated Wire and Cable for the Transmission and Distribution of Electrical Energy. Revision No 1, Dec 1984.

[26] ICEA S-68-516/NEMA WC8-1976, Ethylene-Propylene-Rubber-Insulated Wire and Cable for the Transmission and Distribution of Electrical Energy. Revision No 7, Dec 1984.

[27] IEEE S-135, IEEE/IPCEA [ICEA] Power Cable Ampacities (SH07096).

[28] IEEE Std 48-1975, IEEE Test Procedures and Requirements for High-Voltage AC Cable Terminations.

[29] IEEE Std 525-1978, IEEE Guide for Selection and Installation of Control and Low-Voltage Cable Systems in Substations.

[30] UL 13, Outline of Proposed Investigation of Power-Limited Circuit Cable. July 1978.[31]

[31] UL 854-1986, Safety Standard for Service-Entrance Cables.

[32] UL 910-1985, Standard for Test Method for Fire and Smoke Characteristics of Electrical and Optical Fiber Cables Used in Air-Handling Spaces.

[33] UL 1072-1986, Medium-Voltage Power Cables.

[30] ICEA publications are available from the Insulated Cable Engineers Association, P.O. Box 411, South Yarmouth, MA 02664.

[31] UL publications are available from the Electrical Department, Underwriters Laboratories, 1285 Walt Whitman Road, Melville, NY 11747.

[34] UL 1277-1986, Electrical Power and Control Tray Cables with Optional Optical Fiber Members.

[35] Underground Systems Reference Book. New York: Association of Illuminating Companies, 1957, Chap 10.

12. Busways

12.1 Origin. Busways originated as a result of a request of the automotive industry in Detroit in the late 1920s for an overhead wiring system that would simplify electrical connections for electric motor-driven machines and permit a convenient arrangement of these machines in production lines. From this beginning, busways have grown to become an integral part of the low-voltage distribution system for industrial plants at 600 V and below.

Busways are particularly advantageous when numerous current taps are required. Plugs with circuit breakers or fusible switches may be installed and wired without de-energizing the busway.

Power circuits over 1000 A are usually more economical and require less space with busways than with conduit and wire. Busways may be dismantled and reinstalled in whole or part to accommodate changes in the electrical distribution system layout.

12.2 Busway Contruction. Originally a busway consisted of bare copper conductors supported on inorganic insulators such as porcelain, mounted within a nonventilated steel housing. This type of construction was adequate for the current ratings of 225-600 A then used. As the use of busways expanded and increased loads demanded higher current ratings, the housing was ventilated to provide better cooling at higher capacities. The bus bars were covered with insulation for safety and to permit closer spacing of bars of opposite polarity in order to achieve lower reactance and voltage drop.

In the late 1950s busways were introduced utilizing conduction for heat transfer by placing the insulated conductor in thermal contact with the enclosure. By utilizing conduction, current densities are achieved for totally enclosed busways that are comparable to those previously attained with ventilated busways. Totally enclosed busways of this type have the same current rating regardless of mounting position. A stack of one bus bar per phase is used where each bus bar is up to approximately 7 in wide (1600 A). Higher ratings will use two (3000 A) or three stacks (5000 A). Each stack will contain all three phases and neutral to minimize circuit reactance.

Early busway designs required multiple nuts, bolts, and washers to electrically join adjacent sections. The most recent designs use a single bolt for each stack (with bars up to 7 in wide). All hardware is captive to the busway section when shipped from the factory. Installation labor is greatly reduced with corresponding savings in installation costs.

Busways are available with either copper or aluminum conductors. Compared to copper, aluminum has lower electrical conductivity, less mechanical strength, and upon exposure to the atmosphere quickly forms an insulating film on the surface. For equal current-carrying ability aluminum is lighter in weight and less costly.

For these reasons aluminum conductors will have electroplated contact surfaces (tin or silver) and at electrical joints use Belleville springs and bolting practices that accommodate aluminum's mechanical properties. Copper busway will be physically smaller (cross section) while aluminum busway is lighter in weight and lower in cost. Copper plug-in busway is more tolerant of cycling loads such as welding.

Busway is usually made in 10 ft sections. Since the busway must conform to the building, all possible combinations of elbows, tees, and crosses are available. Feed and tap fittings to other electric equipment, such as switchboards, transformers, motor-control centers, etc, are provided. Plugs for plug-in busway use fusible switches and molded-case circuit breakers. Standard busway current ratings are 20-5000 A for single-phase and three-phase service. Neutral conductors may be supplied if required. Newer designs of busway including plug-in devices can incorporate a ground bar if specified.

Four types of busways are available, complete with fittings and accessories, providing a unified and continuous system of enclosed conductors (Fig 164):

(1) Feeder busway for low-impedance transmission of power

(2) Plug-in busway for easy connection or rearrangement of loads

(3) Lighting busway to provide electric power and mechanical support to fluorescent, high-intensity discharge, and incandescent fixtures

(4) Trolley busway for mobile power *tapoffs* to electric hoists, cranes, portable tools, etc

12.3 Feeder Busway. Feeder busway is used to transmit large blocks of power. It has a very low and balanced circuit reactance for control of voltage at the utilization equipment (Fig 165).

Feeder busway is frequently used between the source of power, such as a distribution transformer or service drop, and the service entrance equipment. Industrial plants use feeder busway from the service equipment to supply large loads directly and to supply smaller current ratings of feeder and plug-in busway, which in turn supply loads through power take-offs or plug-in units.

Available current ratings range from 600-5000 A, 600 V ac. The manufacturer should be consulted for dc ratings. Feeder busway is available in single-phase and three-phase service with 50 and 100% neutral conductor. A ground bus is available with all ratings and types. Available short-circuit current ratings are 50 000-200 000 A, symmetrical rms (see 12.8.2). The voltage drop of low-impedance feeder busway with the entire load at the end of the run ranges from 1-3 V/100 ft, line-to-line, depending upon the type of construction and the current rating (see 12.8.3).

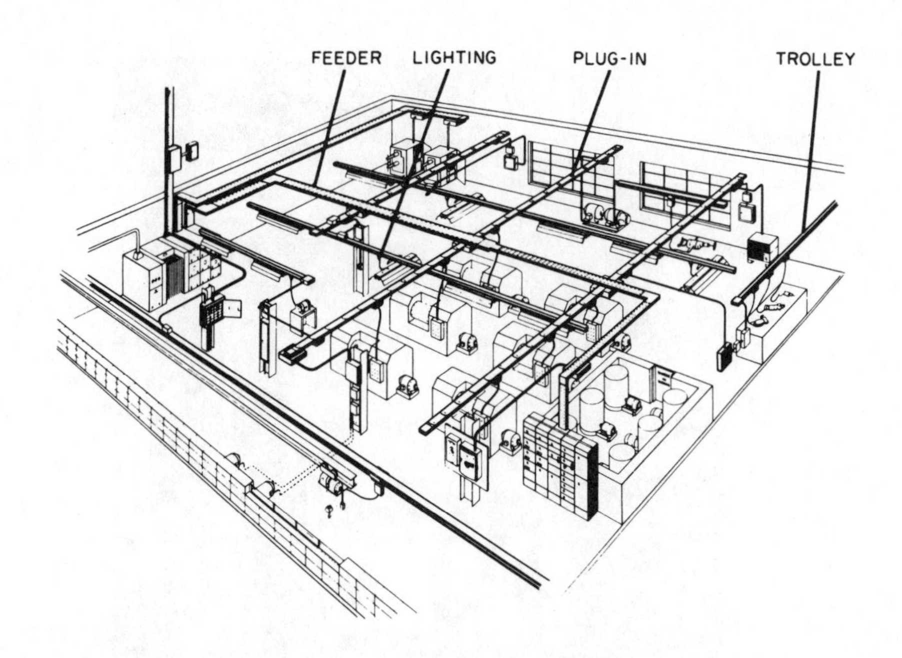

Fig 164
Illustration of Versatility of Busways, Showing Use of Feeder, Plug-In, Lighting, and Trolley Types

Feeder busway is available in indoor and weatherproof (outdoor) construction. Weatherproof construction is designed to shed water (rain). It should be used indoors where the busway may be subjected to water or other liquids. If NEMA "3R" equipment is suitable (see NEMA BU1-1983 [4])[32], weatherproof busway should be used. Busway of any type is not suitable for immersion in water.

12.4 Plug-In Busway. Plug-in busway is used in industrial plants as an overhead system to supply power to utilization equipment. It serves as an elongated switchboard or panelboard running through the area with covered plug-in openings provided at closely spaced intervals to accommodate the plug-in devices placed on the busway near the loads which they supply.

Plug-in tapoff rearrangement is greatly facilitated by the use of flexible bus drop cable. The plug may be removed from the busway together with the bus drop cable and reinstalled with the machine in a minimum of time (Fig 166).

Plug-in devices available include fusible switches, circuit breakers, static voltage protectors, ground indicators, combination starters, lighting contactors, and capacitor plugs.

---

[32] The numbers in brackets correspond to those of the references listed at the end of this chapter; when preceded by B, they correspond to the bibliography at the end of this chapter.

Fig 165
Feeder Busway

Most plug-in busway is totally enclosed with current ratings from 100–4000 A. Usually plug-in and feeder busway sections of the same manufacturer above 600 A have compatible joints, so that they are interchangeable in a run. Plug-in busway may be inserted in a feeder run when a tapoff is desired. Plug-in tapoffs are generally limited to maximum ratings of 800 A.

Short-circuit current ratings vary from 15 000–150 000 A, symmetrical rms (see 12.8.2). The voltage drop ranges from 1–3 V/100 ft, line-to-line, for evenly distributed loading. If the entire load is concentrated at the end of the run, these values double (see 12.8.3).

A neutral bar may be provided for single-phase loads such as lighting. Neutral bars vary from 25–100% of the capacity of the phase bars.

A ground bar is often added for greater system protection and coordination under ground fault conditions. The ground bar provides a low-impedance ground path and also reduces the possibility of arcing at the joint under high-level ground faults. See 12.8.2 for additional details.

Fig 166
Installation View of Small Plug-In Busway Showing Individual Circuit Breaker Power Tapoff and Flexible Bus-Drop Cable

12.5 Lighting Busway. Lighting busway is rated a maximum of 60 A, 300 V-to-ground, with two, three, or four conductors. It may be used on 480Y/277 or 208Y/120 V systems and is specifically designed for use with fluorescent and high-intensity discharge lighting (Fig 167).

Lighting busways provide power to the lighting fixture and also serve as the mechanical support for the fixture. Auxiliary supporting means called *strength beams* are available. Strength beams may be supported at maximum intervals of 16 ft. This permits the strength beam supports to conform to building column spacing. The strength beams provide supports for the lighting busway as required by ANSI/NFPA 70-1987 [2], (National Electric Code [NEC]).

Fluorescent lighting fixtures may be suspended from the busway or they may be ordered with plugs and hangers attached for close coupling of the fixture to the busway. The busway may also be recessed in or surface mounted to dropped ceilings.

Lighting busway is also used to provide power for light industrial applications.

12.6 Trolley Busway. Trolley busway is constructed to receive stationary or movable take-off devices. It is used on a moving production line to supply electric power to a motor or a portable tool moving with a production line, or where operators move back and forth over a range of 10–20 ft to perform their specific operations.

Fig 167
Lighting Busway Supporting and Supplying Power to High-Intensity Discharge Fixture

12.7 Standards. Busways are designed to conform to the following standards:

(1) NEC [2], Article 364
(2) ANSI/UL 857-1981 [3]
(3) NEMA BU1-1983 [4]

ANSI/UL 857-1981 [3] and NEMA BU1-1983 [4] are primarily manufacturing and testing standards. The NEMA standard is generally an extension of the UL standard to areas that UL does not cover. The most important areas are busway parameters resistance R, reactance X, and impedance Z, and short-circuit testing and rating.

The NEC [2] is the most important standard for busway installation. Some of its most important areas are as follows:

(1) Busway may be installed only where located in the open and visible. Installation behind panels is permitted if access is provided and the following conditions are met:

(a) No overcurrent devices are installed on the busway other than for an individual fixture
(b) The space behind the panels is not used for air-handling purposes
(c) The busway is the totally enclosed nonventilating type
(d) The busway is so installed that the joints between sections and fittings are accessible for maintenance purposes

(2) Busway may not be installed where subject to physical damage, corrosive vapors, or in hoistways.

(3) When specifically approved for the purpose, busway may be installed in a hazardous location, outdoors, or in wet or damp locations.

(4) Busway must be supported at intervals not to exceed 5 ft unless otherwise approved. Where specifically approved for the purpose, horizontal busway may be supported at intervals up to 10 ft, and vertical busway may be supported at intervals up to 16 ft.

(5) Busway must be totally enclosed where passing through floors and for a minimum distance of 6 ft above the floor to provide adequate protection from physical damage. It may extend through walls if joints are outside walls.

State and local electrical codes may have specific requirements over and above ANSI/UL 857-1981 [3] and NEC [2]. Appropriate code authorities and manufacturers should be contacted to ensure that requirements are met.

12.8 Selection and Application of Busways. To properly apply busways in an electric power distribution system, some of the more important items to consider are the following.

12.8.1 Current-Carrying Capacity. Busways should be rated on a temperature rise basis to provide safe operation, long life, and reliable service.

Conductor size (cross-sectional area) should not be used as the sole criterion for specifying busway. Busway may have seemingly adequate cross-sectional area and yet have a dangerously high temperature rise. The UL requirement for temperature rise (55 °C) (See ANSI/UL 857-1981 [3]) should be used to specify the maximum temperature rise permitted. Larger cross-sectional areas can be used to provide lower voltage drop and temperature rise.

Although the temperature rise will not vary significantly with changes in ambient temperature, it may be a significant factor in the life of the busway. The limiting factor in most busway designs is the insulation life, and there is a wide range of types of insulating materials used by various manufacturers. If the ambient temperature exceeds 40 °C or a total temperature in excess of 95 °C is expected, then the manufacturer should be consulted.

12.8.2 Short-Circuit Current Rating. The bus bars in busways may be subject to electromagnetic forces of considerable magnitude by a short-circuit current. The generated force per unit length of bus bar is directly proportional to the square of the short-circuit current and is inversely proportional to the spacing between bus bars. Short-circuit current ratings are generally assigned and tested in accordance with NEMA BU1-1983 [4]. The ratings are based on (1) the use of an adequately rated protective device ahead of the busway that will clear the short circuit in 3 cycles and (2) application in a system with short-circuit power factor not less than that given in Table 89.

If the system on which the busway is to be applied has a lower short-circuit power factor (larger X/R ratio) or a protective device with a longer clearing time, the short-circuit current rating may have to be reduced. The manufacturer should then be consulted.

The required short-circuit current rating should be determined by calculating the available short-circuit current and X/R ratio at the point where the input end of the busway is to be connected. The short-circuit current rating of the busway must equal or exceed the available short-circuit current.

The short-circuit current may be reduced by using a current-limiting fuse at the supply end of the busway to cut it off before it reaches maximum value (see Chapter 5).

Short-circuit current ratings are dependent on many factors such as bus bar center line spacing, size, and strength of bus bars and mechanical supports.

Since the ratings are different for each design of bus bar, the manufacturer should be consulted for specific ratings. Short-circuit current ratings should include the ability of the ground return path (housing and ground bar if provided) to carry the rated short-circuit current. Failure of the ground return path to adequately carry this current can result in arcing at joints with attendant fire hazard. The ground-fault current can also be reduced to the point that the overcurrent protective device does not operate.

12.8.3 Voltage Drop. Line-to-neutral voltage drop V_D in busways may be calculated by the following formulas. The exact formulas for concentrated loads at the end of the line are, with V_R known,

$$V_D = \sqrt{(V_R \cos \phi + IR)^2 + (V_R \sin \phi + IX)^2} - V_R \quad \text{(Eq 1)}$$

and with V_S known,

$$V_D = V_S + IR \cos \phi + IX \sin \phi - \sqrt{V_S^2 - (IX \cos \phi - IR \sin \phi)^2} \quad \text{(Eq 2)}$$

where

$$V_R = V_S \frac{Z_L}{Z_S}, \quad V_D = V_S - V_R \quad \text{(Eq 3)}$$

NOTE: Multiply the line-to-neutral voltage drop by $\sqrt{3}$ to obtain the line-to-line voltage drop in three-phase systems. Multiply the line-to-neutral voltage drop by 2 to obtain the line-to-line voltage drop in single-phase systems.

The approximate formulas for concentrated loads at the end of the line are

$$V_D = I(R \cos \phi + X \sin \phi) \quad \text{(Eq 4)}$$

$$V_{pr} = \frac{S(R \cos \phi + X \sin \phi)}{10\, V_k^2} \quad \text{(Eq 5)}$$

Table 84
Busway Ratings as a Function of Short-Circuit Power Factor

Busway Rating (symmetrical rms amperes)	Power Factor	X/R Ratio*
10 000 or less	0.50	1.7
10 001—20 000	0.30	3.2
Above 20 000	0.20	4.9

*X/R is load reactance X divided by load resistance R.

The approximate formula for distributed load on a line is

$$V_{pr} = \frac{S(R \cos \phi + X \sin \phi)\, L}{10\, V_k^2} \left(1 - \frac{L_1}{2L}\right) \qquad \text{(Eq 6)}$$

where

V_D = voltage drop, in volts
V_{pr} = voltage drop, in percent of voltage at sending end
V_S = line-to-neutral voltage at sending end, in volts
V_R = line-to-neutral voltage at receiving end, in volts
ϕ = angle whose cosine is the load power factor
R = resistance of circuit, in ohms per phase
X = reactance of circuit, in ohms per phase
I = load current, in amperes
Z_L = load impedance, in ohms
Z_S = circuit impedance, in ohms, plus load impedance, in ohms, added vectorially
S = three-phase apparent power for three-phase circuits or single-phase apparent power for single-phase circuits, in kilovolt-amperes
V_k = line-to-line voltage, in kilovolts
L_1 = distance from source to desired point, in feet
L = total length of line, in feet

The foregoing formulas for concentrated loads may be verified by a trigonometric analysis of Fig 168. From this figure it can be seen that the approximate formulas are sufficiently accurate for practical purposes. In practical cases the angle between V_R and V_S will be small (much smaller than in Fig 168, which has been exaggerated for illustrative purposes). The error in the approximate formulas diminishes as the angle between V_R and V_S decreases and is zero if that angle is zero. This latter condition will exist when the X/R ratio (or power factor) of the load is equal to the X/R ratio (or power factor) of the circuit through which the load current is flowing.

Figure 168
Diagram Illustrating Voltage Drop and Indicating Error When Approximate Voltage-Drop Formulas Are Used

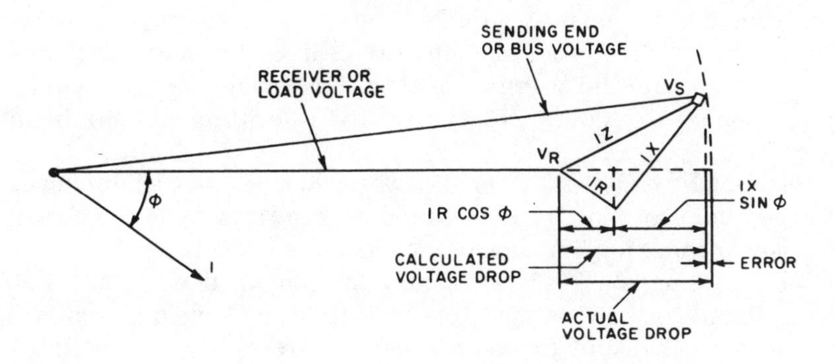

In actual practice, loads may be concentrated at various locations along the feeders, uniformly distributed along the feeder, or any combination of the same. A comparison of the approximate formulas for concentrated end loading and uniform loading will show that a uniformly loaded line will exhibit one half the voltage drop as that due to the same total load concentrated at the end of the line. This aspect of the approximate formula is mathematically exact and entails no approximation. Therefore, in calculations of composite loading involving approximately uniformly loaded sections and concentrated loads, the uniformly loaded sections may be treated as end-loaded sections having one half normal voltage drop of the same total load. Thus the load can be divided into a number of concentrated loads distributed at various distances along the line. The voltage drop in each section may then be calculated for the load which it carries.

Three-phase voltage drops may be determined with reasonable accuracy by the use of Tables 85 and 86. These are typical values for the particular types of busway shown. The voltage drops will be different for other types of busway and will vary slightly by manufacturer within each type. The voltage drop shown is three-phase, line-to-line, per 100 ft, at rated load on a concentrated loading basis for feeder, plug-in, and trolley busway. Lighting busway values are single-phase, distributed loading. For other loading and distances use the formula

$$\text{voltage drop } V_D = \text{table } V_D \left(\frac{\text{actual load}}{\text{rated load}} \right) \left(\frac{\text{actual distance (feet)}}{100 \text{ ft}} \right)$$

The voltage drop for a single-phase load connected to a three-phase busway is 15.5% higher than the value shown in the tables. Typical values of resistance and reactance are shown in Table 87. Resistance is shown at normal room temperature (25 °C). This value should be used in calculating the short-circuit current available in systems since short circuits can occur when busway is lightly loaded or initially energized. To calculate the voltage drop when fully loaded (75 °C), the resistance of copper and aluminum should be multiplied by 1.19.

12.8.4 Thermal Expansion. As load is increased, the bus bar temperature will increase and the bus bars will expand. The lengthwise expansion between no load and full load will range from ½ to 1 inch per 100 ft. The amount of expansion will depend on the total load, the size and location of the tapoffs, and the size and duration of varying loads. To accommodate the expansion, the busway should be mounted using hangers that permit it to move. It may be necessary to insert expansion lengths in the busway run. To locate expansion lengths, the method of support, the location of power take-offs, the degree of movement permissible at each end of the run, and the orientation of the busway must be known. The manufacturer can then make recommendations as to the location and number of expansion lengths.

12.8.5 Building Expansion Joints. Busway, when crossing a building expansion joint, must include provision for accommodating movement of the building structure. Fittings providing for 6 inches of movement are available.

12.8.6 Welding Loads. The busway and the plug-in device must be properly sized when plug-in busway is used to supply power to welding loads. The plug sliding contacts (stabs) and protective device (circuit breaker or fused switch)

Table 85
Voltage-Drop Values for Three-Phase Busways with Copper Bus Bars, in Volts per 100 ft, Line to Line, at Rated Current with Entire Load at End*

Current Rating (amperes)	Load Power Factor (Percent, lagging) 20	30	40	50	60	70	80	90	95	100
Totally Enclosed Feeder Busway										
600	2.28	2.51	2.73	2.93	3.09	3.23	3.31	3.31	3.23	2.83
800	1.75	1.93	2.08	2.23	2.35	2.44	2.49	2.48	2.42	2.10
1000	1.51	1.81	2.11	2.39	2.66	2.92	3.15	3.33	3.39	3.29
1350	1.60	1.87	2.13	2.37	2.60	2.80	2.98	3.11	3.13	2.96
1600	1.90	2.10	2.27	2.43	2.56	2.67	2.73	2.72	2.66	2.31
2000	1.82	2.00	2.16	2.30	2.43	2.52	2.57	2.55	2.49	2.15
2500	1.75	1.91	2.06	2.18	2.29	2.36	2.40	2.37	2.30	1.96
3000	1.96	2.14	2.30	2.43	2.55	2.63	2.67	2.63	2.55	2.17
4000	1.84	2.01	2.16	2.29	2.40	2.49	2.53	2.49	2.42	2.07
5000	1.67	1.83	1.98	2.11	2.22	2.30	2.35	2.33	2.27	1.96
Totally Enclosed Plug-In Busway										
225	1.92	2.08	2.22	2.36	2.46	2.54	2.56	2.52	2.42	2.04
400	2.26	2.40	2.52	2.60	2.66	2.70	2.66	2.54	2.40	1.90
600	4.91	5.03	5.10	5.11	5.04	4.89	4.62	4.11	3.67	2.38
800	5.75	5.91	6.00	6.02	5.96	5.80	5.50	4.92	4.42	2.92
1000	4.77	4.91	4.98	5.02	4.98	4.84	4.60	4.12	3.70	2.46
1350	3.72	3.84	3.92	3.94	3.94	3.84	3.68	3.32	3.01	2.06
1600	3.58	3.70	3.78	3.82	3.80	3.72	3.54	3.22	2.92	2.00
2000	4.67	4.79	4.86	4.86	4.82	4.68	4.42	3.94	3.52	2.30
2500	4.08	4.20	4.26	4.30	4.26	4.14	3.94	3.54	3.18	2.12
3000	3.76	3.87	3.92	3.94	3.90	3.80	3.60	3.24	2.90	1.92
4000	4.64	4.74	4.80	4.79	4.73	4.57	4.30	3.81	3.38	2.15
5000	3.66	3.75	3.78	3.78	3.78	3.62	3.40	3.02	2.70	1.76
Lighting, Single Phase, Distributed Loading										
30	0.84	1.11	1.38	1.65	1.89	2.13	2.40	2.51	2.20	2.75
60	1.08	1.38	1.62	1.98	2.22	2.46	2.70	2.88	3.00	3.00
Trolley										
100	1.16	1.38	1.56	1.74	1.90	2.06	2.20	2.28	2.30	2.18

NOTE: Voltage-drop values are based on bus bar resistance at 75° C (room ambient temperature 25° C plus average conductor temperature at full load of 50° C rise).

*Divide values by 2 for distributed loading.

should have sufficient rating to carry both the continuous and peak welding load. This is normally done by determining the equivalent continuous current of the welder based on the maximum peak welder current, the duration of the welder current, and the duty cycle. Values may be obtained from the welder manufacturer. Loads 600 A and greater require special attention including consideration of bolted taps. As previously stated, copper busway is more tolerant of cycling loads than aluminum busway. When aluminum busway is used, cycling loads should be referred to the manufacturer.

12.9 Layout. Busway must be tailored to the building in which it is installed. Once the basic engineering work has been completed and the busway type, current rating, number of poles, etc determined, a layout should be made for all but the simplest straight runs. The initial step in the layout is to identify and locate the building structure (walls, ceilings, columns, etc) and other equipment that is in the

Table 86
Voltage-Drop Values for Three-Phase Busways with Aluminum Bus Bars, in Volts per 100 ft, Line to Line, at Rated Current with Entire Load at End*

Current Rating	Load Power Factor (Percent, lagging)									
(amperes)	20	30	40	50	60	70	80	90	95	100
Totally Enclosed Feeder Busway										
600	1.64	1.93	2.21	2.48	2.73	2.96	3.16	3.30	3.34	3.17
800	1.69	1.95	2.21	2.44	2.66	2.86	3.03	3.14	3.15	2.94
1000	1.51	1.81	2.11	2.39	2.66	2.92	3.15	3.33	3.39	3.29
1350	1.60	1.87	2.13	2.37	2.60	2.80	2.98	3.11	3.13	2.96
1600	1.70	1.97	2.22	2.45	2.67	2.87	3.04	3.14	3.15	2.94
2000	1.57	1 81	2.03	2.23	2.42	2.59	2.73	2.81	2.81	2.60
2500	1.56	1.78	1.98	2.18	2.35	2.51	2.63	2.70	2.69	2.48
3000	1.64	1.94	2.14	2.37	2.58	2.78	2.94	3.04	3.05	2.85
4000	1.60	1.83	2.04	2.24	2.42	2.59	2.71	2.79	2.78	2.56
Totally Enclosed Plug-In Busway										
100	2.05	2.63	3.20	3.76	4.30	4.83	5.33	5.79	5.98	6.01
225	1.94	2.22	2.49	2.73	2.96	3.15	3.31	3.41	3.40	3.13
400	3.47	3.66	3.81	3.92	3.99	3.99	3.92	3.69	3.45	2.64
600	4.62	4.89	5.12	5.30	5.41	5.45	5.37	5.10	4.80	3.76
800	4.09	4.34	4.54	4.70	4.81	4.84	4.78	4.54	4.28	3.36
1000	3.22	3.43	3.61	3.75	3.85	3.89	3.86	3.70	3.50	2.79
1350	2.92	3.10	3.12	3.36	3.44	3.48	3.44	3.28	3.08	2.44
1600	3.98	4.20	4.38	4.51	4.59	4.61	4.52	4.27	3.99	3.07
2000	3.48	3.68	3.85	3.99	4.07	4.09	4.04	3.83	3.60	2.81
2500	2.83	3.00	3.13	3.24	3.30	3.32	3.27	3.10	2.92	2.27
3000	3.68	3.85	3.99	4.09	4.14	4.12	4.01	3.74	3.47	2.60
4000	3.11	3.27	3.40	3.50	3.55	3.55	3.47	3.26	3.04	2.31

NOTE: Voltage-drop values are based on bus bar resistance at 75° C (room ambient temperature 25° C plus average conductor temperature at full load of 50° C rise).

*Divide values by 2 for distributed loading.

busway route. A layout of the busway to conform to this route is made. Although the preliminary layout (drawings for approval) can be made from architectural drawings, it is essential that field measurements be taken to verify building and busway dimensions prior to the release of the busway for manufacture. Where dimensions are critical, it is recommended that a section be held for field check of dimensions and manufactured after the remainder of the run has been installed. Manufacturers will provide quick delivery on limited numbers of these *field-check* sections.

Busway has great physical and electrical flexibility. It may be tailored to almost any layout requirement. However, some users find it a good practice to limit their busway installations to a minimum number of current ratings and maintain as many 10 ft lengths as possible. This enables them to reuse the busway components to maximum advantage where production line changes, etc, require relocation of the busway.

Another important consideration when laying out busway is coordination with other trades. Since there is a finite time lapse between job measurement and actual installation, other trades may use the busway *clear area* if coordination is lacking. Again, standard components can help since they are more readily available (some-

Table 87
Typical Busway Parameters, Line to Neutral, in Milliohms per 100 ft, 25 °C

Current Rating (amperes)	Feeder Busway Aluminum *R*	Feeder Busway Aluminum *X*	Feeder Busway Copper *R*	Feeder Busway Copper *X*	Plug-In Busway Aluminum *R*	Plug-In Busway Aluminum *X*	Plug-In Busway Copper *R*	Plug-In Busway Copper *X*
100	—	—	—	—	29.1	5.0	—	—
225	—	—	—	—	6.74	3.45	4.44	3.94
400	—	—	—	—	3.20	4.33	2.31	2.76
600	2.56	0.99	2.28	1.68	3.03	3.80	1.92	4.35
800	1.78	0.81	1.27	0.98	2.03	2.52	1.78	3.80
1000	1.59	0.50	1.05	0.82	1.35	1.57	1.20	2.52
1350	1.06	0.44	0.76	0.65	0.88	1.06	0.75	1.44
1600	0.89	0.41	0.70	0.53	0.93	1.24	0.61	1.17
2000	0.63	0.31	0.52	0.41	0.68	0.86	0.56	1.24
2500	0.48	0.25	0.38	0.32	0.44	0.56	0.42	0.86
3000	0.46	0.21	0.35	0.30	0.43	0.62	0.32	0.66
4000	0.31	0.16	0.25	0.21	0.28	0.39	0.26	0.62
5000	—	—	0.19	0.15	—	—	0.17	0.39
Lighting								
30	—	—	—	—	—	—	79.0	3.0
60	—	—	—	—	—	—	51.0	3.0
Trolley								
100	—	—	12.6	4.3	—	—	—	—

NOTE: Resistance values increase as temperature increases. Reactance values are not affected by temperature. The above values are based on conductor temperature of 25° C (normal room temperature) since short circuits may occur when busway is initially energized or lightly loaded. To calculate voltage drop when fully loaded (75° C), multiply resistance of copper and aluminum by 1.19.

times from stock). By reducing the time between final measurement and installation, in addition to proper coordination, the chances of interference from other trades can be reduced to a minimum.

Finally, terminations are a significant part of busway layout considerations. For ratings 600 A and above, direct bused connections to the switchboard, motor-control center, etc, can reduce installation time and problems. For ratings up to 600 A, direct bused terminations are generally not practical or economical. These lower current ratings of busway are usually fed by short cable runs.

12.10 Installation. Busway installs quickly and easily. When compared with other distribution methods, the reduced installation time for busway can result in direct savings on installation costs. In order to ensure maximum safety, reliability, and long life from a busway system, proper installation is a must. The guidelines below can serve as an outline from which to develop a complete installation procedure and timetable.

12.10.1 Procedure Prior to Installation

(1) Manufacturers supply installation drawings on all but the simplest of busway layouts. Study these drawings carefully. Where drawings are not supplied, make your own.

(2) Verify actual components on hand against those shown on installation drawing to be sure that there are no missing items. Drawings identify components by catalog number and location in the installation. Catalog numbers appear on section nameplate and carton label. Location on the installation (item number) will also be on each section.

(3) During storage (prior to installation) all components, even the weatherproof type, should be stored in a clean, dry area and protected from physical damage.

(4) Always read manufacturer's instructions for installation of individual components. If you are still in doubt, ask for more information; never guess.

(5) Electrical testing of individual components prior to installation should be done. Identification of defective pieces prior to installation will save considerable time and money.

(6) Finally, preposition hanger supports (drop rods, etc) and hangers if of the type that can be prepositioned. You are now ready to begin the actual installation of busway components.

12.10.2 Procedure During Installation

(1) Almost all busway components are built with two dissimilar ends that are commonly called *bolt* end and *slot* end. Refer to the installation drawing to properly orient the bolt and slot ends of each component. This is important because it is not possible to properly connect two slot ends or bolt ends.

(2) Lift individual components into position and attach to hangers. It is generally best to begin this process at the end of the busway run that is most rigidly fixed (for example, the switchboards).

(3) Pay particular attention to "TOP" labels and other orientation marks where applicable.

(4) As each new component is installed in position, tighten the joint bolt to proper torque per manufacturer's instructions. Also install any additional joint hardware that may be required.

(5) On plug-in busway installations, attach plug-in units in accordance with manufacturer's instructions and proceed with wiring.

(6) Outdoor busway may require removal of *weep hole* screws and addition of joint shields. Pay particular attention to installation instructions to ensure that all steps are followed.

12.10.3 Procedure After Installation. Be sure to recheck all steps to ensure that you have not forgotten anything. Be particularly sure that all joint bolts have been properly tightened. At this point the busway installation should be almost complete. Before energizing, however, the complete installation should be properly tested.

12.11 Field Testing. The completely installed busway run should be electrically tested prior to being energized. The testing procedure should first verify that the proper phase relationships exist between the busway and associated equipment. This phasing and continuity test can be performed in the same manner as similar tests on other pieces of electric equipment on the job.

All busway installations should be tested with a megohmmeter or high-potential voltage to be sure that excessive leakage paths between phases and ground do not exist. Megohmmeter values depend on the busway construction, type of insulation, size and length of busway, and atmospheric conditions. Acceptable values for a particular busway should be obtained from the manufacturer.

If a megohmmeter is used, it should be rated 1000 V direct current. Normal high-potential test voltages are twice rated voltage plus 1000 V for 1 min. Since this may be above the corona-starting voltage of some busway, frequent testing is undesirable.

For additional details, see NEMA BU 1.1-1981 [5].

12.12 Busways Over 600 V (Metal-Enclosed Bus). Busway over 600 V is referred to as metal-enclosed bus and consists of three types, isolated phase, segregated phase, and nonsegregated phase. Isolated phase and segregated phase are utility-type busways used in power generation stations. Industrial plants outside of power generation areas use nonsegregated phase for connection of transformers and switchgear and interconnection of switchgear lineups. The advantage of metal-enclosed bus over cable is a simpler connection to equipment (no potheads required). It is rarely used to feed individual loads.

12.12.1 Standards. Metal-enclosed bus was covered for the first time in the 1975 NEC. The NEC [2] requires that the metal-enclosed bus nameplate specify its rated

(1) Voltage
(2) Continuous current
(3) Frequency
(4) 60 Hz withstand voltage
(5) Momentary current

The NEC [2] further requires that metal-enclosed bus be constructed and tested in accordance with ANSI/IEEE C37.20-1969 [1].

12.12.2 Ratings. ANSI/IEEE C37.20-1969 [1] specifies the voltage, insulation, and the continuous- and momentary-current levels for metal-enclosed bus (Table 93). The ratings are equal to the corresponding values for metal-enclosed switchgear.

12.12.3 Construction. Metal-enclosed (nonsegregated phase) bus consists of aluminum or copper conductors with bus supports usually of glass polyester or porcelain. Bus bars are insulated with sleeves or by fluid bed process. After installation joints are covered with boots or tape. Metal-enclosed bus is totally enclosed. The enclosure is fabricated from steel in lower continuous-current ratings and aluminum or stainless steel in higher ratings. Normal lengths are 8–10 ft with a cross section of approximately 16 in × 26–36 in, depending on conductor size and spacing. Electrical connection points are electroplated with either silver or tin. Indoor and outdoor (weatherproof) constructions are available.

12.12.4 Field Testing. After installation the metal-enclosed bus should be electrically tested prior to being energized. Phasing and continuity tests can be performed with other associated electric equipment on the job. Megohmmeter tests can be made similar to those described for busway under 600 V. High-potential tests should be conducted at 75% of the values shown in Table 88.

Table 88
Voltage, Insulation, Continuous-Current, and Momentary-Current Ratings of Nonsegregated-Phase Metal-Enclosed Bus

Voltage (kV, rms)			Insulation, Withstand Level (kV)			
Nominal	Rated Maximum	Continuous Current (A)	Power Frequency (rms), 1 min	DC Withstand, 1 min	Impulse	Momentary Current (kA, (asymmetrical)
4.16	4.76	1200	19.0	27.0	60	19-78
13.8	15.00	2000	36.0	50.0	95	19-78
23.0	25.80	3000	60.0	—	125	58
34.5	38.00	—	80.0	—	150	58

Continuous-current ratings are based on a maximum temperature rise of 65 °C of the bus (30 °C if joints are not electroplated). Insulation temperature limits vary with the class of insulating material. Maximum total temperature limits for metal-enclosed bus are based on 40 °C ambient. If the ambient temperature will exceed 40 °C, the manufacturer should be consulted.

The momentary-current rating is the maximum rms total current (including direct-current component) that the metal-enclosed bus can carry for 10 cycles without electrical, thermal, or mechanical damage.

12.13 References

[1] ANSI/IEEE C37.20-1969 (R 1981), IEEE Standard for Switchgear Assemblies Including Metal-Enclosed Bus.

[2] ANSI/NFPA 70-1987, National Electrical Code.

[3] ANSI/UL 857-1981, Safety Standard for Busways and Associated Fittings.

[4] NEMA BU1-1983, Busways.

[5] NEMA BU1.1-1986, Instructions for Safe Handling, Installation, Operation, and Maintenance of Busway and Associated Fitting Rated 600 Volts or Less.

13. Electrical Energy Conservation

13.1 Introduction. Energy became an important design criteria as the cost of electrical energy soared in the 1970s. This chapter describes energy conservation, design, and planning by summarizing ANSI/IEEE Std 739-1984 [B1][33].

The ten topics discussed in this chapter represent three basic facets of energy planning:

(1) Administrative aspects that embody such items as organizing for conservation, embarking on a conservation program, understanding electric rates, managing load and evaluating losses in purchase decisions

(2) Descriptions of electrical devices used in metering and lighting and discussion of equipment efficiencies in relation to the electrical and mechanical environment

(3) On-site generation, which encompasses cogeneration and peak shaving

13.2 Organizing for a Conservation Effort

13.2.1 Obtain Management Approval and Commitment. One key to success in an engineering effort is the in advance approval and blessing from upper management and supervision. This approval is even more important in a conservation plan, because expenditures will generally have no direct effect on production output. However, long-term cost reductions can increase profits.

Goals and guidelines should be established so that management knows what to expect. Furthermore, mutually agreed upon criteria will help an engineer properly direct his/her efforts. There are numerous cases where corporate savings were unnoticed and unappreciated because an engineer failed to notify and work with his management.

The conservation program should include appropriate organizational changes including the establishment of an energy committee composed of engineering, purchasing, accounting, scheduling, production, and labor representatives. In addition,

[33] The numbers in brackets correspond to those of the references listed at the end of this chapter; when preceded by B, they correspond to the bibliography at the end of this chapter.

a management liaison should be included. This committee should have responsibility and commensurate authority to do its job of conserving energy. Clearly defined and accurately written goals (in energy units) and guidelines can properly guide the committee efforts.

Regardless of the breadth of the energy team (which could consist of only one person), the team approach should be used. This means that all affected employees should be consulted before preparing any plans. Results and proposals should then be *completely* communicated to *all* interested and involved employees.

13.2.2 Embarking on an Energy Conservation Program. It is important, from several standpoints, to establish the existing pattern of electrical usage and to identify those areas where energy consumption could be reduced. A monthly history of electric usage is available from the electric bills, and this usage should be carefuly recorded in a format that will facilitate future reference. The following list of items (where appropriate) should be recorded in the electric usage history.

Billing month
Reading date
Days in billing cycle
Kilowatthours (or kilovolt-ampere hours if billed on this basis)
Billing kilowatt demand (or kilovolt-ampere if billed on this basis)
Actual kilowatt demand (or kilovolt-ampere if billed on this basis)
Kilovars (actual and billed)
Kilovar hours (actual and billed)
Power factor
Power bill (broken down into the above categories plus fuel cost)
Production level
Heating or cooling degree days
One additional column for remarks (such as *plant on vacation* for 2 weeks)

In addition, it would be wise to use graphic demand meters to *sketch* plant load patterns for at least several months and preferably an entire year.

A listing of plant operations, plant equipment, and energy conservation opportunities (ECOs) will provide both a history and a basis for evaluating future improvement. The listing of this information along with electric usage is called an energy audit, and this subject will be covered in the next section.

In general there are four categories of ECOs. These four categories are as follows:

(1) *Housekeeping Measures*. Easily performed (and usually low cost) actions that should logically be done (for example, turning lights off when not required, cleaning or changing air filters, cleaning heat exchangers, keeping doors shut, and shutting down redundant motors, pumps, and fans).

(2) *Equipment Modification*. More difficult and usually more expensive actions involving physical changes to the electric system (for example, the addition of solid-state variable speed drives, reducing motor sizes on existing equipment, and modifying heating and cooling systems).

(3) *Process Changes or Better Equipment Utilization*. The restructuring of production schedules or relocating of equipment to reduce energy demand or con-

sumption, or both (for example, combining hot rolling and tempering of steel into one process to eliminate the need to reheat).

(4) *Changes to the Building Shell*. Improving the insulating quality of the building to reduce losses to the outside environment (for example, adding insulation, reducing infiltration, controlling exhaust/intake, etc).

It should be intuitively obvious that housekeeping and low-cost measures should be done immediately. The larger and more expensive ECOs generally take longer to initiate and should often be done after low-cost measures are completed. However, there may be cases where obvious equipment modification improvements can be made concurrently with low-cost improvements.

13.2.3 Energy Needs Versus Energy Uses — The Equipment Energy Audit. The analysis of equipment and its efficiency can be a very simple task requiring only a careful tour of facilities. The analysis can also involve the documentation of all equipment and associated efficiencies. In any case, the results usually have a direct relationship with the effort. Regardless of the intensity of the energy audit, the following five conservation categories should be considered. In all cases there may be other nonenergy related criteria that should be included in an evaluation of the ECO.

(1) *Application* — evaluate the energy effort required to accomplish the task with the objective of using only the amount of input energy needed.

(2) *Utilization* — examine the process or equipment with the objective of changing the way the equipment is being used in order to reduce energy consumption.

(3) *Relocation* — evaluate the overall process or task to determine the best location from an energy standpoint.

(4) *Modification* — improve energy utilization by physical changes to the process or equipment.

(5) *Maintenance* — determine the type and amount of maintenance (cleaning, lubricating or adjusting equipment) that will reduce energy usage or increase efficiency.

The above five categories apply to all types of systems. In some cases more effective use of energy can be achieved by using waste heat or redirecting waste heat. In addition, one should be open to the application of energy in areas not related to the process itself, such as recapturing heat from lighting systems for space conditioning.

The industrial plant can be divided by function into six equipment/process categories. Each function can then be scrutinized in the aforementioned five conservation categories. The six functions are lighting, HVAC (heating, ventilating, and air conditioning) systems, motors and drives, processes (nonmotor such as heat treating), electrical distribution equipment, and the building environmental shell.

A long list of specific actions (or inquiries) is given in ANSI/IEEE Std 739-1984 [B1], but several categories are given in Table 89 to provide a feel for the task required in an energy audit.

13.2.4 Tracking Progress. The energy audit gives revealing information about the energy usage pattern and ECOs. While this information is extremely important, the current status is also important. The establishment of a recording and monitor-

Table 89
Examples of Conservation Categories

Situation	Action	Conservation Category
Hall lighting is 100 fc	Insert dummy or resistance modified tubes	Application/ Utilization
	Remove alternate fixtures	Modification
Drafting lighting is 50 fc (too low)	Move desk to window	Relocation
	Clean fixtures and lamps	Maintenance
	Use task lighting	Modification
Filters on intake air are clogged	Clean or replace filters	Maintenance
	Install recirculating oil bath filters	Modification
	Close up system and remove because it is not needed	Application/ Modification
50 hp motor has 25 hp load	Replace with new 25 hp motor if economically justifiable	Modification/ Application
	Change pulley and increase production and load	Utilization
Heat treating adjacent to office (in warm climate)	Relocate heat treating to rear of plant reducing office A/C requirement (or move the office)	Relocation
24 hp conveyer motor running all day	Install a switching device to automatically control the conveyer motor	Modification

ing system is, therefore, needed to maintain control over the electric usage and to gauge the results of implemented ECOs.

There are four important aspects of tracking progress:

(1) Meters should be installed at various important load centers

(2) These meters should be read and the data recorded on a regular basis (preferably weekly or by shift)

(3) The data should be analyzed to determine the need for action

(4) The information should be kept in terms of energy units on a common base

Since energy usage is frequently a function of production rate, the ratio of kW (kVA) or kWh per unit of output may provide more meaning than strict kW or kWh.

13.2.5 Overall Considerations There are several considerations that set the stage for action on ECOs, and all of them are important:

(1) The designer should recognize the limitation of his capabilities as well as the corporation's to design and implement an ECO. It may be necessary to call in an energy expert or a consultant to prepare the specific design under your guidance.

(2) Careful consideration shall be given to appropriate codes and standards. Some prominent codes and standards are developed and published by the National Fire Protection Association (NFPA), the Environmental Protection Agency (EPA), and the Occupational Safety and Health Administration (OSHA).

(3) The plant energy balance is an important consideration. There may be cases where an ECO action would merely shift the energy source from one point (or fuel) to another. (For example, a building may be heated electrically during the winter

months. As a conservation measure, 15 kW of lighting was turned off. However, this meant that the heaters had to make up the 15 kW of heat that was supplied by the lights. In this case, the lights should have been left on, because the light was free during the winter. On the other hand, lighting adds to A/C load in the summer so that 1 kW less of light will also reduce air conditioning by ½ kW. Hence, energy balance is looking at the entire system or building shell.)

13.3 Dollar Involvement in ECOs—Rates

13.3.1 Introduction. An understanding of utility rates (or tariffs) is important, because the cost of electric service is considered when evaluating an ECO. Once the energy saving is determined, it should be given a dollar value. In most cases, the average cost of electricity does not accurately reflect the savings per kW or kWh. Therefore, this section describes the important terms, concepts, and application of rates.

13.3.2 Rate Textbook. Virtually all rates have two sections: general rules and regulations, and the rate schedules themselves. The written documents are available through the public utility commission or the electric utility. An energy analyst should study the rates of the utility serving the plant. However, it is only necessary to thoroughly understand the particular schedules applicable to the plants involved in the investigation.

The rules and regulations include information regarding billing practices, available voltages, customer responsibilities, voltage regulation, balance and reliability, line extension limits, and temporary service requirements and availability. The rate schedules give the minimum (and maximum) values of usage to qualify for each rate along with the procedure for calculating the cost of electricity. Often, there are charges that apply to all rates, which are listed separately in another section and are called *riders*. Some common rates are residential, commercial, industrial, low load factor, time of day, and area lighting. Some common riders are *fuel-cost* special voltages and charges for extra facilities such as redundant transformers.

13.3.3 Billing Calculations. Each specific rate will contain a means to determine the cost for any or all of the following: kW, kWh, kVA, kVar, and power factor level. The charge may vary with time of day or time of year, and minimum service or customer charges may also be included.

The textbook definition of a kilowatt is a measure of the instantaneous power requirement (that is, the instantaneous rate of energy consumption). The comparable unit of energy is the kilowatthour (kWh). The billing kW is the highest average rate of energy usage over the billing cycle. Usage is generally averaged for a 15-30 min *demand* period. The demand for a 15 min period (¼ h) is determined by multiplying the kWh used in that period by 4.

Some electric rates contain a *ratchet* clause. This ratchet provision sets a minimum level for subsequent billings. This minimum usually continues for 3-11 months. The ratchet is generally applied only to the demand portion of the bill.

13.3.4 Declining Block Rate and Example. To understand the use of the declining block rate in the calculations listed in this section, assume that the A to Z Welding Company used 150 000 kWh (kilowatthours) and had a peak demand of 250 kW. The first 50 kW costs $5 per kW or $250, leaving 200 kW for the remaining blocks. The next block takes 150 kW at $4.50 per kW or $675, leaving only 50 kW for the next block which can accommodate 200. The last 50 kW is then billed at $4 per kW which costs $200. The total demand charge is the sum of the charges from each block, or $1,125, as follows:

Block 1	50 kW · 5.0 $/kW =	$ 250
Block 2	150 kW · 4.5 $/kW =	$ 675
Block 3	50 kW · 4.0 $/kW =	$ 200
Total	250 kW	$ 1125 or $4.50 kW average

The charges for kilowatthours may be done in much the same manner as shown below:

	kWh · Dollars/kWh = Charge	
Block 1	50 000 kWh · 0.07 $/kWh =	$3500
Remaining	= 150 000 – 50 000 = 100 000	
Block 2	50 000 · 0.065	$3250
Remaining	= 100 000 · 50 000 = 50 000	
Block 3	50 000 · 0.060	$3000
Total	150 000 kWh for	$9750 or 6.5¢ per kWh average

The total bill is then $1125 + $9750, or $10 875. The average cost of electricity can be expressed in terms of kilowatts or kilowatthours. The electric cost can be expressed as 43.50 dollars per kW or 7.25 cents per kilowatthour. It is important to note that averages cannot be used to determine energy savings. If the demand of this plant is reduced by 25 kW without an accompanying kWh reduction, the effect on the bill will be seen *only* in Block 3 of the demand charges (this rate of charge is referred to as the tail rate). The correct energy savings then is 25 · 4 or $100. By using the average total cost per kW of $43.5, an erroneous savings of $1088 would be shown.

Most rates include a charge for power factor either

(1) by assuming a power factor in the kW charge
(2) by charging for kVA
(3) by charging for power factors below a certain value or
(4) by charging for kvars (reactive demand).

In any case, the utility is capturing its cost for supplying vars to their customers. The subject of var flow is beyond the scope of this book and those interested should refer to a power system book, such as Stevenson [B17].

13.3.5 Demand Usage Rate. A more complex rate is the demand usage block rate in which the size of certain kWh blocks are determined by the peak demand. This rate allows smaller consumers to take advantage of the lower kWh charge when their energy usage is high. (The *load factor* is a measure of energy usage relative to demand. This factor is the ratio of kilowatthours to kilowatt demand times hours per billing cycle. Its value varies between the limits of 0 and 1.0.) The demand usage rate allows a utility to reduce the number of rates and encourages a more consistent level of electric usage. The concept is best described by the following example.

Suppose that the P and Q Packing Company uses 500 000 kWh and has a peak demand of 1000 kW. The kWh section of their rate schedule is as shown in the following data:

Blocks	Charge
(1) First 50 000 kWh	7 cents per kWh
(2) Next 200 kWh per kW	6 cents per kWh
(3) Next 300 kWh per kW	5 cents per kW
(4) All excess	4 cents per kWh

Since P and Qs demand is 1000 kW, the amount of kWh in Block 2 is 200 · 1000 or 200 000; similarly, the amount in Block 3 is 300 000. Once these block sizes are determined, the following procedure (identical to the previous example) can be used:

Block 1	50 000 · $0.07 =	$ 3500
(500 000 -	50 000 =	450 000 remaining)
Block 2	200 000 · $0.06 =	$12 000
(450 000 -	200 000 =	250 000 remaining)
Block 3	250 000 · $0.05 =	$12 500
Total	500 000 kWh	$28 000

Average cost is 5.6 cents per kWh

13.4 Load Management

13.4.1 Introduction. Any energy conscious engineer or plant manager should attempt to exert control of the plant's energy usage (kWh) and the rate of usage (kW). The simple fact is that no energy is used when equipment is shut off. Therefore, one of the first jobs of an energy engineer is to make sure that unused, redundant, and idling equipment is shut off.

Early demand controllers were tied into the meter pulse system and began shutting down equipment when it appeared that a preset demand would be exceeded. While this procedure can significantly reduce electric costs, there is some question as to the amount of energy saved (or added). A second generation of controllers has increased the effectiveness by removing all nonessential load in addition to keeping the demand under a preset level.

The ensuing sections briefly explain various types of controllers and how to design a proper system. While this discussion does not include any details of the hardware needed to implement an energy conservation program, the chapter does provide sufficient information to properly direct engineers' efforts through the use of publications listed in 13.11, other chapters of this book, and appropriate codes and standards.

13.4.2 Controllers. The control function can be performed by many types of systems ranging from a simple manual system to a sophisticated computer system. The energy engineer should match the system needs with equipment capabilities to determine the optimum choice. In many cases, significant energy, demand, and cost savings can be achieved by prudent operation of equipment or mechanical interlocking.

Semiautomatic controllers are the simplest form of controls. These include time clocks that switch loads based on a predetermined time schedule. Photocells are light-activated controllers that can be used in conjunction with a time clock or other devices. Environmentally controlled switches and sensors are also effective. Devices can either work independently or jointly to control energy usage. However, several considerations should be made when applying a controller. These considerations are as follows:

(1) The operation of the equipment in an automatic mode should not endanger anyone near the equipment or inadvertently interrupt any process.

(2) The controller should be periodically checked to see that it is operating as planned and has not been defeated.

(3) In the cases of time clocks, the time should be checked and the time control adjusted to compensate for changing seasons and conditions.

(4) The controlled equipment should be capable of withstanding the planned number of starts and stops.

There are five types of demand controllers, four of which require a utility demand meter pulse. In any case, one should first recognize that any controller or meter calculates the maximum demand by averaging the kilowatthours over a set interval (a 15 min demand interval would indicate the kilowatthours for 15 min multiplied by four since there are four 15 min periods per h). The five controller schemes are noted below (see ANSI/IEEE Std 739-1984 [B1] for a more complete description).

(1) *Instantaneous*. Controls loads at any time during an interval if the rate of usage exceeds a preset value.

(2) *Ideal Rate*. Controls loads when they exceed the set rate but allows a higher usage at the beginning of the interval.

(3) *Converging Rate*. Has a broader control bandwidth in the beginning of the interval but tightens control at the end of the interval.

(4) *Predictive Rate*. The controller is programmed to predict the usage at the end of the interval by the usage pattern along the interval and switches load to achieve the preset demand level.

(5) *Continuous Interval*. The controller looks at the past usage over a period equal to (or less than) the demand interval. Loads are switched in such a manner that no time period (of an interval's duration) will see an accumulation of kWh that exceed the preset value. This controller needs no utility meter pulse.

Before any of the above controllers can be installed, a load survey should be made. This survey is, in essence, an equipment/process audit and can well be done in conjunction with the audit described in 13.3.

13.4.3 Equipment Audit and Load Profile. Each process and piece of equipment shall be surveyed to find which loads can be switched off and to what extent they can be switched. The engineer shall evaluate any loss of equipment life or mechanical problems associated with switching each load. The survey consists of, but is not limited to, the listing of equipment by the following four categories:

(1) *Critical Equipment*. This equipment is required at all times or it needs to be controlled in the present manner for production, safety, or other reasons.

(2) *Necessary.* While this equipment is required for production (or other reasons), it can be shut down at some measurable financial loss during extreme conditions.

(3) *Deferrable.* This equipment is important but can be turned off for varying periods of time. Some load may even be switched virtually at will provided that some minimum on time is allowed.

(4) *Unnecessary.* This equipment has usually been used or left on even though it is not needed. This equipment should be shut off and periodically checked. Sometimes equipment is used only occasionally and the user fails to de-energize after use, so an indicator or semiautomatic controller may be properly applied.

Once the loads are recorded and analyzed, the proper control method can be established with the help of a load profile. The load profile is established by installing a graphic meter for at least a week (continuous) at various production levels. The profile is then analyzed along with the equipment audit to determine the target demand. Kilowatt recorders are available on a rental basis or for purchase (since continual monitoring is highly desirable). A recording ammeter can also be a good indicator if the power factor is known. The off-peak load pattern can be as revealing as the peak load pattern.

In the absence of load profile, some judgment needs to be made as to the amount of equipment that must be kept running at the estimated time(s) of peak load. This estimate is then made a target or *target demand* for the controller. However, the designer should also plan to control the use of equipment so that it is de-energized when not needed.

13.5 Energy Savings to Dollar Savings

13.5.1 Time Value of Money. Most engineers do not have an unlimited budget and, therefore, often need to make evaluations of various options. The energy savings of one or more projects have to be weighed against their (capital) costs. Since the equipment will usually function for many years and a future savings (or cost) is also involved, the time value of money should generally be considered. In order to properly evaluate an ECO, the installed cost as well as the operating, maintenance, and energy expenses should be determined on an annual basis over the life of equipment. Each annual expenditure (or savings) should be inflated and then discounted to the same base by the appropriate multiplier. If the engineer is unfamiliar with the process, he should at least develop the anticipated costs by year and work with the corporate accountant to determine the *value* of each option to the company.

13.5.2 Evaluating Motor Loss. The energy saving techniques employed in motor design add to its cost. The value of this energy savings is seen in reduced energy costs. The *cost-of-losses* evaluation is the process of determining how much additional investment is justified for each kilowatt of losses saved. This section provides a general, simplified approach. There are several technical papers on the subject that can be used, if more depth on the subject is needed. (See 13.11.)

Most industrial motors have a life of approximately 5–10 years if they are constantly operated at or near their rated horsepower. Therefore, the evaluation of losses should cover a 5-year period. The low-loss motors should have a longer life due to their construction, but insufficient data is available at this time to include the effect of additional motor life in the loss analysis.

The present value of future loss costs is determined by discounting the cost of 1 kW of losses in each of 5 years to year zero. The *present value of an annuity factor* is used for this discounting. The factor is as follows:

$$\text{present worth of an annuity} = \frac{(1 + i)^n - 1}{i\,(1 + i)^n}$$

where

i = discount rate in per unit (a 10% rate = 0.1 per unit)
n = the number of years (5 for motors)

The tail rate (see 13.3.4) should be used in the evaluation of load and no-load losses. Demand costs are not normally significant because most motor demands are diversified and the demand cost is usually insignificant compared to energy costs. However, the annual demand cost is easily calculated by using the formula:

$$\text{cost} = (\text{kW loss}) \cdot (\text{dollars per kW}) \cdot 12 \cdot (\text{diversity effect}).$$

Motor losses are composed of 2 types of losses — no-load losses which stay fairly constant, and load (or copper) losses which vary as the square of the load. Detailed loss information can usually be obtained for large motors (at least several hundred horsepower in size), but only generic information will normally be available for smaller motors. The loss information may consist of total losses at various loads rather than be separated into load and no-load. The cost of no-load losses is simple to determine: it is the product of the energy cost times the number of hours that the motor is operated per year. Load losses are more difficult to determine for two reasons: (1) the load losses vary as the square of the motor load and (2) the motor load needs to be determined. The motor's load cycle can either be measured directly or be calculated. Where total losses are given as a function of load, there is no need to do a no-load loss calculation.

Example. Suppose that a test shows that a motor is used for 3–8 hour shifts per day for 50 weeks every year. During the day, the motor is shut off completely for only 2 hours (between shifts). The remainder of the day, it spends 6 hours with no-load connected, 12 hours at 50% loading, 6 hours at 70% loading, and the remaining 4 hours at 100% loading. The tail rate[34] is \$0.05 per kWh and the discount rate is 20%.

The no-load loss cost is determined by the *on-time*.

$$\text{annual no-load loss cost} = (22\ \text{h/day}) \cdot (7\ \text{days/week}) \cdot (50\ \text{weeks/year}) \cdot 0.05\ \$/\text{kWh}$$

$$\text{annual no-load loss cost} = \$385/\text{kW}$$

The load loss cost is determined by the load cycle which, in this case, is repeated daily.

[34] Tail rate is the energy or demand charge for the last applicable rate block for a particular usage (see section 13.3.4).

(1) Duration	(2) Load	(3) Load^2	(1) · (3) Per Unit Loss
6/24 (0.25)	0	0	0
12/24 (0.50)	0.50	0.25	0.125
6/24 (0.25)	0.70	0.49	0.1225
4/24 (0.1667)	1.00	1.00	0.1667
24/24 (1.0)		Total	0.4142

The per unit losses are 0.4142/kW/day. Therefore, the annual load loss can be calculated as follows:

$$\text{annual load loss cost} = (24\ \text{h}) \cdot (7\ \text{days/week}) \cdot (50\ \text{weeks/year}) \cdot 0.4142 \cdot 0.05\ \$/\text{kWh}$$

$$\text{annual load loss cost} = \$174/\text{kW}$$

The worth of losses for this particular application is found by multiplying the annual costs by the present worth of annuity factor as follows:

Value (over 5 years) of 1 kW reduction of

$$(1)\ \text{no-load loss cost} = \frac{(1+0.2)^5 - 1}{0.2\,(1+0.2)^5} \cdot \$385 = 2.99 \cdot 385 = \$1151/\text{kW}$$

$$(2)\ \text{load loss cost} = \frac{(1+0.2)^5 - 1}{0.2\,(1+0.2)^5} \cdot \$174 = 2.99 \cdot 174 = \$\ 520/\text{kW}$$

Assuming equal load and no load reductions, it would be worth \$1671 to increase the efficiency of a 125 hp motor from 92% to 94% (assuming a 2 kW loss reduction) in this example.

13.5.3 Evaluating Transformer Losses. Transformers can be manufactured with efficiencies as high as 98–99%. Most transformer manufacturers offer a variety of loss designs with associated differences in cost. The manufacturer can determine the optimum design for a given value of losses, which makes it beneficial to include the cost of losses in the bid package. Both load (coil) and no-load (core) loss costs should be included since they each affect design parameters differently.

Transformer losses are determined at 100% load and at a winding temperature of 85 °C for 65 °C rise transformers (75 °C for 55 °C/65 °C rise). The winding loss varies approximately as the square of the load (and vary slightly with operating temperature). The transformer efficiencies at various levels are normally available from the manufacturer.

Had the previous example been a transformer loading situation, the annual load loss value would have been the same (at \$174), but the no-load loss value would increase because transformers are energized 365 days per year for 24 hours per day. The new annual no-load losses would be \$438/kW (24 · 365 · \$0.05). The kVA load and not kW should be used in determining the load losses. Since transformers normally last decades, at least a 10-year evaluation period should be used.

13.5.4 Evaluating Losses in Other Equipment. In general, losses associated with currents are a function of load squared. Magnetic losses in iron core reactors, large magnets, or solenoids are a function of voltage squared. Many experts have decided to increase wire sizes strictly for the purpose of reducing loss costs. For example, a 100 ft run of AWG No 1/0 aluminum has a resistance of 0.021 Ω and a rating of 150 A. The annual cost of losses using the motor example load pattern is as follows (using kW = $I^2R \div 1000$ and the \$174 per kW load less cost).

$$\$174/\text{kW} \cdot \frac{150^2}{1000} \cdot 0.021 = 174 \cdot \frac{22\,500}{1000} \cdot 0.021 = \$82 \text{ per year less cost}$$

For a 3-phase, 3-wire circuit the cost of losses would be 3 times this value or \$246. Use of AWG No 3/0 aluminum at 0.0133 Ω resistance would reduce the 3-phase losses to 0.0133/0.0210 · 246 = \$156 for a savings in excess of \$90/100 circuit ft. See 13.6.3 for more information on feeder oversizing.

13.6 Electrical Equipment and Its Efficient Operation

13.6.1 Losses. All electrical equipment has some type of loss associated with its use. There are 5 different types of losses that should be considered in determining the optimum operating point for a piece of equipment.

(1) *Resistive (copper) losses* are associated with the flow of current. These are generally a function of the square of the current from the equation $P = I^2 \cdot R$. However, the energy engineer has to also recognize the temperature relationship of R, because increased current will invariably increase the operating temperature of a device. Tube devices like thyratrons consume power in their heating elements and their arcs, but the heat is not a square function of current. Solid-state devices have a constant voltage drop when they are conducting; so the power-current function is essentially linear.

(2) *Magnetic losses* are associated with motors, transformers, reactors, regulators, and solenoids. These losses are usually a function of voltage squared (approximately) and consist of hysteresis, eddy current, and mutual induction losses.

(3) *Motion losses* are produced as the equipment operates. These losses include friction loss from bearings, wind, and system restrictions.

(4) *Mechanical losses* are also transferred back to the electric circuit's power requirements. These losses include inefficiencies associated with transmissions, eddy current clutches, and speed control devices (which can even be in the electric circuit).

(5) A combination of factors will cause *additional or unnecessary losses* if a piece of equipment is operated outside of its design limits. Operating above the rated capability can cause overheating (and associated loss costs) as well as destruction of the equipment. Operating the equipment too far below rated capacity wastes capital dollars, causes an increase in the no-load portion of the losses, and lowers the power factor. The key to energy engineering is to match the device to the load and the power supplied to the device.

3.6.2 Efficiency. The textbooks define efficiency as the power (kW) output divided by the power (kW) input at rated output. The percent efficiency is 100 times this value. This method should be discarded in most (but not necessarily all) energy evaluations. The efficiency of a device for any energy engineering effort should be considered over its entire cycle of operation. In an energy evaluation the following expression applies (with the assumption that the output is converted to kWh):

$$\% \text{ Energy Efficiency} = \frac{\text{kWh out (over operating cycle)}}{\text{kWh in (over operating cycle)}} \cdot 100$$

Most equipment is given an efficiency rating at nameplate or full-load conditions. The device is then used under different conditions making the nameplate value of efficiency incorrect for the applied device.

13.6.3 Conductor Oversizing. The cost of losses is becoming sufficiently high to evaluate the installation of wiring that exceeds the ampacity requirement of a particular circuit. In many cases, there will be virtually no change in the cost of the feeder breaker, the conduit, the pull boxes, and the receiving panel. However, in other cases the added cost of equipment upsizing will be significant. The actual costs and availability of capital will determine the actual course of action. See 13.5.4 for a loss calculation example.

13.6.4 Motors. Motors comprise the largest portion of electric energy consumed in plants today. They are a fairly efficient device at rated load. In general, three-phase motors are more efficient than single-phase motors and larger motors are more efficient than smaller ones. There is only minor improvement in efficiency above 200 hp and the *knee* of the efficiency versus size curve occurs at about 10–15 hp. The peak efficiency of a motor occurs at full load with about 105% (of nameplate) balanced voltage at its terminals. However, as load is reduced from nameplate, the optimum efficiency occurs at a lower voltage (see ANSI/IEEE Std 739-1984 [B1] for further detail).

Motor voltage unbalance will increase motor losses due to a negative sequence voltage that causes a rotating magnetic field in the opposite direction of motor rotation. A 2% voltage unbalance will increase losses by 8%, a 3½% unbalance will increase losses by 25%, and a 5% unbalance will increase losses by 50%.

The power factor of most three-phase motors is between 80% and 90% at full load and decreases as load is reduced. The installation of power factor correction (to 95% or so) at the motor terminals will accomplish two tasks. Improved power factor will decrease current requirements thereby reducing I^2R losses in the supply line. More importantly, the use of capacitors at the motor will improve voltage regulation by increasing the voltage level when the motor is used. Large banks of unswitched capacitors can cause problems from several aspects (see ANSI/IEEE Std 739-1984 [B1] for further details) and are, therefore, not recommended as a first choice. If large banks of capacitors exist (or are planned), they should be switched as a function of plant load.

New high-efficiency motors are available and their cost is usually justified. However, each application should be evaluated. These motors achieve a higher efficiency by using higher grade steel, special low friction bearings, added copper windings, closer tolerances, and smaller air gaps. These motors have the added benefit of a longer life because they run cooler than low-efficiency models.

Motor speed control takes on several forms. The earliest methods of speed control involved the use of resistors to reduce the amount of voltage seen by the motors. The losses of this type of system are easily determined. Modern techniques include voltage control with thyristors like SCRs and triacs. These devices are more efficient but supply a somewhat distorted voltage supply, and the devices themselves have losses. The most sophisticated speed control is a variable speed drive (VSD) which varies frequency and voltage level of a synthesized ac voltage wave. The VSD does not synthesize a pure voltage wave so losses occur as a result of the harmonic content. The peak efficiency of the new systems generally occurs at full speed and full load on the motor.

In any case, a rectifier system causes voltage notching on *any* power system. The use of an oscilloscope is recommended in any thyristor application to evaluate the need for filtering at the *source* of the notching. Notching can cause capacitor and other equipment failure when large amounts of capacitance are on the system or the system is at a low-load condition, or both.

13.6.5 Transformers. Transformers are very efficient devices. Their loss evaluation was covered in 13.5.3. The load losses are a function of the square of the load kVA to nameplate kVA ratio (modified for change in resistance as load increases). The no-load losses are a function of voltage squared.

Dry-type transformers are found throughout an industrial complex, and they are usually energized every hour of the day. The losses of these units are significant and they should be switched off when not in use or removed whenever possible (the manufacturer should be consulted since moisture can be accumulated during the *off* time and cause premature failure). However, their losses may be offset by the reduced current achieved by using a higher voltage supply. This discussion does not refer to redistribution centers, which are considered next.

Double-ended substations have been used for many years to increase reliability. A double-ended station is also energy efficient at high loads, although it is less efficient at low loads due to twice the required no-load (core) loss. The decreased load loss from shared load is illustrated as follows:

1 unit of full load = 1 per unit loss

$$\text{2 units at half load} = \left(\frac{1}{2}\right)^2 + \left(\frac{1}{2}\right)^2 = \frac{1}{4} + \frac{1}{4} = \frac{1}{2} \text{ per unit losses}$$

13.6.6 Thyratrons, Ignitrons, and Other Diode Devices. Thyratrons and ignitrons have been used for many years in supplying welders and dc drives. However, the efficiency of these devices is extremely poor in comparison to the new solid-state rectifiers. However, the new devices have losses, too, so their use and application should be carefully considered. Furthermore, the rapid turn-on time and phase control of the new solid-state devices cause the previously mentioned voltage notching, and this phenomena should also be considered.

13.6.7 Capacitors. The use of capacitors can be a significant energy saver if they are properly applied (see 13.6.4). A capacitor bank is also a *load* (a 750 kvar bank draws almost 1000 A at 480 V), so it should be disconnected when var support is not required. If a fuse blows on a large capacitor, an unbalanced voltage will occur along with resultant increases in motor losses. Therefore, the fuse integrity of capacitor banks should be monitored. High harmonic content in the power supply has been known to cause either capacitor failure or unplanned operation of protective devices.

13.6.8 Reactors and Regulators. Reactors are normally installed on a system to reduce fault current. Due to their magnetic fields, they should not be located near steel structures. If they are not designed to carry load (or any large amount of) current, the absence of significant current should be verified (using proper safety practices).

Regulators can be used to balance voltage as well as to supply the required voltage levels for optimum efficiency. However, regulators do have losses that should be considered. The use of switched capacitors for voltage regulation at various feeder and points (or at motors) should be considered from an energy conservation standpoint.

13.6.9 Equipment Overview. Equipment should be operated as near as possible to its nameplace rating, in terms of both load and voltage. The voltage should be well-balanced and regulated in an efficient manner wherever possible. Capacitors should be carefully applied and switched. Unnecessary and underloaded equipment should be removed or replaced. The system should be checked for voltage level, balance, and harmonic contact. Energy-saving devices should be considered if their worth has been scientifically established. Their function should also match the concepts discussed in this section.

13.7 Metering. Metering provides the opportunity to monitor and control the rate of energy consumption in the electrical system. Accurate and complete kW, kWh, and power factor information is needed for a good energy audit as well as for continued measurement and control for an energy conservation program. The metering subject is covered thoroughly in Chapter 10, so no detail is given here. It is worth noting that good metering is virtually worthless without a program of meter reading, data recording, data analysis, and a guide for action. Meters themselves save no energy, but thoughtful use of metering information does save energy and money.

Ideally, metering should be installed as the electric system is being built. However, metering is often an afterthought. There are several cautions that must be observed from any engineering standpoint:

(1) Relay rated PTs and CTs shall not be used for metering because their accuracy is not sufficiently high at normal loads.

(2) The meters should be installed at a point where they can be easily and safely read.

(3) The meter should be out of the way of normal traffic (and even of doors used only on occasion) so that their glass covers, which normally protrude 18 inches from the surface, are not broken.

(4) Clamp on CTs may be required in areas where 24-hour production is required, because the system cannot be shut down long enough to install donut-type current transformers.

(5) Current transformers (and PTs) have insulation voltage ratings that shall exceed the anticipated voltage level of the system on which they are being installed.

(6) All appropriate safety requirements shall be considered before purchasing or installing the equipment.

Portable instruments can be used to provide the input-versus-output energy for various machines and processes. Since the slip is a linear function of the load on a motor from no-load to full-load, a tachometer or strobe tachometer can be used to determine the output horsepower of a motor. The electrical input effect should be translated to nameplate conditions (see ANSI/IEEE Std 739-1984 [B1]) for very accurate information.

Recorded metering information will reveal differences due to the varying energy demands of the electrical system. The metering results should be plotted (as noted in 13.2), and reasons for variation should be sought. Valuable information can be gleaned by reading as frequently as every shift. However, trends tend to be more indicative of improvement or lack of improvement in energy conservation.

Single reading variations may indicate one of the following:

(1) Abnormal production conditions
(2) Faulty or failing equipment
(3) Defeat of energy conserving controls
(4) Meter error
(5) Meter reading error
(6) New production volume
(7) Installation and operation of new equipment

Lack of any changes in kW rate or kWh may indicate that the demand indicator was not reset or that the meter is not being read or is defective. In any case, the meter should be checked and readings verified.

Changes in the trend of usage or usage different from planned targets can signify any of the following:

(1) The targets are too high or too low
(2) Something other than production level affects energy usage
(3) The energy conservation methods are (or are not) working

Regardless of the metering plans, all metering information should be recorded under the following (or similar) headings (see 13.2.2):

(1) Date
(2) Previous meter reading
(3) Present (new) meter reading
(4) Difference
(5) Meter multiplier
(6) kWh consumption
(7) Demand indicator reading
(8) Indicator constant
(9) Peak kW for current (just read) period

13.8 Lighting

13.8.1 Introduction. An energy efficient lighting system is one in which the required amount of light illuminates the subject at the proper level and color with the minimum amount of energy. A lighting system consists of more than just a bulb (lamp) and a fixture above a working station. A well-designed lighting system considers the use of natural light, the proper direction or dispersion of light from the fixture (or both), the effect of reflections off various surfaces, flexibility, cleaning, switchability, dimming, ambient versus task light, etc. In other words, there is more to light efficiency than lumens (light brightness) per watt output of a lamp.

Refer to ANSI/IEEE Std 739-1984 [B1] and to other sources in the Bibliography (13.11) for further information on various light sources and general guidelines on conservation considerations in lighting.

The number of reasonable options is a function of whether a lighting system is being designed for a new building, an existing building, or a renovation project. The engineer should consider only those options that can reasonably be achieved within the restrictions of his budget, the plant's physical characteristics, and other influencing factors.

13.8.2 Types of Lighting. Common lighting systems can be grouped into four different categories — incandescent, fluorescent, high-intensity discharge (HID) and low-pressure sodium (LPS). All of these sources are capable of providing an efficient lighting system in different situations. The *efficiency* of a lamp is more properly designated as efficacy where

$$\text{efficacy} = \frac{\text{light output (in lumens)}}{\text{power input (in watts)}}$$

The efficacy of a lamp depends upon the size of the lamp, the design of the lamp, and the intended application of the lamp. The forthcoming descriptions give a representative number, excluding ballast losses, as well as a description of the light input. All of the following except incandescent light sources require ballasts:

(1) Incandescent lamps have efficacies in the mid-twenties and provide a near perfect white light. They have a rather short life of up to serveral thousand hours at most and come in a variety of sizes and types. The more efficient incandescent lamps include the krypton gas-filled lamps and reflector lamps such as the PAR (outdoor reflector), type R (reflector), and type ER (elliptical reflector).

(2) Fluorescent lamps have efficacies near 80 and are fairly good sources of white light (covering about 80% of the visible spectrum). They have rather long lives at 20 000–30 000 h. Several sizes (from 20–215 W) are available. Energy-efficient lamps are available, and high-efficiency ballasts are being used. Energy-saving ideas for existing installations include the phantom tube and impedance modified lamps, all at some sacrifice of light output.

(3) HID mercury lamps have efficacies in the 60s, but they are not a good source of white light (they cover only about 50% of the visible spectrum), although they are adequate for most applications. Their lives are good at over 20 000 h and they come in a variety of sizes from at least 40–1000 W. A disadvantage, which should be considered in their application, is their 5–7 min start time and 3–6 min restart time.

(4) HID metal halide lamps have efficacies into the 120s and provide a good source of white light (covering 70% of the visible spectrum). The life ranges from 7000–20 000 h, and they come in a variety of sizes from about 150–1500 W. Their one disadvantage is a 3–5 min start time and a 10–15 min restart time after being extinguished, and this characteristic should be considered in their application.

(5) HID high-pressure sodium lamps have efficacies near 140, but they are not a good source of white light (they cover about 21% of the visible spectrum), although the color quality is adequate for many industrial uses. Their life is good at over 20 000 h. They come in sizes from about 50 to over 1000 W. These lamps need a 3–4 min starting time but restart in approximately 1 min.

(6) LPS lamps are the most efficient light sources today at efficacies over 180. However, they provide only one color of light (which covers almost none of the visible light spectrum), so they are primarily used where color is not important. The lamp life is near 18 000 h for virtually all lamps that range from about 20 to almost 200 W. The starting time is approximately 1 min with a nearly instantaneous restart time.

13.8.3 Control. The amount of energy consumed by a lighting system is a function of two parameters (power and time):

$$\text{energy} = \text{power} \cdot (\text{time energized})$$

Therefore, the easiest way to save energy is to turn off the lighting when it is not needed. While this is an easy and obvious statement, achieving the desired results is not necessarily easy. Years of low-cost energy have resulted in the installation of as many lamps as possible per control device and, to eliminate switches, breaker panels have been used to switch lighting. While this is an acceptable design practice (provided that the breakers are approved for the purpose), energy is often wasted. The ultimate system of control would be to remotely control every fixture and program the mode of operation, but at the present time this is impractical. Therefore, the designer should consider the subsequent information in choosing the optimum control system for a particular application. There are manual and automatic control systems and devices.

Before a control function can be used, the lights shall be controllable. At the present time, there are several ways to control fluorescent lighting besides wiring each fixture separately. Two-level ballasts are available, new solid-state dimmers are available, or each ballast (in a fixture) can be wired to a separate phase in a 3-wire single-phase system. In addition, 3-ballast, 3-lamp fixtures are available that will allow 3 lighting levels per fixture, but the overall efficiency suffers because of the 3 ballasts. Solid-state controls are available for dimming ballasted lights, but special ballasts are required and the controls tend to be expensive.

Manual control of a lighting system is often the least expensive but may be the least effective alternative. A manual control requires the designation of an individual who shuts the lights off at a selected time and verifies that the proper level of light is available at the proper time. This type of control is too easy to defeat and requires total reliance on one individual (or group of individuals).

Automatic controls vary from a simple time clock to a sophisticated computer controlled lighting system. The price versus benefit cost analysis will be required for each installation. However, the system should be programmed for normal operation and have a local manual override. Safety considerations shall be made for the possibility of someone being caught in an area when the lights are switched off.

13.8.4 System Considerations. Lighting should be switched in groups (as contrasted to strings) so that areas can be lighted or darkened as conditions change.

The walls and ceilings should be light in color to reflect rather than absorb light rays.

A periodic cleaning and lamp replacing program should be considered in the design stage. Many plants have reduced lamp wattage by as much as 50% with concurrent cleaning and found no change in actual lighting levels.

Maximum use should be made of natural light with careful consideration of heat loss. The heat from ballasts and lamp losses should be recirculated or removed. For example, removal of 1 kW of lighting can reduce air conditioning requirements by ½ kW, but redirecting this heat in the winter would lessen the heating energy requirements.

The high-efficiency lighting sources should be used as often as possible, but the long restart systems should be supplemented with fluorescent or incandescent lighting to prevent a total blackout. Task lighting should be considered especially where intense (100–150 fc) lighting is required.

13.9 Cogeneration. Cogeneration is the process of concurrently producing heat and electricity (or shaft horsepower) in a more efficient manner than if they were produced separately. The most common form of cogeneration used today is a topping cycle in which electricity (or shaft horsepower) is produced, and the exhausted heat is then used to supply process heating.

The use of cogeneration has decreased over the years, but energy conservation and governmental push have increased interest. The high capital commitment required for cogeneration demands careful consideration. The choice of whether to purchase or generate weighs heavily on the following factors:

(1) The cost of electricity versus the cost of prime mover fuel (including receiving and storing facilities)

(2) The current and future availability of prime mover fuel

(3) Standby power supply is normally required

(4) Process heat requirements and electric generation requirements should complement each other in time and magnitude

(5) The process heat and electrical requirements should be around-the-clock

(6) The energy requirements should be fairly high (at least 5 MW)

(7) All governmental benefits (investment tax credits, etc) and requirements (such as pollution control) should be considered

A more complete description of this subject is given in ANSI/IEEE Std 739-1984 [B1].

13.10 Peak Shaving. Many industries are required to have, and exercise, emergency or standby power supplies. These supplies can be used to reduce the electrical demand if they are exercised during peak periods. The use of emergency equipment on a daily basis to reduce demand should be given careful analysis in terms of cost versus (tail rate) savings. Costs should include extra maintenance and loss of life as well as prime mover fuel.

13.11 Bibliography

[B1] ANSI/IEEE Std 739-1984, IEEE Recommended Practice for Energy Conservation and Cost-Effective Planning in Industrial Facilities.

[B2] BONNETT, A. H. Understanding Efficiency in Squirrel-Cage Induction Motors. *IEEE Transactions on Industry Applications*, vol 1A-16, no 4, Jul/Aug 1980, pp 476-483.

[B3] BROWN, R. J. and YANUCK, R. R. *Life Cycle Costing*. Pennsylvania, Commonwealth of Pennsylvania, 1979.

[B4] CUNNINGHAM, R. L. Harmonic Pollution on an Industrial Distribution System. *Conference Record, IEEE Industry Applications Society Annual Meeting, 1980*, no 80CH1575-0, 1980, pp 405-408.

[B5] GIBBONS, W. P. Analysis of Steady-State Transients in Distribution Low Voltage Power Systems With Rectified Loads. *IEEE Transactions on Industry Applications*, vol IA-16, no 1, Jan/Feb 1980, pp 51-59.

[B6] GRANT, E. L., and IRESON, W. G. *Principles of Engineering Economy*, 4th edition. New York: The Ronald Press Company, 1960.

[B7] HARKINS, H. L. Cogeneration for the 1980's. *Conference Record, IEEE Industry Applications Society Annual Meeting, 1978*, no 78CH1346-61A, 1978, pp 1161-1168.

[B8] HELMS, R. N. *Illuminating Engineering for Energy Efficiency*. New York: Prentice-Hall.

[B9] HICKOCK, H. N. Electrical Energy Losses in Power Systems. *IEEE Transactions on Industry Applications*, vol 1A-14, no 5, Sep/Oct 1978, pp 373-386.

[B10] KAUFMAN, J. E., Editor. *IES Lighting Handbook*. New York: Illuminating Engineering Society (IES); Reference Volume, 1984.

[B11] KOVACS, J. P. Economic Considerations of Power Transformer Selection and Operation. *IEEE Transactions on Industry Applications*, vol 1A-16, no 5, Sep/Oct 1980, pp 595-599.

[B12] LINDERS, J. R. Electric Wave Distortions, Their Hidden Costs and Containment. *IEEE Transactions on Industry Applications*, vol 1A-15, no 5, Sep/Oct 1979, pp 458-471.

[B13] NEMA and NECA. *Total Energy Management*. A practical handbook on energy conservation and management, 2nd edition, National Contractors Association, 1979.

[B14] SIECK, R. E. and SCHWEIZER, P. F. An Assessment of Energy Cost Trends on Electrical Equipment Selection for the Petrochemical Industry. *IEEE Transactions on Industry Applications*, vol 1A-16, no 5, Sep/Oct 1980, pp 624-632.

[B15] SMITH, C., Editor. *Efficient Electricity Use*. A reference book on energy management for engineers, architects, planners, and managers, New York: Pergamon Press, 1978.

[B16] STEBBINS, W. L. Implementing an Energy Management Program. *Fiber Producer*, Atlanta: W. R. Smith Publishing Company, Oct 1980, vol 8, no 5, pp 44-47.

[B17] STEVENSON, W. D. *Elements of Power System Analysis*. 4th edition, New York: McGraw-Hill, 1982.

14. Cost Estimating of Industrial Power Systems

14.1 Introduction. Electricity remains the most economical and convenient form of energy for most industrial plant power applications. In order to meet the plant power requirements for the least investment in the electric system, an important stage in planning is the economic comparison of alternative system arrangements. System cost, while important, is but one of several factors to be considered in planning the most suitable distribution system.

Consideration should be given to the concept of *total cost*, or *true cost*. This requires weighing the first cost of the equipment, plus other costs allocated for improving reliability, ease of maintenance and part replacements, safety, and performance. Additional factors such as tax considerations, operational economies, and provision for future improvements or innovations in the manufactured product frequently are critical in the evaluation of competing systems. Under the impacts of cost of power, availability of power, and demand constraints, cogeneration and load control have become increasingly important. Yet there is currently no simple way to forecast precisely the effects of these factors. For a given installation, the engineer should prepare alternate power-distribution schemes that can be reviewed with his/her accounting department to develop the *true cost*. This chapter is restricted to the development of the relative capital cost of power systems.

For the discussion in this chapter, it is assumed that the load survey has been prepared for a new plant or is known for a plant being modernized. The kVA or kW load, area, location, and utility services available to the plant should be known. The nature of the load, size, and location of the major loads, degree of reliability required, flexibility needed for changes, and requirements for expansion should also be determined. After the load requirements are known, a number of economic decisions should be made in planning the distribution system. Among these are the following items, which vary in importance depending upon the size and type of plant under consideration.

14.2 Power Supply. Is a second utility supply source, valuable from a reliability point of view, economically justifiable?

Where the demand for steam is large or where the process produces steam or fuel as a byproduct, what are the merits of cogeneration?

To what extent can conservation techniques be practiced to reduce power requirements?

14.3 Voltage Level. Selection of the distribution voltages is an important factor. Voltage selection is governed by many factors such as available utility supply voltage, cost of equipment, size of transformers, size of motors, location and size of the various loads, interrupting-capacity requirements of the system, length of cable runs, voltage regulation, and the rating of standard equipment.

14.4 Reliability of the Distribution System. The selection of the type of distribution system in an industrial plant resolves itself into an economic consideration of the factors concerning the loss in production and revenue from outages caused by failure of the distribution system. In making the study, remember that the distribution system is but one element in the production chain and should be considered in its proper perspective in the total plant equipment. The possibility of outages in the primary power supply and of consequent mechanical failure of the machine being operated from the electric system should also be considered. The cost of a power outage includes loss of production, spoilage of material under process, cost of re-establishing production, and possible damage to machinery and buildings.

The plant distribution system normally consists of switching and protective devices, transformers, and conductors. Electric equipment can fail from many causes, but the installation of adequate equipment followed by routine preventive maintenance and testing can minimize the number of failures.

14.5 Preparing the Cost Estimate. The capital cost of a power system is the sum of the equipment and material costs, cost of installation, plus miscellaneous other costs incurred in order to provide a complete and ready-to-operate system. In making economic comparisons, it is important that the entire system be included, as each part is economically related to the whole.

The cost estimate will be used by many people, and so the estimator should identify clearly what is included in the estimate and the source or basis for estimating figures. Also, closely related items that are not included in the estimate should be identified lest they be overlooked completely.

When developing relative costs, the estimating engineer can expect to have limited time and information; as a result, he will have to make assumptions from time to time. These assumptions should be both realistic and consistent throughout the estimate or much of the validity of the relative costs may be lost. Basic assumptions should be clearly stated and included as part of the estimate so that people using the estimate know what is included and what is excluded. Typical items to cover are premium time allowance, field engineering, start-up assistance, spare parts, taxes, permits, contingency, foundations, and unusual scheduling or construction conditions.

14.6 Classes of Estimates. There are three basic types of estimates. These vary in accuracy as well as in the time and effort required for preparation. Typically, a preliminary estimate will be prepared first, based on limited information and prepared in a limited time. Next is the engineering estimate, and the most accurate is the detailed estimate. Include sufficient money in the engineering budget to cover the estimating effort.

14.6.1 Preliminary Estimate. Typically, actual cost will range from 15% below to 40% above this estimate. This estimate, which takes the least time to prepare, is highly dependent on good judgment. One approach is to take the known cost of a similar installation and then scale the cost to the size of the system under study, allowing for any conditions peculiar to the new system such as location, new technology, new design concepts, unusual labor productivity, and changes in costs of equipment and labor.

14.6.2 Engineering Estimate. Typically, the actual cost will range from 10% below to 20% above the estimate. This is the kind of estimate shown in the example in this chapter. Obtain prices of major items from vendors or from previous purchases. Use updated data from past jobs for other materials and installation costs. References listed in the Note at the end of this chapter are useful. This technique requires a good understanding of the installation, one-line diagrams, layouts, and a comprehensive list of equipment.

14.6.3 Detailed Estimate. Accuracy should be at least ± 10, perhaps ± 5%. Detailed material requirements are taken off completed drawings. Firm quotations are obtained from vendors. This is the sort of estimate used by contractors when bidding. In most cases, the estimating will be done by experienced estimators using established procedures.

14.7 Equipment and Material Costs. Today's changing market precludes publishing a list of typical costs for equipment and material. Up-to-date costs may be obtained from recent purchases or quotations, quotations for the specific project under study, manufacturers' and distributors' published prices (include all adders to the base price; use the correct discounts), and from estimating guides such as those listed in the Note at the end of this chapter. Typically, major items will be priced as accurately as possible; minor items will be covered by an allowance based on judgment or established percentages.

14.8 Installation Costs. To the extent that installation costs reflect labor productivity and wage rates, the engineer is advised to refer to previous experience of his company, to local contractors, and to current estimating guides. The Note at the end of this chapter lists several such guides, but the list should not be considered exhaustive.

The estimate should include an allowance for premium time or overtime and consideration of crew size and the nonworking foreman. Costs that vary with location, season, and time should be adjusted to reflect anticipated actual costs. For example, work in Alaska tends to be more expensive than work in the contiguous 48 states; cost of winter work frequently varies from work in summer; and work performed months or years in the future probably will have a higher dollar cost than the current cost. Most accounting and estimating departments have established procedures for coping with the time element. The engineer should make adjustments for location and season.

14.9 Other Costs. A contingency item is included to cover costs beyond those defined in the estimate. Contingencies adjust for estimating errors, unforeseen complications, and miscellaneous small tasks, but not for omission of significant items or for change in scope. Contingency values are a judgment decision; they typically range from 5-15% and may differ for equipment, material, and labor.

Escalation, an adjustment for rising costs due to inflation, modifies costs at the time the estimate is made to anticipate costs at the time expenditures are made. Usually the estimating or accounting department has established procedures for coping with this problem.

Salvage value for equipment, cable, etc, should be recognized, but seldom is it credited to the project.

Engineering costs cover complete engineering services, including preliminary plans, estimates, construction drawings and specifications, equipment specifications, review of equipment and construction bids, etc, but does not cover costs for construction supervision or field engineering. Engineering costs should be assigned whether the work is done in-house or by consultants. Typically, engineering costs will run from 8-12% of the construction costs of large, conventional projects. For engineering-intensive jobs such as retrofits, low capital cost jobs, and high-technology projects, the percentage will be much higher.

Special services such as calibrating protective devices, assistance during check-out and start-up, training, special tests, fees, construction power, offices, storage, rental equipment, etc, also should be identified in the estimate.

14.10 Example. The following example shows one technique for developing an *engineering cost estimate*. This example shows technique only. The reader should not use the dollar figures because undoubtedly they will be out of date and probably not directly applicable to his specific project.

Note that costs are segregated to show major items purchased by the owner, equipment and material provided by the contractor, and labor provided by the contractor.

The labor portion of the estimate is developed on a man-hour basis and then multiplied by the labor rate to obtain dollar amounts. This technique facilitates revising the estimate as labor rates change. Also, the man-hour information provides a ready means to check the estimate by comparing time actually spent on the job with time as estimated. For reference on subsequent jobs, the man-hour estimate without the masking effect of labor rates will prove more useful than dollar amounts.

14.11 Design Data. Assume that preliminary design of the power distribution system has been completed, that it conforms to requirements outlined in other chapters of this standard, and that it is described adequately by one-line diagrams (Figs 169, 170, and 171) and a site plan (Fig 172). Figure 173 will be used as a calculation sheet.

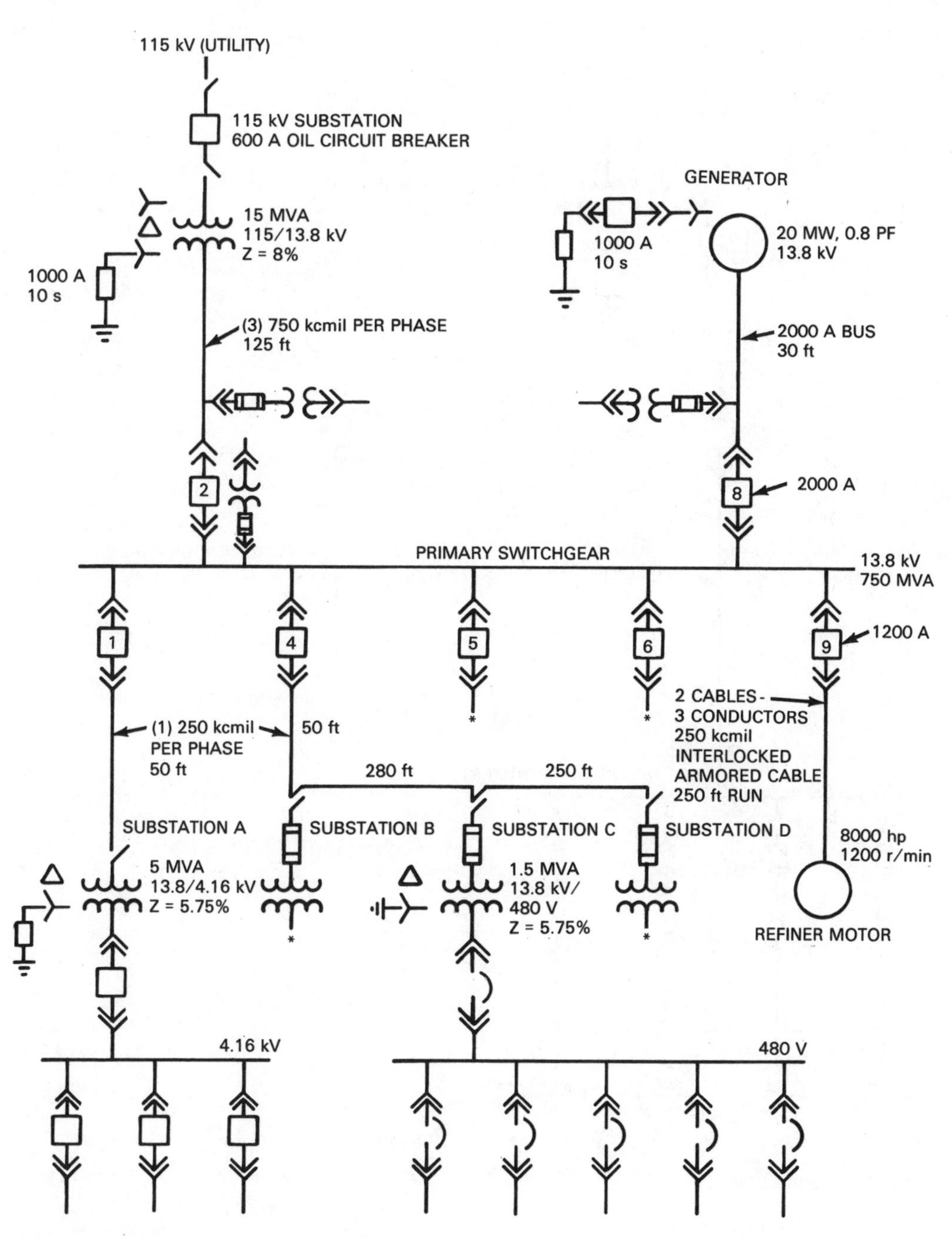

Fig 169
One-Line Diagram

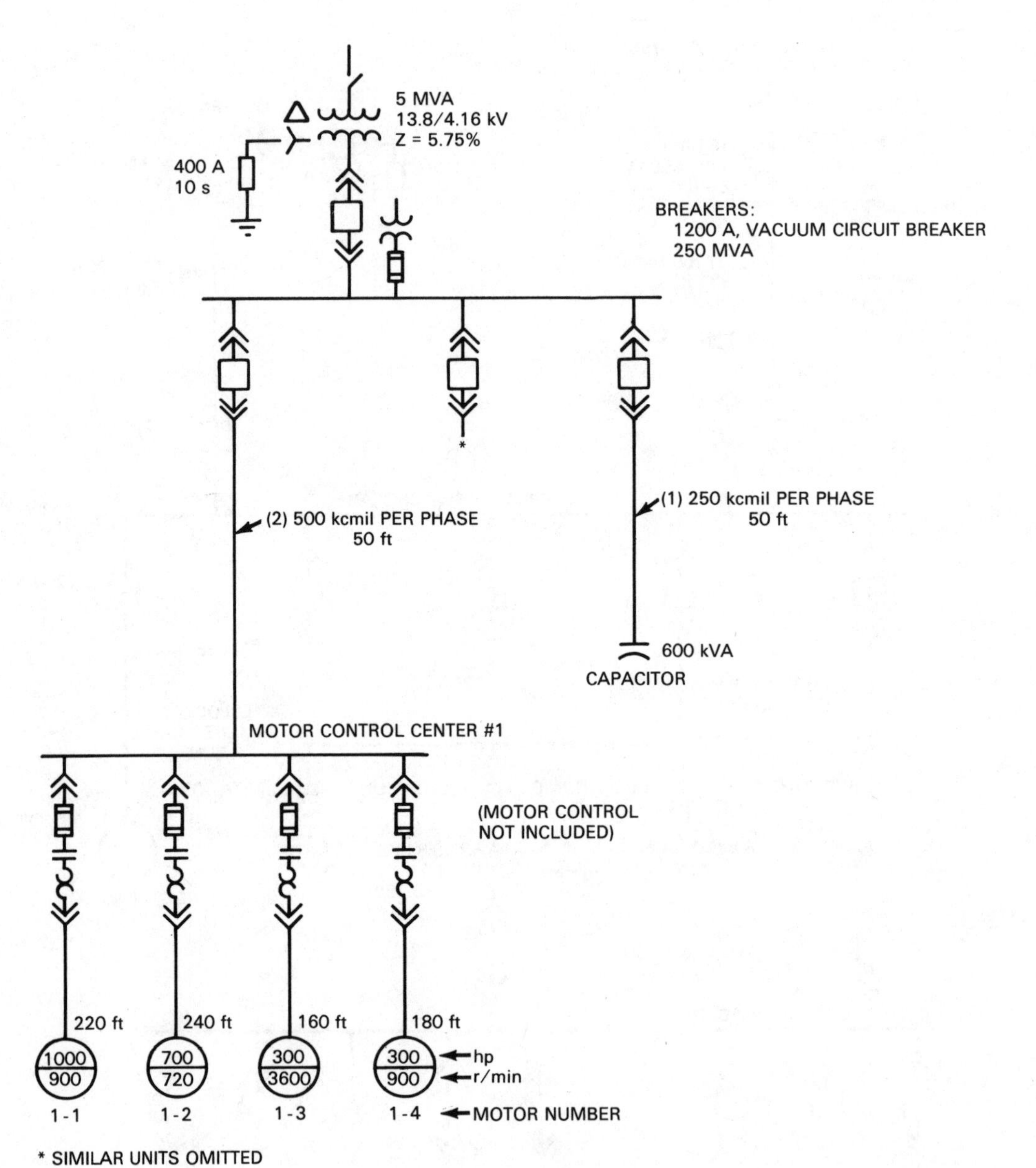

Fig 170
Substation A—5 MVA, 4.16 kV

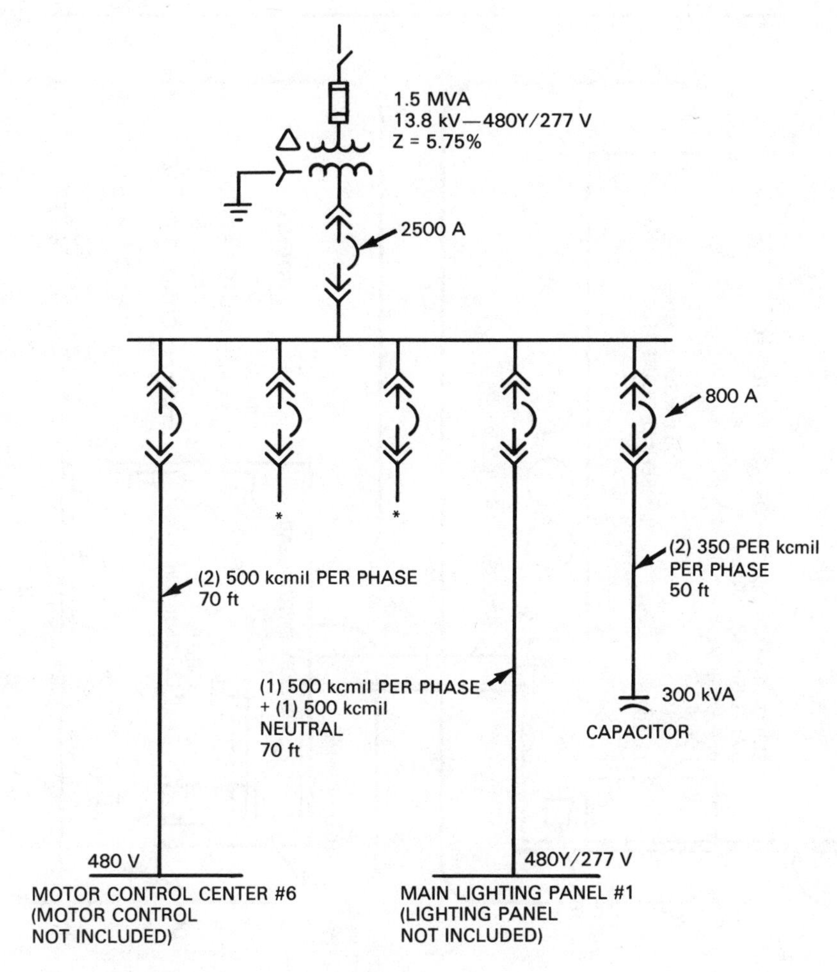

Fig 171
Substation C—1.5 MVA, 480Y/277 V

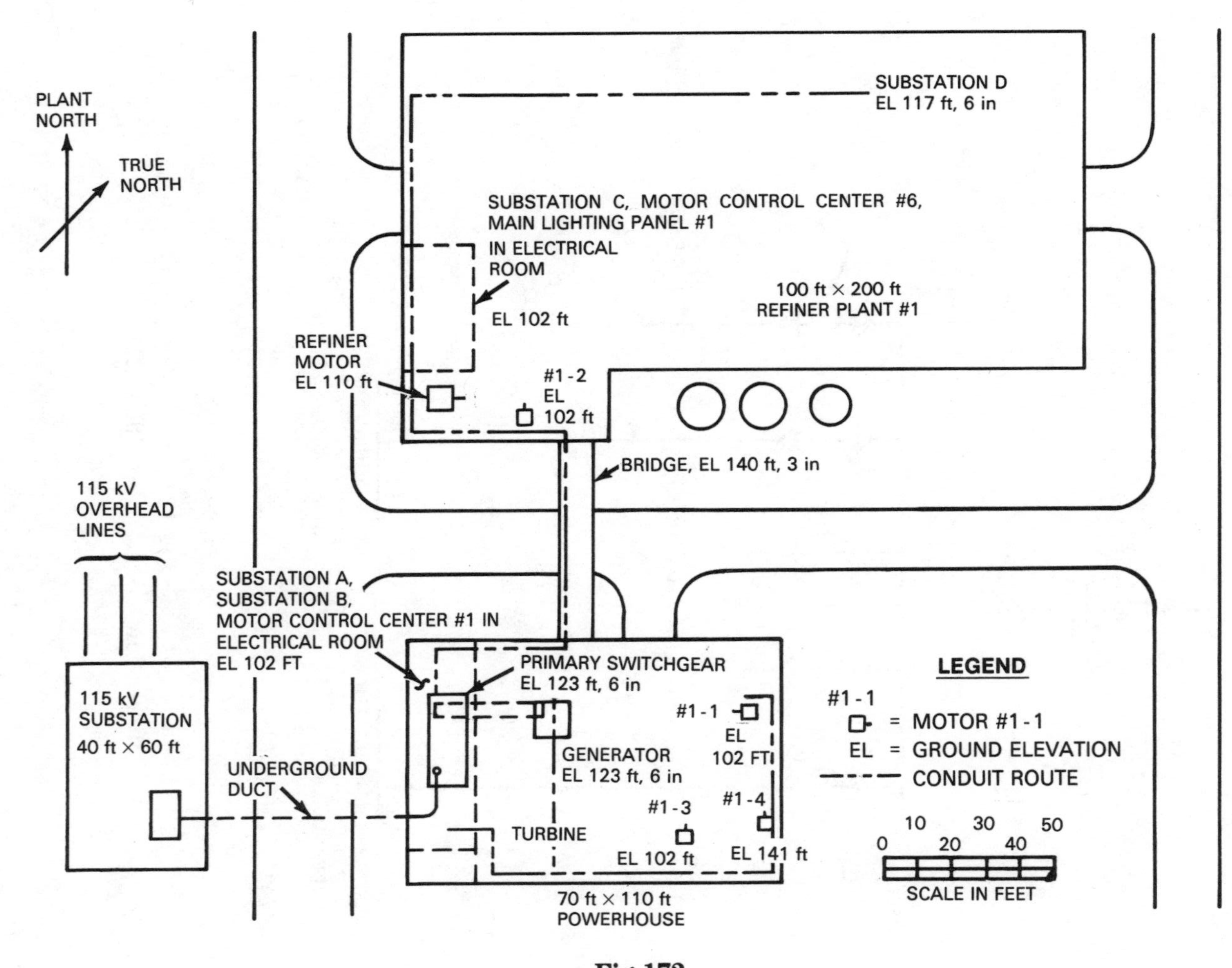

Fig 172
Site Plan

☐ Summary Sheet
☐ Detail Sheet

JOB NAME: ____________________

____

LOCATION: ____________________

____

File No ____________

Sheet ________ of ________

Estimate by ____________

Date ____________

ITEM NO	DESCRIPTION	QUANTITY	MATERIAL			LABOR AT $ /MAN-HOUR		TOTAL $
			UNIT $	$ BY OWNER	$ BY CONTRACTOR	UNIT MAN-HOUR	$ TOTAL	

Fig 173
Cost Estimate Calculation Sheet

14.12 Supporting Data. Maintain a file of quotations and other information that substantiates the cost data (not included for this example, however). Include an allowance in the contractor's material for his overhead expenses and profit.

Decide whether to use direct labor costs for each line item with overhead, profit, etc, added in the summary or to include total costs and profit in the cost per man-hour. In this example, all costs are included in the cost per man-hour.

Cost/man-hour for estimate: Assume one superintendent with three crews, each with four working journeymen and one working foreman. The total crew is 16, with 15 electricians working with the tools. Assume the average work week is 40 hours at straight time and 4 hours at premium time.

Payroll cost/week:

Journeymen:	(12)	·	(40)	@	($ /hr)	=	$
	(12)	·	(4)	@	($ /hr)	=	
Foremen:	(3)	·	(40)	@	($ /hr)	=	
	(3)	·	(4)	@	($ /hr)	=	
Superintendent:			(40)	@	($ /hr)	=	
			(4)	@	($ /hr)	=	________
Subtotal #1, payroll cost/wk						=	$
Add to the payroll cost:							
Direct labor charges ___%						=	$
Indirect labor charges ___%						=	$________
Subtotal #2, payroll, direct and indirect						=	$
Add to Subtotal #2:							
Contractor's profit ___%						=	$________
Subtotal #3, costs and profit						=	$________

Divide subtotal #3 by 15 to obtain the average cost per man-hour for this estimate. For purposes of this example, the average cost is assumed to be $30.00/man-hour.

☑ Summary Sheet
☐ Detail Sheet

JOB NAME: PROPOSED INDUSTRIAL PLANT
POWER SYSTEM
LOCATION: CENTERVILLE, USA

File No. 141-14-6
Sheet 1 of 10
Estimate by SMITH
Date 4 MAY 86

ITEM NO.	DESCRIPTION	QUANTITY	MATERIAL			LABOR AT $30/MAN-HOUR		TOTAL $
			UNIT $	$ BY OWNER	$ BY CONTRACTOR	UNIT MAN-HOUR	$ TOTAL	
	SUMMARY							
	PRIMARY POWER, 13.8 kV							
	OUTDOOR SUBSTATION			147 000	5 600		16 950	169 550
	INCOMING FEEDER			25 200	7 820		20 300	53 320
	PRIMARY SWITCHGEAR			290 000	7 470		15 600	313 070
	GENERATOR—INCLUDED IN MECHANICAL			—	—		—	—
	GENERATOR BUS			14 000	1 000		9 000	24 000
	FEEDERS			20 720	10 175		26 085	56 980
	GROUNDING				6 300		6 600	12 900
	SUBTOTAL, PRIMARY POWER			496 920	38 365		94 535	629 820
	SUBSTATION A, 4.16 kV							
	UNIT SUBSTATION			102 000	3 500		9 300	114 800
	FEEDERS				5 365		5 320	10 685
	CAPACITOR				6 000		750	6 750
	SUBTOTAL, SUBSTATION A			102 000	14 865		15 370	132 235

Fig 174
Sample Cost Estimate Calculation Sheet—Summary

☑ Summary Sheet
☐ Detail Sheet

JOB NAME: PROPOSED INDUSTRIAL PLANT
POWER SYSTEM
LOCATION: CENTERVILLE, USA

File No. 141-14-6
Sheet 2 of 10
Estimate by SMITH
Date 4 MAY 86

ITEM NO.	DESCRIPTION	QUANTITY	MATERIAL			LABOR AT $30/MAN-HOUR		TOTAL $
			UNIT $	$ BY OWNER	$ BY CONTRACTOR	UNIT MAN-HOUR	$ TOTAL	
	SUMMARY (CONT'D)							
	SUBSTATION C, 480 V							
	UNIT SUBSTATION			60 750	2 000		5 900	68 650
	FEEDERS				7 960		7 760	15 720
	CAPACITOR				4 500		600	5 100
	SUBTOTAL, SUBSTATION C			60 750	14 460		14 260	89 470
	SUBSTATION B, SUBTOTAL			60 750	14 460		14 260	89 470
	SUBSTATION D, SUBTOTAL			60 750	14 460		14 260	89 470
	TOTAL FOR POWER SYSTEM			781 170	96 610		152 685	1 030 465
	(NOT INCLUDED IN ESTIMATE:							
	FREIGHT, SPARES, TRAINING, CONTINGENCY, ESCALATION, SALES TAX,							
	PERMITS AND FEES, BOND, FOUNDATIONS, LIGHTING)							

Fig 174 *(Continued)*
Sample Cost Estimate Calculation Sheet—Summary

☐ Summary Sheet
☑ Detail Sheet

JOB NAME: PROPOSED INDUSTRIAL PLANT
POWER SYSTEM
LOCATION: CENTERVILLE, USA

File No. 141-14-6
Sheet 3 of 10
Estimate by SMITH
Date 4 MAY 86

ITEM NO.	DESCRIPTION	QUANTITY	MATERIAL			LABOR AT $30/MAN-HOUR		TOTAL $
			UNIT $	$ BY OWNER	$ BY CONTRACTOR	UNIT MAN-HOUR	$ TOTAL	
	PRIMARY POWER (FIG 169)							
	OUTDOOR SUBSTATION:							
1	TRANSFORMER, 15 MVA, 115/13.8 kV (SEE QUOTE)	1		142 000				142 000
	FOUNDATION—INCLUDED IN STRUCTURAL ESTIMATE			—	—		—	—
	HANDLING, RIGGING, INSTALLATION		LUMP SUM		1000	400	12 000	13 000
	PRIMARY CONNECTIONS		LUMP SUM		2000	100	3 000	5 000
	TESTING (SUBCONTRACT QUOTE + 15%)				1150			1 150
2	GROUNDING RESISTOR, 1000 A, 7.2 Ω, WITH CT	1		5 000				5 000
	SUPPORTING STRUCTURE				1250	15	450	1 700
	INSTALL, CONNECT, TEST				200	50	1 500	1 700
	SUBTOTAL, OUTDOOR SUBSTATION			147 000	5600		16 950	169 550

Fig 175
Sample Cost Estimate Calculation Sheet—Primary Power

☐ Summary Sheet
☑ Detail Sheet

JOB NAME: PROPOSED INDUSTRIAL PLANT
POWER SYSTEM
LOCATION: CENTERVILLE, USA

File No. 141-14-6
Sheet 4 of 10
Estimate by SMITH
Date 4 MAY 86

ITEM NO.	DESCRIPTION	QUANTITY	MATERIAL			LABOR AT $30/MAN-HOUR		TOTAL $
			UNIT $	$ BY OWNER	$ BY CONTRACTOR	UNIT MAN-HOUR	$ TOTAL	
	PRIMARY POWER (CONT'D)							
3	15 kV INCOMING FEEDER:							
	TRENCH AND BACKFILL 18 IN × 42 IN; 10 yd^3 FILL	65 ft	50/yd^3		500	0.15/ft	300	800
	CONCRETE—ENCASED DUCT; (3) = 4 in, (1) = 2 in	65 ft	60/ft		3900	2/ft	3 900	7 800
	CONDUIT RISERS, 4 in ALUMINUM, INCLUDING FITTINGS, HANGERS	210 ft	12/ft		2520	0.60/ft	3 780	6 300
	CABLE, SINGLE-CONDUCTOR 750 kcmil, 15 kV, EPR (9 @ 175 ft)	1575 ft	12/ft	25 200		0.15/ft	8 000	33 200
	TERMINATIONS	18	50/ea		900	8/ea	4 320	5 220
	SUBTOTAL, INCOMING FEEDER			25 200	7820		20 300	53 320
4	PRIMARY SWITCHGEAR:							
a	INDOOR METALCLAD, 15 kV, 750 MVA, (1) INCLUDING BREAKERS AND							
	AUX, (4) FEEDER BREAKERS, (1) MOTOR STARTER BREAKER (SEE							
	SWITCHGEAR QUOTE)	LOT		175 000				175 000
b	ADD FOR GENERATOR, BREAKER AND AUXILIARY, NEUTRAL BREAKER,							
	RELAYING, SYNC, BATTERY (SEE GENERATOR QUOTE)	LOT		115 000				115 000
c	HANDLING, RIGGING				1000	100	3 000	4 000
d	ALIGN, CONNECT, CHECKOUT				720	420	12 600	13 320

Fig 175 *(Continued)*
Sample Cost Estimate Calculation Sheet—Primary Power

☐ Summary Sheet
☑ Detail Sheet

JOB NAME: PROPOSED INDUSTRIAL PLANT
POWER SYSTEM
LOCATION: CENTERVILLE, USA

File No. 141-14-6
Sheet 5 of 10
Estimate by SMITH
Date 4 MAY 86

ITEM NO.	DESCRIPTION	QUANTITY	MATERIAL			LABOR AT $30/MAN-HOUR		TOTAL $
			UNIT $	$ BY OWNER	$ BY CONTRACTOR	UNIT MAN-HOUR	$ TOTAL	
	PRIMARY POWER (CONT'D)							
4e	TESTING (SUBCONTRACT QUOTE + 15%)				5750			5 720
	SUBTOTAL, PRIMARY SWITCHGEAR			290 000	7470		15 600	313 070
5	GENERATOR							
	GENERATOR AND AUXILIARIES, EXCEPT FOR ITEM 4b, INCLUDED							
	IN TURBINE-GENERATOR ESTIMATE; SEE MECHANICAL ESTIMATE							
	INCLUDING ERECTION			—	—		—	—
6	GENERATOR BUS; 2000 A, 15 kV	LOT		14 000	1000	300	9 000	24 000
	15 kV FEEDERS:							
7	SUBSTATION FEEDERS FROM BREAKERS #1, 4 (#5, 6 NOT INCLUDED)							
	3 in ALUMINUM CONDUIT, FITTINGS, HANGERS,	630 ft			5735	0.50/ft	9 450	15 185
	CABLE, SC 250 kcmil, 15 kV, EPR (670 ft RUN INCLUDING ENDS)	2010 ft		9 650		0.08/ft	4 825	14 475
	TERMINATIONS	24			960	6/ea	4 320	5 280

Fig 175 *(Continued)*
Sample Cost Estimate Calculation Sheet—Primary Power

☐ Summary Sheet
☑ Detail Sheet

JOB NAME: PROPOSED INDUSTRIAL PLANT
POWER SYSTEM
LOCATION: CENTERVILLE, USA

File No. 141-14-6
Sheet 6 of 10
Estimate by SMITH
Date 4 MAY 86

ITEM NO.	DESCRIPTION	QUANTITY	MATERIAL			LABOR AT $30/MAN-HOUR		TOTAL $
			UNIT $	$ BY OWNER	$ BY CONTRACTOR	UNIT MAN-HOUR	$ TOTAL	
	PRIMARY POWER (CONT'D)							
8	MOTOR FEEDER							
	3/C, 250 kcmil, 15 kV, INTERLOCKED ARMOR	540/ft	20.5/ft	11 070		0.28/ft	4 550	15 620
	SUPPORTS, BRACKETS, (5 ft CENTERS)	50	40/ea		2 000	1/ea	1 500	3 500
	TERMINATIONS	4	120/ea		480	12/ea	1 440	1 920
9	HI-POT TEST, INCLUDING INCOMING AND GENERATOR BUS				1 000			1 000
	SUBTOTAL, FEEDERS			20 720	10 175		26 085	56 980
10	GROUNDING (INCLUDES GROUND LOOP FOR (2) BUILDINGS,							
	CONNECTIONS TO ALL 15 kV AND 5 kV CLASS EQUIPMENT):							
	4/0 AWG BARE COPPER	2000/ft	2.15/ft		4 300	0.06/ft	3 600	7 900
	CONNECTORS, RODS	LOT			2 000		3 000	5 000
	SUBTOTAL, GROUNDING				6 300		6 600	12 900

Fig 175 *(Continued)*
Sample Cost Estimate Calculation Sheet—Primary Power

☐ Summary Sheet
☑ Detail Sheet

JOB NAME: PROPOSED INDUSTRIAL PLANT
POWER SYSTEM
LOCATION: CENTERVILLE, USA

File No. 141-14-6
Sheet 7 of 10
Estimate by SMITH
Date 4 MAY 86

ITEM NO.	DESCRIPTION	QUANTITY	MATERIAL			LABOR AT $30/MAN-HOUR		TOTAL $
			UNIT $	$ BY OWNER	$ BY CONTRACTOR	UNIT MAN-HOUR	$ TOTAL	
	SUBSTATION A (FIG 170)							
	INDOOR SUBSTATION:							
11a	UNIT SUBSTATION—PRIMARY SWITCH; TRANSFORMER 5 MVA,							
	13.8/4.16 kV; SWITCHGEAR SECTION WITH (4) 2000 A VACUUM CIRCUIT							
	BREAKERS, METERING, RELAYING; TOTAL COST (SEE 2 QUOTES)	1		98 000				98 000
b	HANDLING, RIGGING				500	80	2400	2 900
c	ALIGN, CONNECT, CHECKOUT				500	180	5400	5 900
d	TESTING (SUBCONTRACT + 15%)				1400			1 400
e	FOUNDATION AND ROOM IN STRUCTURAL ESTIMATE, LIGHTING IN							
	BUILDING LIGHTING ESTIMATE			—	—		—	—
12	GROUNDING RESISTOR 400 A, 6 Ω, WITH CT	1		4 000				4 000
	SUPPORTING STRUCTURE				900	15	450	1 350
	INSTALL, CONNECT, TEST				200	35	1050	1 250
	SUBTOTAL, INDOOR SUBSTATION			102 000	3500		9300	114 800

Fig 176
Sample Cost Estimate Calculation Sheet—Substation A

☐ Summary Sheet
☑ Detail Sheet

JOB NAME: PROPOSED INDUSTRIAL PLANT
POWER SYSTEM
LOCATION: CENTERVILLE, USA

File No. 141-14-6
Sheet 8 of 10
Estimate by SMITH
Date 4 MAY 86

ITEM NO.	DESCRIPTION	QUANTITY	MATERIAL UNIT $	MATERIAL $ BY OWNER	MATERIAL $ BY CONTRACTOR	LABOR AT $30/MAN-HOUR UNIT MAN-HOUR	LABOR AT $30/MAN-HOUR $ TOTAL	TOTAL $
	SUBSTATION A (CONT'D)							
	5 kV FEEDERS: (FEEDER #2 NOT INCLUDED)							
13	3 in ALUMINUM CONDUIT, FITTINGS, HANGERS	100 ft	9.1/ft		910	0.5/ft	1500	2 410
	CABLE, SINGLE-CONDUCTOR 500 kcmil, 5 kV, XPE	330 ft	6.8 ft		2245	0.11/ft	1090	3 335
	2½ in ALUMINUM CONDUIT, FITTINGS, HANGERS	50 ft	8.3/ft		415	0.45/ft	675	1 090
	CABLE, SINGLE-CONDUCTOR 250 kcmil, 5 kV, XPE	180 ft	4.7/ft		845	0.08/ft	435	1 280
	TERMINATIONS	18	25/ea		450	3/ea	1620	2 070
14	TESTING (SUBCONTRACT)				500			500
	SUBTOTAL, FEEDERS				5365		5320	10 685
15	CAPACITOR INCLUDING INSTALLATION	1			6000	25	750	6 750

Fig 176 *(Continued)*
Sample Cost Estimate Calculation Sheet—Substation A

☐ Summary Sheet
☑ Detail Sheet

JOB NAME: PROPOSED INDUSTRIAL PLANT
POWER SYSTEM
LOCATION: CENTERVILLE, USA

File No. 141-14-6
Sheet 9 of 10
Estimate by SMITH
Date 4 MAY 86

ITEM NO.	DESCRIPTION	QUANTITY	MATERIAL			LABOR AT $30/MAN-HOUR		TOTAL $
			UNIT $	$ BY OWNER	$ BY CONTRACTOR	UNIT MAN-HOUR	$ TOTAL	
	SUBSTATION C (FIG 171)							
	INDOOR SUBSTATION:							
16a	UNIT SUBSTATION—FUSED PRIMARY SWITCH;							
	TRANSFORMER 1.5 MVA, 13.8 kV/480 V; SWITCHGEAR SECTION WITH							
	2500 A MAIN, (4) 1200 A FEEDER BREAKERS, METERING							
	(USE ACTUAL COST FROM PROJECT 80-107 + 12½%)	1		60 750				60 750
b	HANDLING, RIGGING				500	77	2300	2 800
c	ALIGN, CONNECT, CHECKOUT				500	120	3600	4 100
d	TESTING				1000			1 000
e	FOUNDATION AND ROOM IN STRUCTURAL ESTIMATE;							
	LIGHTING IN BUILDING LIGHTING ESTIMATE			—	—		—	—
	SUBTOTAL, SUBSTATION			60 750	2000		5900	68 650

Fig 177
Sample Cost Estimate Calculation Sheet—Substation C

☐ Summary Sheet
☑ Detail Sheet

JOB NAME: PROPOSED INDUSTRIAL PLANT
POWER SYSTEM
LOCATION: CENTERVILLE, USA

File No. 141-14-6
Sheet 10 of 10
Estimate by SMITH
Date 4 MAY 86

ITEM NO.	DESCRIPTION	QUANTITY	MATERIAL			LABOR AT $30/MAN-HOUR		TOTAL $
			UNIT $	$ BY OWNER	$ BY CONTRACTOR	UNIT MAN-HOUR	$ TOTAL	
	SUBSTATION C (CONT'D)							
	480 V FEEDERS:							
17	2½ IN ALUMINUM CONDUIT, FITTINGS, HANGERS	100 ft	8.3/ft		830	0.45/ft	1350	2 180
	3 IN ALUMINUM CONDUIT, FITTINGS, HANGERS	140 ft	9.1/ft		1275	0.50/ft	2100	3 375
	4 IN ALUMINUM CONDUIT, FITTINGS, HANGERS	70 ft	11/ft		770	0.62/ft	1300	2 070
	SC 250 kcmil, XHW	80 ft	2.3/ft		185	0.08/ft	200	385
	SC 350 kcmil, XHW	360 ft	3.4/ft		1225	0.08/ft	865	2 090
	SC 500 kcmil, XHW	720 ft	5.1/ft		3675	0.09/ft	1945	5 620
	SUBTOTAL, FEEDERS				7960		7760	15 720
18	CAPACITOR INCLUDING INSTALLATION	1			4500	20	600	5 100
	SUBSTATION B AND SUBSTATION D							
	FOR PURPOSES OF THIS EXAMPLE, ASSUME B AND D							
	ARE IDENTICAL TO C.							

Fig 177 *(Continued)*
Sample Cost Estimate Calculation Sheet—Substation C

Note
Selected Sources for Cost-Estimating Information

This is an abbreviated list of references for cost-estimating information. Omission of other current sources is not meant in any way to be judgmental. The reader is urged to consider all available references.

NECA Manual of Labor Units, National Electrical Contractors Association, Inc
7315 Wisconsin Avenue, Bethesda, MD 20814
NECA information is proprietary. Nonmembers of NECA may contact the national office (address above) or the local NECA office for further information.

Mechanical and Electrical Cost Data, R. S. Means Company, Inc
100 Construction Plaza, Kingston, MA 02364
Means offers several different services. Contact Means for further information.

Richardson Rapid Estimating Systems
Richardson Engineering Services, Inc
909 Rancheros Drive, P.O. Box 1055
San Marcos, CA 92069
Richardson offers a wide range of estimating information, including electrical.

Kolstad & Kohnert, Rapid Electrical Estimating and Pricing
McGraw-Hill Book Company, P.O. Box 400, Hightstown, NJ 08520

National Price Service, 4525 West 160th Street, Cleveland, OH 44135

Trade Service Publications, Inc
10996 Torreyana Road, San Diego, CA 92121

U S Department of Labor, Bureau of Labor Statistics
450 Golden Gate Avenue, Box 36017, San Francisco, CA 94102

Appendix
Power System Device Function Numbers[35]

(This Appendix is not a part of ANSI/IEEE Std 141-1986, IEEE Recommended Practice for Electric Power Distribution for Industrial Plants.)

1. master element is the initiating device, such as a control switch, etc, which serves either directly or through such permissive devices as protective and time-delay relays to place an equipment in or out of operation.

NOTE: This number is normally used for a hand-operated device, although it may also be used for an electrical or mechanical device for which no other function number is suitable.

2. time-delay starting or closing relay is a device that functions to give a desired amount of time delay before or after any point of operation in a switching sequence or protective relay system, except as specifically provided by device functions 48, 62, and 79.

3. checking or interlocking relay is a relay that operates in response to the position of a number of other devices (or to a number of predetermined conditions) in an equipment, to allow an operating sequence to proceed, or to stop, or to provide a check of the position of these devices or of these conditions for any purpose.

4. master contactor is a device, generally controlled by device function 1 or the equivalent and the required permissive and protective devices, that serves to make and break the necessary control circuits to place an equipment into operation under the desired conditions and to take it out of operation under other abnormal conditions.

5. stopping device is a control device used primarily to shut down an equipment and hold it out of operation. [This device may be manually or electrically actuated, but excludes the function of electrical lockout (see device function 86) on abnormal conditions.]

---

[35] From ANSI/IEEE C37.2-1979, IEEE Standard Electrical Power System Device Function Numbers.

6. starting circuit breaker is a device whose principal function is to connect a machine to its source of starting voltage.

7. anode circuit breaker is a device used in the anode circuits of a power rectifier for the primary purpose of interrupting the rectifier circuit if an arc back should occur.

8. control power disconnecting device is a disconnecting device, such as a knife switch, circuit breaker, or pull-out fuse block, used for the purpose of respectively connecting and disconnecting the source of control power to and from the control bus or equipment.

NOTE: Control power is considered to include auxiliary power which supplies such apparatus as small motors and heaters.

9. reversing device is a device that is used for the purpose of reversing a machine field or for performing any other reversing functions.

10. unit sequence switch is a switch that is used to change the sequence in which units may be placed in and out of service in multiple-unit equipments.

11. Reserved for future application.

12. overspeed device is usually a direct-connected speed switch which functions on machine overspeed.

13. synchronous-speed device is a device such as a centrifugal-speed switch, a slip-frequency relay, a voltage relay, an undercurrent relay, or any type of device that operates at approximately the synchronous speed of a machine.

14. underspeed device is a device that functions when the speed of a machine falls below a predetermined value.

15. speed or frequency matching device is a device that functions to match and hold the speed or the frequency of a machine or of a system equal to, or approximately equal to, that of another machine, source or system.

16. Reserved for future application.

17. shunting or discharge switch is a switch that serves to open or to close a shunting circuit around any piece of aparatus (except a resistor), such as a machine field, a machine armature, a capacitor, or a reactor.

NOTE: This excludes devices that perform such shunting operations as may be necessary in the process of starting a machine by devices 6 or 42, or their equivalent, and also excludes device function 73 that serves for the switching of resistors.

18. accelerating or decelerating device is a device that is used to close or to cause the closing of circuits which are used to increase or decrease the speed of a machine.

19. starting-to-running transition contactor is a device that operates to initiate or cause the automatic transfer of a machine from the starting to the running power connection.

20. electrically operated valve is an electrically operated, controlled or monitored valve used in a fluid line.

NOTE: The function of the valve may be indicated by the use of the suffixes in 3.3.

21. distance relay is a relay that functions when the circuit admittance, impedance, or reactance increases or decreases beyond a predetermined value.

22. equalizer circuit breaker is a breaker that serves to control or to make and break the equalizer or the current-balancing connections for a machine field, or for regulating equipment, in a multiple-unit installation.

23. temperature control device is a device that functions to raise or lower the temperature of a machine or other apparatus, or of any medium, when its temperature falls below, or rises above, a predetermined value.

NOTE: An example is a thermostat that switches on a space heater in a switchgear assembly when the temperature falls to a desired value as distinguished from a device that is used to provide automatic temperature regulation between close limits and would be designated as device function 90T.

24. Reserved for future application.

25. synchronizing or synchronism-check device is a device that operates when two ac circuits are within the desired limits of frequency, phase angle, and voltage, to permit or to cause the paralleling of these two circuits.

26. apparatus thermal device is a device that functions when the temperature of the shunt field or the amortisseur winding of a machine, or that of a load limiting or load shifting resistor, or of a liquid or other medium exceeds a predetermined value; or if the temperature of the protected apparatus, such as a power rectifier, or of any medium decreases below a predetermined value.

27. undervoltage relay is a relay which operates when its input voltage is less than a predetermined value.

28. flame detector is a device that monitors the presence of the pilot or main flame in such apparatus as a gas turbine or a steam boiler.

29. isolating contactor is a device that is used expressly for disconnecting one circuit from another for the purposes of emergency operation, maintenance, or test.

30. annunciator relay is a nonautomatically reset device that gives a number of separate visual indications upon the functioning of protective devices, and which may also be arranged to perform a lockout function.

31. separate excitation device is a device that connects a circuit, such as the shunt field of a synchronous converter, to a source of separate excitation during the starting sequence; or one that energizes the excitation and ignition circuits of a power rectifier.

32. directional power relay is a relay which operates on a predetermined value of power flow in a given direction, or upon reverse power such as that resulting from the motoring of a generator upon loss of its prime power.

33. position switch is a switch that makes or breaks contact when the main device or piece of apparatus which has no device function number reaches a given position.

34. master sequence device is a device such as a motor-operated multicontact switch, or the equivalent, or a programming device, such as a computer, that establishes or determines the operating sequence of the major devices in an equipment during starting and stopping or during other sequential switching operations.

35. brush-operating or slip-ring short-circuiting device is a device for raising, lowering, or shifting the brushes of a machine, or for short-circuiting its slip rings, or for engaging or disengaging the contacts of a mechanical rectifier.

36. polarity or polarizing voltage device is a device that operates, or permits the operation of, another device on a predetermined polarity only, or verifies the presence of a polarizing voltage in an equipment.

37. undercurrent or underpower relay is a relay that functions when the current or power flow decreases below a predetermined value.

38. bearing protective device is a device that functions on excessive bearing temperature, or on other abnormal mechanical conditions associated with the bearing, such as undue wear, which may eventually result in excessive bearing temperature or failure.

39. mechanical condition monitor is a device that functions upon the occurrence of an abnormal mechanical condition (except that associated with bearings as covered under device function 38), such as excessive vibration, eccentricity, expansion, shock, tilting, or seal failure.

40. field relay is a relay that functions on a given or abnormally low value or failure of machine field current, or on an excessive value of the reactive component of armature current in an ac machine indicating abnormally low field excitation.

41. field circuit breaker is a device that functions to apply or remove the field excitation of a machine.

42. running circuit breaker is a device whose principal function is to connect a machine to its source of running or operating voltage. This function may also be used for a device, such as a contactor, that is used in series with a circuit breaker or other fault protecting means, primarily for frequent opening and closing of the circuit.

43. manual transfer or selector device is a manually operated device that transfers the control circuits in order to modify the plan of operation of the switching equipment or of some of the devices.

44. unit sequence starting relay is a relay that functions to start the next available unit in a multiple-unit equipment upon the failure or nonavailability of the normally preceding unit.

45. atmospheric condition monitor is a device that functions upon the occurrence of an abnormal atmospheric condition, such as damaging fumes, explosive mixtures, smoke, or fire.

46. reverse-phase or phase-balance current relay is a relay that functions when the polyphase currents are of reverse-phase sequence, or when the polyphase currents are unbalanced or contain negative phase-sequence components above a given amount.

47. phase-sequence voltage relay is a relay that functions upon a predetermined value of polyphase voltage in the desired phase sequence.

48. incomplete sequence relay is a relay that generally returns the equipment to the normal, or off, position and locks it out if the normal starting, operating, or stopping sequence is not properly completed within a predetermined time. If the device is used for alarm purposes only, it should preferably be designated as 48A (alarm).

49. machine or transformer thermal relay is a relay that functions when the temperature of a machine armature or other load-carrying winding or element of a machine or the temperature of a power rectifier or power transformer (including a power rectifier transformer) exceeds a predetermined value.

50. instantaneous overcurrent or rate-of-rise relay is a relay that functions instantaneously on an excessive value of current or on an excessive rate of current rise.

51. ac time overcurrent relay is a relay that operates when its ac input current exceeds a predetermined value, and in which the input current and operating time are inversely related through a substantial portion of the performance range.

52. ac circuit breaker is a device that is used to close and interrupt an ac power circuit under normal conditions or to interrupt this circuit under fault or emergency conditions.

53. exciter or dc generator relay is a relay that forces the dc machine field excitation to build up during switching or which functions when the machine voltage has built up to a given value.

54. Reserved for future application.

55. power factor relay is a relay that operates when the power factor in an ac circuit rises above or falls below a predetermined value.

56. field application relay is a relay that automatically controls the application of the field excitation to an ac motor at some predetermined point in the slip cycle.

57. short-circuiting or grounding device is a primary circuit switching device that functions to short circuit or to ground a circuit in response to automatic or manual means.

58. rectification failure relay is a device that functions if one or more anodes of a power rectifier fail to fire, or to detect an arc back, or on failure of a diode to conduct or block properly.

59. overvoltage relay is a relay which operates when its input voltage is more than a predetermined value.

60. voltage or current balance relay is a relay that operates on a given difference in voltage, or current input or output, of two circuits.

61. Reserved for future application.

62. time-delay stopping or opening relay is a time-delay relay that serves in conjunction with the device that initiates the shutdown, stopping, or opening operation in an automatic sequence or protective relay system.

63. pressure switch is a switch which operates on given values, or on a given rate of change, of pressure.

64. ground detector relay is a relay that operates on failure of machine or other apparatus insulation to ground.

NOTE: This function is not applied to a device connected in the secondary circuit of current transformers in a normally grounded power system, where other device numbers with a suffix G or N should be used, that is, 51N for an ac time overcurrent relay connected in the secondary neutral of the current transformers.

65. governor is the assembly of fluid, electrical, or mechanical control equipment used for regulating the flow of water, steam, or other medium to the prime mover for such purposes as starting, holding speed or load, or stopping.

66. notching or jogging device is a device that functions to allow only a specified number of operations of a given device, or equipment, or a specified number of successive operations within a given time of each other. It is also a device that functions to energize a circuit periodically or for fractions of specified time intervals, or that is used to permit intermittent acceleration or jogging of a machine at low speeds for mechanical positioning.

67. ac directional overcurrent relay is a relay that functions on a desired value of ac overcurrent flowing in a predetermined direction.

68. blocking relay is a relay that initiates a pilot signal for blocking of tripping on external faults in a transmission line or in other apparatus under predetermined conditons, or cooperates with other devices to block tripping or to block reclosing on an out-of-step condition or on power swings.

69. permissive control device is generally a two-position device that in one position permits the closing of a circuit breaker, or the placing of an equipment into operation, and in the other position prevents the circuit breaker or the equipment from being operated.

70. rheostat is a variable resistance device used in an electric circuit, which is electrically operated or has other electrical accessories, such as auxiliary, position, or limit switches.

71. level switch is a switch which operates on given values, or on a given rate of change, of level.

72. dc circuit breaker is a circuit breaker that is used to close and interrupt a dc power circuit under normal conditions or to interrupt this circuit under fault or emergency conditions.

73. load-resistor contactor is a contactor that is used to shunt or insert a step of load limiting, shifting, or indicating resistance in a power circuit, or to switch a space heater in circuit, or to switch a light or regenerative load resistor of a power recitifier or other machine in and out of circuit.

74. alarm relay is a relay other than an annunciator, as covered under device function 30, that is used to operate, or to operated in connection with, a visual or audible alarm.

75. position changing mechanism is a mechanism that is used for moving a main device from one position to another in an equipment; as, for example, shifting a removable circuit breaker unit to and from the connected, disconnected, and test positions.

76. dc overcurrent relay is a relay that functions when the current in a dc circuit exceeds a given value.

77. pulse transmitter is used to generate and transmit pulses over a telemetering or pilot-wire circuit to the remote indicating or receiving device.

78. phase-angle measuring or out-of-step protective relay is a relay that functions at a predetermined phase angle between two voltages or between two currents or between voltage and current.

79. ac reclosing relay is a relay that controls the automatic reclosing and locking out of an ac circuit interrupter.

80. flow switch is a switch which operates on given values, or on a given rate of change, of flow.

81. frequency relay is a relay that responds to the frequency of an electrical quantity, operating when the frequency or rate of change of frequency exceeds or is less than a predetermined value.

82. dc reclosing relay is a relay that controls the automatic closing and reclosing of a dc circuit interrrupter, generally in response to load circuit conditons.

83. automatic selective control or transfer relay is a relay that operates to select automatically between certain sources or conditons in an equipment, or performs a transfer operation automatically.

84. operating mechanism is the complete electrical mechanism or servomechanism, including the operating motor, solenoids, position switches, etc, for a tap changer, induction regulator, or any similar piece of apparatus which otherwise has no device function number.

85. carrier or pilot-wire receiver relay is a relay that is operated or restrained by a signal used in connection with carrier-current or dc pilot-wire fault relaying.

86. lockout relay is a hand or electrically reset auxiliary relay that is operated upon the occurrence of abnormal conditons to maintain associated equipment or devices inoperative until it is reset.

87. differential protective relay is a protective relay that functions on a percentage or phase angle or other quantitative difference of two currents or of some other electrical quantities.

88. auxiliary motor or motor generator is one used for operating auxiliary equipment, such as pumps, blowers, exciters, rotating magnetic amplifiers, etc.

89. line switch is a switch used as a disconnecting, load-interrupter, or isolating switch in an ac or dc power circuit, when this device is electrically operated or has electrical accessories, such as an auxiliary switch, magnetic lock, etc.

90. regulating device is a device that functions to regulate a quantity, or quantities, such as voltage, current, power, speed, frequency, temperature, and load, at a certain value or between certain (generally close) limits for machines, tie lines, or other apparatus.

91. voltage directional relay is a relay that operates when the voltage across an open circuit breaker or contactor exceeds a given value in a given direction.

92. voltage and power directional relay is a relay that permits or causes the connection of two circuits when the voltage difference between them exceeds a given value in a predetermined direction and causes these two circuits to be disconnected from each other when the power flowing between them exceeds a given value in the opposite direction.

93. field-changing contactor is a contactor that functions to increase or decrease, in one step, the value of field excitation on a machine.

94. tripping or trip-free relay is a relay that functions to trip a circuit breaker, contactor, or equipment, or to permit immediate tripping by other devices; or to prevent immediate reclosure of a circuit interrupter if it should open automatically even though its closing circuit is maintained closed.

95–99. Used only for specific applications in individual installations where none of the assigned numbered functions from 1 to 94 are suitable.

Index

A

B

C

D

E

F

G

H

I

J

K

L

M

N

O

P

Q

R

S

T

U

V

W

Z